Computational Optical Biomedical Spectroscopy and Imaging

Computational Optical Biomedical Spectroscopy and Imaging

EDITED BY

Sarhan M. Musa

CRC Press is an imprint of the
Taylor & Francis Group, an informa business

CRC Press
Taylor & Francis Group
6000 Broken Sound Parkway NW, Suite 300
Boca Raton, FL 33487-2742

First issued in paperback 2018

ISBN 13: 978-1-138-74850-7 (pbk)
ISBN 13: 978-1-4822-3081-9 (hbk)

Visit the Taylor & Francis Web site at
http://www.taylorandfrancis.com

and the CRC Press Web site at
http://www.crcpress.com

Dedicated to my late father, Mahmoud,

my mother, Fatmeh,

and my wife, Lama.

Contents

Preface

Recent growth in the use of computational optical technology for biomedical research and health care is explosive. New applications are made possible by emerging technologies in lasers, optoelectronic devices, fiber optics, physical and chemical sensors, spectroscopy, and imaging, all of which are being applied to medical research, diagnostics, and therapy. Computational optical biomedical spectroscopy and imaging can be used as a reference and training resource for readers who are interested in learning and doing research in the area of computational optical biomedical spectroscopy and imaging. The readers will become familiar with the useful computational methods and their applications that are used in optical biomedical spectroscopy and imaging. This book can help the reader to compute different algorithms to unravel spectra. It provides the following major benefits: future applications, directions, opportunities, and challenges of optical biomedical spectroscopy and imaging in technical industry, academia, and government; new developments and interdisciplinary research of engineering, science (physics, chemistry, biology), and medicine using computational methods; intended for a broad audience working in the fields of physics, chemistry, biology, medicine, materials science, quantum science, electrical engineering, optical science, computer science, mechanical engineering, chemical engineering, and aerospace engineering; an introduction to the key concepts of computational methods used in optical biomedical spectroscopy and imaging in a manner that is easily digestible to a new beginner in the field; and new directions for researchers in the field of optical biomedical spectroscopy and imaging. This book has 11 chapters and 3 appendices.

Chapter 1 provides a variety of applications of infrared (IR) and Raman microscopic imaging techniques related to phenomena in human hair and skin. In the hair, for example, the authors demonstrate through univariate and multivariate data analyses how to generate infrared (IR) and Raman images providing the spatial distribution of lipids and proteins, protein secondary structure, lipid chain conformational order, and disulfide cross-links. This spectroscopic data provides a correlation among hair chemistry, anatomy, and structural organization. The obtained information provides a basis to understand how cosmetic treatments (hot ironing, bleaching, relaxing, etc.) affect hair protein structures including hair keratin conformational changes (from α-helix to β-sheet) as well as structural damage to disulfide bonds. For studies concerned with the skin, they demonstrate the utility of these techniques in understanding chemically induced molecular structural changes in skin cells. Further, the authors monitor hydration control mechanisms through evaluation of natural moisturizing factor with IR images of corneocyte cells. They also demonstrate through the use of Raman imaging to track the delivery and controlled release of skin care ingredients in the stratum corneum.

Chapter 2 presents the synthesis, characterization, and application of nanoparticles in nanoimaging through reviews of fluorescence bioimaging with applications to chemistry.

Chapter 3 provides the new trends in immunohistochemical, genome, and metabolomics imaging.

Chapter 4 reviews recent developments in the tools and computational techniques that are used to measure and classify cell phenotypes and identifies some of the opportunities and challenges for these approaches.

Chapter 5 reviews functional near-infrared spectroscopy (fNIRS) as well as the physical, mathematical, computational, and instrumentational perspectives of fNIRS.

Also, this chapter provides basic applications of fNIRS in cognitive neuroscience and neurological disorders.

Chapter 6 presents an overview of the existing literature for the automatic identification and characterization of interstitial lung diseases (ILDs) in computed tomography (CT) images. The specific research area has made significant progress in the past two decades providing health-care professionals with valuable tools that are able to improve diagnostic procedures and disease management. For the lung segmentation, new methods were proposed focusing on the processing of cases with ILD pathologies. To this end, the special textural characteristics of the interstitial patterns were considered along with any available anatomical information, often incorporated into a machine-learning framework. Regarding the segmentation of the lung airways, the utilization of the relatively new multidetector CT imaging protocol enabled the use of advanced three-dimensional (3D) region growing techniques that achieved height accuracies and facilitated the ILD pattern identification. Moreover, some methods were proposed for the segmentation of the vascular tree, although there are still several open issues to be addressed. Most of these methods use vessel enhancement filters based on Hessian matrix analysis, whereas some machine-learning approaches were also proposed based on 3D texture features.

Chapter 7 addresses the optical natural fluorescence (ONF) spectra of *Giardia lamblia* cysts excited and detected using new experimental setup. The experimental setup in this chapter is successfully used to study the ONF of *G. lamblia* cysts. The main components consist of optical multimode fibers used as waveguides for illumination and detection, a highly sensitive spectrometer, long-pass filters, and suitable continuous-wave (CW) blue excitation sources. The laser excitations are blue shifted (458 and 401 nm). At 458 nm excitation, the peak of the ONF for *G. lamblia* is located at 525 nm; at 401 nm excitation, the peak of the ONF is located at 496 nm; and at 496 nm excitation, the peak of the ONF is located at 532 nm approximately.

Chapter 8 presents the applied computational optical biomedical spectroscopy to enhance the efficiency of biomedical imaging materials, which is potentially used to develop novel imaging techniques for early detection, screening, diagnosis, and image-guided treatment of life-threatening diseases and cancer with higher resolution and lower doses.

Chapter 9 provides a different computational approach as applied to various nanoplasmonic structures for simulation of mainly the resonance response in the spectral domain to be applied to optical biomedical spectroscopy and imaging.

Chapter 10 introduces nanoimaging system design principles and representative techniques and their applications in the areas of bioscience and medical diagnostics. Selected nanospectroscopic techniques as applied to medical imaging and radiotherapy are discussed. The principles of imaging mass spectroscopy, one of the newest metabolic imaging techniques, are introduced and discussed. Also, new optical polarimetric metrics, namely, the polarimetric exploratory data analysis (pEDA), which combines polarimetry with histogram data analysis, aimed at providing enhanced discrimination among cancerous tissues and cells, is also discussed. A preclinical example of enhanced discrimination among different lung cancer types is presented based on the pEDA, and a computational analysis of the polarimetric principles is offered and discussed.

Chapter 11 presents a brief background and review on medical imaging instrumentation and techniques.

Finally, the book concludes with appendices. Appendix A shows common material and physical constants, with the consideration that the constant values of materials vary from one published source to another due to many varieties of most materials and conductivity is sensitive to temperature, impurities, moisture content, as well as the dependence of relative permittivity and permeability on temperature and humidity and the like. Appendix B provides equations for photon energy, frequency, wavelength, and electromagnetic spectrum, including the approximation of common optical wavelength ranges of light. In addition, it provides a figure for wavelengths of commercially available lasers.

Appendix C provides common symbols and useful mathematical formulae.

Sarhan M. Musa
Prairie View A&M University

MATLAB® is a registered trademark of The MathWorks, Inc. For product information, please contact:

The MathWorks, Inc.
3 Apple Hill Drive
Natick, MA 01760-2098, USA
Tel: +1 508 647 7000
Fax: +1 508 647 7001
E-mail: info@mathworks.com
Web: www.mathworks.com

Acknowledgments

I express my sincere appreciation and gratitude to all the contributors of this book. I thank Brain Gaskin and James Gaskin for their wonderful hearts and for being great American neighbors. I acknowledge the outstanding help and support of the team at Taylor & Francis Group/CRC Press in preparing this book, especially from Nora Konopka, Michele Smith, Michael Slaughter, Kari Budyk, and Robert Sims.

I thank Professors John Burghduff and Mary Jane Ferguson for their support, understanding, and being great friends. I also thank Dr. Fadi Alameddine for taking care of my mother's health very well.

I thank Dr. Kendall T. Harris, my college dean, for his continued support. Finally, the book would never have seen the light of day without the constant support, love, and patience of my family.

Editor

Sarhan M. Musa, BSc, MSc, PhD, is currently an associate professor in the Department of Engineering Technology, Roy G. Perry College of Engineering, Prairie View A&M University, Prairie View, Texas. He is the founding director of Prairie View A&M University AVAYA Networking Academy (PVNA), Prairie View, Texas. He is a frequent invited speaker on computational optical spectroscopy and imaging, has provided consultation for multiple organizations nationally and internationally, and has written and edited several books. Dr. Musa is a senior member of the Institute of Electrical and Electronics Engineers (IEEE) and is also a Local Telecommunications Division (Sprint) and a Boeing Welliver Fellow.

Contributors

M. Anthimopoulos
ARTORG Center for Biomedical Engineering Research
University of Bern
Bern, Switzerland

S. M. Ashraf
Materials Research Laboratory
Department of Chemistry
Jamia Millia Islamia
New Delhi, India

Mahua Bera
Department of Applied Optics and Photonics
University of Calcutta
Kolkata, India

Kaushik Brahmachari
Department of Applied Optics and Photonics
University of Calcutta
Kolkata, India

K. Broderick
Microsystems Technology Laboratory
Massachusetts Institute of Technology
Cambridge, Massachusetts

H. Chen
School of Physics and Engineering
Tongji University
Shanghai, People's Republic of China

A. Christe
Department of Diagnostic, Interventional and Pediatric Radiology
Bern University Hospital "Inselspital"
Bern, Switzerland

S. Christodoulidis
ARTORG Center for Biomedical Engineering Research
University of Bern
Bern, Switzerland

and

School of Electrical and Computer Engineering
Aristotle University of Thessaloniki
Thessaloniki, Greece

Richard Conroy
National Institute of Biomedical Imaging and Bioengineering
Bethesda, Maryland

Aditi Deshpande
Department of Biomedical Engineering
The University of Akron
Akron, Ohio

T. Farrahi
Charles L. Brown Department of Electrical Engineering
University of Virginia
Charlottesville, Virginia

Sharmila Ghosh
Department of Applied Optics and Photonics
University of Calcutta
Kolkata, India

George C. Giakos
Department of Electrical and Computer Engineering
Manhattan College
Riverdale, New York

Kendall T. Harris
Department of Mechanical Engineering
Roy G. Perry College of Engineering
Prairie View A&M University
Prairie View, Texas

R. Koglin
Department of Biomedical Engineering
The University of Akron
Akron, Ohio

Jing Li
Faculty of Health Sciences
Avenida Padre Tomás Pereira
Taipa, Macau SAR, People's Republic of China

Yinan Li
Department of Electrical and Computer Engineering
The University of Akron
Akron, Ohio

Xiaohong Lin
Bioimaging Core
Faculty of Health Sciences
University of Macau
Taipa, Macau SAR, People's Republic of China

B. Liu
School of Physics and Engineering
Tongji University
Shanghai, People's Republic of China

G. Livanos
Department of Electronic and Computer Engineering
Technical University of Crete
Chania, Greece

Fengmei Lu
Faculty of Health Sciences
Avenida Padre Tomás Pereira
Taipa, Macau SAR, People's Republic of China

Stefanie Marotta
Department of Electrical and Computer Engineering
The University of Akron
Akron, Ohio

Roger L. McMullen
Ashland Specialty Ingredients
Ashland, Inc.
Bridgewater, New Jersey

Richard Mendelsohn
Department of Chemistry
Rutgers University
Newark, New Jersey

S. Mougiakakou
ARTORG Center for Biomedical Engineering Research
University of Bern
Bern, Switzerland

Osama M. Musa
Ashland Specialty Ingredients
Ashland, Inc.
Bridgewater, New Jersey

Sarhan M. Musa
Department of Engineering Technology
Roy G. Perry College of Engineering
Prairie View A&M University
Prairie View, Texas

Ying Na
Hangzhou Dianzi University
Hangzhou, People's Republic of China

C. Narayan
The University of Akron
Akron, Ohio

Vinay Pai
National Institute of Biomedical Imaging and Bioengineering
Bethesda, Maryland

P. Pignalosa
Department of Electrical and Computer Engineering
University of Michigan
Ann Arbor, Michigan

and

Massachusetts Institute of Technology
Cambridge, Massachusetts

T. Quang
Department of Biomedical Engineering
The University of Akron
Akron, Ohio

Sukla Rajak
Department of Applied Optics and Photonics
University of Calcutta
Kolkata, India

Mina Ray
Department of Applied Optics and Photonics
University of Calcutta
Kolkata, India

Ufana Riaz
Materials Research Laboratory
Department of Chemistry
Jamia Millia Islamia
New Delhi, India

Suman Shrestha
Department of Radiology
University of Massachusetts Medical School
Worcester, MA

Viroj Wiwanitkit
Tropical Medicine
Hainan Medical University
Hainan, People's Republic of China

Y. Yi
Department of Electrical and Computer Engineering
University of Michigan
Ann Arbor, Michigan

and

Massachusetts Institute of Technology
Cambridge, Massachusetts

Zhen Yuan
Faculty of Health Sciences
Avenida Padre Tomás Pereira
Taipa, Macau SAR, People's Republic of China

M. Zervakis
Department of Electronic and Computer Engineering
Technical University of Crete
Chania, Greece

Guojin Zhang
Ashland Specialty Ingredients
Ashland, Inc.
Bridgewater, New Jersey

1

Applications of Vibrational Spectroscopic Imaging in Personal Care Studies

Guojin Zhang, Roger L. McMullen, Richard Mendelsohn, and Osama M. Musa

CONTENTS

1.1 Introduction

In the last decade, infrared (IR) and Raman microscopic imaging techniques have gained recognition not only in academic research laboratories but also in solving a variety of real-world problems confronting various industries. A major advantage of spectroscopic imaging techniques is the availability of direct molecular structural information inherent in IR and Raman spectra. The vast spectra are collected quickly by advanced imaging techniques, and the rich chemical-specific information is extracted with chemometrics. In this chapter, we demonstrate a variety of applications of both techniques related to phenomena in human hair and skin. In the hair, for example, we demonstrate through univariate and multivariate data analyses how to generate IR and Raman images containing the spatial distribution of lipids and proteins, protein secondary structure, lipid chain conformational order, and disulfide cross-links. This spectroscopic data provides a correlation of hair chemistry, anatomy, and structural organization. The obtained information provides a basis for us to understand how cosmetic treatments (hot ironing, bleaching, relaxing, etc.) affect hair protein structures, including hair keratin conformational changes (from α-helix to β-sheet) as well as structural damage to disulfide bonds. For studies concerned with the skin, we demonstrate the utility of these techniques in understanding chemically induced molecular structural changes in skin cells. Further, we are able to monitor hydration control mechanisms through evaluation of natural moisturizing factor (NMF) with IR images of corneocyte cells. Spectroscopic imaging also allows us to conduct skin permeability studies, which we demonstrate through the use of Raman imaging to track the delivery and controlled release of skin care ingredients in the stratum corneum (SC).

1.2 Characterization of Hair Chemistry, Microanatomy, and Structural Organization by IR and Raman Imaging

1.2.1 Hair Structure and Function

Human hair is a complex biological material consisting of distinct morphological components, each composed of several chemical species. The cuticle, the outermost layer of the hair, is composed of 5–10 layers of thin, flat, but circumferentially curved, overlapping cuticle cells and provides a chemically resistant region surrounding the cortex. The cortex, the major component of the hair mass, contains elongated keratinized cortical cells and the cell membrane complex. Cortical cells are spindle-shaped and full of intermediate filaments (crystalline phase) embedded in an amorphous matrix. The medulla is the innermost structure and generally occupies only a small percentage of the mass. In some fine hair, the medulla is completely absent; however, it is almost always found in Asian hair and other fibers.

1.2.2 IR and Raman Imaging of Hair Fibers

1.2.2.1 Overview of Experimental Protocols and Band Assignments

IR and Raman spectroscopy have been extensively used to probe hair chemical composition and keratin structure for years without supplying any spatially resolved information [1–5]. Recently published work in Ashland's Measurement Science laboratory demonstrates that IR imaging in conjunction with multivariate analysis has the enhanced capability to detect altered spatial regions in the hair sample [6]. The spatially resolved molecular information is comparable with that obtained by an attenuated total reflection (ATR) and synchrotron source-based spectrometer, which provides better spatial resolution [7,8]. Figure 1.1 illustrates typical protocols utilized to obtain hair cross sections. A 1 cm-long hair bundle was

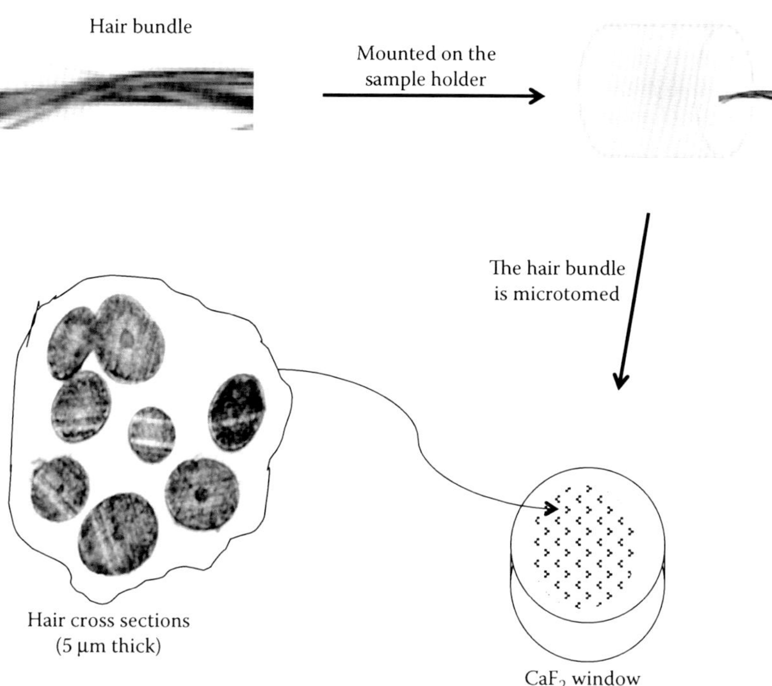

FIGURE 1.1
A typical procedure of hair sectioning for IR imaging.

cut from the middle of European dark brown hair tresses and mounted on the top of a sample holder by embedding it in ice. The hair bundle was then microtomed at –30°C into 5 μm-thick cross sections with a cryostat. Hair cross sections were collected onto CaF_2 windows. This preparation technique avoids any possibility of contamination with embedding or fixing medium. IR images are generally acquired with a 6.25 μm step size, eight scans for each spectrum, and 8 cm^{-1} spectral resolution. The main advantage of IR imaging is its capability of rapid data acquisition. Therefore, this approach permits sampling of much larger areas at a higher signal-to-noise ratio, however, with a lower resolution. Figure 1.2 demonstrates the nature of the information available in a typical IR imaging experiment of hair cross sections. Figure 1.2a contains typical IR spectra obtained from the medulla, cortex, and cuticle. The corresponding band assignments according to the literature are listed in Table 1.1 [9–12]. These vibrational modes coincide with protein secondary structure and lipid conformational order. Figure 1.2c and d depicts IR spatial images of protein distribution and the relative concentrations of lipid. The protein distribution was generated by integrating a protein band (Amide I, 1650 cm^{-1}), and the lipid concentration was imaged by taking the ratio of the lipid band at 2850 cm^{-1} to a band at 2960 cm^{-1} attributed to C–H asymmetric stretching of CH_3 moieties in amino acid side chains of proteins. The images clearly demonstrate that simple functional group mapping proves useful for tracking the concentrations of important chemical species in hair cross sections.

Raman spectroscopy complements the molecular vibration information available from IR spectroscopy and likewise contains detailed information about the structure, chemistry, and interactions of molecules of biomedical interest. Raman imaging has a higher spatial resolution and is suitable for the characterization of localized biochemical processes

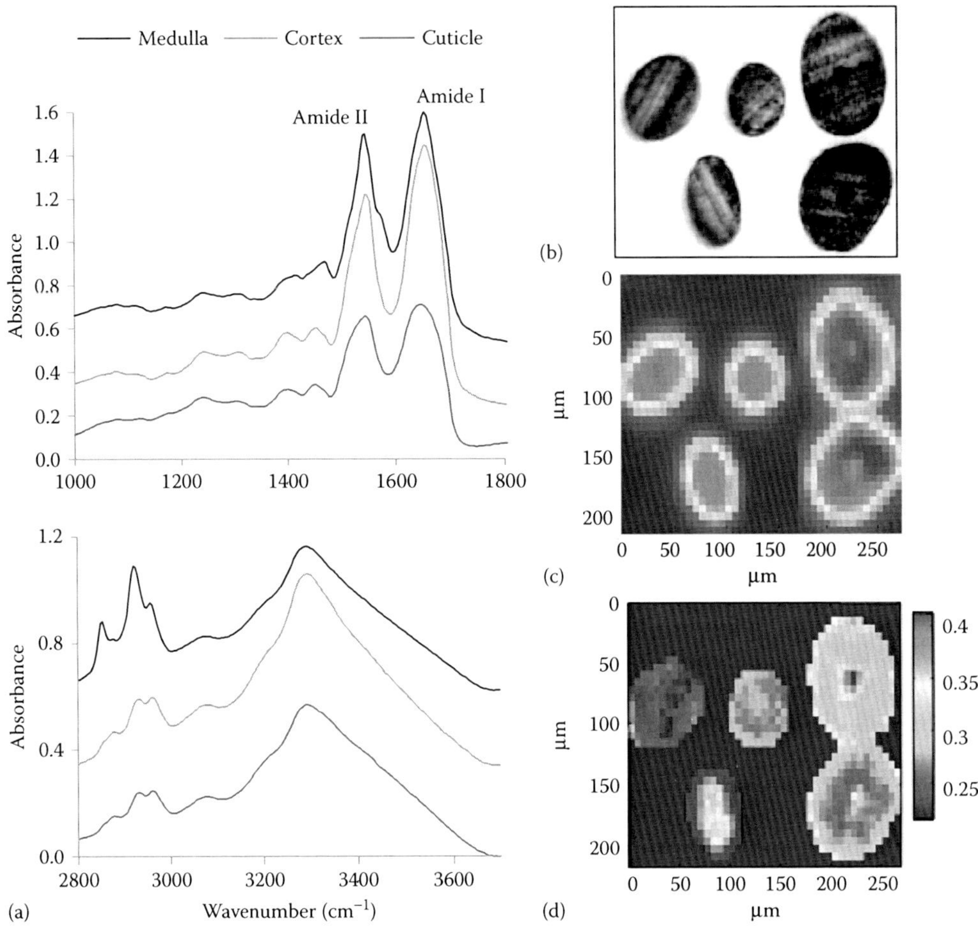

FIGURE 1.2
(See color insert.) Optical and IR imaging of hair cross sections. (a) Representative IR spectra obtained from the medulla, cuticle, and cortex in the spectral regions of 1000–1800 and 2800–3700 cm^{-1}. Each spectrum was obtained from a single pixel with dimensions of 6.25 μm × 6.25 μm. (b) Optical image of microtomed hair cross sections. (c) IR image of protein distribution. (d) IR image of lipid distribution. Image was obtained from integration peak area ratios of 2850 cm^{-1} (C–H stretching of CH_2 groups in lipid chains) to 2960 cm^{-1} (C–H stretching of CH_3 terminal groups in proteins).

[12–14]. In addition, Raman microscopy has the benefit of permitting confocal measurements of molecular structure or chemistry. Thus, the physical sectioning of hair fibers is not required to obtain molecular chemistry and structural information within the fiber. This provides opportunities for the nondestructive examination of hair fibers under many different conditions. However, the high melanin granule content in pigmented hair results in the absorption of laser light and subsequent sample destruction and/or fluorescence. Therefore, Raman spectroscopic analysis is limited to hair fibers with very little color. In the described experiment, a series of Raman spectra were acquired along a confocal line from the hair surface to 80 μm deep with a 2 μm step size. Figure 1.3a displays typical spectra acquired from the cortex with the corresponding assignments displayed in Table 1.1. A very useful utility of Raman microscopy is the capability to probe the conformation

TABLE 1.1

IR and Raman Band Assignments Corresponding to Lipid and Protein Conformation in the Hair and Skin

IR (cm^{-1})	Raman (cm^{-1})	Assignment and Approximate Description of Vibrational Mode
3278		Amide A, bonded N–H stretching
3070		Amide B, first overtone, Amide II at 1547 cm^{-1}
2957	2958	$\nu(CH_3)$ asymmetric stretching
2918	2883	$\nu(CH_2)$ asymmetric stretching from ordered lipid
2872	2876	$\nu(CH_3)$ symmetric stretching protein with minor from lipid contribution
2850	2852	$\nu(CH_2)$ symmetric stretching mostly from ordered lipid
	1685	Amide I, random coil, undefined structure
1650	1652	Amide I, 80% $\nu(C=O)$, α-helix
1630	1668	Amide I, β-sheet
1547	1543	Amide II, 60% δ(NH) and 40% ν(CN), α-helix
1540		Amide II, random coil
1517		Amide II, β-sheet
	1440	$\delta(CH_2)$ scissoring
~1395		$\nu(COO^-)$ symmetric
	1274	Amide III, α-helix, disordered structure
1247	1243	Amide III, β-sheet

δ, deformation; ν, stretch.

of disulfide cross-links in hair fibers [15], a parameter not accessible by IR spectroscopy. The importance of disulfide cross-links for the stabilization of hair protein structures is well known. In the Raman spectra shown in Figure 1.3a, the band at 509 cm^{-1} is assigned to the S–S stretching mode. To evaluate the spatial variation of S–S cross-linking across hair fibers, spectra obtained from the confocal lines were normalized to the peak area of a band at ~1448 cm^{-1} (the C–H scissoring mode from both CH_2 and CH_3 groups). The map of the integrated area of the band at 509 cm^{-1} is shown in Figure 1.3b, indicating that proteins in the cuticle region contain a higher level of S–S cross-links, which are consistent with the cuticle containing a higher content of half-cystine than the cortex [16]. It is noticeable that the S–S cross-links are unevenly distributed through different layers of the cuticle.

1.2.2.2 Microanatomy Characterization of Hair Fibers by IR Imaging in Conjunction with Factor Analysis

1.2.2.2.1 Factor Analysis

The spectroscopic images include a wide range/amount of spectra (between several hundreds and 15,000) with large variable features from one pixel to another due to sample heterogeneities. Efficiently extracting information from such large quantities of spectral data is a big challenge. Principal component analysis (PCA) is a multivariate approach, which is able to condense the information into a small set of dimensions with minimum loss of information. However, the loading vectors from PCA calculations are not generally pure component spectra. Principal components (PCs) are ranked in terms of the percentage of the total variance that they explain. Therefore, it is difficult to use PCs to establish the spectra–structure relationship. If the spectral image is proportional to an absorbance

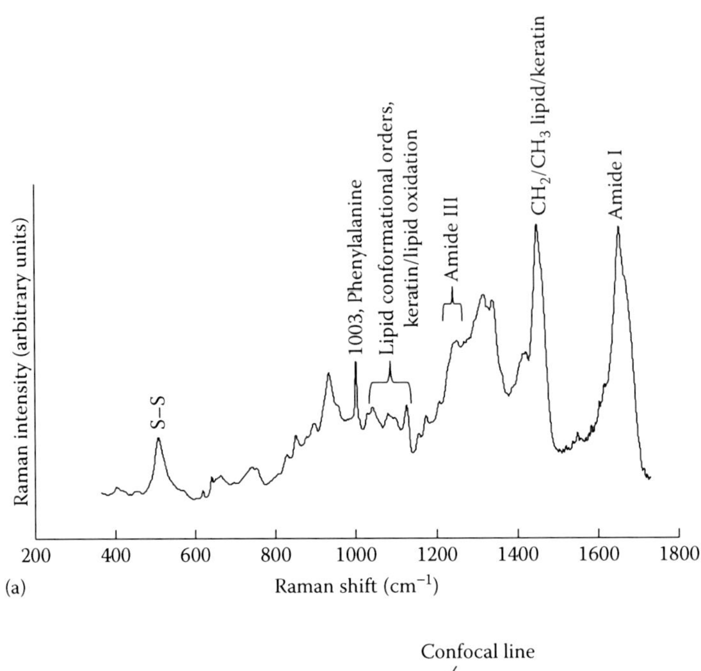

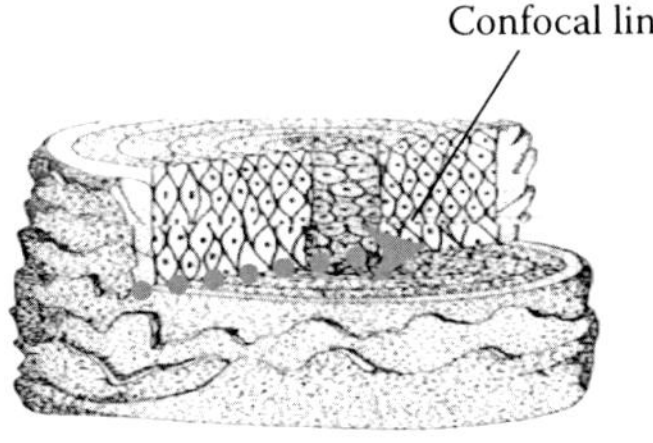

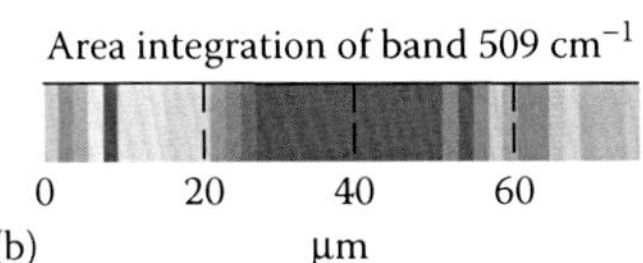

FIGURE 1.3
Raman images of chemical constituents in a confocal line of the hair fiber. (a) A typical Raman spectrum acquired from the cortical region. (b) *Upper*: Schematic diagram of confocal Raman acquisition within the hair fiber. *Lower*: Raman image of S–S cross-link distribution along a confocal line. The map was created by integrating the peak areas of band 509 cm^{-1}, the S–S stretching mode of hair keratin fibers. Before peak integration, Raman spectra were normalized by the band intensity of 1448 cm^{-1}. Dark shades of gray in the image correspond to lower levels of disulfide bonds while brighter shades of gray are indicative of higher levels of disulfide bonds.

image, then to the extent that Beer's law is obeyed, the intensity at the (i,j)th pixel and kth wavelength can be expanded as follows:

$$I_{ijk} \approx \sum_{m} C_m e_{mk} \tag{1.1}$$

where:

C_m is the concentration of species m

e_{mk} is the molar extinction coefficient of species m at the kth wavelength of the spectrum

If we *unfold* the image cube into a two-dimensional array, X, with pixel spectra along each row, we can then express that matrix as a sum of outer products of concentration vectors and pure component spectra:

$$X = \sum_{m} C_m e_m \tag{1.2}$$

In the above equation, each C_m is a $p \times 1$ vector (p is the number of pixels in the image), each component of which represents the concentration of species m at a particular pixel, and e_m is a $1 \times k$ (k is the number of channels in each pixel spectrum) spectrum of species m. Note that PCA produces a factorization of the data matrix:

$$X = \sum_{j} S_j L_j \tag{1.3}$$

where:
S_j are the scores
L_j are the corresponding loadings

If the data matrix can be factorized in terms of concentrations and spectra, then it will be possible to find transformations between L_j and e_m. The corresponding score images will then represent the concentration profiles of various components. Factor analysis seeks transformations from the abstract PCA loading vectors to spectra loadings, which are able to explain spectra–structure relationships. Factor analysis was conducted by ISys 3.1 software developed by Malvern Instruments Ltd, Malvern, UK. It is performed in a three-step sequence: determination of the number of factors, diagonalization of the original covariance matrix, and score segregation.

In choosing how many factors to utilize, a useful first step is to carry out PCA because the number of components needed is often a good guide to the number of factors in factor analysis. The number of PCs required is judged according to some criteria, such as a *scree plot*, or the percentage of the total variance explained by the number of components. In both hair and skin studies, the molecular information from spectra is used to identify biological structures within the tissue. The number of factors is also assessed with skin and hair biological structure models. As noted above, the hair consists of the cuticle, the cortex, and the medulla. If exogenous agents or additional treatments are applied to the substrate, more factors will be introduced. Although the choice of factor numbers is subjective, the ability of the method to determine the correlation structure and the ease of interpreting factors arising from overlapped component spectra make factor analysis a powerful multivariate statistical approach.

Score segregation begins by normalizing the scores by factor numbers so that their intensities range from 0.0 to 1.0, then by performing vector normalization. The score is generated by matrix multiplication, $S = X \times L$, where X is a two-dimensional array with pixel spectra along each row and L is the transpose of the matrix of normalized loading. Score segregation is an iterative technique, which relies on the convergence of the loading vectors as measured by the loading difference error. The factor loadings are calculated according to $(S'S)^{-1}S'X$, unless $S'S$ is not invertible, in which iteration stops. Under this condition, a factor has to be deleted or added before continuing. At the end of factor analysis, one set of factor loadings and score images is generated. Factor loadings are assumed to represent spectra from the location at which they are determined, whereas scores indicate the relative contributions of various factors from particular locations.

1.2.2.2.2 IR Imaging of Microanatomy of Hair

The benefit of using both factor analysis and existing spectra–structure correlations to characterize the hair structure is demonstrated by the following example: An IR image acquired from microtomed virgin hair sections (5 μm thick) was analyzed by multivariate factor analysis performed in separate spectral regions 1000–1476, 1476–1700, and 2832–3700 cm^{-1}. The detailed method is described in a previous publication [6]. The corresponding optical micrograph is shown in Figure 1.4a. Four factor loadings from each spectral region are shown in Figure 1.4b and labeled as f1–f4. Representative score images corresponding to the four factor loadings in the spectral region 1480–1700 cm^{-1} are presented in Figure 1.4c, with the color bar marked from deep red for the highest score to dark blue for the lowest score. Score images of factor loadings from the remaining spectral region of 1000–1476 and 2832–3700 cm^{-1} show essentially the same results (not shown here). Score distributions depict correlations between factor loadings and raw spectra directly obtained from these measurements. Factor loadings generated from factor analysis are

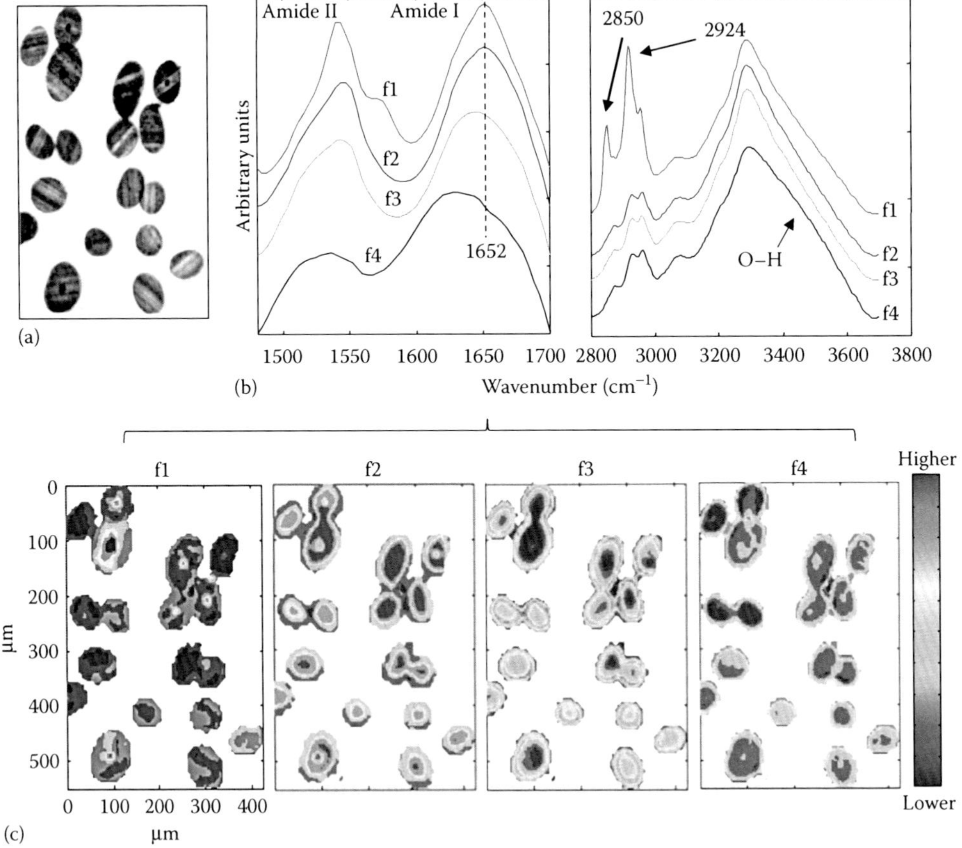

FIGURE 1.4
(See color insert.) (a) An optical micrograph of microtomed (5 μm thick) hair cross sections. Factor analysis was conducted on an IR image acquired from hair cross sections. (b) The four distinct factor loadings in the regions of 1480–1700 and 2830–3700 cm^{-1} generated by the ISys score segregation algorithm are offset and labeled as f1–f4. (c) The spatial distribution of factor scores for each of the loadings in the spectral region of 1480–1700 cm^{-1} as marked. Dark blue indicates the lowest score, whereas green, yellow, orange, and red indicate progressively higher scores. Factor loadings and score images have been assigned to different micro-regions in hair as described in the text.

assumed to represent spectral features of the raw spectra acquired from areas with higher scores. As shown in Figure 1.4c, the score distribution images clearly delineate the known anatomical regions of the hair, that is, f1 corresponds to the medulla, f2 to the cortex, and f4 to the cuticle. The remaining factor (f3) highlights the transition region between the cortex and the cuticle. In the same way, the factor loadings contain spectral features specific to each hair region. Briefly, in the factor loadings of the spectral region from 2830 to 3700 cm^{-1}, f1 presents strong lipid bands at 2850 and 2924 cm^{-1}, indicating generally higher lipid content in the medulla area. The band centered at 3400 cm^{-1} is assigned to O–H stretching, and the increased intensity at 3400 cm^{-1} shown in f4 reflects relatively higher content of OH groups in the epicuticle area. The high level of OH groups may be attributed to water absorbed on the hair surface and/or a high level of serine residues in the cuticle. In the factor loadings of the spectral region from 1480 to 1750 cm^{-1}, the frequencies of Amide I and Amide II shown in f1 and f2 indicate predominantly an α-helical structural conformation in the cortex and medulla regions. The broadening contour of Amide I and the relatively increased intensities at ~1630 cm^{-1} (Amide I) and 1516 cm^{-1} (Amide II) shown in f3 and f4 indicate more β-sheet contribution in the cuticle area.

As demonstrated above, although the factor analysis condenses a large amount of spectra into several factor loadings, the factor loadings resemble real spectra and do represent the molecular structural features of the raw spectra from the location at which they are measured. Score images spatially delineate micro-regions within hair sections according to their underlying molecular structure. It is feasible to utilize such a multivariate algorithm to efficiently examine the altered spatial regions of hair samples of biomedical interest. An understanding of the spatially resolved spectroscopy of hair components paves the way to the investigation of molecular alteration of hair proteins and lipids resulting from environmental stresses and disease status.

1.2.3 Investigating Hair Damage Resulting from Common Cosmetic Treatments with IR and Raman Imaging

1.2.3.1 Hot Iron Treatment

Hot flat irons are one of the most popular thermal styling appliances used to create straight styles and operate at temperatures in excess of 200°C, which cause damage to hair keratin. The keratin structural modifications from thermal treatment were examined by IR and Raman imaging. It was found that thermal treatment of hair causes the conversion of the α-helical protein to β-sheet in addition to protein loss, which results from 12 minutes of thermal exposure at 232°C (short 12-second intermittent heat cycles) [17]. Figure 1.5a shows the typical IR spectra and their second derivative analysis corresponding to the cortical region of undamaged European dark brown hair from 1480 to 1700 cm^{-1} (Amide I and II). The second derivative curve displays the minor component of β-sheet (Amide I at 1630 cm^{-1}, Amide II at 1516 cm^{-1}) and a major α-helical structure (Amide I at 1650 cm^{-1}, Amide II at 1548 cm^{-1}) for undamaged hair. To compare the changes of these two compositions after thermal treatment, the ratio of the peak intensities at 1516 cm^{-1} (attributed to the β-sheet formation) and 1548 cm^{-1} (α-helix band) was used to quantify the conversion of α-helical to β-sheet conformation. These ratio maps of hair cross sections are shown in Figure 1.5b. It can be seen that the outermost layers of hair have higher β-sheet levels than the cortex in virgin hair [Figure 1.5b(i)]. The β-sheet content of hair after thermal treatment increases and becomes more pronounced in the outside layer of thermally treated hair as the heating commences at the surface [Figure 1.5b(ii)].

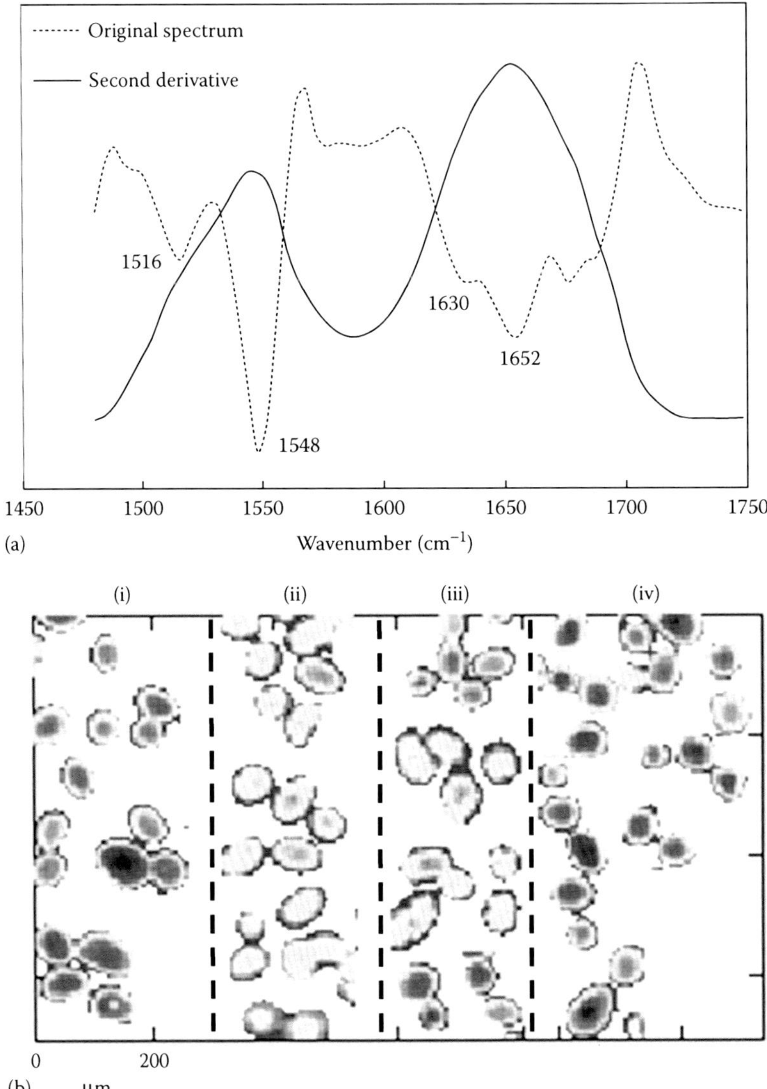

FIGURE 1.5
(See color insert.) (a) IR spectra and the second derivative curve in the amide region of 1480–1700 cm^{-1}. The spectrum was obtained from the cortical region of undamaged European dark brown hair. The assignments of marked peaks are described in Table 1.1. (b) β-Sheet distributions in hair cross sections. The color coding is red > orange > yellow > green > blue. The maps are constructed from the ratio of peak intensities at 1516 cm^{-1} (attributed to β-sheet formation) and 1548 cm^{-1} (α-helix band). Hair samples include European dark brown hair (i), thermally treated hair without polymer pretreatment (ii), thermally treated hair with pretreatment of 0.5% hydroxyethylcellulose (iii), and thermally treated hair with pretreatment of 1% poly(vinylpyrrolidone-*co*-acrylic acid-*co*-laurylmethacrylate) + 0.5% hydroxyethylcellulose (iv).

The protein molecular structural changes from thermal treatment can also be examined by confocal Raman spectroscopy. Figure 1.6a shows Raman spectra obtained from the cortical regions of both thermally treated and untreated hair fibers. Compared to untreated hair, the Raman spectrum corroborates the β-sheet assignment with the more pronounced appearance of a high-frequency Amide I component at about 1669 cm^{-1} and a strong Amide III component at approximately 1240 cm^{-1} in thermally treated hair. The broadening on the high-frequency side of the Amide I band in the Raman spectrum obtained from thermally treated hair fibers indicates the formation of more random coil structure in thermally treated hair.

Disulfide cross-links are very important for stabilizing the hair protein structure. The changes in disulfide bonds resulting from thermal treatment were probed in the spectral range of 470–550 cm^{-1}, as shown in Figure 1.6b. The relative concentrations of disulfide bonds within the hair fibers are mapped by the band intensity ratio of the S–S stretching mode at 509 cm^{-1} to the band at 1004 cm^{-1} assigned to phenylalanine of keratin (Figure 1.6c). The results indicate that thermal treatment cleaves the disulfide bonds as the intensity of S–S stretching significantly decreases in thermally treated hair. A new band arises at 487 cm^{-1} after the thermal treatment. The clear assignment of this band has not been resolved yet. However, it has been reported that the position of the S–S stretching mode varies with different disulfide conformers [4,18]. This newly formed band may relate to the conformational transformation of the disulfide band during thermal treatment.

To alleviate native protein structural damage, a thermal protection route was developed by Ashland (Bridgewater, NJ), aiming to put polymer barriers on the hair surface prior to hot ironing cycles without affecting styling. A variety of polymers have been tested [17]. Examples included in Figure 1.5b(iv) indicate that pretreatment with poly(vinylpyrrolidone-*co*-acrylic acid-*co*-laurylmethacrylate) is able to effectively prevent the conversion of α-helical to β-sheet conformation with the intensity ratio close to that of nonthermally treated hair. However, hydroxyethylcellulose, a thickener usually used in shampoo formulations, does not offer any thermal protection to hair fibers [Figure 1.5b(ii)].

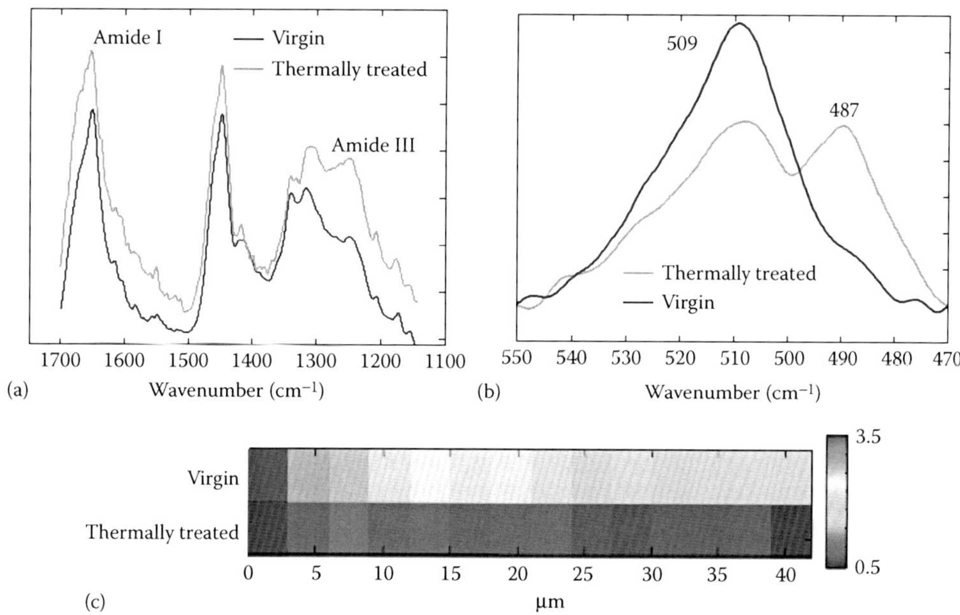

FIGURE 1.6
(See color insert.) Representative Raman spectra in the spectral region of (a) 1150–1700 cm^{-1} and (b) 470–550 cm^{-1}. The Raman spectra were obtained from the cortical region of both thermally treated and virgin hair fibers. (c) Raman image of disulfide bond distribution along a confocal line. The confocal line was acquired from the hair surface to 42 μm deep into the fiber structure. The schematic diagram of confocal Raman acquisition is explained in Figure 1.3b. The map was created by the intensity ratio of the band at 509 cm^{-1}, arising from the S–S stretching mode, to the band at 1004 cm^{-1} (phenylalanine in keratin).

1.2.3.2 Hair Bleaching Treatment

Hair bleaching is another common cosmetic treatment which involves lightening hair color as well as hair dyeing. It is well known that significant damage to hair is derived from oxidation due to bleaching. Figure 1.7a presents typical IR spectra obtained from the cortical regions of virgin and bleached hair. As shown in the spectra, the distinct band at 1040 cm^{-1} in bleached hair is assigned to S=O stretching of sulfonates, indicating oxidation of cysteine after the hair bleaching treatment. The relative sulfonate concentration was imaged from the integrated area ratio of the band at 1040 cm^{-1} to Amide I (Figure 1.7b). It is obvious that sulfonate is distributed across the entire hair section, indicating damage in protein structure occurs in the whole hair fiber, but is more pronounced in the outside layers of bleached hair, where more half-cystine residues are normally found (in virgin hair). The concentration of sulfonates in the cuticle is up to 3 times higher than that in the cortex.

1.2.3.3 Hair Relaxing

Hair relaxing is a hair styling process by which hair curl is eliminated or straightened. Although the formulation is hair-type dependent, hair fibers are exposed to alkaline conditions during the relaxing process. In this study, hair fibers were treated with 2% NaOH solution for 10 minutes. Average spectra were obtained from 20 hair cross sections of both virgin and treated hair. Mean spectra along with spectral difference in the region of 900–1500 cm^{-1} are displayed in Figure 1.8a and b. The most prominent feature in the spectra is the increased intensity in the band at 1404 cm^{-1} in the spectrum acquired from alkali-treated hair cross sections. The band at 1404 cm^{-1} has been assigned to carboxylate symmetric stretching. The area integration images of Amide I and COO^- groups for both virgin hair and alkali-treated hair are displayed in Figure 1.8c. Alkali treatment cleaves the peptide amide bond resulting in a significant increase of COO^- groups across the hair fiber. The COO^- groups increase ~60%, whereas Amide I bonds decrease ~30%. This result indicates that not all COO^- groups are generated from amide bond cleavage of peptide

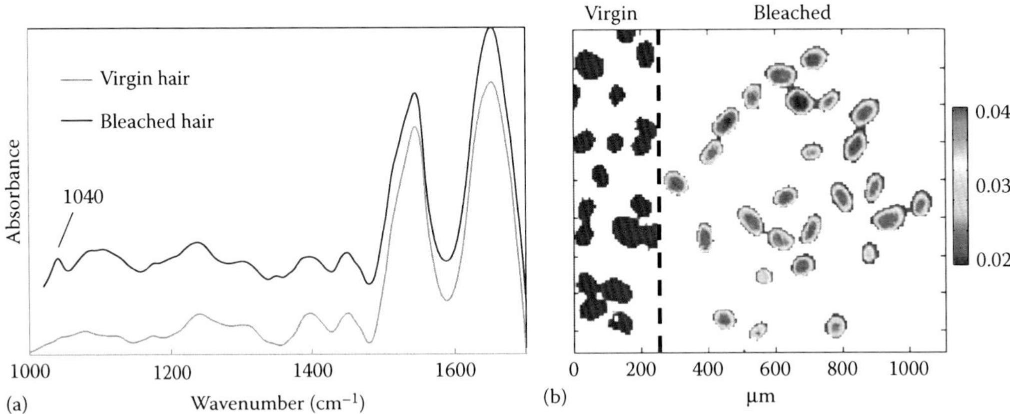

FIGURE 1.7
(See color insert.) (a) IR spectra acquired from the cortical regions of virgin and bleached hair in the spectral regions of 1000–1700 cm^{-1}. The marked band at 1040 cm^{-1} in the spectrum of bleached hair arises from S=O stretching mode of the sulfonate groups. (b) IR images of sulfonate distribution in hair cross sections (5 µm thick) microtomed from virgin and bleached hair fibers.

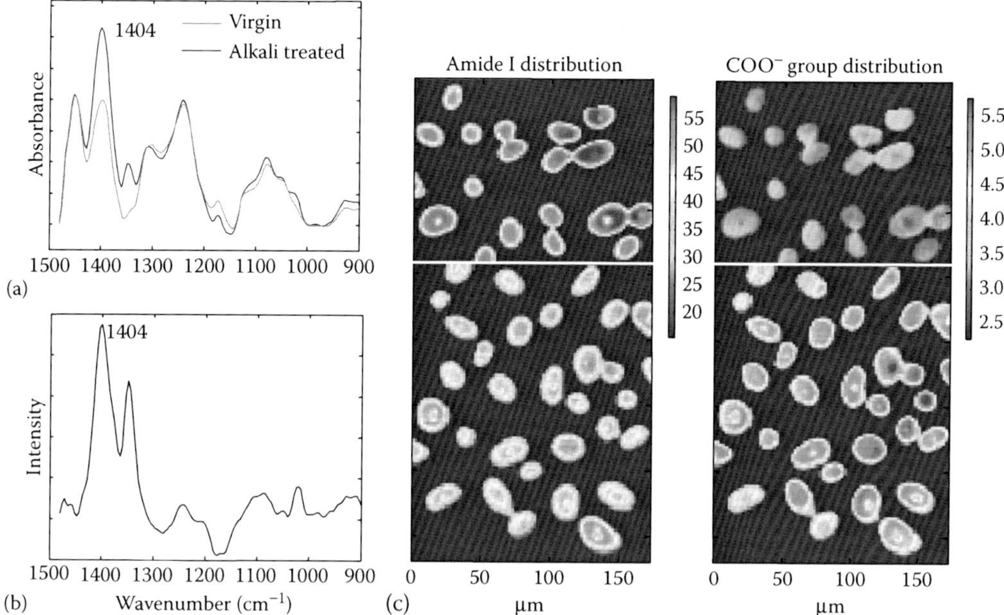

FIGURE 1.8
(See color insert.) (a) Mean IR spectra acquired from 20 hair cross sections microtomed from virgin and alkali-treated hair in the spectral region of 900–1500 cm^{-1}. (b) The difference spectrum obtained by subtracting the mean spectrum of virgin from the mean spectrum of alkali-treated hair. (c) IR images of Amide I and COO$^-$ group distributions. *Upper*: virgin hair. *Lower*: alkali-treated hair.

backbones during the relaxing process. Half of newly formed COO$^-$ groups may result from other sources, for example, hydrolysis of acidic amino acid groups such as aspartic and glutamic acid residues [19].

1.2.4 Investigation of Lipid Biophysical Properties of Hair Fibers

Hair contains free, integral, and covalently attached lipids. Free lipids mainly supplied from sebaceous glands, primarily consist of free fatty acids and neutral fat (esters, waxes, hydrocarbons, and alcohols). They reside on the surface of hair as well as in its internal structure. Integral lipids are not covalently bonded to hair but serve a structural function as components of the cell membrane complex between cuticle and cortical cells. Covalently attached lipids, in particular 18-methyleicosanoic acid, are found on the surface of cuticle cells [20,21]. In a recent study, researchers found that removal of the free and integral lipids does not have an impact on hair stiffness and elasticity; however, it does change the binding capacity of hair with rinse-off ingredients [22]. The lipid distribution in hair can be easily mapped by both IR and Raman imaging. Figure 1.9a shows the total lipid distribution across hair sections. This image was constructed from IR imaging by integrating the band areas and taking the ratio of the peak at 2850 cm^{-1}, resulting from C–H stretching of CH_2 groups in lipids, to Amide I. It is observed that lipid concentrations vary among different hair fibers. However, we find that the medulla and the cuticle always have relatively higher concentrations of lipids. Raman complements the information that IR provides. In addition to lipid distribution, Raman spectra also contain a detailed information regarding lipid conformational order (fluidity) of lipid chains [23,24]. A representative Raman spectrum obtained from the medulla area is displayed in Figure 1.9d. The lipid bands at 1125 and 1080 cm^{-1} are both assigned to C–C skeletal stretching. The former band, in particular, is indicative of an all-*trans* acyl chain conformation and has been extensively used for monitoring lipid conformational order (fluidity). The peak intensity ratio of 1125/1080 cm^{-1} along a confocal line from the hair surface to the interior region of the fiber is shown in Figure 1.9e. The map shows that lipids in the cuticle region are considerably more conformationally ordered than lipids in the medulla. The ordered lipids probably pertain to structural lipids that form the cell membrane complex of hair, both covalently and noncovalently bound. The disordered lipids, however, are free lipids, many of which probably originate from sebum. Hair lipid was extracted by a method with increasing polarity of solvents: it was first extracted with *n*-hexane for 4 hours, then with *t*-butanol for 4 hours, and finally with a mixture of chloroform and methanol (70:30, v/v) for 6 hours. The lipid distributions from the extracted hair are displayed in Figure 1.9b and c. After hexane extraction, the surface lipid, about 13% of total lipids, is removed, but lipids within the internal structure of hair remain. Continuous chloroform/methanol extraction removes most of the internal lipids, about 98% of total lipids. As solvent extraction only removes the noncovalently bonded lipid in the hair, it is estimated that covalently bonded lipid is less than 2% of total lipids.

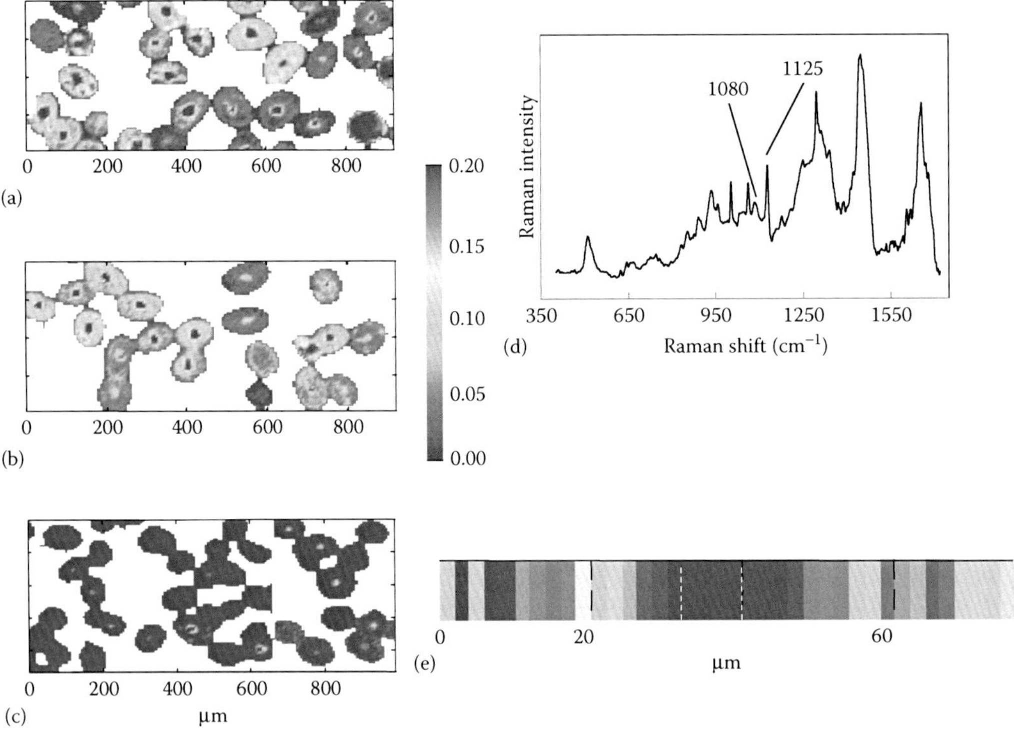

FIGURE 1.9
IR images of lipid distribution in hair cross sections of (a) virgin hair, (b) hair extracted with hexane, and (c) hair extracted with hexane and then with a mixture of chloroform and methanol (70:30). (d) Raman spectrum acquired from the medulla region of a virgin hair fiber in the spectral region of 400–1700 cm^{-1}. The marked bands at 1125 and 1180 cm^{-1} are assigned to C–C skeletal stretching in lipids, the former band from all-*trans* acyl chains and the latter from random conformations. (e) Raman image of conformationally ordered lipids in a confocal line. The schematic diagram of confocal Raman acquisition is explained in Figure 1.3b. The Raman map was constructed from the intensity ratio of the band at 1125 to 1080 cm^{-1}.

1.3 Spectroscopic Imaging of Skin: From Single Cells to Intact Tissue

1.3.1 Skin Structure and Function

The skin is the largest organ of human body. It plays many vital roles in both regulating heat and water loss from the body and preventing the ingress of noxious chemicals or microorganisms. Typical skin structure is depicted in Figure 1.10. It contains three layers in order from the outermost to the innermost as is follows: the SC, the epidermis, and the dermis. Although the SC is thin (~10–20 μm), it provides the main barrier to permeability and serves to maintain water homeostasis. Underlying the SC is the viable epidermis, which possesses a complex multilayered structure. The principal cell in the viable epidermis is the keratinocyte, whose main function is to generate SC. During this process, the cells migrate upward; meanwhile, keratinocytes differentiate and synthesize keratins. Eventually, the keratinocytes become flattened in the stratum granulosum layer as enzymes in this layer degrade viable cell components, such as nuclei and organelles. In normal, healthy skin, the thickness of the epidermis ranges from 50 to 150 μm. The dermis, however, consists of tough connective tissues and a variety of specialized structures. Its thickness ranges from 0.6 to 3 mm and collagen accounts for ~75% of its dry weight.

The region of the skin of major importance in cosmetic science and pharmacology research is the outermost layer, the SC. It has a unique structure and provides the skin with its permeation barrier function. The traditional *bricks and mortar* model provides an interpretation of the SC organization in which corneocytes of this layer (the *bricks*) are embedded in a continuous matrix of highly ordered intercellular lipid phase (the *mortar*). Corneocytes are anucleated cells with aligned keratin filaments wrapped inside a cell envelope (CE).

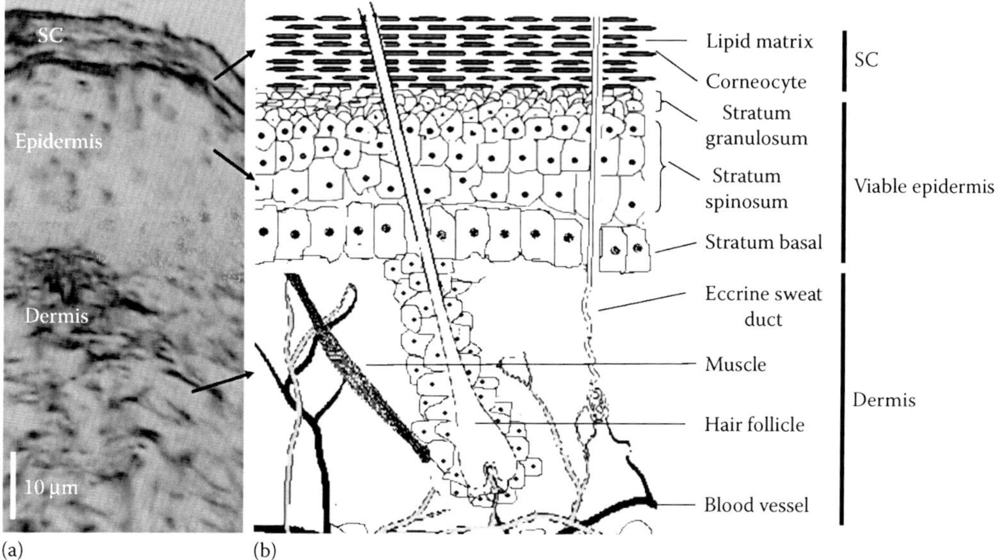

FIGURE 1.10
Structure of the skin. (a) An optical micrograph of a 5 μm-thick microtomed unstained pig skin section depicting the SC, underlying epidermis, and dermis regions. (b) A schematic cross section through the skin with anatomical features and subregions labeled.

1.3.2 IR Microscopy and Imaging of Corneocytes

This section describes the applications of IR microscopy and imaging to the study of isolated human corneocytes. The approach is suitable for monitoring the molecular structure, chemical composition, and biological maturation process of corneocytes.

1.3.2.1 Investigating Molecular Structures and Compositions of Isolated Human Corneocytes by IR Microscopy

Corneocytes consist of closely packed arrays of keratin filaments inside a highly insoluble, cross-linked protein envelope, referred to as the CE. Isolation of corneocytes was achieved by tape-stripping human forearm skin. The corneocyte-attached tape was then flushed with hexane. The hexane/corneocyte suspension was then sonicated for ~2 minutes to break up desmosomes and disrupt lipid cohesion between individual corneocytes. A small aliquot of this suspension was deposited on CaF_2 windows, and IR spectra were acquired after the solvent evaporated. A typical optical image of an isolated corneocyte is displayed in Figure 1.11a(i), and the corresponding IR spectrum is presented in Figure 1.11b(i). Refer to Table 1.1 for the assignments of vibrational modes. To resolve the protein molecular structure of corneocytes, IR spectra were also acquired from the extracted human epidermal keratin [Figure 1.11b(ii)] and the isolated human SC [Figure 1.11b(iii)]. As shown in Figure 1.11b, corneocytes share similar spectral features with keratin in the Amide I and II regions, which reflect the fact that proteins in the corneocytes are dominated by α-helical conformation. The SC presents a slightly different contour in the Amide II band compared to individual corneocytes, which is attributed to more N–H and C–N vibrations from intercellular ceramides. The CE, shown in Figure 1.11a(ii), was isolated by extracting the dried corneocytes in 2% sodium dodecyl sulfate (SDS) with 50 mM β-mercaptoethanol for 1.5 hours at 60°C, centrifuged at 30,000 rpm, and then rinsed with 0.1% SDS solution to remove the β-mercaptoethanol. A small aliquot of CE/0.1% SDS suspension was deposited on the CaF_2 window. The CE on the window was gently washed with distilled water to remove the remaining SDS and dried under vacuum overnight for IR measurements. The mean IR spectrum obtained from 12 randomly picked CEs is presented in Figure 1.11c(i) (dashed line). The CE mainly consists of two components: defined structural proteins and covalently attached lipids. The latter binds the envelope protein from the exterior surface of the envelope to the intercellular lipid domains. It was previously reported that CE proteins are predominantly in the β-sheet conformation [25]. However, as shown in Figure 1.11c, there is no substantial evidence that CE proteins are dominated by β-sheets since no band appears at or near 1630 cm^{-1}. The normalized spectra (by the peak intensity of the Amide I band) in the region of 1480–1750 cm^{-1} are displayed in the inset of Figure 1.11c. The CE has a broader Amide I peak compared to the corneocyte, a possible consequence in which the CE contains a variety of proteins in various conformations [26]. In particular, the stronger Amide II band at 1540 cm^{-1} may suggest that there are random coils in the CE. Generally speaking, the thickness of a corneocyte is ~500 nm. Assuming that the CE has the same extinction coefficient as the corneocyte, it is estimated from the Amide I intensity, according to Beer's law, that the thickness of the corneocyte envelope is ~25–30 nm.

Both corneocytes and CEs present strong lipid bands at ~2850 and 2920 cm^{-1}. The lipids in the envelope are slightly disordered as the frequency of the CH_2 stretching mode is shifted by 1 cm^{-1}. This could be caused by the isolation process in which keratin filaments are removed and the bonding between keratin and envelope lipid is disrupted.

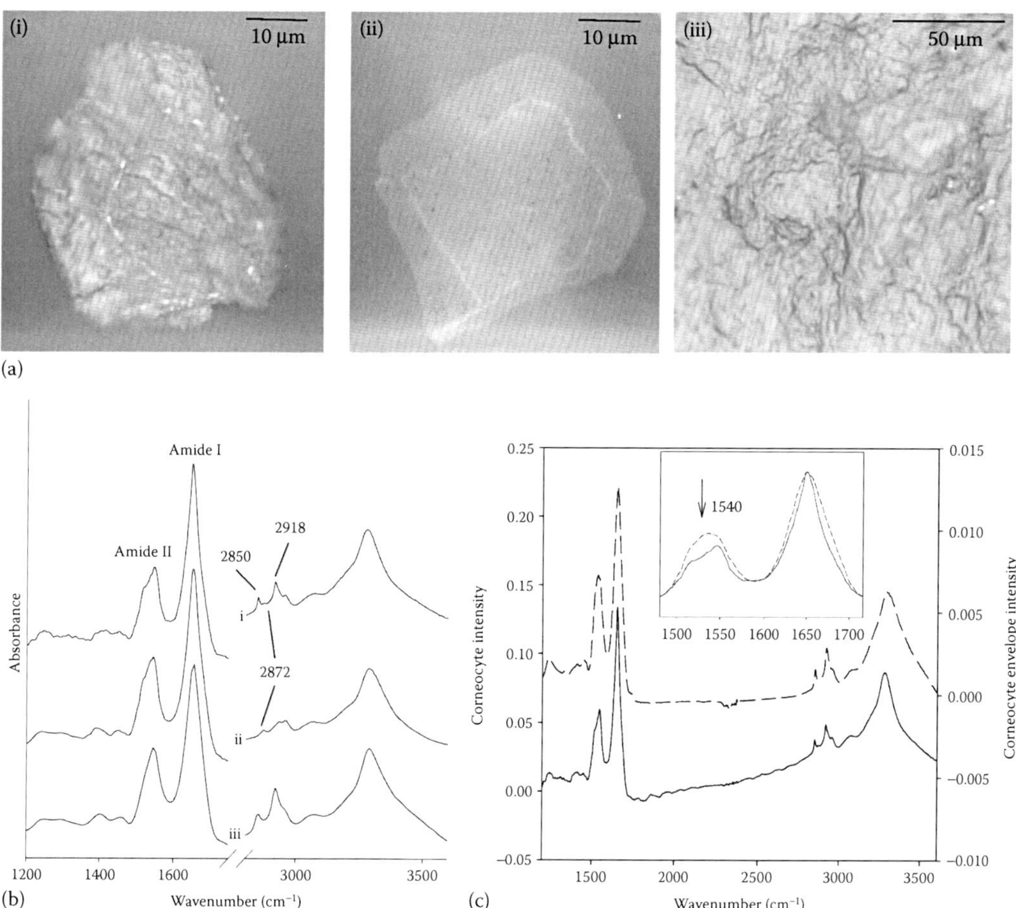

FIGURE 1.11
(a) Optical image of corneocyte (i), CE (ii), and isolated stratum corneum (iii). In (i) and (ii), the bar = 10 μm. (b) IR spectra of corneocyte (i), epidermal keratin (ii), and stratum corneum (iii). (c) IR spectra of CE (---) and corneocyte (——). Inset: IR spectra of corneocyte and CE in the spectral region of 1480–1750 cm^{-1}. The spectra are normalized to the Amide I band intensity.

1.3.2.2 Understanding Solvent-Induced Protein Structural Changes in Corneocytes by IR and Raman Microscopy

1.3.2.2.1 Dimethyl Sulfoxide

Dimethyl sulfoxide (DMSO) is well known as an efficient permeation enhancer. It has successfully ferried many drugs across the SC including morphine sulfate, penicillin, steroids, and cortisone. Its mechanism is generally thought to involve DMSO alteration of the protein conformation in the SC. Most studies have been completed on intact SC [27]. Vibrational spectroscopy is a great technique for the evaluation of keratin secondary structure; however, it can be hampered by the native ceramide contribution to the Amide I and II bands. This concern is mostly overcome by acquiring spectra of isolated corneocytes in which the ceramide contribution to the amide modes is substantially reduced. In this study, isolated corneocytes are directly treated with DMSO. The IR spectrum of an isolated corneocyte treated with pure DMSO for 1 minute is presented in Figure 1.12b. The most significant change in the IR spectra is the appearance of an intense low-frequency Amide I component observed for the DMSO-treated sample at 1626 cm^{-1} in conjunction with a weak high-frequency shoulder at ~1695 cm^{-1}. This pattern is diagnostic for formation of an antiparallel β-sheet structure. A redistribution of the maximum intensity in the IR Amide II band from 1550 to 1517 cm^{-1} for the same two spectra is also consistent with β-sheet formation. The mechanism by which DMSO converts keratin conformation from α-helix to β-sheet is still under debate. DMSO is known to form an association complex with water stronger than that formed between water molecules alone [28]. It is thought that DMSO interrupts hydrogen bonds between NH and CO groups in the peptide backbone, thereby displacing the bound water necessary for maintaining secondary structure. Interestingly, upon rehydration of DMSO-treated corneocytes, the IR spectra of corneocytes are very similar to those of untreated corneocytes (Figure 1.12a and c). This result provides direct evidence that the conformational change induced by DMSO is generally reversible. When water molecules competitively bind with DMSO, the keratin conformation reverts to its native state upon the removal of DMSO. This work demonstrates that although DMSO performs its enhancement function by denaturing skin proteins, the denaturing process is generally reversible when water is present. In this case, appropriate concentrations of DMSO will not cause irreversible damage to the skin barrier. This is a desirable property for penetration enhancers.

1.3.2.2.2 Chloroform/Methanol

In studies of native protein structure in the SC, the mixture of chloroform and methanol (2:1, v/v) is a very common solvent used to remove lipids. The current work indicates that this procedure is not innocuous and could induce irreversible changes in protein structure. Isolated corneocytes were continuously extracted with chloroform/methanol

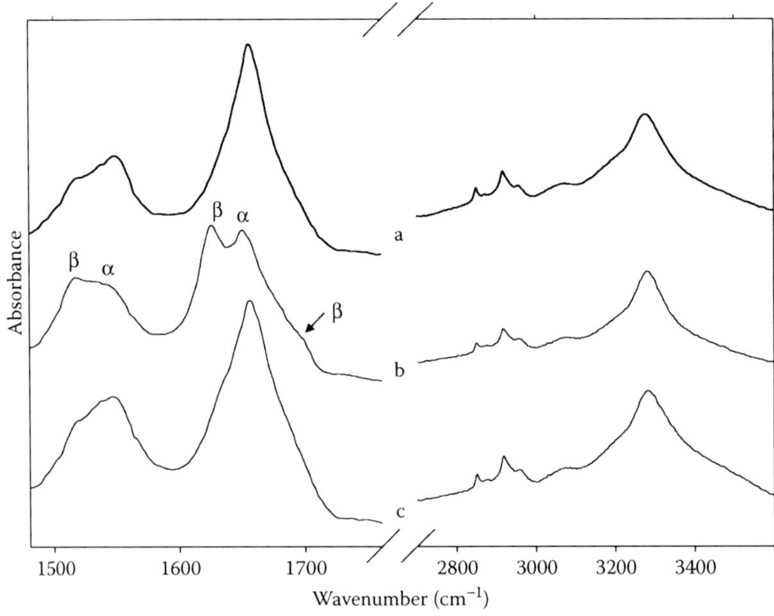

FIGURE 1.12
DMSO-induced conformational changes in the keratin of single corneocytes. IR spectra were acquired from an isolated corneocyte prior to (a) and following (b) immersion in DMSO and after an overnight rehydration period (c). α-Helical and β-sheet components of the Amide I and II bands are noted.

for 68 hours. This procedure removed most free lipids in the corneocytes. As shown in Figure 1.13(i), lipid characteristic bands at 2850 and 2920 cm^{-1} disappear [Figure 1.13a(i) and b(i)]. In the amide spectral region of 1480–1700 cm^{-1}, several changes are apparent in the IR and Raman spectra of corneocytes upon exposure to solvents. The low-frequency Amide I component in the IR spectrum [Figure 1.13b(i)] at 1626 cm^{-1} and the appearance of an Amide III band at 1240 cm^{-1} in the Raman spectrum [Figure 1.13b(ii)] provide strong evidence that the chloroform/methanol mixture induces β-sheet formation. The broadening on the high-frequency side of the Amide I band in the Raman spectra is much more obvious after chloroform/methanol treatment. The Amide I in this region (~1685 cm^{-1}) is assigned to non-hydrogen-bonded disordered structures [29]. After rehydration, α-helical conformation is partially recovered, while part of the β-sheet converts to undefined or random conformation as the Amide I band in both IR and Raman spectra becomes broader.

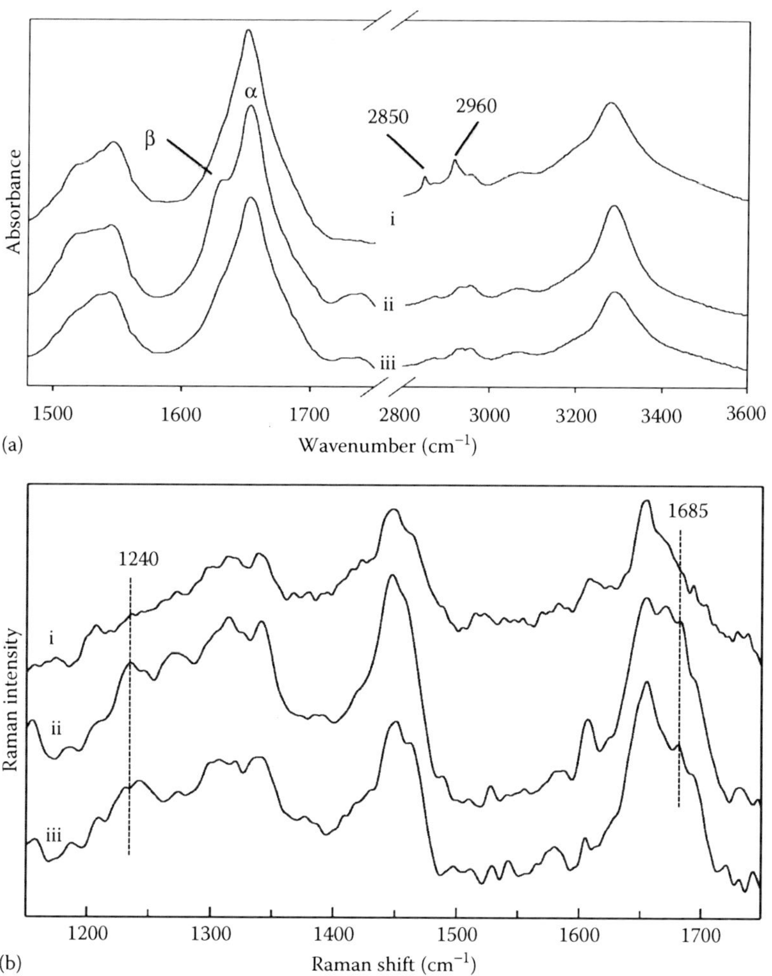

FIGURE 1.13
Chloroform/methanol-induced conformational changes in the keratin of single corneocytes. IR (a) and Raman spectra (b), respectively, of an isolated corneocyte prior to (i) and following (ii) immersion in chloroform/methanol and after an overnight rehydration period (c) (top to bottom, respectively). α-Helical and β-sheet components of the Amide I and II bands are noted in (a). One component of the Amide III band is marked at 1240 cm^{-1} in (b).

Compositional comparison of α-keratin with β-keratin indicates that the α-helix keratin contains larger amounts of glutamic and asparatic acid residues than the β-keratin. These two residues together constitute 23% of amino acid residues in the SC [30]. One factor that may contribute to the conformational changes is that methanol esterifies glutamic or aspartic acid side chains, which could likely occur during the chloroform/methanol solvent extraction. Figure 1.14 presents the IR spectra of pure human epidermal keratin before and after the chloroform/methanol treatment and following rehydration. The band at 2872 cm^{-1} is assigned to C–H symmetric stretching of CH_3 groups, usually resulting from protein terminals. Based on the intensity of this band in the untreated keratin spectrum, the other two spectra were normalized. The spectral regions of 1188–1480 and 1480–1780 cm^{-1} are presented as insets. After chloroform/methanol treatment, the ester band C=O at 1720–1740 cm^{-1} increases along with the β-sheet band at 1626 cm^{-1}, whereas the band intensity at 1404 cm^{-1} and the underlying band at 1575 cm^{-1}, assigned to symmetric and asymmetric stretching of, respectively, $-COO^-$ groups in glutamate and aspartate decrease (Figure 1.14, inset). The result indicates that methanol esterifies glutamate and aspartate amino acid side chains. As is well known, α-keratin is a left-handed coiled–coiled structure, which is maintained not only by the hydrogen bonds between amide protons and carbonyl oxygen atoms but also by the interaction between side chains in amino acid residues. The esterification induced by methanol significantly interrupts the specific side chain interactions existing in keratin and leads to the formation of some β-sheets and a certain level of undefined secondary structures.

1.3.2.3 Characterizing Corneocyte Maturation by IR Spectroscopic Imaging

The SC is biologically very active. There are many types of enzymes identified in this layer [31,32]. For example, in cornified cells, filaggrin, which is associated with the corneocyte keratin filaments in the deepest layers of the SC, is continuously hydrolyzed into free

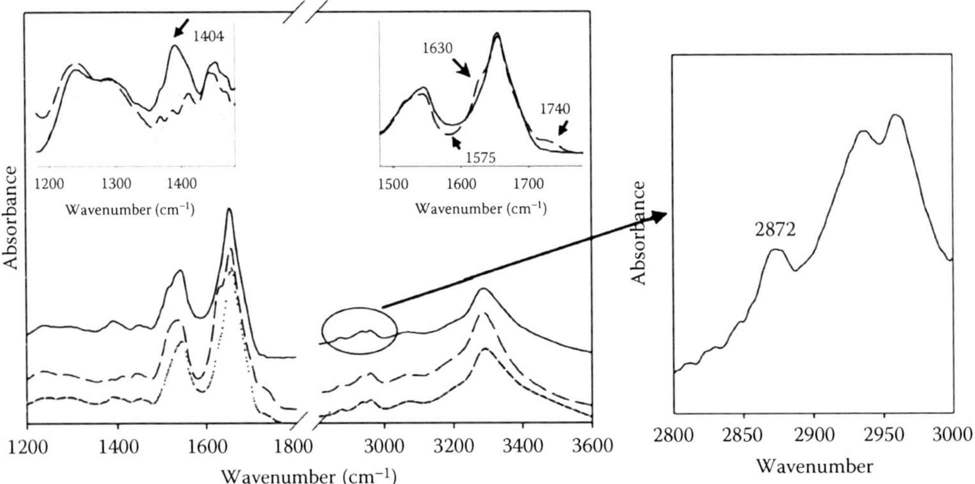

FIGURE 1.14
Chloroform/methanol–induced conformational changes in pure epidermal keratin. IR spectra of epidermal keratin prior to (——) and after immersion in chloroform/methanol for 2 hours (----), and following an overnight rehydration (·····). The spectra are normalized by the intensity of the band at 2872 cm^{-1}. Insets are the normalized spectra in the regions of 1188–1480 and 1480–1780 cm^{-1}.

amino acids by protease enzymes. These highly concentrated hydrophilic amino acids, termed collectively the *natural moisturizing factor,* contribute to the flexibility of the cornified layer and are very important for the retention of water in corneocytes [33]. In the past few years, our laboratory has done a number of studies in collaboration with Rutgers University (Newark, NJ) aiming to characterize the important properties of corneocytes [12,34]. In this section, we give an example demonstrating the applications of IR imaging of corneocyte maturation by monitoring NMF concentration in the corneocytes.

The distinct layers of human corneocytes were collected by sequentially tape-stripping the inner forearm area where the impact of environmental stress is minimized (the third and eleventh tape-stripped layers are shown in this study). IR images were acquired from multiple corneocytes in which a 4 × 4 pixel area located in the center portion of the image of each corneocyte was spatially masked and analyzed. The spatial size for each pixel is 6.25 μm × 6.25 μm. A mean spectrum was calculated from the masked images acquired from 36 corneocytes, isolated from each layer and displayed in Figure 1.15. Significant differences are observed between the two mean spectra, the most prominent being the increased intensity in the band at 1404 cm^{-1} (due to carboxylate symmetric stretching) in the spectrum acquired from corneocytes isolated from the deeper layer. We identified that these carboxylate groups result from the NMF, the breakdown products of filaggrin, for example, amino acid salts and ionized carboxylic acid derivatives [12]. The features observed in the difference spectrum (subtracting third layer from eleventh layer) of corneocytes are quite comparable to those in the spectrum of the dried NMF film. Qualitatively, the concentration of NMF in corneocytes appears to increase significantly with the SC depth for the outermost (~11) layers. This work demonstrates the feasibility of acquiring

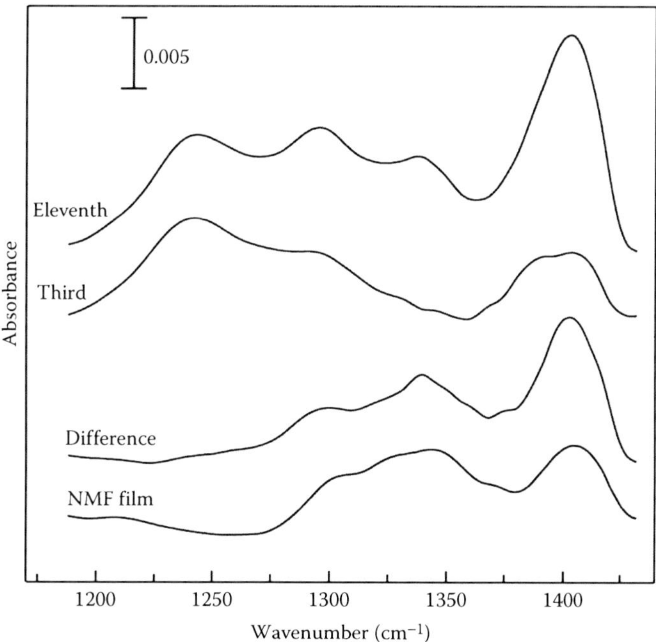

FIGURE 1.15
Mean IR imaging spectra of multiple corneocytes from two depths in human stratum corneum compared to an IR spectrum of a model NMF film. Mean IR spectra (1180–1430 cm^{-1} region) of 36 corneocytes isolated from the eleventh and third tape-stripped layers (top two spectra) along with the difference spectrum (eleventh layer minus third layer) and a spectrum of the model NMF film.

molecular level biochemical and structural information from single corneocytes sampled at different points in the maturation and exfoliation process. In healthy skin, the complex process of complete cell turnover takes ~30 days. Most notably, this work reveals the depth dependence of NMF concentration in the outer ~6 μm of the SC. NMF components are thought to facilitate critical biochemical events in the SC. Perturbations in NMF concentration are linked to aging, surfactant exposure, and disease states of varying severity. Thus, the quantitative analysis of the NMF concentration gradient in the SC provides meaningful information to the medical, pharmaceutical, and cosmetic industries.

1.3.3 Raman Spectroscopic Imaging of Salicylic Acid Delivery into the Skin by Different Vehicles

In general, traditional methods of evaluating permeation and metabolism of exogenous materials topically applied to the skin are usually carried out by either perturbing the skin structure or exposing the skin to nonphysiological conditions. One common method involves the use of a diffusion cell, in which a buffer solution is in constant contact with the dermis (acceptor phase), possibly extracting skin constituents and changing the physiological state of the skin. Subsequent high-performance liquid chromatography (HPLC) and mass spectrometry studies of the acceptor phase, to identify and quantify permeants, provides high selectivity and sensitivity; however, information pertaining to the spatial distribution of the exogenous substances in the skin and possible perturbations to the structure of endogenous skin components remain unknown. Tape-stripping the uppermost cornified layers of the skin in the SC is another common protocol. Depth-dependent concentration profiles and an evaluation of endogenous skin structure can be obtained when tape-stripping is followed by a technique such as ATR infrared spectroscopy (ATR-IR). However, the sampling is limited to the SC and spatial heterogeneity in the plane parallel to the skin surface cannot be determined.

The applicability of confocal Raman microspectroscopy as a noninvasive optical approach for studies of localized biomedical process and the delivery and metabolism of prodrugs/drugs has been previously demonstrated [13,34–36]. The unique advantage of this technique is its capability to provide direct spatially resolved concentration and molecular structural information while keeping the sample intact under essentially *in vivo* conditions. By optically dissecting the skin with a spatial resolution of ~1 μm in the x and y directions (parallel to the skin surface) and 2–3 μm in the z direction (depth in skin), the delivery and metabolism of exogenous materials in the skin may be mapped as a confocal line or a lateral plane under the surface. In this study, this novel approach will be applied to track the permeation of salicylic acid, a multifunctional cosmetic ingredient that is used to treat ailments of the skin such as acne, dermatitis, and psoriasis.

In this analysis, a 5% lightly cross-linked poly(vinylpyrrolidone) (PVP) gel, manufactured by Ashland, was used as a delivery vehicle and compared to a water-based system. The concentration of salicylic acid was 2%. During the experiment, a salicylic acid–5% PVP gel and a salicylic acid solution were applied to the skin surface and incubated for 1 or 2.5 hours at ~100% humidity and room temperature. The treated skin sample was then blotted to remove excessive formulation and equilibrated in a closed sampling cell for a specified amount of time (0.5 hour for lateral map and 1 hour for confocal line) to allow the sample to settle and minimize the shrinkage of the skin during the measurement. Confocal lines were acquired down to 30 μm deep with a 3 μm step size. The lateral Raman maps of 50 μm by 40 μm were obtained in the lateral direction at ~6–8 μm under the skin surface with a 10 μm step size.

The spectrum of untreated pig skin acquired at a depth of ~6 μm beneath the SC surface, along with spectral region of salicylic acid solution and PVP gel, is presented in Figure 1.16a. Salicylic acid has a Raman marker band at 1033 cm^{-1}, and this band partially overlaps with a native skin band at 1030 cm^{-1}. To quantify salicylic acid concentration in the skin, the integrated area ratios of the band at ~1033 cm^{-1} to the skin phenylalanine band at 1004 cm^{-1}

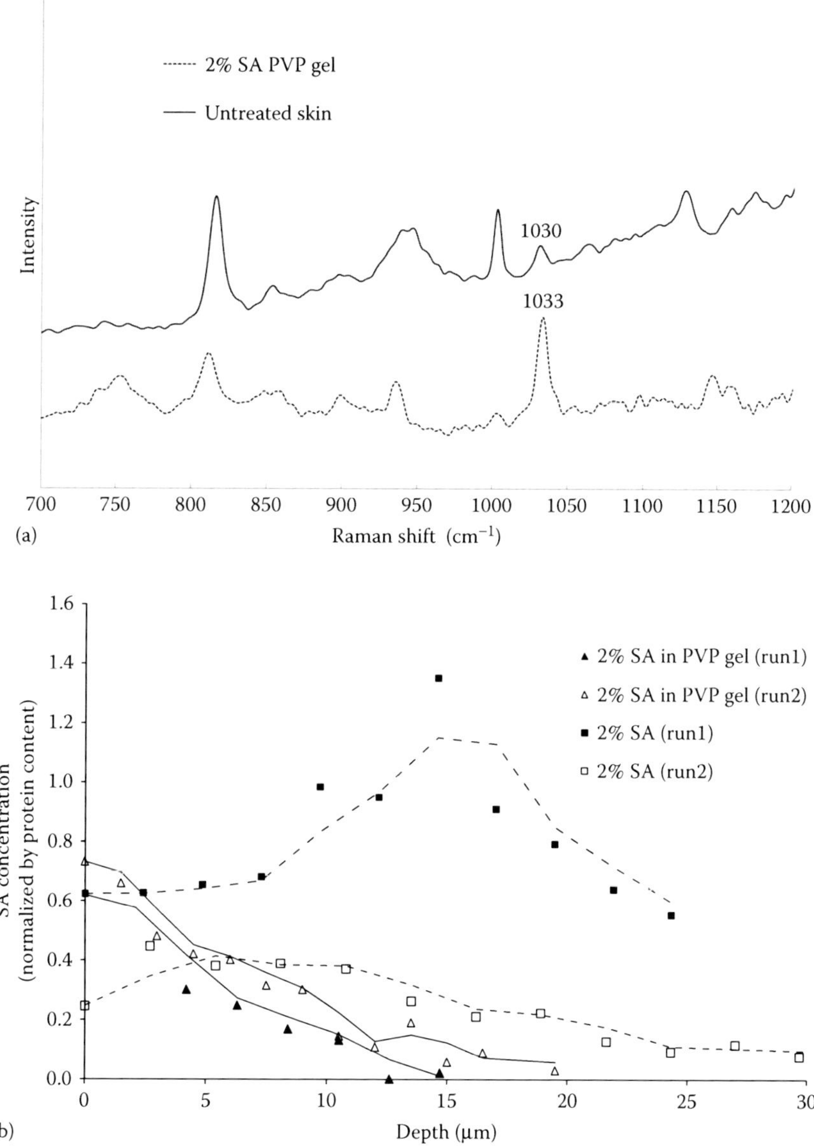

FIGURE 1.16
(See color insert.) (a) Representative Raman spectrum of untreated pig skin acquired at a depth of ~6 μm beneath the SC surface, along with the spectrum of 2% salicylic acid (SA) in PVP gel. (b) The relative concentration of SA change in skin with depth after 2% SA was applied on the skin surface for 1 hour with different vehicles. (c) SA distributions inside the stratum corneum. Each Raman map was acquired laterally at ~6–8 μm below the skin surface. The skin was treated with 2% SA in a PVP gel for 1 hour (i), 2% SA for 1 hour (ii), 2% SA in a PVP gel for 2.5 hours (iii), and 2% SA for 2.5 hours (iv).

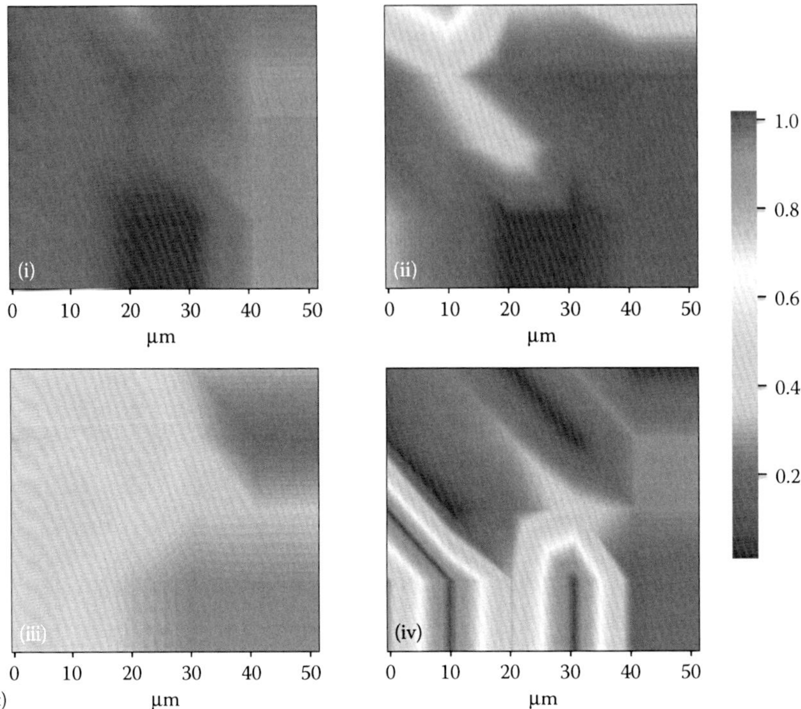

FIGURE 1.16
Continued

were obtained from both untreated and treated samples. By subtracting the average ratio obtained from the untreated skin samples from the ratios (1033–1004 cm^{-1}) obtained from the treated samples, salicylic acid concentrations were estimated. Figure 1.16b depicts salicylic acid concentration changes in the skin with depth after 1 hour application. The salicylic acid concentrations are calculated from the spectra acquired from confocal lines. With the application of the salicylic acid water solution, salicylic acid penetrated through the SC and entered the viable epidermis. Larger variations of salicylic acid concentrations in the skin were observed between different confocal lines, reflecting the unevenness of salicylic acid distribution inside the skin. In the case of 5% PVP as a delivery vehicle, salicylic acid mostly resides inside the SC or only reaches the boundary of viable epidermis. To evaluate the temporal and spatial variations of salicylic acid distribution inside the skin, lateral Raman maps were obtained at ~6–8 μm under the skin surface. The salicylic acid concentration maps are displayed in Figure 1.16b. The average concentration of salicylic acid within the mapped area was calculated from each map and displayed in Figure 1.16c. Lateral maps of 1 hour salicylic acid application indicate that the water solution delivered more salicylic acid into the skin compared with the 5% PVP gel. The PVP gel is a lightly cross-linked polymer with high molecular weight and is not able to permeate into the skin. This result indicates that the PVP gel controls the release of salicylic acid from polymer and decreases the rate of salicylic acid permeation. After 2.5 hour application, lateral maps showed that the water solution and the 5% PVP gel delivered overall approximately equal amounts of salicylic acid into the skin; however, salicylic acid distributions within the SC are very different. Salicylic acid delivered by 5% PVP is more evenly distributed inside the SC than the salicylic acid solution.

This investigation demonstrates the ability of confocal Raman imaging to monitor the temporal and spatial distribution of exogenous agent permeation into the skin with different delivery systems. The inherent strength of the techniques applied herein allows for the analysis of targeted active delivery without the introduction of potentially perturbing probes.

References

1. A. Kuzuhara, "Analysis of structural changes in permanent waved human hair using Raman Spectroscopy," *Biopolymers*, 85(3), 274–283 (2006).
2. A. Kuzuhara, "Analysis of structural change in keratin fibers resulting from chemical treatments using Raman spectroscopy," *Biopolymers*, 77(6), 335–344 (2005).
3. J. Strassburger, M. Breuer, "Quantitative Fourier transform infrared spectroscopy of oxidized hair," *J. Soc. Comet. Chem.*, 36(1), 61–74 (1985).
4. N. Nishikawa, Y. Tanizawa, S. Tanaka, Y. Horiguchi, T. Asakura, "Structural change of keratin protein in human hair by permanent waving treatment," *Polymer*, 39, 3835–3840 (1998).
5. V.F. Kalasinski, "Biomedical applications of infrared and Raman microscopy," *Appl. Spectrosc. Rev.*, 31(3), 193–249 (1996).
6. G. Zhang, L. Senak, D.J. Moore, "Measuring changes in chemistry, composition and molecular structure within hair fibers by infrared and Raman spectroscopic imaging," *J. Biomed. Opt.*, 16(5), 056009 (2011).
7. K.L.A. Chan, S.G. Kazarian, A. Mavraki, D.R. Williams, "Fourier transform infrared imaging of human hair with a high spatial resolution without the use of a synchrotron," *Appl. Spectrosc.*, 59(2), 149–155 (2005).
8. P. Dumas, L. Miller, "The use of synchrotron infrared microspectroscopy in biological and biomedical investigations," *Vibt. Spectrosc.*, 32(1), 3–21 (2003).
9. A. Barth, "The infrared adsorption of amino acid side chains," *Prog. Biophys. Mol. Biol.*, 74, 141–173 (2000).
10. L.K. Tamm, S.A Tatulian, "Infrared spectroscopy of proteins and peptides in lipid biolayers," *Quart. Rev. Biophys.*, 30, 365–429 (1997).
11. L.J. Bellamy, *The Infrared Spectra of Complex Molecules*, Chapman & Hall: London (1975).
12. G. Zhang, D.J. Moore, C.R. Flach, R. Mendelsohn, "Vibrational microscopy and imaging of skin: From single cells to intact tissue," *Anal. Bioanal. Chem.*, 387, 1591–1599 (2007).
13. K.L.A. Chan, G. Zhang, M. Tomic-Canic, O. Stojadinovic, B. Lee, C.R. Flach, R. Mendelsohn, "A coordinated approach to cutaneous wound healing: Vibrational microscopy and molecular biology," *J. Cell Mol. Med.*, 12(5B), 2145–2154 (2008).
14. P.J. Caspers, G.W. Lucassen, E.A. Carter, H.A. Bruining, G.J. Puppels, "In vivo confocal Raman microspectroscopy of the skin: Noninvasive determination of molecular concentration profiles," *J. Invest. Dermatol.*, 116(3), 434–442 (2001).
15. S. Schlucker, K.R. Strehle, J.J. DiGiovanna, K.H. Kraemer, I.W. Levin, "Conformational differences in protein disulfide linkages between normal hair and hair from subjects with trichothildystrophy: A quantitative analysis by Raman microspectroscopy," *Biopolymers*, 82(6), 615–622 (2006).
16. C.R. Robbins, *Chemical and Physical Behavior of Human Hair*, 5th ed., pp. 122–126, Springer-Verlag: New York (2012).
17. Y. Zhou, R. Rigoletto, D. Koelmel, G. Zhang, T.W Gillece, L. Foltis, D.J. Moore, X. Xu, C. Sun, "The effect of various cosmetic pretreatments on protecting hair from thermal damage by hot flat ironing," *J. Cosmet. Sci.*, 62, 265–282 (2011).
18. H.E. Van Wart, H.A. Scheraga, "Raman spectra of cystine-related disulfides effect of rotational isomerism about carbon-sulfur bonds on sulfur-sulfur stretching frequencies," *J. Phys. Chem.*, 80(16), 1812–1823 (1976).

19. C.R. Robbins, *Chemical and Physical Behavior of Human Hair*, 5th ed., pp. 274–275, Springer-Verlag: New York (2012).
20. C.R. Robbins, "The cell membrane complex: Three related but different cellular cohesion components of mammalian hair fibers," *J. Cosmet. Sci.*, 60, 437–465 (2009).
21. L.N. Jones, D.E. Rivett, "18-Methyleicosanoic acid in the structure and formation of mammalian hair fibres," *Micron*, 28, 469–485 (1997).
22. R.L. McMullen, D. Laura, S. Chen, D. Koelmel, G. Zhang, T. Gillece, "Determination of the physicochemical properties of delipidized hair," *J. Cosmet. Sci.*, 64, 355–370 (2013).
23. R. Mendelsohn, "Laser-Raman spectroscopic study of egg lecithin and egg lecithin-cholesterol mixtures," *Biochim. Biophys. Acta*, 290, 15–21 (1972).
24. C. Xiao, C.R. Flach, M. Marcott, R. Mendelsohn, "Uncertainties in depth determination and comparison of multivariate with univariate analysis in confocal Raman studies of a laminated polymer and skin," *Appl. Spectrosc.*, 58(4), 382–389 (2004).
25. N.D. Lazo, J.G. Meine, D.T. Downing, "Lipids are covalently attached to rigid corneocyte protein envelopes existing predominantly as β-sheets: A solid-state nuclear magnetic resonance study," *J. Invest. Dermatol.*, 105, 296–300 (1995).
26. N.D. Lazo, D.T Downing, "A mixture of α-helical and 3_{10}-helical conformations for involucrin in the human epidermal corneocyte envelope provides a scaffold for the attachment of both lipids and proteins," *J. Biol. Chem.*, 274, 37340–37344 (1999).
27. R.P. Oertel, "Protein conformational changes induced in human stratum corneum by organic sulfoxides: An infrared spectroscopic investigation," *Biopolymers*, 16(10), 2329–2345 (1977).
28. A.N.C. Anigbogu, A.C. Williams, B.W. Barry, H.G.M. Edwards, "Fourier transform Raman spectroscopy of interactions between the penetration enhancer dimethyl sulfoxide and human stratum corneum," *Int. J. Pharm.*, 125, 265–282 (1995).
29. P. Yager, B.P. Gaber, "Membranes" in *Biological Applications of Raman Spectroscopy*, Ed. T.G. Spiro, pp. 203–261, John Wiley and Sons: New York (1987).
30. C.K. Mathew, K.E. Van Holde, K.G. Ahern, *Biochemistry*, 3rd ed., Prentice Hall: Englewood Cliffs, NJ (2000).
31. E. Candi, R. Schmidt, G. Melino, "The cornified envelope: A model of cell death in the skin," *Nat. Rev. Mol. Cell Biol.*, 6, 328–340 (2005).
32. E. Ekholm, M. Brattsand, T. Egelrud, "Stratum corneum tryptic enzyme in normal epidermis: A missing link in the desquamation process," *J. Invest. Dermatol.*, 114, 56–63 (2000).
33. A.V. Rawlings, I.R. Scott, C.R. Harding, P.A. Bowser, "Stratum corneum moisturization at the molecular level," *J. Invest. Dermatol.*, 103, 731–740 (1994).
34. G. Zhang, D.J. Moore, R. Mendelsohn, C.R. Flach, "Vibrational microspectroscopy and imaging of molecular composition and structure during human corneocyte maturation," *J. Invest. Dermatol.*, 126(5), 1088–1094 (2006).
35. G. Zhang, D.J. Moore, B. Sloan, C.R. Flach, R. Mendelsohn, "Imaging the prodrug-to-drug transformation of a 5-fluorouracil derivative in skin by confocal Raman microscopy," *J. Invest. Dermatol.*, 127, 1205–1209 (2007).
36. P.J. Caspers, G.W. Lucassen, E.A. Carter, H.A. Bruining, G.J. Puppels, "In vivo confocal Raman microspectroscopy of the skin: Noninvasive determination of molecular concentration profiles," *J. Invest. Dermatol.*, 116(3), 434–442 (2001).

2

Fluorescence Bioimaging with Applications to Chemistry

Ufana Riaz and S.M. Ashraf

CONTENTS

2.1 Introduction

Optical imaging has till date been the most versatile visualization tool in diverse research areas. Microscopy still remains a diagnostic and a flexible visualization tool, which has been merged with new technologies [1–5]. It has applications spanning from the decoding of the human genome and high-throughput screening to noninvasive imaging of functional and molecular contrast in intact tissues [6,7]. One of the fundamental reasons to use optical imaging in research is the wealth of contrast mechanisms that can be offered through the use of physical properties of light (i.e., polarization, interference, etc.) and the ability to capitalize on a wide range of light–tissue interactions and their corresponding photophysical and photochemical mechanisms and processes occurring at the molecular level such as multiphoton absorption, second-harmonic generation, and fluorescence. This technique also offers a convenient technology for experimentation as most of the components can be assembled at the laboratory scale and can be made portable or compact. Thus, high quality of optical components and high detection sensitivity can be achieved at a very moderate cost. The utilization of such technology offers a highly versatile platform to probe at scales spanning from the molecular to the system level, which can give information and deeper insight into different areas of physical, chemical, and biological research. Although optical microscopy has shown advantages in enhancing contrast details due to variations in the index of refraction, these transmitted light techniques fail to provide any means to distinguish individual objects or identify small organelles other than by their shape or optical density. To overcome these limitations, staining agents and techniques that depended on poorly understood interactions were developed by histologists. Moreover, they were also unable to allow the detection of objects that were smaller than the diffraction limit or that did not exhibit sufficient contrast. Fluorescence microscopy has overcome all these limitations by rejecting the excitation light, leaving only the visible light as the source of emission. The development of immunocytochemistry during the nineteenth century gave this technique its full potential, which was initially limited to fixed samples, and was later on also applied to live cell imaging around the 1980s [8]. Confocal laser scanning microscopy (CLSM) is also seen as an indispensable tool for three-dimensional (3D) imaging with capabilities extended to multiphoton excitation processes and lifetime imaging, and also to improve its 3D resolution. Single-molecule spectroscopy (SMS) has recently shed light on inter- and intramolecular interactions. Advances in the detector technology extended the fluorescence microscopy to its ultimate level of sensitivity. Single fluorescent molecules can now be detected in a living cell and localized with nanometer precision in real time along with improvements in image versatility, sensitivity, and resolution. Fluorescence was also used as a technique to probe the dynamics, conformational changes, and interactions of single molecules. This modern development has evolved fluorescence microscopy from a purely imaging technique to nanospectroscopy and has extended the application of the microscope to structural biology, biochemistry, and biophysics, providing new tools for the proteomics era.

Fluorescence microscopy and imaging have received wide attention due to the increasing availability of fluorescent proteins (FPs), dyes, and probes that enable the noninvasive study of gene expression, protein function, protein–protein interactions, and a large number of cellular processes [9,10]. There is an increasing list of fluorescent imaging techniques that offer microscopic resolutions and methods that operate at resolutions beyond the diffraction limit offering single-molecule sensitivity [11–14]. This technique is also gaining momentum as an imaging method for whole-body tissue molecular imaging method. Fluorescence rejects the excitation signal and allows only detection of the fainter emission light using filters and dichroic mirrors that do not lead to any resolution improvement, but can detect subresolution objects as contrast spots of fluorescent light with a diffraction-limited size. The development of digital, high-sensitivity cameras, and successive excitation and detection of different fluorophores has now made it possible to obtain multicolor images with good signal-to-noise ratio (SNR) [15]. Several of these methods have been proposed and implemented in recent years, in both wide-field and point detection geometries with the aim to recover the contribution of each fluorophore to the recorded emission at different wavelengths for each individual pixel [16,17]. As fluorescence is an incoherent process and emission spectra are well defined, intensities from different fluorophores simply add at each wavelength of the emission spectrum. By analyzing the emission spectra, we can interpret the contribution of each individual fluorophore by a simple matrix inversion [11,18–20].

2.2 Fundamental Principle of Fluorescence Spectroscopy

Fluorescence is the phenomenon of photon emission following absorption of one or more photon(s) by a molecule or a fluorophore that returns to its ground state. A series of experimental and theoretical studies by several investigators in the early twentieth century uncovered most of today's known properties of advance fluorescence techniques [21–23]. Excitation and emission processes in a typical molecule are represented by a Jablonski diagram (Figure 2.1), which depicts the initial, final, and intermediate electronic and vibrational states of the molecule. In general, fast intramolecular vibrational relaxations result in emitted photons having a lower energy. The emitted photon is detected within a typical delay (lifetime) after absorption of the excitation photon, which depends on the species studied and its local environment. Organic dyes have typical lifetimes from several tens of picoseconds to several nanoseconds. Longer lifetimes are obtained for fluorescent semiconductor nanocrystals (NCs) (tens of nanoseconds), organometallic compounds (hundreds of nanoseconds), and lanthanide complexes (up to milliseconds). The probability distribution of emission times usually depends on exponential, characterized by a lifetime $\tau = \Gamma^{-1} = (k_r + k_{nr})^{-1}$, where k_r and k_{nr} are the radiative and nonradiative decay rates, respectively, and depend on the local environment via perturbations of the intramolecular transition matrix elements [14]. In its excited state, a molecule has a probability to end up in a nonemitting triplet state during microseconds to milliseconds, resulting in dark states [4]. The efficiency of photon absorption is proportional to $(\vec{E} \cdot \vec{\mu})^2$, where $\vec{E}$ represents the local electric field and $\vec{\mu}$ is the absorption dipole moment of the fluorophore [24]. The orientation of the molecule's absorption dipole can be determined by the emitted fluorescence as a function of the orientation of the linear polarization of the excitation light, which allows the determination of the spatial orientation of the fluorophore [23–26]. Fluctuations faster than the fluorescence lifetime lead to a depolarized emission; fluctuations taking place over timescales longer than the fluorescence lifetime but shorter than the integration time lead to a depolarized emission.

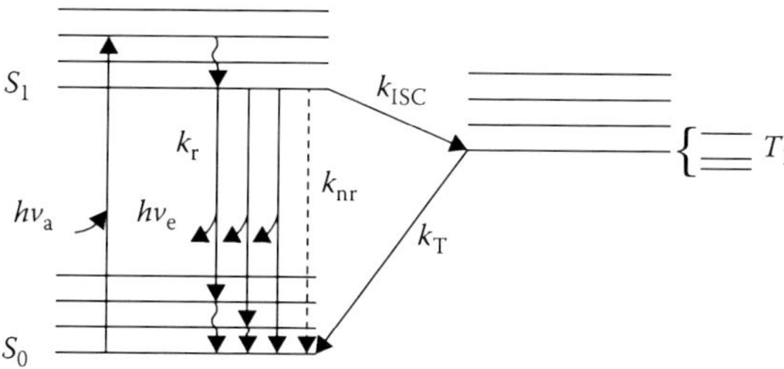

FIGURE 2.1
Jablonski diagram for fluorescence. k_{ISC}, intersystem crosslinking rate; k_T, transition state rate; k_r and k_{nr}, radiative and nonradiative decay rates; S_0, ground state; S_1, excited state; T_1, transition state. (From Michalet, X. et al., *Annu. Rev. Biophys. Biomol. Struct.*, 32, 161–182, 2003. With permission.)

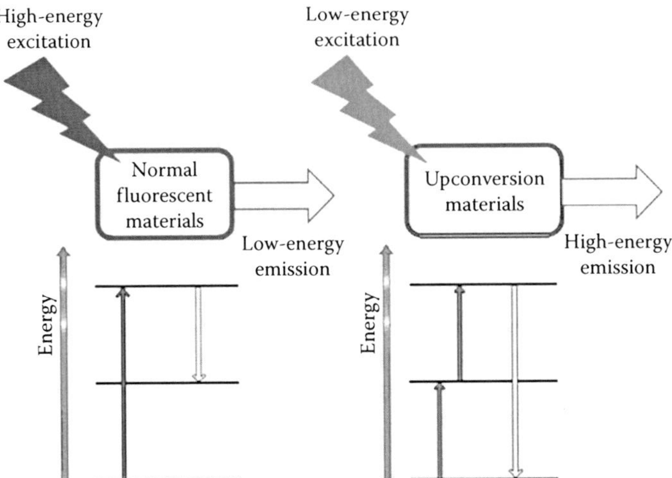

FIGURE 2.2
Energy-level diagram of the emission processes of normal fluorescent materials and upconversion materials. (Reprinted from *Prog. Polym. Sci.* 39, Chen, M. and Yin, M., Design and development of fluorescent nanostructures for bioimaging, 365–395, Copyright 2014, with permission from Elsevier.)

Lately, upconversion (UC) nonmaterials have gained significant importance. The phenomenon of *upconversion* was defined as a nonlinear optical process in which the sequential absorption of two or more photons leads to the emission of a single photon at a shorter wavelength, which is shown in Figure 2.2. Their major advantages are as follows:

- They minimize the optical damage to cell tissues.
- They exhibit high tissue penetration.
- Background interference of radiation can be effectively avoided.
- They provide excellent photostability and low toxicity as bioimaging agents.

2.3 Instrumentation: Quantification and Sensitivity of the Spectrophotometer

The commercialization and development of fluorescence imaging systems has progressed rapidly over the past decade and a wide variety of approaches have been used. A review of commercial systems shows that the technology holds a wide range of approaches and capabilities. The range of choices in the field is driven by factors such as cost, data types required, ease of use, ability to couple to other imaging systems, and scientific collaborations. Figure 2.3 provides an overview of the hardware required for different types of imaging data. All *in vivo* fluorescence imaging systems are designed around the idea that a portion of the fluorescence emission spectrum of the molecule of interest can be separated spectrally from the excitation and/or background light signals. This is achieved by filtering strategies, such as dielectric interference filters, liquid crystal tunable filters, and

	Steady state	Frequency domain	Time domain
Signal biochemical properties	Intensity Concentration and quantum yield	Intensity and phase Lifetime concentration and quantum yield	Time-gated signal and point-spread function Lifetime concentration and quantum yield
Excitation sources	Virtually any light source Filament and gas lamps LED Laser diode Gas laser Solid-state laser	Frequency-modulated source LED Laser diode Other modulated sources	Pulsed source Laser diode Tunable lasers
Detection instrumentation	Virtually any light detector CCD ICCD EMCCD PMT APD	Homo- or heterodyned detection ICCD EMCCD PMT APD	Time-correlated single-photon counting and time-gated detection ICCD EMCCD PMT APD
Filtering techniques	Interference or absorption filters (filter wheels) Liquid crystal tunable filters Spectrograph gratings Dichroic mirrors		

FIGURE 2.3
Three major hardware design strategies for *in vivo* fluorescence imaging are shown in the columns of the table. Comparison of the imaging approaches is provided in terms of the main four components needed for each. EMCCD, electron magnetic chiral dichroism; ICCD, intensified charge-coupled device. (Reprinted from *Photochem. Photobiol. B*, 98, Leblond, F. et al., Preclinical whole-body fluorescence imaging: Review of instruments, methods and applications, 77–94, Copyright 2010, with permission from Elsevier.)

spectrograph gratings. The filtering capacity of a system is a major fundamental limitation that governs the lower limits of detection of fluorophores.

For imaging applications based on cameras (Figure 2.4), many filtering strategies such as interference filters can at best provide 2–3 orders of magnitude of suppression of the excitation light, so the remainder of the excitation light is left to leak through to the detector and ultimately provides a background signal. These attenuation factors are lower by several orders of magnitude than typical values provided for high-end filters by manufacturers.

For imaging applications based on cameras, the angle of incidence is not normal on average, which explains the relatively limited attenuation factors of these instruments. In situations where light is collected from a point or a small area on the surface (Figure 2.3), the filters can be placed in a near focal geometry, allowing improved filtering of fluorescent light. When designing an imaging system, a reasonable strategy for maximizing signal intensity could consist in irradiating tissue with light such that the fluorescent molecules of interest are excited right at the peak wavelength of their absorption spectrum. Generally, the choice of the excitation wavelengths must be balanced with the ability of the detection system to filter and reject the excitation light in the measured signal. Consequently,

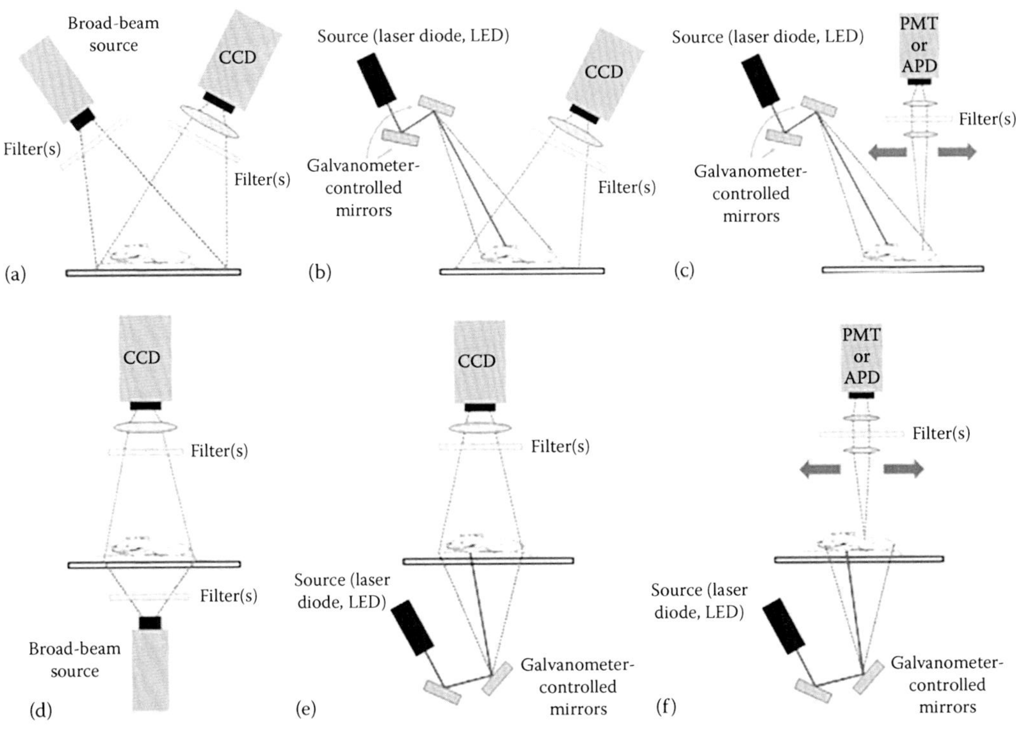

FIGURE 2.4
Schematic rendering of different methods that can be used for whole-body fluorescence imaging. The first row shows epi-illumination geometries: (a) broad beam illumination with wide-field camera detection; (b) raster-scan illumination with wide-field camera detection; (c) raster-scan illumination and detection. The lower row of images shows the corresponding transillumination configurations. Not shown in the figure are configurations optimized for tomography imaging and fiber-based planar configurations. (Reprinted from *Photochem. Photobiol. B*, 98, Leblond, F. et al., Preclinical whole-body fluorescence imaging: Review of instruments, methods and applications, 77–94, Copyright 2010, with permission from Elsevier.)

fluorophores with small Stokes shifts, such as most organic fluorophores, are usually excited at wavelengths shorter than wavelengths situated toward the right on the absorption peak, in order to allow better wavelength separation between the incident light and the emitted light. While some excitation sources allow flexibility in terms of illumination wavelengths, others are fixed, and thus the fluorophores to be used must be considered in the design or purchase phase of the system. Using grating-based spectrographs, liquid crystal tunable filters, or filter wheels with several band-pass filters over the emission range, pre-recorded basis spectra can be used to decouple the autofluorescence signal from the fluorescence of the probe. The spectrally resolved data provide a platform for imaging several fluorophores simultaneously, using the same spectral unmixing algorithms. Excitation can be achieved by a wide variety of light sources including filtered filament or gas discharge lamps (tungsten and xenon), light-emitting diodes (LEDs), or any laser system, including gas, crystal, and diode lasers. Versatile imaging instruments, in terms of number of available wavelengths, can be designed using expansive systems such as tunable lasers (e.g., Ti:sapphire laser), but relatively inexpensive broadband lamps are used to provide a flexible platform. However, the isotropic emission of lamps limits the ability to focus a large amount of power on the tissue surface and limits their use to applications having large fiber optic cable delivery or direct lens-based delivery of the light. Lamps require relatively long warm-up times and are considerably less stable than solid-state devices. Intensity variations of lamps can be in the range of 10%–50% over an imaging session, whereas the stability of diode lasers can typically be in 0.1% range if they are low power or temperature controlled. Lamp-based systems are not suitable for time or frequency domain measurements that require precise and rapid modulation of the source intensity.

Quantification signifies the potential to provide datasets from which the correct concentration or density of molecules can be achieved. In the commercial and laboratory systems, *in vivo* fluorescence imaging is not capable of providing quantitative information. It can only be approached by combining light transport modeling, data normalization, and calibration. The closest to actual quantification is attained by using normalized datasets, either in the ratio of emission to excitation [10,27] or in the wavelength ratio of the fluorophore at different wavelengths [13]. The approach of normalized emission/excitation ratio for fluorescence tomography works well because the signals travel similar path length of the tissue. The importance of this normalization is to cancel out geometrical and heterogeneity effects that are uncontrolled experimentally, because they enter the signal typically as multiplicative factors, and are almost equal in magnitude for each of the two signals.

Spatial resolution, sensitivity, and quantification potential are features that a manufacturer looks for prior to acquiring a fluorescence instrument. Sensitivity is an important criterion that refers to the concentration or density of fluorescent molecules that can be locally detected in a material. The signal from the target is always polluted by nonspecific signals coming from autofluorescence of the material to be investigated, excitation light leakage through the filters, as well as other signals in the form of unbound exogenous dyes or contrast agents present in the material. Typically, a large part of the autofluorescence comes from the nonbound or nonspecifically bound components. This implies that for an imager to be efficient, the detection system should have acceptable noise characteristics and large dynamic range to allow background removal techniques to be applied to the signal. If the dynamic range is too small, it is likely that in applications where targets are associated with smaller contrast than autofluorescence will dominate the detectable image. At a minimum, an *in vivo* system should come equipped with a cooled charge-coupled device (CCD) camera with large dynamic range. Some systems allow for the possibility to perform bioluminescence imaging. In this case, there is no autofluorescence, but signal levels are typically 2–3 orders of magnitude smaller than for fluorescence imaging.

2.4 Types of Fluorescent Imaging

Direct imaging administers an engineered fluorescent probe that targets a specific moiety such as a receptor or an enzyme. Fluorescent probes for direct imaging are categorized as active or activatable.

2.4.1 Active Probes

Active probes are essentially fluorochromes that are attached to an affinity ligand specific for a certain target. This paradigm is similar to probe design practices seen in nuclear imaging, except that a fluorochrome is used in the place of the isotope such as monoclonal antibodies and antibody fragments [28], modified or synthetic peptides [29], and labeled small molecules [30]. They fluoresce even if they are not bound to the intended target and yield nonspecific background signals, unless long circulating times are allowed to efficiently remove the nonbound probe from circulation.

2.4.2 Activatable Probes

Activatable probes carry quenched fluorochromes that are arranged in close proximity to each other so that they self-quench or they are placed next to a quencher using enzyme-specific peptide sequences [31]. These peptide sequences can be cleaved in the presence of the enzyme, thus freeing the fluorochromes that can then emit light upon excitation [32–34].

In contrast to active probes, activatable probes minimize background signals because they are essentially dark in the absence of the target and can improve contrast and detection sensitivity. Fluorescence probes target specific cellular and subcellular events, and this ability differentiates them from nonspecific dyes, such as indocyanine green, which reveals generic functional characteristics such as vascular volume and permeability. Fluorescence probes typically consist of the active component, which interacts with the target such as the affinity ligand, the reporting component such as fluorescent dye or quantum dot (QD) used, and possibly a delivery vehicle like a biocompatible polymer. The important characteristic in the design of active and activatable probes for *in vivo* studies is the use of fluorochromes that operate in the near-infrared (NIR) spectrum of optical energy. This is due to the low light absorption that tissue exhibits in this spectral window, which makes light penetration of several centimeters possible.

2.4.3 Indirect Fluorescence Imaging

Indirect imaging is a strategy utilized for *in vitro* reporting assays and is utilized to study gene expression and gene regulation. The most common practice is the introduction of a transgene in the cell. The transgene encodes for an FP, which acts as an intrinsically produced reporter probe, and the transcription of the gene leads to the production of the FP, which can then be detected with optical imaging methods [35]. Therefore, gene regulation is imaged indirectly by visualizing and quantifying the presence of FPs in tissues. Cells can be stably transfected to express FPs and report on their position for cell-trafficking studies. The transgene can also be placed under promoters for studying gene regulation. Fusing the FP encoding gene to a gene of interest offers an opportunity for visualizing virtually every protein *in vivo*, which yields a chimeric protein that maintains the functionality of the original protein but is tagged with the FP so that it can be visualized *in vivo*. The protein of interest can also be transcribed and separately translated by the FP under control of the same promoter using a transgene containing an internal ribosomal entry site between the genes encoding for the FP and the gene of interest [36].

Several different FP approaches were developed to allow interrogation of protein–protein interactions through the utilization of fluorescence resonance energy transfer (FRET) techniques or protein function [37]. Various gene strategies were used for imaging modalities, such as positron emission tomography (PET) or magnetic resonance imaging, for example, when transcription of the reporter gene leads to upregulation of a receptor or enzyme. This exhibits trapping and increased accumulation of an extrinsically administered reporter probe [38–40]. Such methods are not commonly used for *in vivo* fluorescence imaging, although examples were reported, for instance, for β-galactosidase-based fluorescent probe activation [40]. The FPs most commonly used are enhanced mutants of the green FP (GFP) isolated from the jellyfish *Aequorea victoria*. Red-shifted proteins are beneficial for microscopy and small animal imaging because tissue yields reduced autofluorescence at longer wavelengths [41–45]. Therefore, a better contrast can be achieved in the far red and NIR (>600 nm). In addition, significantly less absorption of light in the far red and NIR is observed from the tissue compared to visible wavelengths; therefore, higher detection sensitivity can be achieved in this spectral region. Although the best FPs reported so far still require excitation within the highly absorbing visible region (<600 nm), further red-shifted FPs are needed with high efficiency and low toxicity in developmental biology, cancer and stem cell research immunology, and drug discovery [46–48]. Reporter gene imaging is a generalizable platform where few well-validated reporter gene and reporter probe pairs can be used to image many different molecular and genetic processes. Indirect optical imaging is also widely performed using bioluminescence imaging, in which case light is intrinsically generated in tissues through chemiluminescent reactions [26,48].

2.5 Fluorescence Detection

Fluorescence can be classified based on the excitation and emission schemes. Wide-field detection schemes that use epifluorescence illumination with lamps [49], defocused laser excitation, or total internal reflection excitation. The detectors used are back-thinned CCDs with quantum efficiencies (QEs) up to 90%, to a few full frames per second [50]. The new electron-multiplying CCD technology permits enhanced frame rates, with transfer rates as high as 10 MPixels/s with single-molecule sensitivity [51]. Point detection schemes include CLSM and near-field scanning optical microscopy (NSOM). The excitation volume has a radius of the order of the excitation wavelength for CLSM and of the tapered fiber core (~100 nm) for NSOM [52,53]. Commercial CLSM use a pair of galvanometer-mounted mirrors or acousto-optical deflectors to move the excitation laser beam across the sample and photomultiplier tubes (PMTs) for photon detection, which shows the advantage of great speed but does not provide enough sensitivity for SMS as SMS requires a slower, stage-scanning method and sensitive avalanche photodiodes (APDs). Indeed, until recently, PMT had a QE <20% in the visible spectrum, against ~70% for silicon APD. Progress in photocathode technology (using GaAsP) increases the QE of PMT, making them attractive detectors for beam-scanning CLSM because of their larger sensitive area as suggested by Synge [54]. The underlying molecules are sensitive only to the near-field contribution of the transmitted laser electric field, which extends over distances smaller than the wavelength, resulting in a higher resolution image collected by a microscope objective lens. The use of lasers as excitation sources limits the range of accessible excitation wavelengths and provides higher intensities with pulsed excitation along with ultrafast lasers. They can be used in the NIR range to perform two-photon fluorescence excitation [55]. Even though spatial resolution is not necessarily enhanced and fluorophore photobleaching is increased [56], the advantages of this technique are as follows:

- Out-of-focus bleaching is reduced, and sample penetration is increased.
- The reduced absorption of NIR radiation allows thick, live tissues to be imaged with little damage to the environment.

The technique has thus found an extensive application in neuroimaging and deep tissue imaging [57]. In this technique, wide-field imaging is necessary for particle-tracking studies that reduce the amount of time needed to accumulate a statistically significant number of observations. It is significantly relevant for experiments for irreversible processes controlled by modification of an external parameter. Point detection geometries allow fluorescence time traces of immobilized molecules with high temporal resolution, as well as fluorescence lifetime information [58], although they are slow for imaging single molecules. It is also extensively used for the study of freely diffusing molecules in solution or embedded in fluid membrane in combination with fluorescence correlation spectroscopy [59].

2.6 Imaging Modes: Intensity and Spectrum

With the advent of digital, high-sensitivity cameras, successive excitation and detection of different fluorophores make it possible to obtain multicolor images with good SNR [60]. Standard fluorophores have broad emission spectra, which cause the spectral separation by band-pass interference filters imperfect and impractical in the case of several colors. Hence, it is advantageous to use spectral imaging methods that collect all emitted light for further processing in both wide-field and point detection geometries [61,62]. The aim is to obtain the contribution of each fluorophore to the recorded emission at different wavelengths for each individual pixel. Although fluorescence is an incoherent process and emission spectra are well defined, intensities from different fluorophores simply add at each wavelength of the emission spectrum. Knowing the emission spectra, we can recover the contribution of each individual fluorophore by spectral changes upon variation of the local concentration of an ionic species [63,64]. There are technologies developed for visualizing fluorescence, and each of them is discussed in Sections 2.6.1 through 2.6.3.

2.6.1 *In Vivo* Imaging

The progress with microscopic methodologies for *in vivo* imaging, particularly intravital confocal, two-photon, and multiphoton microscopies, has yielded great insights into biology by imaging fluorescence with high resolution (0.5–3.0 μm) at depths of several hundred μm under the surface [65,66]. Intravital microscopy enables the characterization of the efficiency of fluorescence developed for macroscopic molecular imaging. This is fundamentally based on observing dynamically the fluorescent probe microdistribution, the particulars of binding or activation, and subsequent immune and clearance responses as a function of time [66]. These technologies were utilized in flexible fiber probes that are now used with endoscopic methods for obtaining images of sites that were previously inaccessible. Some representative images from normal human alveoli and tumoral vessels obtained from a mouse prostate are shown in Figure 2.5.

Wang et al. [15] developed an *in vivo* confocal microscope employing a novel dual-axis architecture using two low numerical aperture (NA) objectives oriented with the illumination and collection beams crossed at an angle, as shown in Figure 2.6, which results in a significant reduction of the axial resolution and allows for long working distances.

2.6.2 Planar Imaging

The most common method to record fluorescence deeper from tissues is associated with illuminating tissue with a plane wave, that is, an expanded light beam, and then collecting fluorescence signals emitted toward the camera. These methods can be generally referred to as planar methods and can be applied in epi-illumination/transillumination mode.

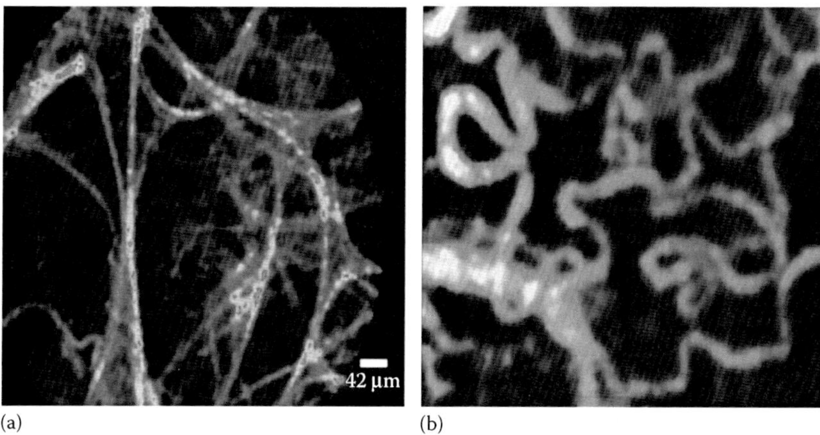

FIGURE 2.5
(See color insert.) Confocal images obtained *in vivo* with a flexible fiber probe of 650 μm. (a) Normal human alveoli: Visualization of normal distal lung, with distinct alveolar microarchitecture. The signal shown is tissue autofluorescence; no dye was applied in this case. (b) *In vivo* angiogenesis imaging: Visualization of tumoral vessels in a mouse prostate after FITC-dextran (500 kDa) injection in the tail. The site was accessed through a microincision in the skin at the site of the tumor. Field of view is 400 × 280 μm. (From Ntziachristos, V., *Annu. Rev. Biomed. Eng.*, 8, 1–33, 2006. With permission.)

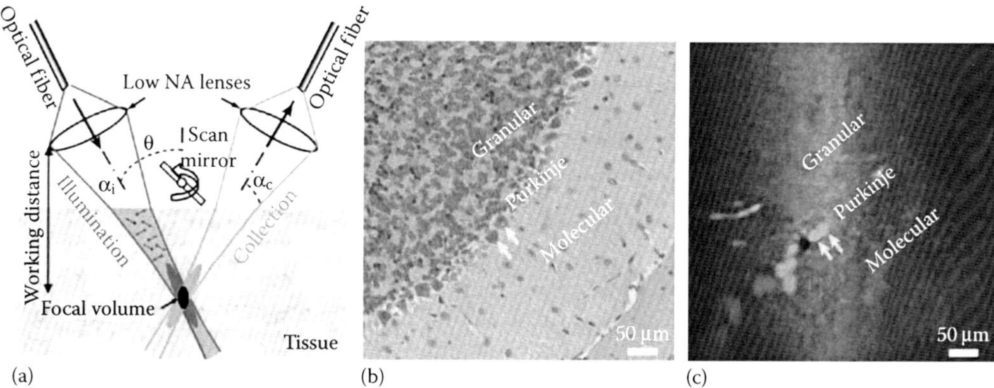

FIGURE 2.6
(See color insert.) Dual-axis confocal microscope for *in vivo* imaging. (a) Schematic of architecture: Two low NA objectives are oriented with the illumination and collection beams crossed at an angle so that the focal volume is defined at the intersection of the photon beams, offering a significant reduction of the axial resolution, long working distances, large dynamic range, and rejection of light scattered along the illumination path. (b) Fluorescence images from the cerebellum of a transgenic mouse that expresses GFP driven by a β-actin-CMV (Cytomegalovirus) promoter. The image was collected at an axial depth of 30 μm and the scale bar is 50 μm. (c) Corresponding histology showing the Purkinje cell bodies, marked by the arrows, aligned side by side in a row, which separate the molecular from the internal granular layer. (From Ntziachristos, V., *Annu. Rev. Biomed. Eng.*, 8, 1–33, 2006. With permission.)

2.6.2.1 Epi-Illumination Imaging

In epi-illumination imaging, the image is noninvasively captured over the surface and subsurface fluorescence activity from the entire animal. The technique collects the emitted light from the same side of the tissue and is also called fluorescence reflectance imaging. Owing to the diffusive nature of photons in the tissue, the light that reaches the

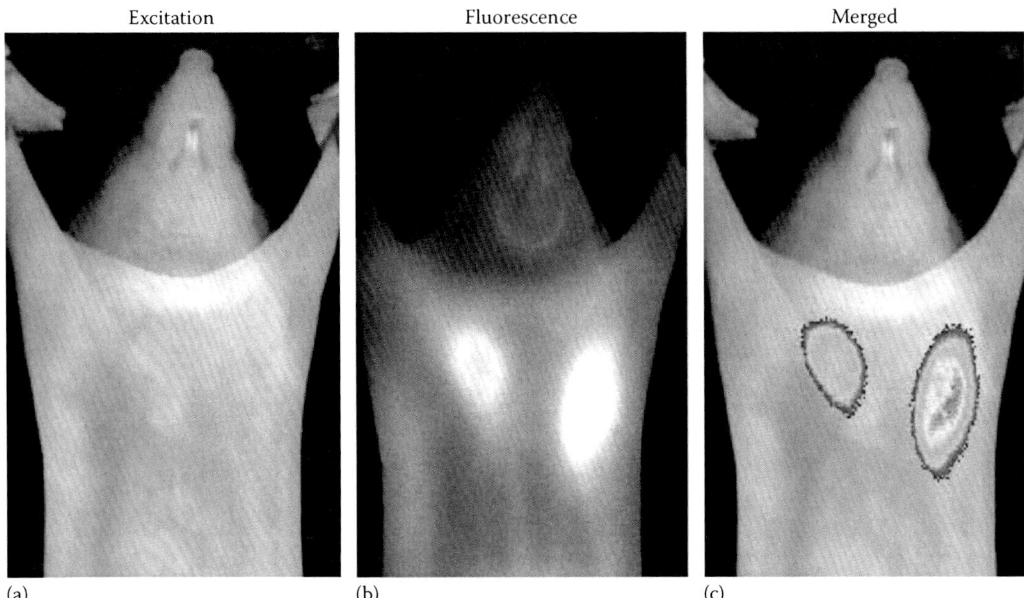

FIGURE 2.7
(See color insert.) *In vivo* epi-illumination imaging of cathepsin activity from a nude female mouse with two HT1080 tumors implanted subcutaneously. (a) Image obtained at the emission wavelength; (b) fluorescence image; (c) merged image, that is, superposition of the fluorescence image shown in color on the excitation image. A threshold was applied on the fluorescence image to remove low-intensity background signals and allow for the simultaneous visualization of (a) and (b). (From Ntziachristos, V., *Annu. Rev. Biomed. Eng.*, 8, 1–33, 2006. With permission; Courtesy of Stephen Windsor; the cathepsin-sensitive probe was kindly provided by Dr. Ching Tung, both with the Center for Molecular Imaging Research, Massachusetts General Hospital and Harvard Medical School, Boston, Massachusetts.)

surface propagates only a few millimeters under the surface, and if an appropriate wavelength is applied, it can excite not only superficial but also subsurface fluorochromes. The fluorescence emitted can be captured by CCD camera using appropriate filters.

Figure 2.7 shows an *in vivo* imaging of protease upregulation in subcutaneous HT1080 tumors subcutaneously implanted in a female nude mouse using cathepsin-sensitive probe. Epi-illumination methodologies combine simplicity of development and operation and aid in significant advancements in the field of fluorescence molecular imaging [67,68]. Some significant drawbacks associated with them are that they cannot resolve depth and cannot account for nonlinear dependencies of the signal detected on depth [69]. Although the fluorescence intensity in this case depends linearly on fluorochrome concentration, it has a strong nonlinear dependence on lesion depth and on the optical properties of the lesion.

2.6.2.2 Transillumination Imaging

Transillumination developed in 1974 is a technique used in place of planar imaging where the source and the detector are placed on the opposite sides of tissues, and the relative attenuation of fluorescence emitted is recorded [70]. This technique is now employed with advanced illumination, detection, and scanning techniques, and has recently received attention in dental research, imaging cardiac muscle activity, and small animal imaging [71–74]. Transillumination images yield similar nonlinear dependencies to epi-illumination images. A notable feature of transillumination is that the volume of interest is entirely sampled because light propagates through it. Normalized transillumination measurements at the emission wavelength are divided by geometrically identical transillumination measurements at the excitation wavelength and can yield certain benefits of enhanced quantification and contrast [74].

2.6.3 Tomographic Imaging

Optical tomography is defined as 3D reconstruction of the internal distribution of fluorochromes or chromophores in tissues based on light measurements collected at the tissue boundary. The principle of operation is similar to X-ray computed tomography (CT) where the tissue is illuminated at different points and the collected light is used in combination with a mathematical formulation to describe photon propagation in tissues. Major distinction of optical tomography compared with tomographic methods based on high-energy rays is that photons in the NIR or visible are highly scattered by tissues, which yields a nonlinear dependence of the photon field ϕ detected on tissue optical properties and the source–detector distance that is based on the following equation [16]:

$$\phi \approx \frac{\exp(-ikr)}{r}$$

where:
r is the source–detector distance assuming a point source and a point detector

$$k = \left(\frac{-c\mu_a + i\omega}{cD}\right)^{1/2}$$

Here, k is the propagation wave number of the photon wave that depends on the absorption coefficient μ_a, the diffusion coefficient D, the speed of light c in tissue, and the modulation frequency ω of the photon beam that illuminates the tissue. This equation describes a generic dependence that illuminates the complex nature of photon attenuation in tissues. The characteristics of optical tomography is that it is generally based on physical models of photon propagation, and therefore, it not only yields 3D imaging and *deep tissue* imaging but also offers true quantification of optical contrast. Most of the mathematical problems used in tomography attempt to model the photon propagation in tissues as a diffusive process, that is, utilize solutions of the diffusion equation. As higher number of spatial frequencies can be sampled with a point source, most solutions are obtained for a point source and a point detector, and form the basis for explaining the expected response of each detector for an assumed medium.

After intensity and spectrum, fluorescence lifetime is the next most useful observable fluorescence emission property for imaging. As fluorophores excited by a pulsed laser emit fluorescence photons after a few nanoseconds, lifetime is extremely sensitive to the molecule's environment. The purpose of fluorescence lifetime imaging microscopy (FLIM) is to measure the fluorescence lifetime at each point of a sample to detect the core environment of a known single fluorophore and is especially relevant for FPs that cannot easily be separated spectrally [75–77]. The technique is applied in two ways [75]:

1. One uses a time-domain measurement.
2. The other uses a frequency-domain measurement.

In the first case, the scanning process is time consuming, whereas in the second case, the camera detects only photons emitted during a fixed time window after the laser pulse. The time-gated detection can only distinguish between fluorophores of well-separated lifetimes and necessitates acquiring two sets of images. This technique has widely exploited for lanthanide chelates and metal ligand complexes imaging [78,79]. These fluorophores have much longer lifetimes than do the autofluorescence of cell proteins, which make them attractive probes for high sensitivity. The frequency-domain approach is based on radiofrequency modulation of the laser intensity and the image intensifier gain, either in phase or out of phase. This process can be extremely time consuming and usually results in photobleaching. Applications to live intracellular imaging have shown the power of this imaging technique. The advantages of two-photon microscopy can be combined with FLIM in the frequency domain [80], time-gated FLIM [81], or time-correlated single-photon counting FLIM and FRET imaging using fluorescence donor/acceptor photobleaching [82]. FRET provides a concentration independent measurement and minimizes illumination [83]. A combination of intensity and lifetime observations can benefit FRET study giving access to dynamic interactions between FRET pairs on a cell-wide scale [84].

2.7 Fluorescent Labeling

Fluorescent organic molecules [85], FPs [86,87], conjugated polymers [88], light-harvesting complexes [89], dendrimers [90], and semiconductor NCs [91] were extensively studied at cryogenic as well as at room temperature. Each of these systems exhibits fluorescence based on specific processes, which can be quite different. In NCs, ultraviolet (UV)-visible photon absorption by a semiconductor compound leads to the creation of an electron–hole pair (exciton). The pair recombines within few tens of nanoseconds, emitting a visible photon whose wavelength depends on the NC diameter owing to quantum confinement effects. Statistical labeling is mainly used for imaging purposes restricted to preliminary stages of assay development, as for the FRET-based analysis of *staphylococcal nuclease* dynamics [93,94]. Site-specific labeling is a necessity when precise distance or orientation information is required [95]. A careful choice of labeling chemistry, optimization of labeling reaction, and rigorous characterization of the labeled biomolecules, site-specificity, and retention of functionality are necessary requirements. Some molecules do not require labeling due to the presence of fluorescent moieties in their inherent structure as in the case of proteins with tryptophan residues, enzymes using NADH (Nicotinamide adenine dinucleotide), or flavins as cofactors.

2.8 Organic Dyes as Fluorescent Labels

The optical properties of organic dyes depend on the electronic transition(s) involved and can be fine-tuned by elaborate design strategies if the structure–property relationship is known for a given class of the dye [95,96]. The emission of organic dyes typically originates from either an optical transition delocalized over the whole chromophore or intramolecular charge transfer transitions [95]. Fluoresceins, rhodamines, most 4,4′-difluoro-4-bora-3α,4α-diaza-*s*-indacenes, and cyanines are resonant dyes that are slightly structured having comparatively narrow absorption and emission bands, high molar absorption coefficients, and moderate-to-high fluorescence quantum yields. The poor separation of the

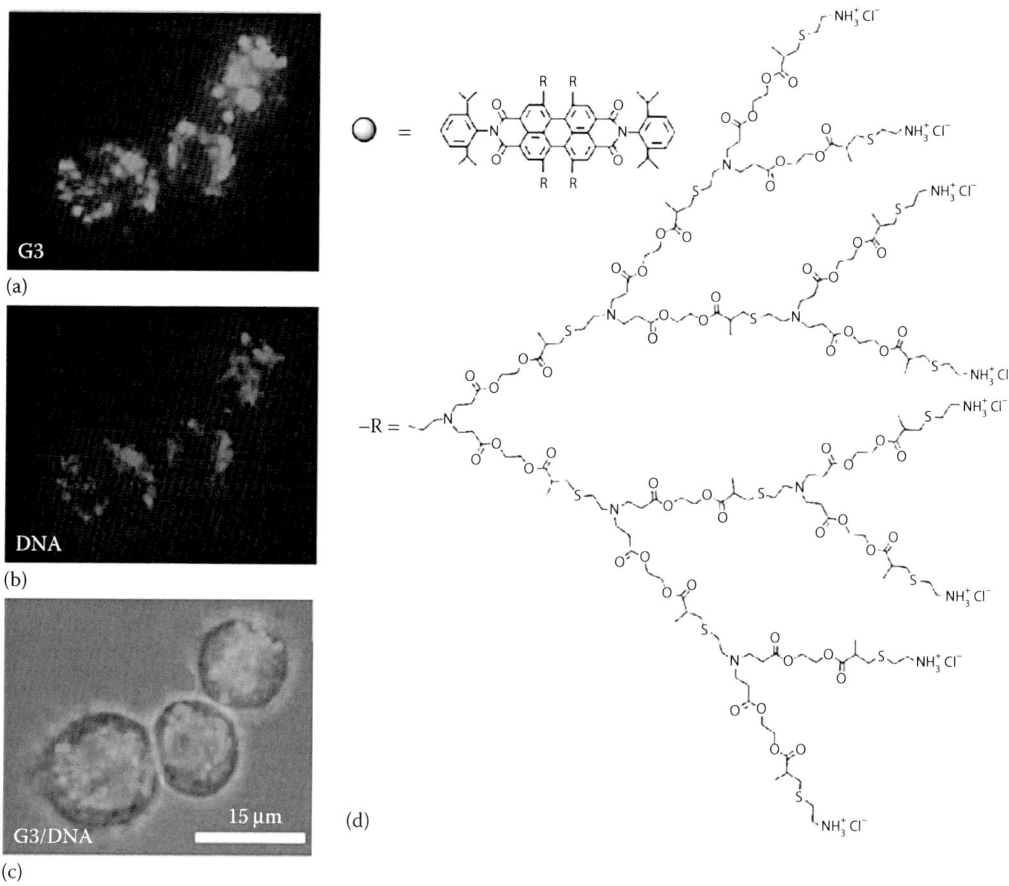

FIGURE 2.8
Fluorescence images (a–c) of the G3–DNA complex inside of cells after 48 h of incubation. (a) G3 fluorescence image; (b) CXR reference dye labeled DNA; (c) merged channels of (a) and (b); (d) structure of perylenediimide (PDI)-cored dendrimers. (Reprinted from *Chem. Commun.*, 49, Xu, Z. et al., Fluorescent water-solubleperylene-diimide-cored cationic dendrimers: Synthesis, optical properties, and cell uptake, 3646–3648, Copyright 2013, with permission from Elsevier.)

absorption and emission spectra favors cross talk between different dye molecules. Dyes such as coumarins have a well-separated structure, less absorption and emission bands in polar solvents, and a larger Stokes shift, the size of which depends on solvent or matrix polarity (Figure 2.8). Their molar absorption coefficients and fluorescence quantum yields are generally smaller than those of dyes with a resonant emission.

2.9 QDs as Fluorescent Labels

Nanometer-sized crystalline particles, also called quantum dots, are composed of periodic groups of II–VI compound semiconductors such as CdSe and III–V compound semiconductors such as InP materials. They are high fluorescence emitters with size-dependent emission wavelengths (Figure 2.9). They possess extreme brightness and resistance to

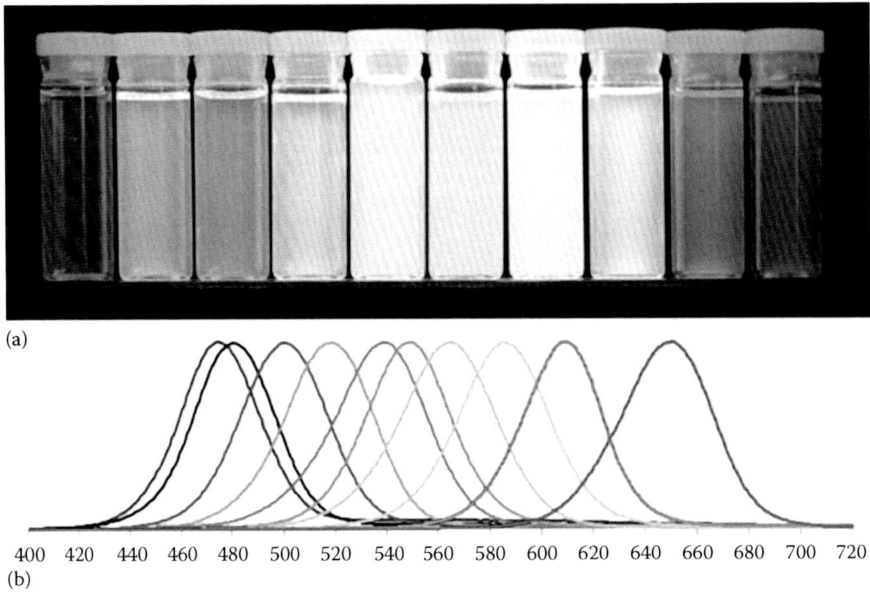

FIGURE 2.9
(a) Colloidal suspension of QDs with different sizes viewed under long-wave UV illumination; (b) narrow emission spectra of related QDs with different sizes. (Reprinted from *Nano Today*, 4, Zrazhevskiy, P. and Gao, X., Multifunctional quantum dots for personalized medicine, 414–428, Copyright 2009, with permission from Elsevier.)

photobleaching, which enables the use of very low intensity lasers over extended time periods, making them especially useful for live cell imaging, such as consecutive acquisition of *z*-stacks for high-resolution 3D reconstructions over time (four-dimensional imaging). The intense brightness of QDs is helpful in single-particle detection and the increasing as well as tunable emission wavelength and distinct emission spectra highly facilitates data acquisition and analysis of multiple tagged molecules [97–104]. QDs have several drastically enhanced properties compared to organic fluorophores, one of which is their unique optical spectra. Organic dyes have typically narrow absorption spectra, which means that they can only be excited within a narrow wavelength. They also have asymmetric emission spectra broadened by a red tail, whereas QDs have broad absorption spectra, enabling excitation by a wide range of wavelengths, and their emission spectra are symmetric and narrow. Hence, multicolor NCs of different sizes can be excited by a single wavelength shorter than their emission wavelengths, with minimum signal overlap. QDs are also very stable light emitters owing to their inorganic composition and are less susceptible to photobleaching than organic dye molecules [102]. The photostability of QDs was compared with commonly used fluorophores, such as rhodamine and fluorescein [103–105]. This extreme photostability makes QDs very attractive probes for imaging thick cells and tissues over long time periods. The two-photon cross section of QDs is significantly higher than that of organic fluorophores, making them quite well suited for examination of thick specimens and *in vivo* imaging using multiphoton excitation, and their fluorescence lifetime is 10–40 ns, which is significantly longer than typical organic dyes that decay on the order of a few nanoseconds [106–109]. When combined with pulsed laser and time-gated detection, the use of QD labels can produce images with greatly reduced levels of background noise. There are also disadvantageous photophysical properties of QDs such as blinking; that is, QDs randomly alternate between an emitting state and a nonemitting state. Till date, there has been limited evidence that QD blinking can be suppressed on some timescale by passivating the QD surface [107]. It has also been reported that QDs reveal photobrightening, which means that the fluorescence intensity increases upon excitation. Although in most cases this property can be advantageous, it is problematic in fluorescence quantization studies. Both blinking and photobrightening are linked to mobile charges on the surfaces of the dots and are considered as limitations of QDs [110].

2.9.1 Biocompatibility

The core–shell QDs synthesized are only soluble in nonpolar solvents because of their hydrophobic surface layer, whereas for their application as probes for examination of biological specimens, the surface must be hydrophilic. One of the easiest approaches is to exchange the hydrophobic surfactant molecules with bifunctional molecules, which are hydrophilic on one side and hydrophobic on the other side and which bind to the ZnS shell. Most often, thiols (–SH) are used as anchoring groups on the ZnS surface and carboxyl (–COOH) groups are used as the hydrophilic ends. Many biological applications of QDs were achieved by using mercapto-hydrocarbonic acids (SH–...–COOH) to make QDs water soluble [111,112]. The long-term stability of QDs depends on the bond between thiols and ZnS [113,114]. Another alternative approach that was used is to grow a silica shell around the particle, also called surface silanization [115]. The trimethoxysilane groups can be cross-linked by the formation of siloxane bonds and the most frequently used are aminopropylsilanes (APS), phosphosilanes, and polyethylene glycol-silane [115]. Because the silica shells are highly cross-linked, silanized QDs are extremely stable. Other solubilization methods include coating the surface with amphiphilic polymers [116], phospholipid micelles [117], dithiothreitol [118], organic dendron [119,120], and oligomeric ligands [121]. The hydrophobic tails of the polymer intercalate with the surfactant molecules and the hydrophilic groups stick out to ensure water solubility of the particle. But this method increases the final size of the particles after coating (19–25 nm), which places restrictions on many biological applications [116].

2.9.2 Toxicity

There have been several concerns about the toxicity of QDs, especially when they are used to study live cells and animals because they contain elements such as Cd and Se. When properly capped by both ZnS and hydrophilic shells, no acute and obvious CdSe QD toxicity was detected in studies pertaining cell proliferation [122,123] and animal models [124]. The toxicity of QDs was also investigated in *Xenopus* embryos, where injection of low concentrations of QDs showed no adverse effects, although some abnormalities were noted with increasing concentrations, but it was unclear whether they resulted from the QDs or changes in the osmotic equilibrium of the cell. Primary hepatocyte cultures were also examined because the liver is the major target of Cd injury [123]. Cytotoxicity was observed when Cd^{2+} was released. When the QD surface coating was not stable, exposing the CdSe to oxidization by air or UV damage [123]. Hence, strategies to protect the QD surfaces from oxidative environments are critical for live cell imaging. Coating the QD surface with ZnS eliminates toxicity owing to air exposure but provides only partial protection of the core from UV exposure. Larger molecules, such as bovine serum albumin and polymer–streptavidin, further slow down the photooxidation process. There is an urgent need for further investigations into QD toxicity in live animals and safety for diagnostic applications.

2.10 Comparison of Fluorescence Properties of QDs with Organic Dyes

In comparison to organic dyes, QDs have narrow emission band of mostly symmetric shape. The spectral position of absorption and emission is tunable by particle size (the so-called quantum size effect; Figure 2.10). The width of the emission peak, in particular, is mainly determined by QD size distribution. The broad absorption allows free selection of the excitation wavelength, and the size-dependent molar absorption coefficients at the first absorption band of QDs are generally large compared to organic dyes. Typical molar absorption coefficients are 100,000–1,000,000 M^{-1} cm^{-1} [117,124–129], whereas for dyes molar absorption coefficients at the main (long-wavelength) absorption maximum are about 25,000–250,000 M^{-1} cm^{-1} [106,107]. Fluorescence quantum yields of properly surface-passivated QDs are in most cases high in the visible light range, the visible–NIR wavelength as well as the NIR wavelength (Table 2.1) [112–117]. By contrast, organic dyes have fluorescence quantum yields that are high in the visible light range but are at best moderate in the NIR wavelength range [115]. Moreover, compared to organic dyes, another favorable feature of QDs is the typically very large two-photon action cross section [121]. The fluorescence lifetimes of organic dyes are about 5 ns in the visible light and 1 ns in the NIR wavelengths, but the typically mono-exponential decay kinetics enable straightforward dye identification from measurements of fluorescence lifetimes, making dyes suitable for applications involving lifetime measurements. In the case of QDs, the comparatively long lifetimes enable straightforward temporal discrimination of the signal from cellular autofluorescence and scattered excitation light, enhancing the sensitivity wavelength and time-dependent, bi- or multiexponential QD decay behavior [111,130,131].

2.11 Fluorescent Conjugated Polymer Nanoparticles for Bioimaging

Despite these unique advantages, fluorescence methods are generally restricted to small animal studies and not to the clinic diagnostics due to photon-limiting interferences (scattering, absorption, and autofluorescence) occurring in biological media. These limitations can be dealt with if a breakthrough in the signal intensity is achieved by utilizing exogenous NIR fluorescent probes that are orders of magnitude brighter than organic dyes [107,132]. Conjugated polymers hold great potential since they possess large absorptivities due to their high chromophore density. Although their bioapplications have remained at the early stages of *in vitro* and *ex vivo* studies till date, their densely packed nanoparticulate formulation holds great potential for *in vivo* use owing to their organic nature of chemical constitution and water dispersibility as well as their superior light absorbing ability [133,134]. The successful *in vivo* application of conjugated polymer nanoparticles in sentinel lymph node (SLN) mapping and biopsy is widely used for various types of cancer staging and surgery [129]. In spite of the advantageous absorption properties, however, the fluorescence quantum yield of conjugated polymers, particularly in the red or NIR region, drops significantly by solidification such as nanoparticle formation, as typically observed for most organic chromophores at high concentration or in the aggregated state [111]. An exceptional case is the cyano-substituted derivatives of poly(*p*-phenylenevinylene) that, in the aggregated film state, exhibit an efficient interchain excitonic photoluminescence in the long-wavelength region with fairly high quantum yields. The simultaneous achievement of high nanoscopic chromophore density and

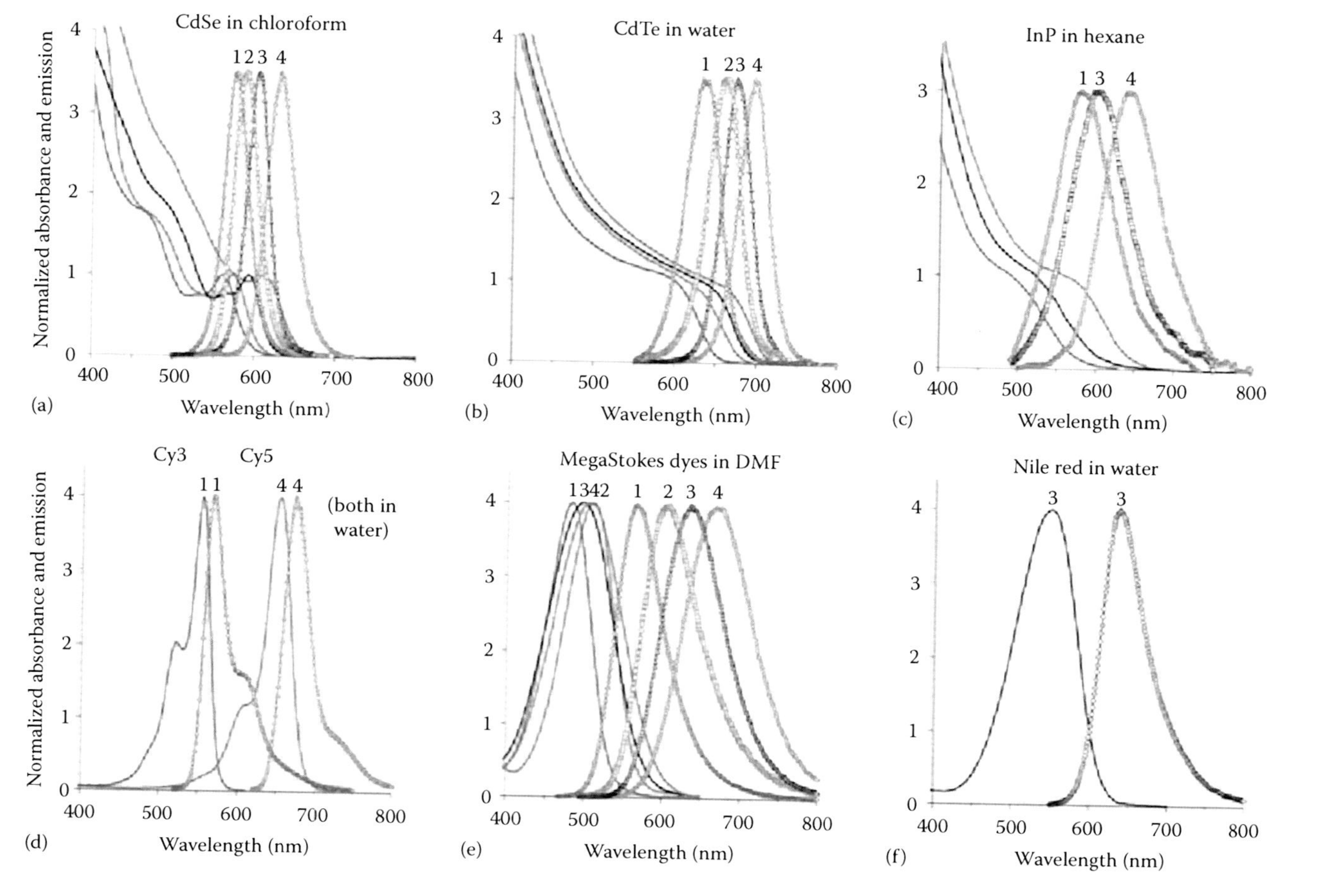

FIGURE 2.10
Spectra of QDs and organic dyes. Absorption (lines) and emission (symbols) spectra of representative QDs (a–c) and organic dyes (d–f) color coded by size ($1 < 2 < 3 < 4$). Mega Stokes dyes were designed for spectral multiplexing in dimethylformamide (DMF). (Reprinted by permission from Macmillan Publisher Ltd., *Nat. Methods*, Resch-Genger, U. et al., 5, 763–775, 2008, copyright 2008.)

TABLE 2.1

Comparison of Organic Dyes and QDs

Property	Organic Dye	QD[a]
Absorption spectra	Discrete bands: FWHM, 35[b] to 80–100 nm[c]	Steady increase toward UV wavelengths starting from absorption onset; enables free selection of excitation wavelength
Molar absorption coefficient	2.5×10^4–2.5×10^5 M^{-1} cm^{-1} (at long-wavelength absorption maximum)	10^5–10^6 M^{-1} cm^{-1} at first exitonic absorption peak, increasing toward UV wavelengths; larger (longer wavelength) QDs generally have higher absorption
Emission spectra	Asymmetric, often tailing to long-wavelength side: FWHM, 35[b] to 70–100 nm[c]	Symmetric, Gaussian profile; FWHM, 30–90 nm
Stokes shift	Normally <50 nm[b], up to >150 nm[c]	Typically <50 nm for visible wavelength-emitting QDs
Quantum yield	0.5–1.0 (visible[d]), 0.05–0.25 (NIR[d])	0.1–0.8 (visible), 0.2–0.7 (NIR)
Fluorescence lifetimes	1–10 ns, mono-exponential decay	10–100 ns, typically multiexponential decay
Two-photon action cross section	1×10^{-52}–50×10^{-48} cm^4 s per photon (typically about 1×10^{-49} cm^4 s per photon)	2×10^{-47}–4.7×10^{-46} cm^4 s per photon
Solubility or dispersibility	Control by substitution pattern	Control via surface chemistry (ligands)
Binding to biomolecules	Via function groups following established protocols; often several dyes bind to a single biomolecule labeling-induced effects on spectroscopic properties of reporter studied for many common dyes	Via ligand chemistry; few protocols available; several biomolecules bind to a single QD; very little information available on labeling-induced effects
Size	~0.5 nm; molecule	6–60 nm (hydrodynamic diameter); colloid
Thermal stability	Depends on dye class; can be critical for NIR wavelength dyes	High; depends on shell or ligands
Photochemical stability	Sufficient for many applications (visible wavelength), but can be insufficient for high-light flux applications; often problematic for NIR wavelength dyes	High (visible and NIR wavelengths); orders of magnitude higher than that of organic dyes; can reveal photobrightening

Toxicity	From very low to high; depends on dye	Little known yet (heavy metal leakage must be prevented, potential nanotoxicity)
Reproducibility of labels (optical and chemical properties)	Good, owing to defined molecular structure and established methods of characterization; available from commercial sources	Limited by complex structure and surface chemistry; limited data available; few commercial systems available
Applicability to single-molecule analysis	Moderate; limited by photobleaching	Good; limited by blinking
FRET	Well-described FRET pairs; mostly single-donor–single-acceptor configurations; enables optimization of reporter properties	Few examples; single-donor–multiple-acceptor configurations possible; limitation of FRET efficiency due to nanometer size of QD coating
Spectral multiplexing	Possible, three colors (MegaStokes dyes), four colors (energy-transfer cassettes)	Ideal for multicolor experiments; up to five colors demonstrated
Lifetime multiplexing	Possible	Lifetime discrimination between QDs not yet shown; possible between QDs and organic dyes
Signal amplification	Established techniques	Unsuitable for many enzyme-based techniques; other techniques remain to be adapted and/or established

Source: Macmillan Publishers Ltd., *Nat. Methods*, Resch-Genger, U. et al., 5, 763–775, 2008, copyright 2008.

Note: Properties of organic dyes depend on dye class and are tunable via a substitution pattern. Properties of QDs depend on material, size, size distribution, and surface chemistry.

Unless stated otherwise, all values were determined in water for organic dyes and in organic solvents for ODs, and refer to the free dye or OD.

[a] Emission wavelength regions for QD materials (approximate): CdSe, 470–660 nm; CdTe, 520–750 nm; InP, 620–720 nm; PbS, >900 nm; PbSe, >1000 nm.

[b] Dyes with resonant emission such as fluoresceins, rhodamines, and cyamines.

[c] CT dyes.

[d] Definition of special regions used here: visible, 400–700 nm; NIR, >700 nm.

high fluorescence efficiency can lead to a significant breakthrough in the signal output from individual nanoprobes. Conjugated nanoparticles concentrated with dyes exhibiting aggregation possessed enhanced fluorescence that can be used as bright nanoprobes for two-photon fluorescence microscopy. Surfactant-stabilized cyanovinylene-backboned polymer dots (cvPDs) were directly synthesized via *in situ* colloidal polymerization in the aqueous phase without using any harmful organic solvents (Figure 2.11). The resulting fluorescence color of the cvPDs was readily tuned throughout the broad spectral window by varying the aromatic structure of the monomers (Figure 2.8). Typical of common

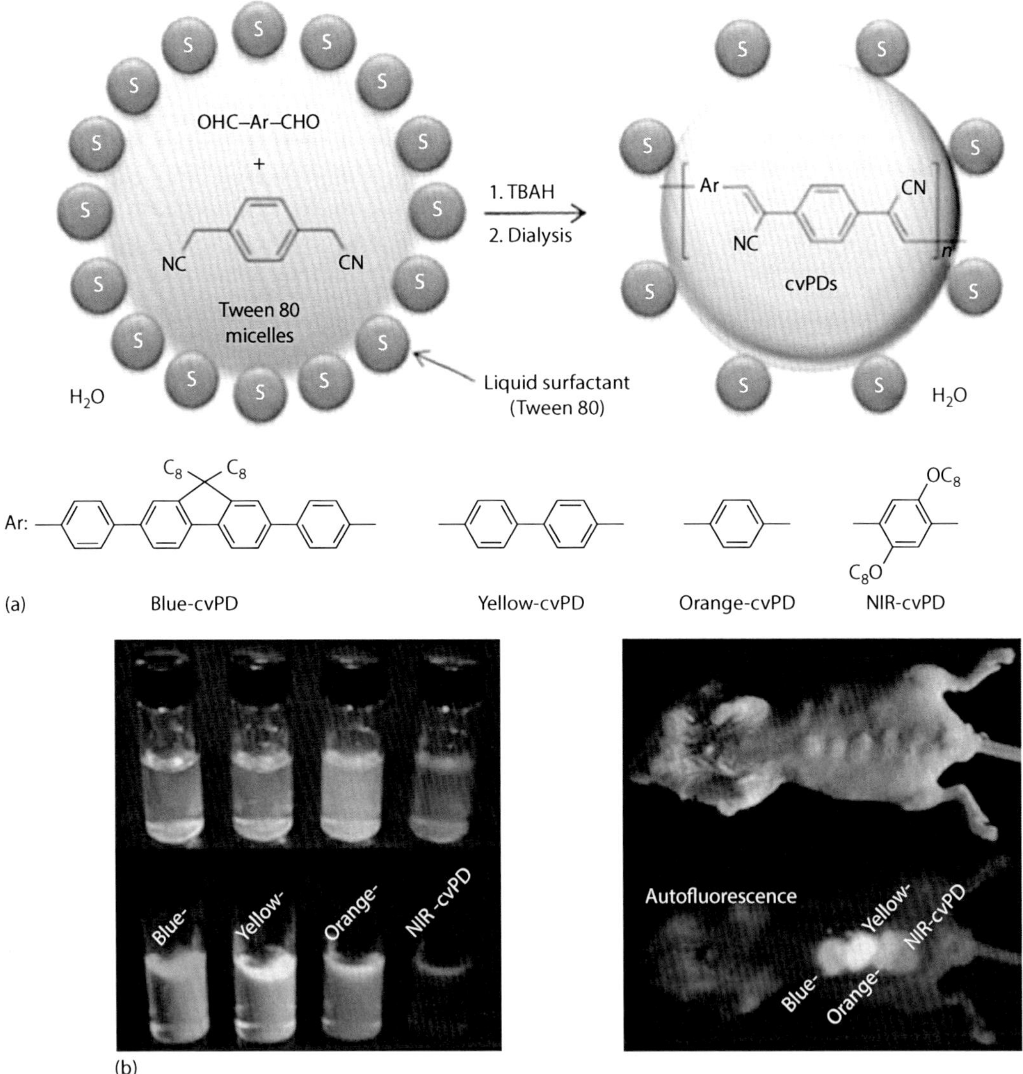

FIGURE 2.11
(See color insert.) (a) Schematic diagram depicting colloidal synthesis of cvPDs through tetrabutylammonium hydroxide (TBAH)-catalyzed Knoevenagel condensation in the hydrophobic core of solvent-free aqueous micelles. (b) True-color photographs of water-dispersed cvPDs (left) and a cvPD-injected live mouse (right) under room light (top) and UV excitation at 365 nm for fluorescence (bottom). (Kim, S. et al., *Chem. Commun.*, 46, 1617–1619, 2010. Reproduced by permission of The Royal Society of Chemistry.)

conjugated polymers, cvPDs have large Stokes shifts with minimal spectral overlaps between the absorption and emission bands, which minimize fluorescence loss by self-absorption in the chromophore-concentrated condition. Similar to QDs, broad absorption profiles of cvPDs allow single-wavelength excitation of multicolor fluorescence, useful for simultaneous imaging of plural molecular targets (Figure 2.11) [107]. Even under UV excitation, the fluorescence spots of subcutaneously injected cvPDs were clearly distinguished from the background in terms of color and brightness, indicating their potential for multicolor *in vivo* imaging (Figure 2.12) [133–143].

Conjugated polymers are known to possess high absorption coefficients and high fluorescence efficiency, revealing their wide range of applications in optoelectronic thin-film devices [144,145]. The extraordinary light-gathering power of conjugated polymers is evidenced by the first reported direct determination of the optical absorption cross section of a single molecule at room temperature, in which an optical cross section of the conjugated polymer poly[2-methoxy-5-(2-ethylhexyloxy)-1,4-phenylenevinylene] (MEH-PPV) in the vicinity of 10^7 M^{-1} cm^{-1} was determined [146]. Highly fluorescent nanoparticles consisting of one or more hydrophobic conjugated polymers were developed and investigated [147,148]. These CPdot nanoparticles are a new class of highly fluorescent probes with potential applications for biosensing and imaging. Detection of two-photon excited fluorescence of single nanoparticles was demonstrated using relatively low-power lasers, demonstrating the great potential of these CPdots for multiphoton fluorescence imaging applications. The conjugated polymers employed are the polyfluorene derivative poly(9,9-dihexylfluorenyl-2,7-diyl) [PDHF; average molecular weight (MW) 55,000], the copolymer poly{(9,9-dioctyl-2,7-divinylenefluorenylene)-*alt-co*-[2-methoxy-5-(2-ethylhexyloxy)-1,4-phenylene]} (PFPV; average MW 270,000), and the

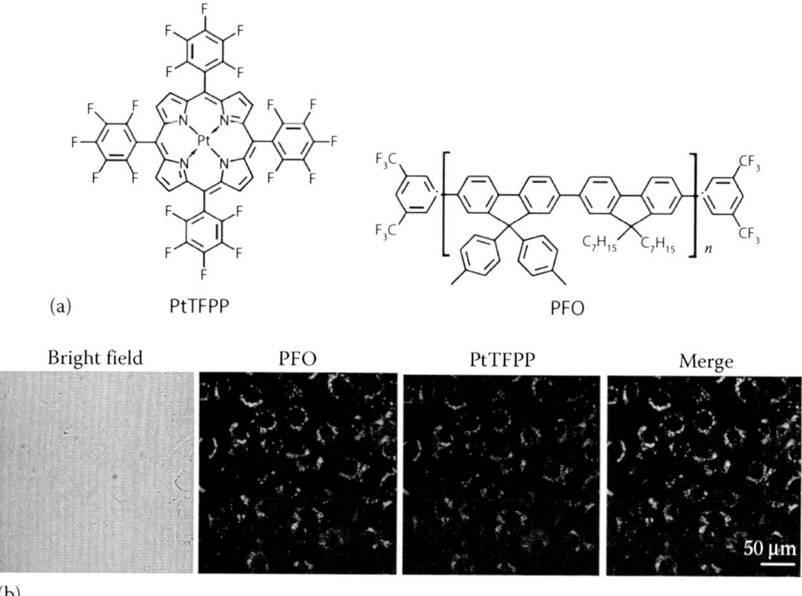

FIGURE 2.12
(a) Chemical structures of phosphorescent PtTFPP reporter and PFO; (b) fluorescent images of MEF cells stained with MM2 (10 g mL^{-1}, 16 h). PtTFPP, tetrakis(pentafluorophenyl)porphyrin platinum; PFO, Polyfluorene; MEF, metal-enhanced fluorescence; MM2, mm^2. (Kondrashina A.V. et al.: A phosphorescent nanoparticle-based probe for sensing and imaging of (intra) cellular oxygen in multiple detection modalities. *Adv. Funct. Mater.* 2012. 22. 4931–4939. Copyright Wiley-VCH Verlag GmbH & Co. KGaA. Reproduced with permission.)

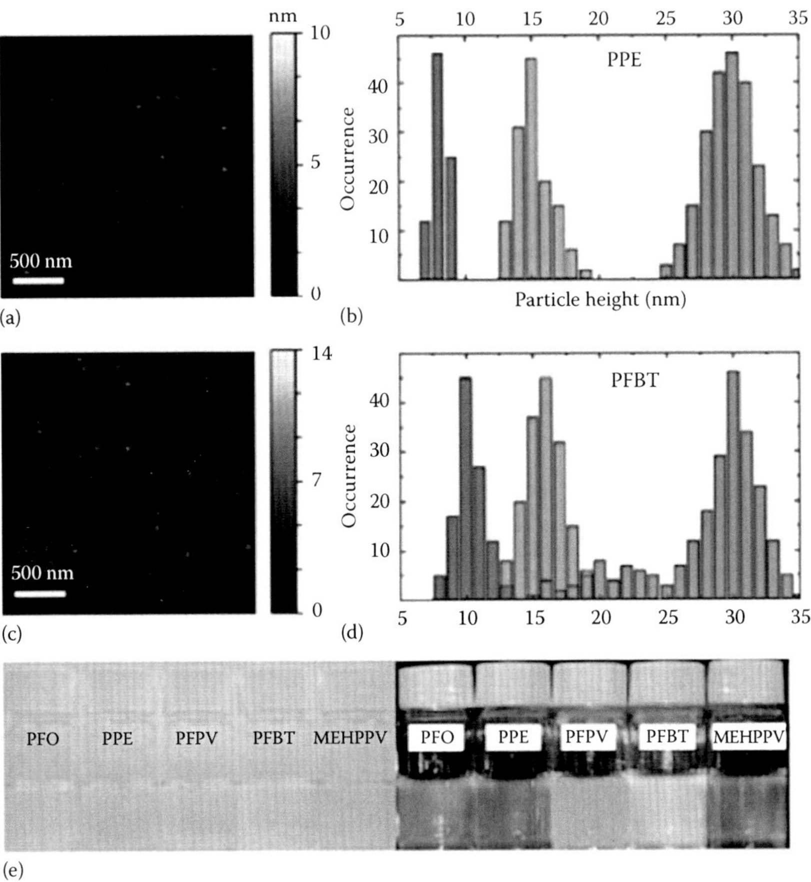

FIGURE 2.13
(a) Typical atomic force microscopy (AFM) image of small PPE dots; (b) histograms of particle height of the PPE dots prepared with different precursor concentration; (c) typical AFM image of small PFBT dots; (d) histograms of particle height of the PFBT dots prepared with different precursor concentrations; (e) photographs of aqueous CPN suspensions under room light (left) and UV light (right) illumination. PPE, Poly(phenylene ethynylene); PFBT, Poly(fluorene-alt-benzothiadiazole). (Reprinted with permission from Wu, C., *ACS Nano*, 2, 2415, 2008, Copyright 2008, American Chemical Society.)

polyphenylenevinylene derivative MEH-PPV (average MW 200,000) (Figure 2.13) [147,148]. Their size can be controlled over the range of 5–50 nm by varying the concentration of the precursor solution. Unlike inorganic semiconductor nanoparticles, the particle size does not affect the shape of the absorption and fluorescence spectra of CPdots. Instead, an increase in particle size largely results in an increase in the optical cross section per particle [147,148]. Single conjugated polymer molecules typically exhibit complex photophysics such as fluorescence intermittence (blinking) and photon antibunching [88,149].

As fluorescent probes for imaging or single-particle tracking, the relatively steady fluorescence of CPdots compares favorably to that of QDs, which typically exhibit pronounced blinking on timescales of milliseconds to hundreds of seconds (Figure 2.14). Analyses of single-particle kinetics traces indicate that ~10^6 photons per particle (~10 nm diameter) were detected prior to photobleaching. This is lower than the photostability under one-photon excitation (~10^7 photons detected), consistent with prior observations that single fluorophores exhibit lower photostability under two-photon excitation than under one-photon

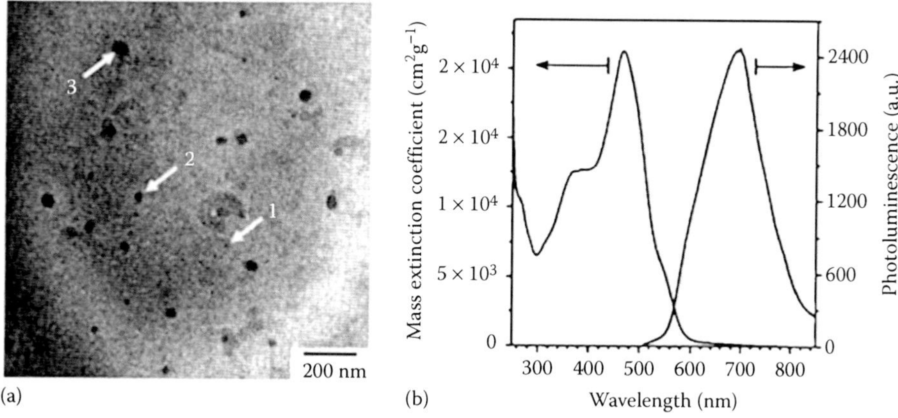

FIGURE 2.14
(a) Transmission electron microscopy (TEM) image of NIR-cvPDs. Nanoparticles representing different size distributions are indicated: (1) 7.2 ± 0.9, (2) 25.7 ± 4.3, and (3) 47.8 ± 5.0 nm (mean ± standard deviation). (b) Mass extinction coefficient and photoluminescence spectra of water-dispersed NIR-cvPDs. Single particle fluorescence kinetics traces. (Kim, S. et al., *Chem. Commun.*, 46, 1617–1619, 2010. Reproduced by permission of The Royal Society of Chemistry.)

excitation. Silica encapsulation would likely improve photostability [147]. Demonstration of single-particle imaging using relatively low laser excitation levels demonstrates the potential utility of CPdots for multiphoton fluorescence microscopy applications and raises the possibility of developing inexpensive NIR diode lasers for two-photon excited fluorescence imaging (Figure 2.15).

Figure 2.15 shows microscopic images of live BALB/c and fixed 3T3 cells stained by the conjugated polymer nanoparticles (CPNs) overnight. The results showed that the CPNs were cell permeable and accumulate exclusively in the cytosol without any considerable inhibition of cell viability. In addition, it was demonstrated that CPNs exhibit high resistance to photobleaching, in contrast to commercially available dyes.

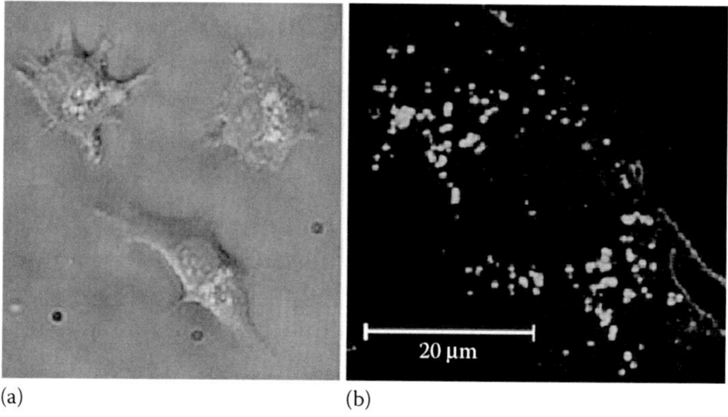

FIGURE 2.15
Fluorescence images of live (a) and fixed (b) cells. (a) BALB/C 3T3 cells were incubated sequentially with CPNs. The image is a composite of two micrographs using GFP (for CPNs) and 4′,6-diamidino-2-phenylindole (DAPI)/Hoechst/aminomethyl coumarin acetate (AMCA) (for Hoechst) filter sets. (b) Live BALB/C 3T3 cells were incubated with CPNs and fixed for confocal microscopic study. (Moon, J.H. et al.: Live-cell-permeable poly[p-phenylene ethynylene]. *Angew. Chem. Int. Ed.* 2007. 46. 8223–8225. Copyright Wiley-VCH Verlag GmbH & Co. KGaA. Reproduced with permission.)

2.12 Experimental Validation of Monte Carlo Modeling of Fluorescence

The Monte Carlo (MC) method has been developed to simulate light propagation in tissues for nearly two decades [150]. This computational modeling tool can provide insight into the design of experimental setups for optical measurements from human tissues, for understanding light distribution in human tissues, and for validation of analytical models of light transport. Also several parameters can be modeled including illumination and collection geometries, 3D light distribution in tissues, and various types of light–tissue interactions such as absorption, scattering, and even fluorescence. One of the important prerequisites to utilizing an MC code is that it should be validated before use against a gold standard, as a test dataset generated from an analytical model of light transport or experimental measurements on tissue phantom models made up of a mixture of absorbers, scatterers, and fluorophores. Significant numbers of studies were carried out to evaluate the effect of excitation and emission geometries, sample geometries, as well as absorption and scattering on tissue fluorescence [151–156]. MC modeling was also employed to relate the bulk tissue fluorescence spectrum to the fluorescence originating from different layers within the tissue, and for verification of analytical models of tissue fluorescence [157–160].

Keijzer et al. [151] simulated fluorescent light transport in turbid media using MC modeling, and showed that fluorescence spectra measured from turbid media depend on the geometry of excitation light delivery and emission light collection. Welch et al. [153] used MC simulations to evaluate the effect of scattering, absorption, boundary conditions, geometry of the tissue sample, and quantum yield of tissue fluorophores on the fluorescence spectrum measured from tissues, and showed that MC modeling provides a realistic method for interpreting the effect of tissue sample geometries on the remitted fluorescence. Avriller et al. [154] used a fast MC simulation to correct the distortion in tissue fluorescence spectra that arises from different separations between illumination and collection fibers. Jianan et al. [155] used MC simulations to identify appropriate illumination and collection geometries, at which the effect of optical properties on tissue fluorescence spectra could be minimized, and found that maximizing the overlap between illumination and collection areas minimized the effect of absorption on the measured tissue fluorescence. Pfefer et al. [156] examined the effect of optical fiber diameter, fiber–tissue spacing, and fiber NA on fluorescence spectroscopy using a single optical fiber probe and investigated that increasing the fiber diameter spacing increased the mean photon path length of collected fluorescent photons and produced a transition from a superficial to a deeper and more homogeneous probing volume in tissues. Increasing the NA results in an increase in fluorescence intensity, but the path lengths of collected fluorescent photons were insignificantly affected for NAs less than 0.8. Hyde et al. [160] compared the spatially resolved fluorescence obtained using MC simulations to that obtained using diffusion theory and that obtained in phantom model experiments, and revealed that the experimentally measured fluorescence was significantly higher than that predicted by MC simulations or diffusion theory at detector positions close to the source ~1 mm, while the results were in agreement when the source–detector separations were around 10 mm. Liu and Ramanujam [161] proposed the design of a variable aperture probe for depth-dependent fluorescence measurements from tissues using MC simulations to show that the fluorescence measured using completely overlapping illumination and collection apertures with variable diameters that were related to the depth of a fluorescent target in a turbid medium epithelial tissue.

Although MC modeling was widely used to simulate the effect of illumination and collection geometries and sample geometries, and the effect of absorption and scattering on fluorescent light transport in tissues, the experimental validation of MC simulations for fluorescent light propagation in turbid media was carried out only to a limited extent. Pogue and Burke [162] compared MC simulations and experimental measurements of fluorescence as a function of fluorophore concentration in a turbid medium, measured with a 1-mm-diameter fiber bundle composed of several fibers with a diameter of 100 mm for a low and a high absorption coefficient. Although they showed similar trends in both computational and experimental results, they did not carry out a quantitative comparison of the two approaches. Liu et al. [163] experimentally verified MC modeling of fluorescence and diffuse reflectance measurements in turbid, tissue phantom models. He performed two series of simulations and experiments, in which one optical parameter (absorption or scattering coefficient) was varied, whereas the other was fixed, to assess the effect of the absorption coefficient (μ_a) and scattering coefficient (μ_s) on the fluorescence and diffuse reflectance measurements in a turbid medium. The conversions accounted for the differences between the definitions of the absorption coefficient and fluorescence quantum yield of fluorophores in a tissue phantom model and those in an MC simulation. The findings indicated good agreement between the simulated and experimentally measured results in most cases. This dataset served as a systematic validation of MC modeling of fluorescent light propagation in tissues. The simulations are carried out for a wide range of absorption and scattering coefficients as well as the ratios of scattering coefficient to absorption coefficient, and thus would be applicable to tissue optical properties over a wide wavelength range (UV–visible/NIR). The fiber optic probe geometries that are modeled in this study include those commonly used for measuring fluorescence from tissues in practice. A 3D weighted photon MC code was written in standard C programming and was modified to simulate fluorescence. The original code was used for diffuse reflectance simulations. Up to 5 million photons were launched in each simulation at random uniformly distributed locations over a range of angles defined by an NA of 0.22 and over a circular or ringlike illumination area at the top surface of the medium. A rejection scheme was used to determine whether or not the absorbed fraction of a photon packet was emitted as a fluorescent photon in the case of fluorescence simulation. The fluorescence or diffuse reflectance escaping the medium was collected over a circular area defined by the collection diameter and an NA of 0.22. The refractive index of the medium above the model was set to 1.452 to simulate an optical fiber, and that below the model was set to 1.0.

To simulate fluorescence and diffuse reflectance measurements with fiber optic probe (Figure 2.16a), the illumination diameter was set to 200 mm and single collection fiber with a diameter of 100 mm was used. Although the actual fiber optic probe had nine collection fibers, the comparison of the simulations to the experimental measurements was not affected, because the profiles of the normalized fluorescence/reflectance versus absorption/scattering coefficients were used, rather than the absolute intensities. The center-to-center distance was set to be 187.5 mm. In the case of the fiber optic probe (see Figure 2.16b for the cross-sectional view), photons were incident over a ring area, which was defined by an inner diameter of 1.52 mm and an outer diameter of 2.18 mm. All the photons exiting within a circular area defined by an inner diameter of 1.52 mm and a cone defined by an NA of 0.22 were collected. In the case of a fiber optic probe (see Figure 2.16c for the cross-sectional view), the illumination area was defined by a diameter of 1180 mm, which was obtained by calculating the distance from the geometrical center of the fiber bundle to the outer edge of the outermost illumination fiber. Three collection fibers were used, each with a diameter of 200 mm and at different distances from the center of the illumination area. The three center-to-center distances were 735, 980, and 1225 mm.

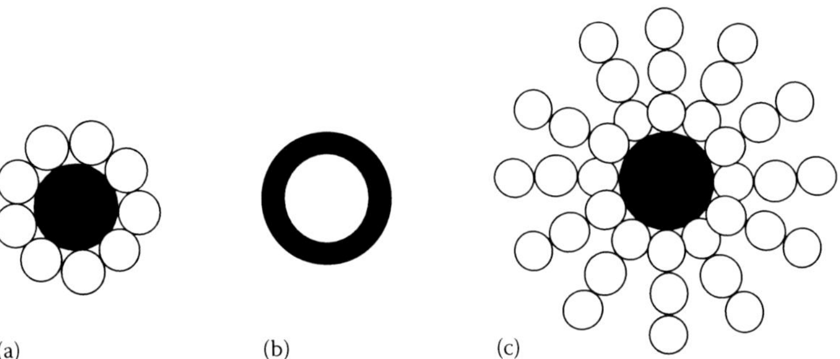

FIGURE 2.16
Cross-sectional views of the common ends of three fiber optic probes employed to measure the fluorescence, and in most cases the diffuse reflectance, of the tissue phantom models. The filled area represents (a) the illumination fiber(s) and (b) the unfilled area represents the collection fiber(s). (c) The collection fibers are round, although they appear slightly elliptical. (From Liu, Q., *J. Biomed. Opt.*, 8, 223–236, 2003. With permission.)

Deng and Gu [164] carried out detailed investigation into the penetration depth through brain cortex tissue in multiphoton fluorescence microscopy according to the double-layer MC model. The dependence of image resolution and signal level on the focal depth in a double-layer human cortex structure was analyzed and the effect of the NA of an imaging objective on these two parameters was investigated. The effect of the gray matter thickness on image resolution and signal level was also explored.

The effective point spread function (EPSF) was used to investigate transverse resolution (Γ) and signal level (η) along the focal depth under multiphoton fluorescence microscopy. The EPSF through a turbid medium under a microscope was explained by the distribution of photons in the focal region, which propagated through a turbid medium, the aperture of a detection objective, and other optical apertures. Two steps were involved to derive an EPSF at a given focal plane in microscopic fluorescence imaging. First, an excited photon distribution $I_{ex}(r)$ where r is the radial distance from the focus (r) was calculated using MC simulation, and a fluorescence photon distribution $p_n(r)$ was produced according to the weighting function under one-photon (1p), two-photon (2p), and three-photon (3p) excitation:

$$p_n(r) = \alpha_n I_{ex}^n(r)$$

where:
$n = 1, 2,$ and 3 corresponding to 1p, 2p, and 3p excitation, respectively
$\alpha_n = 1$ (taken in this case)

In the second step of the MC simulation, fluorescence photons excited by the above equation were monitored, and those fluorescence photons at r reaching the detector led to a photon distribution $h_n(r)$, which is referred to as the EPSF. The image intensity of a thin object with a fluorescence strength function $O(x,y)$ was given by a convolution relation:

$$I_n(x,y) = \int\int_{-\infty}^{\infty} h_n + \left\{\left[(x)^2 + (y)^2\right]^{1/2}\right\} \times O(x-x', y-y')\mathrm{d}x'\mathrm{d}y'$$

The Fresnel formulas were used to determine the internal reflection or transmittance of the photons on the boundaries and the interface. Illumination at 10^7 photons were used to

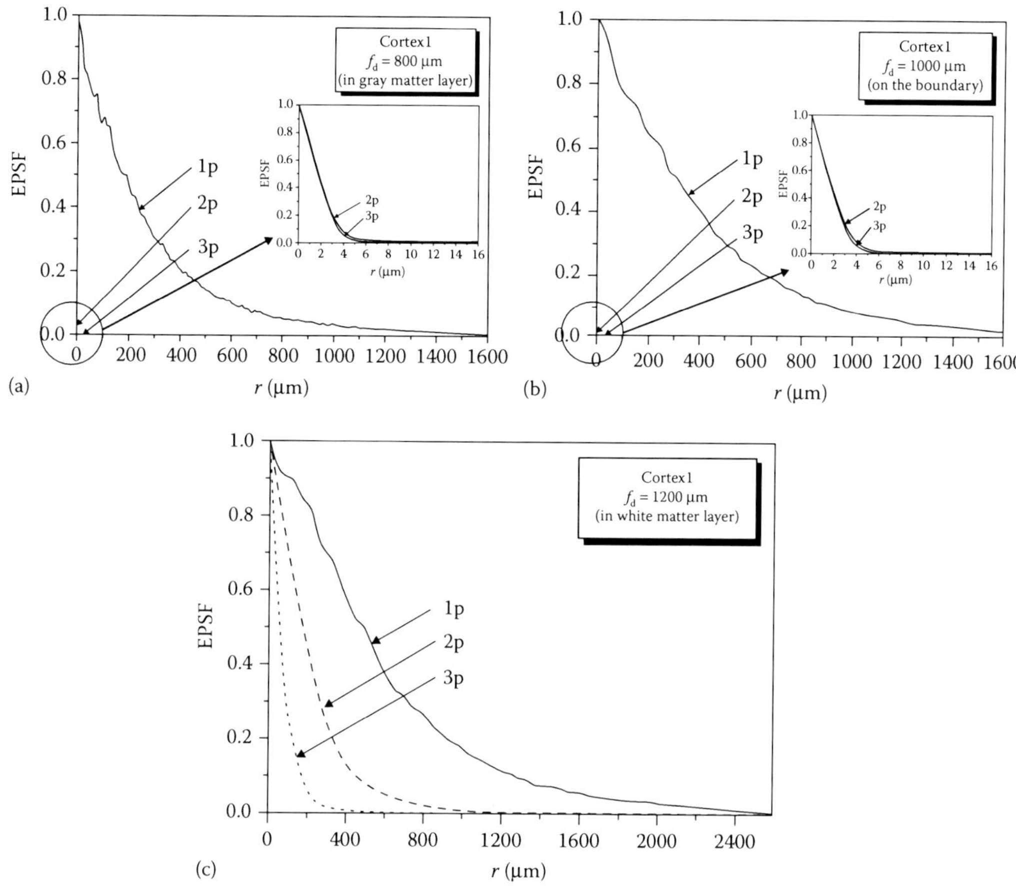

FIGURE 2.17
Comparison of 1p, 2p, and 3p fluorescence EPSFs in cortex1 (NA is 0.25): (a) At the focal depth of 800 μm (within the gray matter layer); (b) at the focal depth of 1000 μm (on the boundary); (c) at the focal depth of 1200 μm (within the white matter layer). (From Deng, X. and Gu, M., *Appl. Opt.*, 42, 3321–3329, 2003. With permission of Optical Society of America.)

ensure the accuracy of simulation results under 1p, 2p, and 3p excitation, The detector size was chosen to be large enough (infinite) to collect all the fluorescence photons incident on the objective lens and covered by the collection lens. The lenses for excitation and collection were taken to be identical. The 1p, 2p, and 3p fluorescence EPSFs at the focal depths of 800, 1000, and 1200 μm in cortex 1 are depicted in Figure 2.17 for an objective NA of 0.25. The focal depth f_d was defined to be the distance between the surface of the gray matter layer to the focal plane. It is clear that the focal depths of 800, 1000, and 1200 μm mean that the focus is within the gray matter layer, at the boundary, and within the white matter layer. It was observed in all three cases that the EPSF under 1p excitation was the broadest, and its difference from those under 2p and 3p excitation was significant, especially within the gray matter layer (Figure 2.17a) and on the boundary (Figure 2.17b). The EPSFs under 2p and 3p excitation within the gray matter layer (Figure 2.17a) were sharp and narrow, and the difference between them was indistinguishable. At the boundary, the difference of the EPSF between 2p and 3p excitation was slightly large (Figure 2.17b), showing narrower EPSF under 3p excitation. The EPSF under 1p, 2p, and 3p excitation dramatically changed

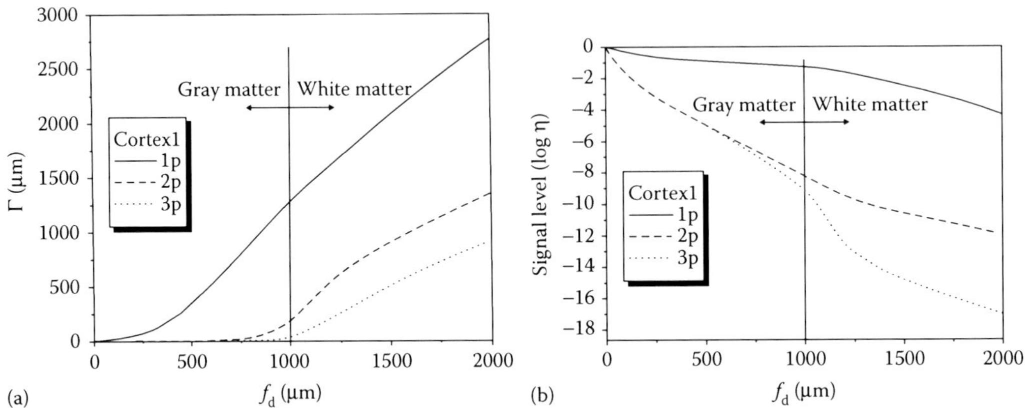

FIGURE 2.18
(a) Transverse resolution and (b) signal level as a function of the focal depth in cortex1 under 1p, 2p, and 3p excitation (NA is 0.25). (From Deng, X. and Gu, M., *Appl. Opt.*, 42, 3321–3329, 2003. With permission of Optical Society of America.)

to broad shape (Figure 2.17c) as the white matter layer had a shorter scattering mean free path length (l) compared with that in the gray matter layer, which resulted in stronger scattering effect and accordingly a dominant contribution of scattered photons in image formation.

Image resolution (Γ) and signal level (η) under 1p, 2p, and 3p excitation along the entire cortex1 thickness were noticed in Figure 2.18a and b. Within a shallow depth in gray matter (f_d < 250 μm), image resolution under 1p excitation was observed to be better than 100 μm (Figure 2.12a). Image resolution under 2p and 3p excitation kept the near diffraction-limited value in the gray matter layer. The diffraction-limited resolution under 2p excitation was maintained almost up to the depth of 1000 μm. However, under 2p excitation, the diffraction-limited resolution was maintained only within a depth of 750 μm because by comparing the scattering mean free path lengths, the scattering mean free path length under 2p and 3p excitation within the gray matter layer was much longer than that under 1p excitation. When the focus moved from the gray matter layer to the white matter layer, the resolution degraded rapidly under 2p and 3p excitation because of the significant reduction of the scattering mean free path length from the gray matter layer to the white matter layer. The 3p excitation resulted in the best image resolution in the white matter layer, showing that in 3p excitation it was approximately 1.5 times better than that under 2p excitation. For the effect of the objective NA, the use of the lower NA objective was better for image resolution when scattered photons were dominant. However, the influence of the NA objective was insignificant and neglected, especially under 2p and 3p excitation cases. Under 1p excitation, the image resolution was strongly affected by the use of a high NA objective in a medium with a small scattering mean free path length and a small anisotropy value.

Tromberg et al. [165] developed a numerical model to simulate the effects of tissue optical properties, objective NA, and instrument performance on two-photon-excited fluorescence imaging of turbid samples. Model data were compared with measurements of fluorescent microspheres in a tissue like scattering phantom. The results showed that the measured two-photon-excited signal decayed exponentially with increasing focal depth. The overall decay constant was a function of absorption and scattering parameters at both excitation and emission wavelengths. The MC model was split into two halves: excitation of two-photon fluorescence and collection of the emitted fluorescent photons. In calculating the two photon excitation, photons were launched into the tissue in focused beam geometry. Photons were uniformly distributed at the objective lens, and the initial direction of each photon was determined from the focal depth and the starting point on the objective lens. The initial directions were varied so that, in the absence of tissue scattering, the intensity profile at the focus would approximate a diffraction-limited spot. Aberrations caused by the objective lens were not taken into account in the model, since the aim was to quantify the dominant effects of multiple light scattering in the tissue. As each photon propagated through the medium, the coordinates $\mathbf{r} = (x,y,z)$ of all interaction points were recorded and stored for postprocessing. Each photon was propagated until it was either absorbed, reflected out of the medium, or had a path length that exceeded a preset value. Although the model included single-photon absorption through the absorption coefficient, μ_a, it was assumed that multiphoton absorption did not alter the distribution of excitation light within the medium. Once the simulation was run and the entire photon trajectories for all photons were stored, the spatial and temporal distribution of photons, $G(\mathbf{r},t)$, due to an infinitely short pulse was predetermined. To determine the spatial and temporal distribution of photons due to a laser pulse of specified pulse width and energy, the temporal impulse response at each spatial point within the medium was convolved with the laser pulse profile:

$$I_{ex}(\mathbf{r},t) = \int_{-\infty}^{\infty} G(\mathbf{r},t')f(t-t')\,dt'$$

where:

$G(\mathbf{r},t)$ is the impulse response
$f(t)$ is the temporal laser pulse profile

A Gaussian temporal profile was assumed throughout this study. In the equation, $I_{ex}(\mathbf{r},t)$ value represents the spatial and temporal distribution of the excitation light in the medium. The spatial distribution of generated two-photon fluorescence sources within the medium, $F_{ex}(r)$, was related to the square of the instantaneous intensity of the excitation light:

$$F_{ex}(r) = \frac{1}{2}\phi\sigma C(r)\int_{-\infty}^{\infty} I_{ex}^2(r,t)\,dt$$

where:

ϕ is the fluorescence QE
σ is the two-photon absorption cross section
$C(r)$ is the spatially dependent fluorophore concentration

The fraction of the generated two-photon fluorescence was collected by the detector and the second half of the MC simulation was used. Fluorescent photons were launched from within the sample with a spatial distribution given by $F_{ex}(r)$ and isotropic initial directions. The optical properties of the sample were those at the emission wavelength. The fluorescent photons were propagated until they were either absorbed or exited the medium. For those photons exiting the top surface of the sample, a geometrical ray trace was performed to determine whether the photon was collected by the objective lens. The fraction of fluorescent photons reaching the detector, $F_{em}(\mathbf{r})$, was determined with the second half of the simulation; the total number of photons reaching the detector due to a single laser pulse focused at a depth, z_f, were calculated from the product of the generation and collection of fluorescence:

$$s(z_f) = \eta \int F_{ex}(r) F_{em}(\mathbf{r}) dr$$

where:

$F_{em}(\mathbf{r})$ is the fraction of fluorescent photons generated at $\mathbf{r}$ that reach the surface within the acceptance angle of the objective lens

η describes the collection efficiency of the optics, filters, and detector QE

With the two-part simulation, the dependence of the optical properties of the medium on the generation and collection of two-photon fluorescence at increasing focal depths was studied for each set of optical properties. The resulting detector signal as a function of focal depth, $S(z_f)$, was fitted to a function of the form:

$$s(z_f) = S_o \exp\left[(-b_{ex}\mu_t^{ex} + b_{em}\mu_t^{em})z_f\right]$$

where:

μ_t^{ex} and μ_t^{em} signify the total attenuation coefficients ($\mu_t = \mu_s + \mu_a$) at the excitation and emission wavelengths, respectively

b_{ex} and b_{em} are parameters determined by means of fitting the computed decay of the generation and collection of two-photon fluorescence

The term S_o is the amount of fluorescence generated in a nonscattering medium for the same NA, fluorophore concentration, and QE

The fluorophore concentration, $C(\mathbf{r})$, was assumed to be uniform over the entire sample in all simulations. The model showed that the two-photon-excited fluorescence signal was dependent on the properties of the medium and the instrument. Fluorescence was found to decay exponentially with a slope determined by the scattering and absorption coefficients of the medium at the excitation and emission wavelengths. The optimal objective NA was found to vary with focal depth, illustrating the trade-off between higher instantaneous excitation intensity at high NAs and the increased probability of scattering for photons entering turbid samples at high incidence angles. The model predictions for the decay of the two-photon signal with depth were comparable with the measured decay in a turbid phantom. On the basis of the predicted form of the two-photon signal decay with depth and a minimum acceptable SNR, the maximum imaging depth was predicted for a complete set of experimental parameters and was found to be approximately 412 mm, assuming a fluorophore with characteristics similar to chloroaluminum-sulfonated phthalocyanine.

Ma et al. [166] adopted a simplified approach to simulate the fluorescence signal from a fluorophore submerged inside a turbid medium using the MC method. A single MC simulation of the excitation light that is computationally less expensive was directly used for well-validated nonfluorescence photon migration MC codes. Fluorescence signals from a mouse tissue like phantom were computed using both the simplified MC simulation and the diffusion approximation. The relative difference of signal intensity was found to be at most 30% for a fluorophore placed in the medium at various depths and horizontally midway between a source and a detector pair separated by 3 mm. The detected fluorescence signal $F(\vec{r},t)$ from a point fluorophore is a convolution of the following [167,168]:

$$F(\vec{r},t)=\iiint H_x(\vec{r}-\vec{r_s},t''')\left[\sum_i \frac{A_i}{\tau_i}\exp\left(-\frac{t''-t'''}{\tau_i}\right)\right]\times E_m(\vec{r_d}-\vec{r},t'-t''))S(t-t')dt'dt''dt'''$$
$$=H_x(\vec{r}-\vec{r_s},t)\otimes\left[\sum_i \frac{A_i}{\tau_i}\exp\left(-\frac{t}{\tau_i}\right)\right]\otimes E_m(\vec{r_d}-\vec{r},t)\otimes S(t) \quad (2.1)$$

where:

the $\otimes$ sign represents the convolution integral

$H_x(\vec{r}-\vec{r_s},t)$ is the light propagated from the excitation source at position $\vec{r_s}$ to the fluorophore at $\vec{r}$

$E_m(\vec{r_d}-\vec{r},t)$ is the fluorescent light propagated from the fluorophore at $\vec{r}$ to the detector at $\vec{r_d}$

$S(t)$ is the system impulse response function (IRF)

The terms in square brackets represent fluorescent decay after excitation, where A_i and τ_i are the amplitude and lifetime of the ith fluorophore component

A_i is related to the concentration, extinction coefficient, and quantum yield of the fluorophore. Since this study focuses on the issue of light propagation in tissue, which is described by $H_x(\vec{r}-\vec{r_s},t)$ and $E_m(\vec{r_d}-\vec{r},t)$, it is convenient to take out the terms not related to photon propagation. In other words, we simply assume that

$$\left[\sum_i \frac{A_i}{\tau_i}\exp\left(-\frac{t}{\tau_i}\right)\right]\otimes S(t)=1 \quad (2.2)$$

and the quantity we examine is

$$F_o(\vec{r},t)=H_x(\vec{r}-\vec{r_s},t)\otimes E_m(\vec{r_d}-\vec{r},t) \quad (2.3)$$

This corresponds to instant fluorescence lifetime and system IRF. Here, we compare $F_o(\vec{r},t)$ when $H_x(\vec{r}-\vec{r_s},t)$ and $E_m(\vec{r_d}-\vec{r},t)$ are obtained by the diffusion approximation (DA) and the simplified fluorescent MC simulation. In practice, fluorescence lifetime and system IRF are always finite. To adapt that, the results presented here can be generalized by convolving with the fluorescent decay and the system IRF as shown:

$$F_o(\vec{r},t)=F_o(\vec{r},t)\otimes\left[\sum_i \frac{A_i}{\tau_i}\exp\left(-\frac{t}{\tau_i}\right)\right]\otimes S(t)$$

The photons were launched at the source location $\vec{r_s}$ and some were absorbed, whereas others were scattered and only a portion of them reached the detector. The fluorophore functioned like a detector to receive excitation photons in the first step and a source to emit fluorescent photons in the second step. The statistical distribution of photons absorbed in each voxel was computed and converted to photon influence by dividing it by the absorption coefficient μ_a. This is the first step of fluorescence, corresponding to the calculation of $H_x(\vec{r}-\vec{r_s},t)$ under DA. Then the photons absorbed at $\vec{r_s}$ were converted to fluorescence and their propagation was simulated in a similar manner as in the first step. Due to light attenuation, the number of photons that reach position $\vec{r_s}$ decreased exponentially with the distance relative to the position of the light source $|\vec{r}-\vec{r_s}|$. The results of MC simulation for excitation light $H_x(\vec{r}-\vec{r_s},t)$ to represent the fluorescent light propagation for a detector at any position $E_m(\vec{r_d}-\vec{r},t)$ could be applicable, provided the optical characteristics of the

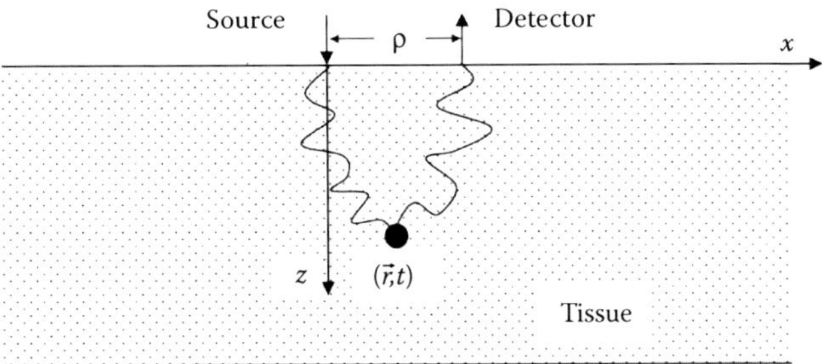

FIGURE 2.19
Schematic of fluorescence spectroscopy in turbid medium such as biological tissue when reflection geometry is used. (From Ma, G. et al., *Appl. Opt.*, 46, 1686–1692, 2007. With permission of Optical Society of America.)

media between source and fluorophore and fluorophore and detector were identical and the distances from source to fluorophore and from fluorophore to detector were the same. Specifically to the configuration shown in Figure 2.19, the excitation photons were injected at $\vec{r}_s$ (0,0,0) and the detector is at $\vec{r}_d$ (3,0,0). The photon influence of excitation light $H_x(\vec{r}-\vec{r}_s,t)$ at $\vec{r}$ (x,y,z) was first computed by the nonfluorescence MC simulation. The photon fluence $E_m(\vec{r}_d-\vec{r},t)$ detected at $\vec{r}_d$ for the fluorescent signal coming from $\vec{r}$ was observed to be the same as the excitation photon fluence $H_x(\vec{r}-\vec{r}_s,t)$ at $\vec{r}$ when injected at $\vec{r}_d$. Since the medium is a homogeneous slab, the distribution of excitation photons for the source at $\vec{r}_s$ (3,0,0) was obtained from that of the source at $\vec{r}_s$ (0,0,0) by shifting the coordinates only 3 mm along the x-axis. Hence, the measurable fluorescent signal was easily calculated from a single MC simulation of the excitation light. The method utilized well-validated nonfluorescence MC codes and saved the computation time and could also be used to compute fluorescence signal from tissues in most applications as long as the specified conditions were satisfied [169,170]. By comparing the results from the DA with MC simulation, the diffusion approximation was an insignificant issue when used to interpret the fluorescence signal from mouse tissue. Large errors are found only when applying the DA at shallow depths, which are at most 30% for the cases. DA provided a computationally fast approach to calculate the fluorescence signal that could potentially be extended to account for inhomogeneity and irregular shape of mouse tissues using its numerical solution based on finite-element mesh.

Chen and Intes [171] presented a comparison study on the computational efficiency of three MC-based methods for time-domain fluorescence molecular tomography. The methods investigated were used to generate time-gated Jacobians, which were the perturbation MC (pMC) method, the adjoint MC (aMC) method, and the midway MC (mMC) method. The effects of the different parameters on the computation time and statistics reliability were evaluated and the methods were applied to a set of experimental data for tomographic application. It was investigated that the parameters affected the computational time for the three methods differently (linearly, quadratically, or not significantly) and the noise level of the Jacobian varied when the parameters changed. The experimental results in preclinical settings demonstrated the feasibility of using both aMC and pMC methods for time-resolved whole-body studies in small animals within a few hours. The aMC method was proved efficient when the sources and detectors used were small relative to the target volume, such as the point sources and detectors widely used in tomography. Due to the Jacobian being the product of the source field and the detection field, this method suffered from high variance on the boundaries. The forward aMC method was readily expanded to the time-domain method by introducing a time variable t. According to the transport theory, the fluorescence intensity measured at r_d and time t for an impulsive excitation at r_s and $t_0 = 0$ was given as follows:

$$U_F = (r_s, r_d, t) = \int_{\Omega} dr \int_0^t dt' \int_0^{t'} dt'' G^x(r_s, r, t' - t'') \times G^m(r_s, r_d, t'') \eta(r) e^{-(t-t')/\tau} \quad (2.4)$$

where:

the integration domain Ω is defined as the entire imaging volume
$\eta(r)$ is the yield distribution
G^x and G^m are the time-dependent background Green's functions for light propagation at the excitation and emission wavelengths, which can be solved analytically or numerically

Equation 2.4 can be written in a more concise linear form:

$$U_F = (r_s, r_d, t) = \int_{\Omega} dr W(r_s, r_d, r, t) \eta(r) \quad (2.5)$$

$$W(r_s, r_d, r, t) \eta(r) = \int_0^t dt' e^{-(t-t')/\tau} \int_0^{t'} dt'' \times G^x(r_s, r, t' - t'') G^m(r_s, r_d, t'') \quad (2.6)$$

where:

$W(r_s, r_d, r, t)$ is the weight function of the measurement
$U_F = (r_s, r_d, t)$ with respect to $\eta(r)$

The double convolution in time was required to calculate the full weight function, which affected the time efficiency of this method. The MC method was also employed to calculate the Green's functions $G(x,m)$ instead of the ubiquitous diffusion equation at the excitation and emission wavelengths. To calculate G^x (r_s, r, t), a forward MC simulation with particles starting at a source and traveling toward the detector was applied. Conversely, the aMC simulation calculated $G_n^m(r, r_d, t'')$ following backward propagating photons from the detector to the target volume based on the general reciprocity theorem and originates when the transport from r to the r_d is replaced by an adjoint transport from r_d to r. The time-gated pMC method for fluorescence tomography was used by directly utilizing the photon path information in the forward MC method for fluorescence generation. The double convolution was taken as follows:

$$W(r_s, r_d, r, t') = \int_0^t dt' W'(r_s, r_d, r, t') e^{-(t-t')/\tau} \quad (2.7)$$

and

$$W'(r_s, r_d, r, t') = \int_0^{t'} dt'' G_n^x(r_s, r, t' - t'') G_n^m(r, r_d, t'') \quad (2.8)$$

In the pMC method, assuming the same optical properties at the excitation and emission wavelengths, the background weight function can be calculated implicitly based on

$$W\left(r_s, r_d, r, t'\right) = \sum_{i=1}^{n} w_i^x\left(r_s, r_d, t'\right) u_a^x(r) l_i(r) \tag{2.9}$$

where:

n is the total number of photons propagating from r_s, passing through r, and detected by the detector r_d at time t

w_i is the detected weight of the ith ($i = 1,\ldots,n$) photon

$l_i(r)$ is the path length that the photon passes at r

Thus, the weight function was the weighted average of the photon paths at each subregion and time bin. If the absorption coefficients were different between the excitation and emission wavelengths, Equation 2.9 was modified to accommodate the difference by adding a correction term:

$$W(r_s, r_d, r, t') = \sum_{i=1}^{n} w_i^x(r_s, r_d, t') \mu_a^x(r) l_i(r) \times \exp - \left(\sum_{j=p_i+1}^{q_i} \Delta\mu_a(r_j) \right) \tag{2.10}$$

where:

$\Delta\mu_a = \Delta\mu_a^m - \Delta\mu_a^x$ is the difference between the absorption coefficients at emission and excitation wavelengths

r_j ($j = 1,\ldots,q_i$) are the regions that the ith photon passes from r_s to r_d and the p_ith region is r

Since the MC simulation stores the path histories for each subregion within the volume of interest, any distribution of absorption coefficient at the excitation and emission wavelengths could be rapidly simulated. Moreover, this formulation allowed to compute fluorescence Jacobians for all the detectors and time simultaneously in a computationally efficient manner by allocating the memory to store the time-resolved weight matrix for each detector. In tomographic settings, dense source–detector pairs were employed. Figure 2.20 shows the time costs to generate a full set of Jacobians for reconstruction for all the gates (the aMC method with 10^8 photons and the pMC method with 10^9 photons). The convolution time for the adjoint method became dominant when the number of source–detector pairs increased, which implied that, even if the forward MC simulations were accelerated by using parallel computing, the required convolution took long time, provided the number of source–detector pairs was large. The total time for the pMC method was found to be less than that for the aMC method when $N_s = N_d > 150$. When N_d was much greater than N_s, the two methods show nearly identical time efficiency, but the pMC method shows superior efficiency. However, this is the case when the Jacobian for all the gates was generated. The time cost for convolution could be reduced up to 90% leading to computational times almost equivalent to that for only forward simulations (dotted lines in Figure 2.20), where the aMC method was found to be more efficient than the pMC method for the source–detector combinations. Moreover, the computation for the convolution in the aMC method was performed under MATLAB® and was optimized for greater computational efficiency. The convolution time was reduced significantly by optimized computational methods, such as parallel acceleration using multicore central processing unit (CPU) or graphics processing unit (GPU). Hence, for

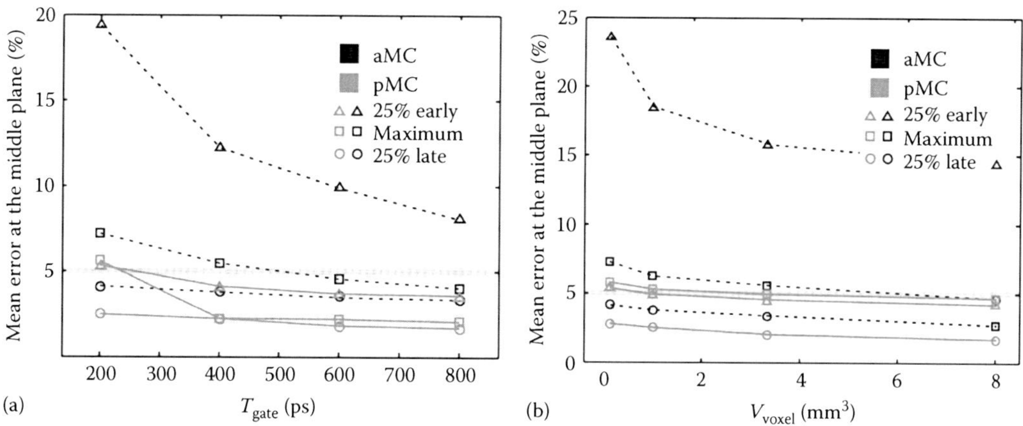

FIGURE 2.20
The mean error at the middle plane of the Jacobian using different T_{gate} (a) and V_{voxel} (b). (From Chen, J. and Intes, X., *Med. Phys.*, 38, 5788–5798, 2011. With permission of American Association of Physicists in Medicine.)

tomographic systems with high spatial and temporal resolution, the pMC method was more efficient than the aMC method, while for simpler fiber-based systems with the capability of acquiring less resolved signals, the aMC method was more suitable. The influence of the number of photons on statistics was found to be different for the aMC and pMC methods.

2.13 Conclusion

The measurement of fluorescence via computational methods offers the possibility for characterizing very small domains on the surface or in the interior of the cell such as the cytoplasmic viscosity of living cells. Moreover, computational methods provide additional and complementary information, such as the nature of the fluorophore and its influence on the target molecule reflecting heterogeneity in the molecular population and in the size, shape, and internal motions of the fluorophore–biomolecule conjugate. Real molecules often exhibit molecular asymmetry, anisotropic rotational modes, and multiple lifetimes, and in the cellular environment, molecular heterogeneity requires in-depth investigation to explore the real interactions occurring at the *in vivo* level. An adequate computational formulation of such systems is a necessity. The extensive efforts on the development of fluorescent nanoparticles have attractive features for their applications in bioimaging. Their specific structural designs, synthetic strategies, and the size effects on the optical properties improve their fluorescence intensity, specificity in targeting, biocompatibility, and nontoxicity. Although many fluorescent nanoparticles have been synthesized, there are still much to be done before they could be widely used as fluorescent probes in clinical tests. With further advances in the design and synthesis of multifunctional fluorescent materials with high quality, the widespread application of nanomaterials could be expected not only in advanced bioimaging but also in ultrasensitive molecular diagnosis, novel light-emitting nanodevices, and intracellular drug delivery.

Acknowledgment

The corresponding author Dr. Ufana Riaz wishes to acknowledge the Department of Science and Technology (DST), India, for funding a major research project under the DST-SERB program.

References

1. Adachi K, Yasuda R, Noji H, Itoh H, Harada Y et al. 2000. Stepping rotation of F1-ATPase visualized through angle-resolved single-fluorophore imaging. *Proc. Natl. Acad. Sci. U.S.A.* 97:7243–7247.
2. Agard DA, Sedat JW. 1983. Three-dimensional architecture of a polytene nucleus. *Nature* 302:676–681.
3. Alivisatos AP. 1996. Semiconductor clusters, nanocrystals, and quantum dots. *Science* 271:933–937.
4. Ambrose WP, Moerner WE. 1991. Fluorescence spectroscopy and spectral diffusion of single impurity molecules in a crystal. *Nature* 349:225–227.
5. Ambrose WP, Goodwin PM, Martin JC, Keller RA. 1994. Alterations of single molecule fluorescence lifetimes in near-field optical microscopy. *Science* 265:364–367.
6. Basché T, Moerner WE, Orrit M, Talon H. 1992. Photon antibunching in the fluorescence of a single dye molecule trapped in a solid. *Phys. Rev. Lett.* 69:1516–1519.
7. Becker W, Bergmann A, Konig K, Tirlapur U. 2001. Picosecond fluorescence lifetime microscopy by TCSPC imaging. *Proc. SPIE* 4262:414–419.
8. Wang Y-L, Taylor DL, eds. 1989. *Fluorescence Microscopy of Living Cells in Culture.* Parts A and B. San Diego, CA: Academic. Vols. 29, 30.
9. Tsien RY. 2005. Building and breeding molecules to spy on cells and tumors. *FEBS Lett.* 579:927–932.
10. Funovics M, Weissleder R, Tung CH. 2003. Protease sensors for bioimaging. *Anal. Bioanal. Chem.* 377:956–963.
11. Ragan TM, Huang H, So PTC. 2003. In vivo and ex vivo tissue applications of two-photon microscopy. *Methods Enzymol.* 361:481–505.
12. Condeelis J, Segall JE. 2003. Intravital imaging of cell movement in tumours. *Nat. Rev. Cancer* 3:921–930.
13. Michaelis J, Hettich C, Mlynek J, Sandoghdar V. 2000. Optical microscopy using a single-molecule light source. *Nature* 405:325–328.
14. Hell SW. 2003. Toward fluorescence nanoscopy. *Nat. Biotechnol.* 21:1347–1355.
15. Wang TD, Contag CH, Mandella MJ, Chan NY, Kino GS. 2003. Dual-axes confocal microscopy with post-objective scanning and low-coherence heterodyne detection. *Opt. Lett.* 28:1915–1917.
16. Farrell TJ, Patterson MS, Wilson B. 1992. A diffusion-theory model of spatially resolved, steady-state diffuse reflectance for the noninvasive determination of tissue optical-properties in vivo. *Med. Phys.* 19:879–888.
17. Taroni P, Danesini G, Torricelli A, Pifferi A, Spinelli L, Cubeddu R. 2004. Clinical trial of time-resolved scanning optical mammography at 4 wavelengths between 683 and 975 nm. *J. Biomed. Opt.* 9:464–473.
18. Kak A, Slaney M. 1988. *Principles of Computerized Tomographic Imaging.* New York: IEEE Press.
19. Oleary MA, Boas DA, Chance B, Yodh AG. 1995. Experimental images of heterogeneous turbid media by frequency-domain diffusing-photon tomography. *Opt. Lett.* 20:426–428.

20. Oleary MA, Boas DA, Li XD, Chance B, Yodh AG. 1996. Fluorescence lifetime imaging in turbid media. *Opt. Lett.* 21:158–160.
21. Buchalla W, Lennon AM, van der Veen MH, Stookey GK. 2002. Optimal camera and illumination angulations for detection of interproximal caries using quantitative light-induced fluorescence. *Caries Res.* 36:320–326.
22. Eppstein MJ, Hawrysz DJ, Godavarty A, Sevick-Muraca EM. 2002. Three dimensional, Bayesian image reconstruction from sparse and noisy data sets: Near-infrared fluorescence tomography. *Proc. Nat. Acad. Sci. U.S.A.* 99:9619–9624.
23. Ntziachristos V, Yodh AG, Schnall M, Chance B. 2000. Concurrent MRI and diffuse optical tomography of breast after indocyanine green enhancement. *Proc. Natl. Acad.Sci. U.S.A.* 97:2767–2772.
24. Jiang HB, Paulsen KD, Osterberg UL, Pogue BW, Patterson MS. 1996. Optical image reconstruction using frequency-domain data: Simulations and experiments. *J. Opt. Soc. Am. A Opt. Image Sci. Vis.* 13:253–266.
25. Griekspoor A, Zwart W, Neefjes J. 2005. Presenting antigen presentation in living cells using biophysical techniques. *Curr. Opin. Microbiol.* 8:338–343.
26. Paris S, Sesboue R. 2004. Metastasis models: The green fluorescent revolution? *Carcinogenesis* 25:2285–2292.
27. Rasmussen A, Deckert V. 2005. New dimension in nano-imaging: Breaking through the diffraction limit with scanning near-field optical microscopy. *Anal. Bioanal. Chem.* 381:165–172.
28. Neri D, Carnemolla B, Nissim A, Leprini A, Querze G et al. 1997. Targeting by affinity-matured recombinant antibody fragments of an angiogenesis associated fibronectin isoform. *Nat. Biotechnol.* 15:1271–1275.
29. Achilefu S, Bloch S, Markiewicz MA, Zhong TX, Ye YP et al. 2005. Synergistic effects of light-emitting probes and peptides for targeting and monitoring integrin expression. *Proc. Nat. Acad. Sci. U.S.A.* 102:7976–7981.
30. Moon WK, Lin YH, O'Loughlin T, Tang Y, Kim DE et al. 2003. Enhanced tumor detection using a folate receptor-targeted near-infrared fluorochrome conjugate. *Bioconjug. Chem.* 14:539–545.
31. Weissleder R, Tung CH, Mahmood U, Bogdanov A. 1999. In vivo imaging of tumors with protease-activated near-infrared fluorescent probes. *Nat. Biotechnol.* 17:375–378.
32. Tung CH. 2004. Fluorescent peptide probes for in vivo diagnostic imaging. *Biopolymers* 76:391–403.
33. Bremer C, Tung CH, Weissleder R. 2001. In vivo molecular target assessment of matrix metalloproteinase inhibition. *Nat. Med.* 7:743–748.
34. Wunder A, Tung CH, Muller-Ladner U, Weissleder R, Mahmood U. 2004. In vivo imaging of protease activity in arthritis—A novel approach for monitoring treatment response. *Arthritis Rheum.* 50:2459–2465.
35. Tsien RY. 1998. The green fluorescent protein. *Annu. Rev. Biochem.* 67:509–544.
36. Mohrs M, Shinkai K, Mohrs K, Locksley RM. 2001. Analysis of type 2 immunity in vivo with a bicistronic IL-4 reporter. *Immunity* 15:303–311.
37. Chamberlain C, Hahn KM. 2000. Watching proteins in the wild: Fluorescence methods to study protein dynamics in living cells. *Traffic* 1:755–762.
38. Ichikawa T, Hogemann D, Saeki Y, Tyminski E, Terada K et al. 2002. MRI of transgene expression: Correlation to therapeutic gene expression. *Neoplasia* 4:523–530.
39. Tjuvajev JG, Chen SH, Joshi A, Joshi R, Guo ZS et al. 1999. Imaging adenoviral-mediated herpes virus thymidine kinase gene transfer and expression in vivo. *Cancer Res.* 59:5186–5193.
40. Tung CH, Zeng Q, Shah K, Kim DE, Schellingerhout D, Weissleder R. 2004. In vivo imaging of beta-galactosidase activity using far red fluorescent switch. *Cancer Res.* 64:1579–1583.
41. Matz MV, Fradkov AF, Labas YA, Savitsky AP, Zaraisky AG et al. 1999. Fluorescent proteins from nonbioluminescent Anthozoa species. *Nat. Biotechnol.* 17:969–973. Erratum. 1999. *Nat. Biotechnol.* 17(12):1227.
42. Zhang J, Campbell RE, Ting AY, Tsien RY. 2002. Creating new fluorescent probes for cell biology. *Nat. Rev. Mole. Cell Biol.* 3:906–918.

43. Wang L, Jackson WC, Steinbach PA, Tsien RY. 2004. Evolution of new nonantibody proteins via iterative somatic hypermutation. *Proc. Nat. Acad. Sci. U.S.A.*101:16745–16749.
44. Muller MG, Georgakoudi I, Zhang Q, Wu J, Feld MS. 2001. Intrinsic fluorescence spectroscopy in turbid media: Disentagling effects of scattering and absorption. *Appl. Opt.* 40:4633–4646.
45. Wobus AM, Boheler KR. 2005. Embryonic stem cells: Prospects for developmental biology and cell therapy. *Physiol. Rev.* 85:635–678.
46. Hoffman RM. 2002. Green fluorescent protein imaging of tumour growth, metastasis, and angiogenesis in mouse models. *Lancet Oncol.* 3:546–556.
47. Dewhirst MW, Shan S, Cao YT, Moeller B, Yuan F, Li CY. 2002. Intravital fluorescence facilitates measurement of multiple physiologic functions and gene expression in tumors of live animals. *Dis. Markers* 18:293–311.
48. Shah K, Jacobs A, Breakefield XO, Weissleder R. 2004. Molecular imaging of gene therapy for cancer. *Gene Ther.* 11:1175–1187.
49. Grinvald A, Haas E, Steinberg IZ. 1972. Evaluation of the distribution of distances between energy donors and acceptors by fluorescence decay. *Proc. Natl. Acad. Sci. U.S.A.* 69:2273–2277.
50. Schmidt T, Schütz GJ, Baumgartner W, Gruber HJ, Schindler H. 1996. Imaging of single molecule diffusion. *Proc. Natl. Acad. Sci. U.S.A.* 93:2926–2929.
51. Mackay CD, Tubbs RN, Bell R, Burt D, Jerram P, Moody I. 2001. Sub-electron read noise at MHz pixel rates. *Proc. SPIE* 4306:289–298.
52. Enderle T, Ha T, Ogletree DF, Chemla DS, Magowan C, Weiss S. 1997. Membrane specific mapping and colocalization of malarial and host skeletal proteins in the *Plasmodium falciparum* infected erythrocyte by dual-color near-field scanning optical microscopy. *Proc. Natl. Acad. Sci. U.S.A.* 94:520–525.
53. Pawley JB, ed. 1995. *Handbook of Biological Confocal Microscopy*. New York: Plenum.
54. Synge EH. 1928. A suggested method for extending microscopic resolution into the ultra-microscopic region. *Philos. Mag.* 6:356–362.
55. Denk W, Strickler JH, Webb WW. 1990. Two-photon laser scanning fluorescence microscopy. *Science* 248:73–76.
56. Patterson GH, Piston DW. 2000. Photobleaching in two-photon excitation microscopy. *Biophys. J.* 78:2159–2162.
57. Mainen ZF, Maletic-Savatic M, Shi SH, Hayashi Y, Malinow R, Svoboda K. 1999. Two-photon imaging in living brain slices. *Methods* 18:231–239.
58. Macklin JJ, Trautman JK, Harris TD, Brus LE. 1996. Imaging and time-resolved spectroscopy of single molecules at an interface. *Science* 272:255–258.
59. Widengren J, Rigler R. 1998. Fluorescence correlation spectroscopy as a tool to investigate chemical reactions in solutions and on cell surfaces. *Cell. Mol. Biol.* 44:857–879.
60. Hiraoka Y, Sedat JW, Agard DA. 1987. The use of charge-coupled device for quantitative optical microscopy of biological structures. *Science* 238:36–41.
61. Lacoste TD, Michalet X, Pinaud F, Chemla DS, Alivisatos AP, Weiss S. 2000. Ultrahigh-resolution multicolor colocalization of single fluorescent probes. *Proc. Natl. Acad. Sci. U.S.A.* 97:9461–9466.
62. Malik Z, Cabib D, Buckwald RA, Talmi A, Garini Y, Lipson SG. 1996. Fourier transform multipixel spectroscopy for quantitative cytology. *J. Microsc.* 182:133–140.
63. Haugland RP. 2002. *Handbook of Fluorescent Probes and Research Products*. Eugene, OR: Molecular Probes. 966 pp.
64. Brasselet S, Moerner WE. 2000. Fluorescence behavior of single-molecule pH sensors. *Single Mol.* 1:17–23.
65. Wei XB, Runnels JM, Lin CP. 2003. Selective uptake of indocyanine green by reticulocytes in circulation. *Invest. Ophthalmol. Vis. Sci.* 44:4489–4496.
66. Bogdanov AA, Lin CP, Simonova M, Matuszewski L, Weissleder R. 2002. Cellular activation of the self-quenched fluorescent reporter probe in tumor microenvironment. *Neoplasia* 4:228–236.
67. Mahmood U, Tung C, Bogdanov A, Weissleder R. 1999. Near infrared optical imaging system to detect tumor protease activity. *Radiology* 213:866–870.

68. Yang M, Baranov E, Jiang P, Sun FX, Li XM et al. 2000. Whole-body optical imaging of green fluorescent protein-expressing tumors and metastases. *Proc. Natl. Acad. Sci. U.S.A.* 97:1206–1211.
69. Ntziachristos V, Ripoll J, Wang LHV, Weissleder R. 2005. Looking and listening to light: The evolution of whole-body photonic imaging. *Nat. Biotechnol.* 23:313–320.
70. Cutler M. 1929. Transillumination as an aid in the diagnosis of breast lesions. *Surg. Gynecol. Obstet.* 48:721–729.
71. Franceschini MA, Moesta KT, Fantini S, Gaida G, Gratton E et al. 1997. Frequency-domain techniques enhance optical mammography: Initial clinical results. *Proc. Nat. Acad. Sci. U.S.A.* 94:6468–6473.
72. Grosenick D, Moesta KT, Wabnitz H, Mucke J, Stroszczynski C et al. 2003. Time-domain optical mammography: Initial clinical results on detection and characterization of breast tumors. *Appl. Opt.* 42:3170–3186.
73. Baxter WT, Mironov SF, Zaitsev AV, Jalife J, Pertsov AM. 2001. Visualizing excitation waves inside cardiac muscle using transillumination. *Biophys. J.* 80:516–530.
74. Ntziachristos V, Turner G, Dunham J, Windsor S, Soubret A et al. 2005. Planar fluorescence imaging using normalized data. *J. Biomed. Opt.* 10(6):064007.
75. Pogue BW, Poplack SP, McBride TO, Wells WA, Osterman KS et al. 2001. Quantitative hemoglobin tomography with diffuse dear-infrared spectroscopy: Pilot results in the breast. *Radiology* 218:261–266.
76. Eppstein MJ, Dougherty DE, Hawrysz DJ, Sevick-Muraca EM. 2001. Three-dimensional Bayesian optical image reconstruction with domain decomposition. *IEEE Trans. Med. Imaging* 20:147–163.
77. Stryer L, Haugland RP. 1967. Energy transfer: A spectroscopic ruler. *Proc. Natl. Acad. Sci. U.S.A.* 58:719–726.
78. Terpetschnig E, Szmacinski H, Malak H, Lakowicz JR. 1995. Metal-ligand complexes as a new class of long-lived fluorophores for protein hydrodynamics. *Biophys. J.* 68:342–350.
79. Vereb G, Jares-Erijman E, Selvin PR, Jovin TM. 1998. Temporally and spectrally resolved imaging microscopy of lanthanide chelates. *Biophys. J.* 74:2210–2222.
80. Trautman JK, Macklin JJ, Brus LE, Betzig E. 1994. Near-field spectroscopy of single molecules at room temperature. *Science* 369:40–42.
81. Sytsma J, Vroom JM, DeGrauw CJ, Gerritsen HC. 1998. Time-gated fluorescence lifetime imaging and microvolume spectroscopy using two-photon excitation. *J. Microsc.* 191:39–51.
82. Gadella TWJ, Jovin TM. 1995. Oligomerization of epidermal growth factor receptors on A431 cells studied by time-resolved fluorescence imaging microscopy—A stereochemical model for tyrosine kinase receptor activation. *J. Cell. Biol.* 129:1543–1558.
83. Gadella TW, Jr, van der Krogt GN, Bisseling T. 1999. GFP-based FRET microscopy in living plant cells. *Trends Plant Sci.* 4:287–291.
84. Hanley QS, Arndt-Jovin DJ, Jovin TM. 2002. Spectrally resolved fluorescence lifetime imaging microscopy. *Appl. Spectrosc.* 56:155–166.
85. Betzig E, Chichester RJ. 1993. Single molecules observed by near-field scanning optical microscopy. *Science* 262:1422–1425.
86. Dickson RM, Cubitt AB, Tsien RY, Moerner WE. 1997. On/off blinking and switching behaviour of single molecules of green fluorescent protein. *Nature* 388:355–358.
87. Lu HP, Xun L, Xie XS. 1998. Single-molecule enzymatic dynamics. *Science* 282:1877–1882.
88. Vanden Bout DA, Yip W-T, Hu D, Fu D-K, Swager TM, Barbara PF. 1997. Discrete intensity jumps and intramolecular electronic energy transfer in the spectroscopy of single conjugated polymer molecules. *Science* 277:1074–1077.
89. Wu M, Goodwin PM, Ambrose WP, Keller RA. 1996. Photochemistry and fluorescence emission dynamics of single molecules in solution-B-phycoerythrin. *J. Phys. Chem.* 100:17406–17409.
90. Hofkens J, Schroeyers W, Loos D, Cotlet M, Köhn F et al. 2001. Triplet states as non-radiative traps in multichromophoric entities: Single molecule spectroscopy of an artificial and natural antenna system. *Spectrosc. Acta A* 57:2093–2107.

91. Empedocles SA, Norris DJ, Bawendi MG. 1996. Photoluminescence spectroscopy of single CdSe nanocrystallite quantum dots. *Phys. Rev. Lett.* 77:3873–3876.
92. Waggoner A. 1995. Covalent labeling of proteins and nucleic acids with fluorophores. *Methods Enzymol.* 246:362–373.
93. Ha T, Ting AY, Liang J, Caldwell WB, Deniz AA et al. 1999. Single-molecule fluorescence spectroscopy of enzyme conformational dynamics and cleavage mechanism. *Proc. Natl. Acad. Sci. U.S.A.* 96:893.
94. Deniz AA, Laurence TA, Beligere GS, Dahan M, Martin AB et al. 2000. Single-molecule protein folding: Diffusion fluorescence resonance energy transfer studies of the denaturation of chymotrypsin inhibitor 2. *Proc. Natl. Acad. Sci. U.S.A.* 97:5179–5184.
95. Mason WT. 1999. *Fluorescent and Luminescent Probes for Biological Activity*, 2nd edn. London: Academic Press.
96. Dähne S, Resch-Genger U, Wolfbeis OS, eds. 1998. Near-infrared dyes for high technology applications. NATO ASI Series, 3. *Hightechnology*. Dordrecht, The Netherlands: Kluwer Academic Publishers. Vol. 52.
97. Alivisatos P. 2004. The use of nanocrystals in biological detection. *Nat. Biotechnol.* 22:47–52.
98. Parak WJ, Gerion D, Pellegrino T, Zanchet D, Micheel C et al. 2003. Biological applications of colloidal nanocrystals. *Nanotechnology* 14:R15–R27.
99. Gao X, Nie S. 2003. Molecular profiling of single cells and tissue specimens with quantum dots. *Trends Biotechnol.* 21:371–373.
100. Michalet X, Pinaud F, Lacoste TD, Dahan M, Bruchez M et al. 2001. Properties of fluorescent semiconductor nanocrystals and their applications to biological labeling. *Single Mol.* 2:261–276.
101. Sutherland AJ. 2002. Quantum dots as luminescent probes in biological systems. *Curr. Opin. Solid State Mater. Sci.* 6:365–370.
102. Bruchez M, Moronne M, Gin P, Weiss S, Alivisatos AP. 1998. Semiconductor nanocrystals as fluorescent biological labels. *Science* 281:2013–2016.
103. Wu X, Liu H, Liu J, Haley KN, Treadway JA et al. 2003. Immunofluorescent labeling of cancer marker Her2 and other cellular targets with semiconductor quantum dots. *Nat. Biotechnol.* 21:41–46.
104. Xiao Y, Barker PE. 2004. Semiconductor nanocrystal probes for human metaphase chromosomes. *Nucleic Acids Res.* 32:e28.
105. Ness JM, Akhtar RS, Latham CB, Roth KA. 2003. Combined tyramide signal amplification and quantum dots for sensitive and photostable immunofluorescence detection. *J. Histochem. Cytochem.* 51:981–987.
106. Larson DR, Zipfel WR, Williams RM, Clark SW, Bruchez MP et al. 2003. Water-soluble quantum dots for multiphoton fluorescence imaging in vivo. *Science* 300:1434–1436.
107. Gao X, Cui Y, Levenson RM, Chung LWK, Nie S. 2004. In vivo cancer targeting and imaging with semiconductor quantum dots. *Nat. Biotechnol.* 22:969–976.
108. Lounis B, Bechtel HA, Gerion D, Alivisatos AP, Moerner WE. 2000. Photon antibunching in single CdSe/ZnS quantum dot fluorescence. *Chem. Phys. Lett.* 329:399–404.
109. Dahan M, Laurence T, Pinaud F, Chemla DS, Alivisatos AP et al. 2001. Time-gated biological imaging by use of colloidal quantum dots. *Opt. Lett.* 26:825–827.
110. Hohng S, Ha T. 2004. Near-complete suppression of quantum dot blinking in ambient conditions. *J. Am. Chem. Soc.* 126:1324–1325.
111. Aldana J, Wang YA, Peng XG. 2001. Photochemical instability of CdSe nanocrystals coated by hydrophilic thiols. *J. Am. Chem. Soc.* 123:8844–8845.
112. Willard DM, Carillo LL, Jung J, Van Orden A. 2001. CdSe-ZnS quantum dots as resonance energy transfer donors in a model protein-protein binding assay. *Nano Lett.* 1:469–474.
113. Mattoussi H, Mauro JM, Goldman ER, Anderson GP, Sundar VC et al. 2000. Self-assembly of CdSe-ZnS quantum dot bioconjugates using an engineered recombinant protein. *J. Am. Chem. Soc.* 122:12142–12150.
114. Mattoussi H, Mauro JM, Goldman ER, Green TM, Anderson GP et al. 2001. Bioconjugation of highly luminescent colloidal CdSe-ZnS quantum dots with an engineered two-domain recombinant protein. *Phys. Status Solidi B Basic Res.* 224:277–283.

115. Gerion D, Pinaud F, Williams SC, Parak WJ, Zanchet D et al. 2001. Synthesis and properties of biocompatible water-soluble silica-coated CdSe/ZnS semiconductor quantum dots. *J. Phys. Chem. B* 105:8861–8871.
116. Pellegrino T, Manna L, Kudera S, Liedl T, Koktysh D et al. 2004. Hydrophobic nanocrystals coated with an amphiphilic polymer shell: A general route to water soluble nanocrystals. *Nano Lett.* 4:703–707.
117. Dubertret B, Skourides P, Norris DJ, Noireaux V, Brivanlou AH, Libchaber A. 2002. In vivo imaging of quantum dots encapsulated in phospholipid micelles. *Science* 298:1759–1762.
118. Pathak S, Choi SK, Arnheim N, Thompson ME. 2001. Hydroxylated quantum dots as luminescent probes for in situ hybridization. *J. Am. Chem. Soc.* 123:4103–4104.
119. Wang YA, Li JJ, Chen HY, Peng XG. 2002. Stabilization of inorganic nanocrystals by organic dendrons. *J. Am. Chem. Soc.* 124:2293–2298.
120. Guo W, Li JJ, Wang YA, Peng XG. 2003. Conjugation chemistry and bioapplications of semiconductor box nanocrystals prepared via dendrimer bridging. *Chem. Mater.* 15:3125–3133.
121. Kim S, Bawendi MG. 2003. Oligomeric ligands for luminescent and stable nanocrystal quantum dots. *J. Am. Chem. Soc.* 125:14652–14653.
122. Parak WJ, Boudreau R, Le Gros M, Gerion D, Zanchet D et al. 2002. Cell motility and metastatic potential studies based on quantum dot imaging of phagokinetic tracks. *Adv. Mater.* 14:882–885.
123. Derfus AM, Chan WCW, Bhatia SN. 2004. Probing the cytotoxicity of semiconductor quantum dots. *Nano Lett.* 4:11–18.
124. Akerman ME, Chan WC, Laakkonen P, Bhatia SN, Ruoslahti E. 2002. Nanocrystal targeting in vivo. *Proc. Natl. Acad. Sci. U.S.A.* 99:12617–12621.
125. Hanaki K, Momo A, Oku T, Komoto A, Maenosono S et al. 2003. Semiconductor quantum dot/albumin complex is a long-life and highly photostable endosome marker. *Biochem. Biophys. Res. Commun.* 302:496–501.
126. Kloepfer JA, Mielke RE, Wong MS, Nealson KH, Stucky G, Nadeau JL. 2003. Quantum dots as strain- and metabolism-specific microbiological labels. *Appl. Environ. Microbiol.* 69:4205–4213.
127. Chen CC, Yet CP, Wang HN, Chao CY. 1999. Self-assembly of monolayers of cadmium selenide nanocrystals with dual color emission. *Langmuir* 15:6845–6850.
128. Mitchell GP, Mirkin CA, Letsinger RL. 1999. Programmed assembly of DNA functionalized quantum dots. *J. Am. Chem. Soc.* 121:8122–8123.
129. Sun B, Xie W, Yi G, Chen D, Zhou Y, Cheng J. 2001. Microminiaturized immunoassays using quantum dots as fluorescent label by laser confocal scanning fluorescence detection. *J. Immunol. Method* 249:85–89.
130. Winter JO, Liu TY, Korgel BA, Schmidt CE. 2001. Recognition molecule directed interfacing between semiconductor quantum dots and nerve cells. *Adv. Mater.* 13:1673–1677.
131. Zhang K, Chang H, Fu A, Alivisatos, AP, Yang H. 2006. Continuous distribution of emission states from single CdSe/ZnS QDs. *Nano Lett.* 6:843–847.
132. Park K, Lee S, Kang E, Kim K, Choi K, Kwon IC. 2009. New generation of multifunctional nanoparticles for cancer imaging and therapy. *Adv. Funct. Mater.*19:1553–1566.
133. Liu B, Bazan GC. 2005. Methods for strand-specific DNA detection with cationic conjugated polymers suitable for incorporation into DNA chips and microarrays. *Proc. Natl. Acad. Sci. U.S.A.* 102:589–593.
134. Moon JH, McDaniel W, MacLean P, Hancock LF. 2007. Live-cell-permeable poly(p-phenylene ethynylene). *Angew. Chem., Int. Ed.* 46:8223–8225.
135. Wu C, Bull B, Christensen K, McNeill J. 2009. Ratiometric single-nanoparticle oxygen sensors for biological imaging. *Angew. Chem., Int. Ed.* 48:2741–2745.
136. Sigurdson CJ, Nilsson KPR, Hornemann S, Manco G, Polymenidou M, Schwarz P, Leclerc M, Hammarstro P, Wuthrich K, Aguzzi A. 2007. Prion strain discrimination using luminescent conjugated polymers. *Nat. Methods* 4:1023–1030.
137. Wu C, Bull B, Szymanski C, Christensen K, McNeill J. 2008. Multicolor conjugated polymer dots for biological fluorescence imaging. *ACS Nano* 2:2415–2423.

138. Jakub JW, Pendas S, Reintgen DS. 2003. Current status of sentinel lymph node mapping and biopsy: Facts and controversies. *Oncologist* 8:59–68.
139. Samuel IDW, Rumbles G, Collison CJ. 1995. Efficient interchain photoluminescence in a high-electron-affinity conjugated polymer. *Phys. Rev. B: Condens. Matter* 52:R11573–R11576.
140. Kim S, Pudavar HE, Bonoiu A, Prasad PN. 2007. Aggregation-enhanced fluorescence in organically modified silica nanoparticles: A novel approach toward high-signal-output nanoprobes for two-photon fluorescence bioimaging. *Adv. Mater.* 19:3791–3795.
141. Kim S, Huang H, Pudavar HE, Cui Y, Prasad PN. 2007. Intraparticle energy transfer and fluorescence photoconversion in nanoparticles: An optical highlighter nanoprobe for two-photon bioimaging. *Chem. Mater.* 19:5650–5656.
142. Kim S, Ohulchanskyy TY, Pudavar HE, Pandey RK, Prasad PN. 2007. Organically modified silica nanoparticles co-encapsulating photosensitizing drug and aggregation-enhanced two-photon absorbing fluorescent dye aggregates for two-photon photodynamic therapy. *J. Am. Chem. Soc.* 129:2669–2675.
143. Landfester K, Montenegro R, Scherf U, Gunter R, Asawapirom U, Patil S, Neher D, Kietzke T. 2002. Semiconducting polymer nanospheres in aqueous dispersion prepared by a miniemulsion process. *Adv. Mater.* 14:651–655.
144. Yu G, Gao J, Hummelen JC, Wudl F, Heeger AJ. 1995. Optimization of conjugated-polymer-based bulk heterojunctions. *Science* 270:1789–1791.
145. Friend RH, Gymer RW, Holmes AB, Burroughes JH, Marks RN, Taliani C, Bradley DDC, Dos Santos DA, Bredas JL, Loglund M, Salaneck WR. 1999. Electroluminescence in conjugated polymers. *Nature* 397:121–128.
146. Szymanski C, Wu C, Hooper J, Salazar MA, Perdomo A, Dukes A, McNeill JD. 2005. Single molecule nanoparticles of the conjugated polymer MEH–PPV, preparation and characterization by near-field scanning optical microscopy. *J. Phys. Chem. B* 109:8543–8546.
147. Wu C, Szymanski C, McNeill J. 2006. Preparation and encapsulation of highly fluorescent conjugated polymer nanoparticles. *Langmuir* 22:2956–2960.
148. Wu C, Peng H, Jiang Y, McNeill J. 2006. Energy transfer mediated fluorescence from blended conjugated polymer nanoparticles. *J. Phys. Chem. B* 110:14148–14154.
149. Grey JK, Kim DY, Norris BC, Miller WL, Barbara PF. 2006. Size-dependent spectroscopic properties of conjugated polymer nanoparticles. *J. Phys. Chem. B* 110:25568–25572.
150. Wilson BC, Adam G. 1983. A Monte Carlo model for the absorption and flux distributions of light in tissue. *Med. Phys.* 10:824–830.
151. Keijzer M, Kortum RRR, Jacques SL, Feld MS. 1989. Fluorescence spectroscopy of turbid media: Autofluorescence of the human aorta. *Appl. Opt.* 28:4286–4292.
152. Pogue BW, Hasan T. 1996. Fluorophore quantitation in tissue-simulating media with confocal detection. *IEEE J. Sel. Top. Quantum Electron.* 2:959–964.
153. Welch AJ, Gardner C, Kortum RR, Chan E, Criswell G, Pfefer J, Warren S. 1997. Propagation of fluorescent light. *Lasersm. Surg. Med.* 21:166–178.
154. Avrillier S, Tinet E, Ettori D, Tualle JM, Gelebart B. 1998. Influence of the emission-reception geometry in laser-induced fluorescence spectra from turbid media. *Appl. Opt.* 37:2781–2787.
155. Qu JY, Huang Z, Jianwen H. 2000. Excitation-and-collection geometry insensitive fluorescence imaging of tissue-simulating turbid media. *Appl. Opt.* 39:3344–3356.
156. Pfefer TJ, Schomacker KT, Ediger MN, Nishioka NS. 2001. Light propagation in tissue during fluorescence spectroscopy with single-fiber probes. *IEEE J. Sel. Top. Quantum Electron.* 7:1004–1012.
157. Zheng W, Huang Z, Xie S, Li B, Krishnan SM, Chia TC. 2001. Autofluorescence spectrum of human bronchial tissue by Monte Carlo modelling. *Acta Photon. Sin.* 30:669–674.
158. Wu J, Feld MS, Rava RP. 1993. Analytical model for extracting intrinsic fluorescence from a turbid media. *Appl. Opt.* 32:3585–3595.
159. Muller MG, Georgakoudi I, Qingguo Z, Jun W, Feld MS. 2001. Intrinsic fluorescence spectroscopy in turbid media: Disentangling effects of scattering and absorption. *Appl. Opt.* 40:4633–4646.

160. Hyde DE, Farrell TJ, Patterson MS, Wilson BC. 2001. A diffusion theory model of spatially resolved fluorescence from depth-dependent fluorophore concentrations. *Phys. Med. Biol.* 46:369–383.
161. Liu Q, Ramanujam N. 2002. Relationship between depth of a target in a turbid medium and fluorescence measured by a variable-aperture method. *Opt. Lett.* 27:104–106.
162. Pogue BW, Burke G. 1998. Fiber-optic bundle design for quantitative fluorescence measurement from tissue. *Appl. Opt.* 37:7429–7436.
163. Liu Q, Zhu C, Ramanujam N. 2003. Experimental validation of Monte Carlo modeling of fluorescence in tissues in the UV-visible spectrum. *J. Biomed. Opt.* 8(2):223–236.
164. Deng X, Gu M. 2003. Penetration depth of single-, two-, and three-photon fluorescence microscopic imaging through human cortex structures: Monte Carlo simulation. *Appl. Opt.* 42(16): 3321–3329.
165. Dunn AK, Wallace VP, Coleno M, Berns MW, Tromberg BJ. 2000. Influence of optical properties on two-photon fluorescence imaging in turbid samples. *Appl. Opt.* 39(7):1194–1201.
166. Ma G, Delorme J-F, Gallant P, Boas DA. 2007. Comparison of simplified Monte Carlo simulation and diffusion approximation for the fluorescence signal from phantoms with typical mouse tissue optical properties. *Appl. Opt.* 46(10):1686–1692.
167. Patterson MS, Pogue B. 1994. Mathematical model for time-resolved and frequency-domain fluorescence spectroscopy in biological tissues. *Appl. Opt.* 33:1963–1974.
168. Sevick-Muraca EM, Burch CL. 1994.Origin of phosphorescence signals reemitted from tissues. *Opt. Lett.* 19:1928–1930.
169. Hayakawa CK, Spanier J, Bevilacqua F, Dunn AK, You JS, Tromberg BJ, Venugopalan V. 2001. Perturbation Monte Carlo methods to solve inverse photon migration problems in heterogeneous tissues. *Opt. Lett.* 26:1335–1337.
170. Kumar YP, Vasu RM. 2004. Reconstruction of optical properties of low-scattering tissue using derivative estimated through perturbation Monte-Carlo method. *J. Biomed. Opt.* 9:1002–1012.
171. Chen J, Intesa X. 2011. Comparison of Monte Carlo methods for fluorescence molecular tomography—Computational efficiency. *Med. Phys.* 38(10):5788–5798.
172. Michalet X, Kapanidis AN, Laurence T, Pinaud F, Doose S, Pflughoefft M, Weiss S. 2003. Jablonski diagram for fluorescence. *Annu. Rev. Biophys. Biomol. Struct.* 32:161–182.
173. Chen M, Yin M. 2014. Design and development of fluorescent nanostructures for bioimaging. *Prog. Polym. Sci.* 39(2):365–395.
174. Leblond F, Davis SC, Valdés PA, Pogue BW. 2010. Pre-clinical whole-body fluorescence imaging: Review of instruments, methods and applications. *Photochem. Photobiol. B.* 98(1):77–94.
175. Ntziachristos V. Confocal images obtained in vivo with a flexible fiber probe of 650 µm. *Annu. Rev. Biomed. Eng.* 8:1–33.
176. Xu Z, He B, Shen J, Yang W, Yin M. 2013. Fluorescent water-soluble perylenediimide-cored cationic dendrimers: Synthesis, optical properties, and cell uptake. *Chem. Commun.* 49:3646–3648.
177. Zrazhevskiy P, Gao X. 2009. Multifunctional quantum dots for personalized medicine. *Nano Today* 4:414–428.
178. Resch-Genger U, Grabolle M, Cavaliere-Jaricot S, Nitschke R, Nann T. 2008. Quantum dots versus organic dyes as fluorescent labels. *Nat. Methods* 5(9):763–775.
179. Kim S, Lim C-K, Na J, Lee Y-D, Kim K, Choi K, Leary JF, Kwon IC. 2010. Conjugated polymer nanoparticles for biomedical in vivo imaging. *RSC Chem. Commun.* 46:1617–1619.
180. Kondrashina AV, Dmitriev RI, Borisov SM, Klimant I, O'Brien I, Nolan YM, Zhdanov AV, Papkovsky DB. 2012. A phosphorescent nanoparticle-based probe for sensing and imaging of (intra)cellular oxygen in multiple detection modalities. *Adv. Funct. Mater.* 22:4931–4939.

3

New Trends in Immunohistochemical, Genome, and Metabolomics Imaging

G. Livanos, Aditi Deshpande, C. Narayan, Ying Na, T. Quang, T. Farrahi, R. Koglin, Suman Shrestha, M. Zervakis, and George C. Giakos

CONTENTS

3.1 Introduction

Digital diagnostic pathology has become one of the most valuable and convenient advancements in technology over the past few years. It allows us to acquire, store, and analyze pathological information from the images of histological and immunohistochemical glass slides, which are scanned to create digital slides. This becomes the primary method of practicing pathology with the ability to image the entire glass slide known as *whole-slide imaging* (WSI). Optical imaging enables the visualization of immunohistochemically stained tissue samples by creating digitized slides that can be studied on a computer for characterization and diagnosis. Images of whole slides can be captured using standard multispectral imaging or fluorescence. Genomic assays are created using the basic image features and parameters with the use of digital microscopy [1]. Digital pathology provides reliability and reproducibility which traditional pathology techniques could not. Till date, however, the adoption of digital platforms by the pathology community as a whole has been slow, and the applications of WSI systems in pathology have been limited to education, research, and specific niches in clinical practice. Much work remains to be done before WSI technology for diagnostic purposes can be widely adopted.

The digitization of pathology enables the pathologist to view and analyze samples and data from any location in the world. Using this *tele-pathology* approach, the data in the form of images can be transmitted easily via the Internet, also decreasing manual labor in the laboratory. This digitization of pathology eliminates the limitations and shortcomings of the conventional manual *microscopy* or pathology such as having a single field of view and not being able to adjust the image acquisition quality as much (exposure, magnification, etc.). It also makes the storage and retrieval of the specimen much easier compared to the cumbersome traditional methods. These digital slides can be accessed from anywhere and at any time, and the images can also be enhanced, edited, or processed to extract diagnostically relevant information.

Immunohistochemistry (IHC) is used in the visualization of biomarker expression in histological samples and their qualitative analysis (Figure 3.1). We know that certain antibodies bind to specific antigens in the body, formed by the cell and tissue constructions and the intercellular substances. This principle is utilized to detect antigens (proteins serving as biomarkers) for diagnostic purposes. The word *immuno* refers to the antibodies being used and *histo* is derived from histology, pertaining to tissues. Along with diagnostic purposes to identify abnormal cell growth related to cancer or other conditions, IHC is also widely used in researches concerning the study of the presence and distribution of proteins and other biomarkers in the body. The diagnostic uses of IHC include studying tumor morphology and detecting metastasis of diseases, detecting and identifying viruses and microbacteria, and assessing the functional and metabolic stage of tissues. Precise evaluation and accurate IHC analysis are achieved by the use of a highly sensitive and high-resolution, calibrated imaging system consisting of a camera (stand-alone or with a microscope), which can accurately reproduce the color and the intensity pattern of the IHC slide, and this reproducibility should be consistent slide to slide. The acquired images can be studied, enhanced, and stored using one of the many software applications available for IHC imaging systems.

Cancer constitutes one of the most frequent causes of death, yet it can be efficiently treated and cured when it is diagnosed in the first stages. Over the past few years, medical advances have led to the identification of numerous tumor biomarkers, facilitating the understanding of the molecular basis of tumor progression and treatment response. Prognostic markers aim to objectively estimate the patient's overall outcome, whereas predictive markers focus on the objective evaluation of the possible benefits from a specific clinical intervention.

Oncogenes are notable both for their role in the pathogenesis of various types of cancer and for their selection as a target of treatment. Overexpression of such receptors in human tissues is associated with increased disease recurrence, poorer relapse-free survival, and worse prognosis. The evaluation of protein overexpression has been proved very helpful, as it can become a preliminary prognostic and predictive factor for detection and therapy of

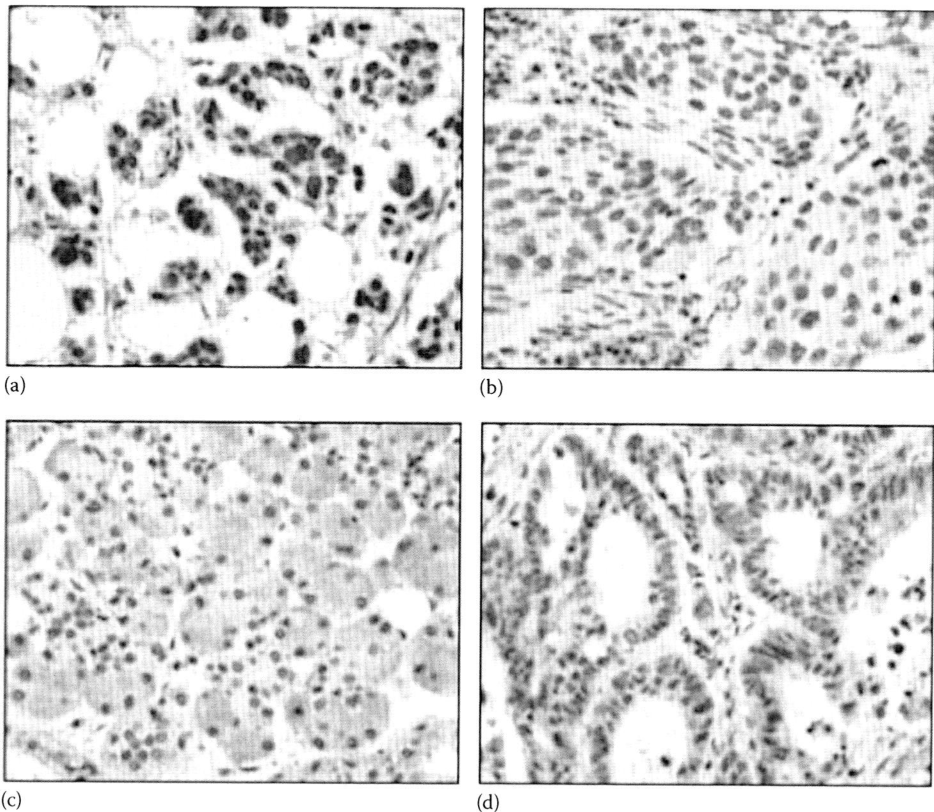

FIGURE 3.1
(See color insert.) A typical digital IHC image: (a) Breast carcinoma; (b) lung carcinoma; (c) pancreas tissue; (d) colon carcinoma.

various diseases and malignancies. As a representative example, the HER2/neu oncogene is known for its participation in breast cancer evolution mechanisms. It is considered to be overexpressed in tumors with much higher level than the relatively low degree in normal tissue. Because of its prognostic role as well as its ability to predict response to trastuzumab (a specialized drug), breast tumors are routinely checked for overexpression of HER2/neu.

IHC was established as the most common and approachable method for detecting protein overexpression. It is a laboratorial process during which tissue samples are affected by antibodies in order to stain the membranes of tumor cells with a specific color and highlight the presence of specific proteins in cells or tissues. Qualitative and quantitative tissue characterization and evaluation was achieved using IHC on frozen and archival tissues by detecting specific antigens or cells based on an antigen–antibody reaction, which facilitates the identification of a large number of proteins, enzymes, and tissue structures. As regards the specific application, IHC *stains* the membranes of HER2-overexpressed cells, monitoring them with a characteristic color (brown or red depending on the staining protocol adopted by the laboratory expert) in the extracted microscope images. When reporting results, the degree of HER2 protein overexpression measured is scored according to the intensity of membrane staining and the percentage of tumor cells stained.

Although the immunohistological method is of relatively low cost, easy to be standardized, simple, widely available, and applicable in numerous specialists' laboratories, its interpretation relies mainly on subjective visual estimates by an expert, yielding only semiquantitative results, which are open to interobserver variation. Yet, the interpretation of such results is subjective and causes certain inconsistencies upon the diagnosis, as the result is highly dependent on the experience of the specialist and the quality of the tissue preparation stage. Consequently, there is a need to establish a more objective and generally accepted method to qualify and quantify the results produced by IHC. Interpolating image processing techniques and computer-assisted microscopy into the experts' experience and knowledge proves to be the most efficient tool for tissue characterization through microscopy and spectroscopy.

Image segmentation aims at extracting attributes of interest from an image considering its common properties, such as discontinuities and similarities, within the different object classes forming the captured scene. Advanced image analysis techniques are adopted in order to accurately segment the cells within the sample images and precisely extract their membrane contour and degree of staining. In this chapter, fundamental notions of image processing theory will be presented, as well as their application/implementation to automated quantitative evaluation of tissue samples status, emphasizing on the algorithmic aspects of each technique, its advantages, and the possibilities for further development and enrichment. Color deconvolution through model conversion and thresholding enhances the pixel intensity differences between the regions of interest in the test images; image clustering reveals the key segments for the evaluation procedure; edge following and linking via the active contours algorithm combined with mathematical morphology and distance transforms are performed in order to extract the complete border of the cell membranes and produce numerical data referring to the shape, size, and color intensity of the segmented regions. In the following text, these key categories of methodologies and their application to IHC image analysis will be presented in a simple and compact way, focusing on the prospects and future challenges for their improvement and direct incorporation to laboratory measurements and industry. The novelty introduced in this chapter lies on the fusion of fundamental approaches incorporated in a powerful detecting system, where each successive stage is supported and initialized by the previous level of processing. In this modular way, the analysis derives detailed numerical data on tissue anatomy and status (overexpression, number of cells, average cell area, intermediate tissue area, position of each cell, etc.) along with the cell characteristics (shape, size, compactness, borders, etc.). In addition, the proposed procedure is automatically adapted to different staining protocols, the morphology of cells, or the selection of the areas of interest to be processed.

Genome refers to the complete set of DNA of an organism, containing the hereditary information. Genomics is the branch of genetics that involves applying DNA sequencing methods and other recombinant DNA techniques to analyze the functions and structures of genomes. Some of the most important fields being researched today are to understand the complex biological mechanisms behind cognitive and responsive behavior of various kinds. The field of molecular genetics has seen tremendous advancements over the years and behavioral changes due to varying gene sequences and other biological mechanisms can be studied extensively due to the currently available functional and metabolic neuroimaging. Efforts are being made to identify and understand variations in the standard *reference* human gene sequence [2,3], which in turn causes variations in human biology and behavior patterns. This has also enabled us to study biological functional polymorphisms, which occur in nature due to genetic variations and sometimes due to adaptation. The various metabolic imaging modalities such as positron emission tomography (PET), functional magnetic resonance imaging (MRI), and electroencephalogram (EEG) are being used to identify, study, and characterize the functional genomics present in the brain [4].

This is a rapidly progressing field and worldwide several people are being scanned for genotyping and genome-wide scans [5], mostly utilizing PET imaging and structural as well as functional MRI. Due to the extensive research, the sample size is now large enough to identify and confirm the changes in the brain function caused by specific gene sequences. These noninvasive imaging techniques for the study of genomics have made it possible to determine the role of specific genes and proteins that can be related to various diseases. This study of protein and gene expression in the body can lead to a better understanding of many complicated diseases and their treatment options.

3.2 Immunohistochemistry and Whole Slide Imaging

IHC is the localization of antigens in tissue sections by the use of labeled antibodies as specific reagents through antigen–antibody interactions that are visualized by a marker such as fluorescent dye, enzyme, radioactive element, or colloidal gold [6]. A series of advancements in the technology, particularly during late 1980s and 1990s, led to the development of antigen retrieval techniques to make IHC possible on nearly all archival tissues. This, coupled with sensitive detection systems, and better antibodies have made this technique routine in surgical pathology and research. With the expansion and development of IHC technique, enzyme labels have been introduced. Other labels include radioactive elements, and the immunoreaction can be visualized by autoradiography. Since IHC involves specific antigen–antibody reaction, it has apparent advantage over traditionally used special enzyme staining techniques that identify only a limited number of proteins, enzymes, and tissue structures. Therefore, IHC has become a crucial technique and is widely used in many medical research laboratories as well as clinical diagnostics.

Image processing for feature extraction and statistical analysis can be utilized for evaluating human tissue, which is crucial for cancer diagnosis, characterization, and treatment. In Section 3.3, we present a compact analysis of the fundamental aspects of this imaging approach along with representative evaluation algorithms originated from international bibliography. Each technique uniquely contributes to extracting useful information. By suitably selecting the appropriate parameters and fusing these methodologies in a united scheme, we can build an effective, automated, and functional imaging system for the assessment of protein overexpression. By utilizing microscope imaging of tissue slides, we are capable of studying tissue at a cellular level, obtaining information about its morphology, the shape characteristics (shape, perimeter, area) of cells, and the distinct composite segments of each cell (cell nuclei, cell membranes, connective tissue).

The simplest method to study cells and tissues is the use of photon microscope. Thin slices of tissues are laid on glass tiles, stained with the appropriate pigment, lightened by common light, and observed in the microscope. The use of electron microscope improves discrimination ability, overcoming the drawback of photon microscope to visualize thin and small structures inside the cells. IHC and fluorescence *in situ* hybridization (FISH) are techniques that have proved extremely useful tools for improving the visual result of microscopy. IHC utilizes antibodies against specific cellular elements, whereas FISH aids in the study of DNA or RNA [7], extracting significant information at the molecular level. In this way, structures that could not be visible with other techniques are now observable in the photon and electron microscopes.

For digital imaging of IHC slides, proper staining and preparation of the sample is extremely crucial so that it preserves the tissue morphology and structure. This is followed

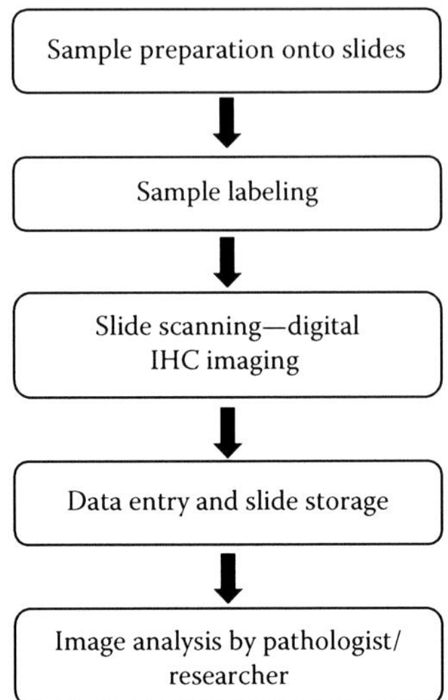

FIGURE 3.2
Step-by-step process of IHC and digital pathology.

by the imaging and storage of the slides digitally. The step-by-step process for IHC imaging is described in Sections 3.2.1 through 3.2.5 (Figure 3.2).

3.2.1 Sample Preparation

The tissue samples to be stained should be fixed well before complete depletion of oxygen occurs in them. The sample preparation influences the quality of the image capture and thus the obtained digital slides to a large extent. This makes it very essential to optimize the process of tissue preparation for ideal fine-quality staining, good-quality glass without any marks or dust, and timely tissue fixation. If there are any folds in the tissue or if certain sections are thicker than the rest, improper scanning can occur, rendering poor-quality digital slides, making them blurry. Thus, it is very important to have clean slides with no pockets of air or dirt trapped in them, no scratches or fingerprints present, no broken or chipped sections should be present, no adhesive sticking out, and so on [8].

Correct positioning of the tissue on the glass slide (ideally in the center) is also important to make the imaging/scanning of the slide easier, and a set of slides should have consistency in the placing and orientation of the specimen. Usually, the slides contain a label slip on the edge, indicating the details of the slide and its contents from the manufacturer. Any excess glue used to stick this label can interfere with the tissue imaging and distract the focus of the camera. Tissues can be embedded in paraffin or just contain frozen sections or cell smears.

The staining in all the slides should be consistently done to acquire accurate images. For paraffin embedding, the tissues must be immersed in the fixative immediately after dissection and the fixation time can vary from 4 to 24 hours. Over-fixation can cause a masking of the antigen, so the time should be monitored carefully. This tissue is first dehydrated before adding the paraffin wax. One of the benefits of frozen tissue samples is the time saved by omitting the initial fixation step required for paraffin-embedded tissue. Frozen tissue can be prepared by immersing the tissue in liquid nitrogen or isopentane, or by burying it in dry ice. For frozen samples, a short fixation is done after freezing and sectioning.

3.2.2 Sample Labeling/Staining

In this process, for the purpose of detection, antibodies can be grouped as primary or secondary reagents such that the primary antibodies are *raised* to detect a specific type of antigen and are mostly unlabeled, whereas the secondary antibodies are those that are usually bound to certain linker molecules, which further attract reporter molecules.

The detection method is one of the governing factors of the type of reporter molecules, the most popular being chromogenic and fluorescence detection, which are in turn mediated by either an enzyme or a fluorophores. When chromogenic reporters are used, an enzyme label reacts with a substrate to return a deeply colored product, which is analyzed using an ordinary light microscope. Although there are plenty of enzyme substrates, alkaline phosphatase and horseradish peroxidase are the most frequently used labels for protein detection.

3.2.2.1 Target Antigen Detection Methods

The direct method just consists of one step, in which a labeled antibody reacts directly with the antigen in tissue sections. This method employs only one antibody and is simple and fast, but the sensitivity is lower due to the signal being amplified lesser, as opposed to indirect methods. The indirect method involves an unlabeled primary antibody (comprising the first layer) that adheres to the target antigen in the tissue and a labeled secondary antibody (second layer) that reacts with the aforementioned primary antibody. This method is more sensitive than the direct method due to a greater signal amplification.

Another highly efficient labeling technique that is commonly used these days is fluorescence microscopy. The specimen is illuminated with light of a specific wavelength, which is absorbed by the fluorophores, causing them to emit light of longer wavelengths (i.e., of a different color than the absorbed light). The illumination light is separated from the much weaker emitted fluorescence through the use of a spectral emission filter. Most fluorescence microscopes in use are epifluorescence microscopes, where excitation of the fluorophore and detection of the fluorescence are done through the same light path (i.e., through the objective).

For fluorescence microscopy, the sample must be fluorescent. A fluorescent sample can be created by various methods such as labeling using a fluorescent stain or by the expression of a fluorescent protein, which may or may not be naturally present in the sample. The intrinsic fluorescence of a sample (i.e., autofluorescence) can be used. Biological fluorescent stains can also be employed, which are inherently fluorescent and bind to certain molecules of interest. Nucleic acid stains such as Hoechst, DAPI (4′,6-diamidino-2-phenylindole), DRAQ5, and DRAQ7, which all bind to the DNA can be used for the labeling of the nucleus. Drugs and toxins are also used as fluorescent stains as they have the capacity to bind to certain structures specifically.

Improvements in genetic understanding and advanced techniques being developed to modify DNA allow scientists to genetically alter proteins to carry a fluorescent protein reporter. This allows researchers to directly select their protein of interest and make it fluorescent and easily track its location. The green fluorescent protein (GFP) is a protein composed of 238 amino acid residues and exhibits green fluorescence when exposed to light in the blue-to–ultraviolet (UV) range. The GFP gene is frequently used as a reporter of expression in biological studies.

3.2.3 Slide Scanning/Imaging

An imaging system comprises the sample to be imaged, a light source that illuminates the sample, filters if needed, and detectors to capture the transmitted or the backscattered light depending on the geometry of the experiment. Any charge-coupled device or complementary metal–oxide–semiconductor cameras are generally used for image capturing. The calibration of any imaging system as well as the scanning parameters is important in transforming a glass slide into a high-quality digital slide. The most important and critical property of any imaging system is its image quality. The reproducibility of color and intensity is extremely necessary in order to acquire consistent and accurate images of the slides. There are various causes for problems in calibration such as problems with the light source, camera gains, macro camera offsets, location of the slide on the mount, and optical misalignment. The *virtual slide* can be assembled in various ways, depending on the particular scanner being used (tiling, line scanning, dual sensor scanning, dynamic focusing, or array scanning) [1]. The resulting digital images acquired can be examined any time at any location and studied using an interactive digital pathology software on the computer screen.

A large number of scanners and cameras are available from various brands, offering faster image capture and better image quality. However, there are still some issues that need attention before these systems can be implemented clinically and for diagnostic purposes. Stability, focus, and consistency of image quality can be problematic with the early virtual slide scanners. Also, uniform illumination is also a major factor in image quality; nonuniform illumination produces highly inconsistent and unclear images. Quality of the slide, focusing, and compression at the scanner or camera end have an influence on image quality as well. At the user end, it is not easy to improve the focus and the compression, apart from the compression ratio. The vital components in all slide scanners are light microscope optics and light source (illumination systems). Most vendors and brands use single-axis optical instruments, but array microscopes have a much larger field of view that offers the possibility of high-speed scanning. Resolution depends on the wavelength of light used and the numerical aperture (NA) of the lens system [resolution = (f) wavelength/2NA]. When the incident light is not properly filtered, there is a tendency for the wavelength to shift to longer values (more red) because of the characteristics of the light source in common use. Most of the currently available microscopes correct for this with a neutral density filter for brightness and a blue filter (depending on the light source) for color correction. Some scanners have more than one objective lens, whereas some scanners have a zoom/magnifying lens. The recent integration of light-emitting diodes for illumination in certain newly designed scanners offers a striking alternative approach.

Another important factor in the quality of the digital images are the specifications of the display screen, which include its physical size, pixel dimension affecting resolution, noise, temporal response, pixel defects, luminance response, and ambient lighting.

3.2.4 Data Entry and Slide Storage

After the slides are scanned and the digital images are acquired, these slide images are stored according to the convenience of the user. Certain available software applications can be used to *read* the image using a bar code and store all the images in a labeled database. The storage, accessing, and sharing of slide images have become easier and faster due to the invent of digital pathology.

3.2.5 Image Analysis, Processing, and Diagnosis

A pathologist or researcher studies the saved images to analyze the tissue for presence of abnormalities and to characterize the irregularities present in the tissue as cancer or some other kind of malformation. Various image processing techniques can also be utilized to enhance the images in order to improve the visualization of the regions of interest.

3.3 Delineation of Tumor Margins

In this study, a detailed analysis of digital scanning of slides obtained using fluorescence microscopy and their further processing and analysis is presented [9]. This work presents the design aspects of a digital fluorescence imaging system implemented with unsupervised machine learning techniques aimed at enhancing the imaging of Gli36Δ5 (glioblastoma multiforme) tumor margins. The samples were illuminated by a 488-nm argon laser attached to the microscope, and the resulting autofluorescence images were captured, and then processed using unsupervised clustering-based image processing techniques in conjunction with edge detection algorithms.

Optical microscopy images of the brain tumor samples were recorded using a Zeiss laser scanning microscope, at a 5× magnification. For fluorescence images, the samples were interrogated with a 488-nm Argon laser attached to the microscope and the resulting autofluorescence images were captured in the green wavelength range with the band center wavelength at 530 nm.

The brain tumor samples used in this study, consisting of Gli36Δ5 tumor cells, are on 10-μm tissue slices cut from the first 2 mm of the anterior brain section. There are five slides with three sections each, corresponding to the depth of the slice from the beginning of the anterior brain section (starting from 10 to 150 μm, in 10-μm slices). The tumor is located on the upper edge of these slices and these tumor cells express GFP. Fluorescence imaging is used to detect cancer as certain biomarker fluorophores are expressed much more strongly in tumor cells than in healthy ones, aiding imaging for discrimination purposes. The most common fluorophores in tissues have their absorption and emission spectrum range from UV to visible. Optical fluorescence microscopy is one of the best techniques for high-quality histological imaging as it does not require the samples to be sectioned or stained. This method is not detrimental to the tissue samples and is time efficient.

To aid enhanced discrimination of the tumor and its margin, a three-step image processing algorithm was designed and implemented (Figure 3.3).

Without going into extensive details, the techniques to enhance the demarcation of tumor boundaries are briefly explained in Sections 3.3.1 through 3.3.3.

3.3.1 Otsu Thresholding

Otsu thresholding is a nonparametric and unsupervised technique and the algorithm assumes that the image consists of pixels belonging to two classes (background and object), that is, the image has a bimodal histogram. It then iteratively calculates an optimal threshold value to segment the images. This technique selects a threshold value such that the intraclass variance (variance among members of one class) is minimum and, consequently, the interclass variance (variance between members of the two different classes) is maximum. This ensures efficient segmentation of the images into two sections, background and object, as variance is a good measure of the homogeneity of a region. This technique is applied on the fluorescent images.

The intraclass variance is then calculated as the weighted sum of the variances of each of the two classes, where the *weight* is taken as the probability of the pixel belonging to that particular class, and probabilities/weights are calculated for each pixel corresponding to different intensity levels. Hence, this method works by calculating what fraction of the total number of pixels belongs to the *background* class and what fraction to the *object* class.

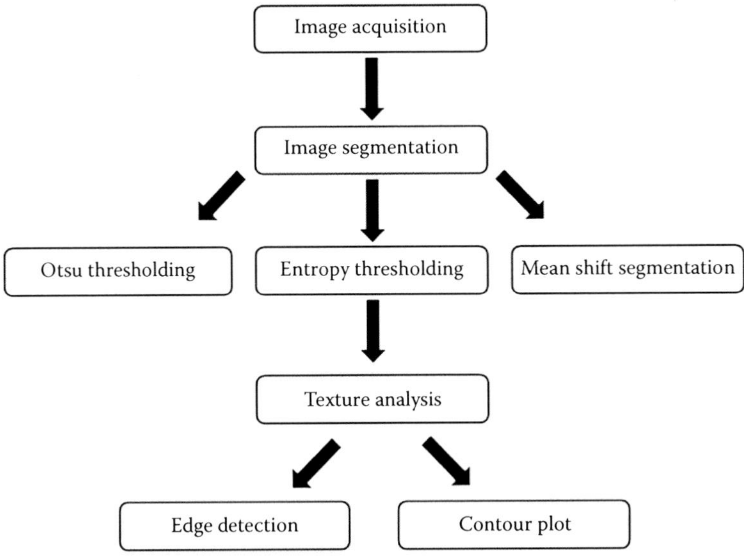

FIGURE 3.3
Step-by-step process followed for enhanced imaging of tumor margins.

If we consider an image with L gray levels in all, its normalized histogram consists of, for each gray level x, the frequency with which this gray level occurs in the image, represented by p_x. If we assume that the image we are dealing with consists of pixels belonging to two classes—object and background—and that the object is bright on a dark background, then the fraction of pixels belonging to the two different classes can be described as follows:

$$\theta(t) = \sum_{x=1}^{t} p_x$$

where:

$\theta(t)$ represents the fraction of pixels belonging to the background class

t corresponds to the threshold value, above which the intensity levels of the pixels will correspond to the object class (below which are the background pixels). Similarly, the fraction of pixels belonging to the object class can be represented as follows:

$$1 - \theta(t) = \sum_{x=t+1}^{L} p_x$$

where:

$\theta(t)$ corresponds to the fraction of pixels belonging to the background class

L is the total number of gray levels or intensity values

t is the threshold value of the intensity, above which the pixels belong to the object class

The total variance, as explained above, is calculated by summing the variance within the two classes (object and background) represented as σ_w^2 and the variance between the two classes σ_B^2. The mean value of the object class is represented as μ_o and the mean of the background class is represented by μ_b. The same notation is followed for the variances:

$$\sigma_T^2 = \theta(t)\sigma_b^2 + [1-\theta(t)]\sigma_o^2 + (\mu_b - \mu)^2\theta(t) + (\mu_o - \mu)^2[1-\theta(t)]$$

As described earlier, this algorithm works by minimizing the variance within the classes (σ_W^2) and maximizing the variance between the two classes ($\sigma_B{}^2$). After mathematical calculations, we get

$$\sigma_B^2(t) = \frac{[\mu(t) - \mu\theta(t)]^2}{\theta(t)[1-\theta(t)]}$$

This value is maximized by starting with a small *guess* value and iteratively modifying the threshold value so that the maximum variance between the two classes is obtained.

3.3.2 Entropy Thresholding

Another technique that has been applied in this study for the thresholding of images is *entropy thresholding*. This is one of the standard and routinely used methods in which the entropy (based on information theory, also known as *Shannon entropy*) of the background is maximized along with the probability distribution of the object. This entropy measures the randomness of the uncertainty of a random variable. The goal of this technique is to separate the histogram (probability distribution) of the image into two different (independent) distributions based on a particular and optimal threshold value. This maximizes the entropies of both the distributions, yielding maximum information.

Let $p_1, p_2, \ldots, p_n$ be the probability distributions of the various intensity levels (gray levels) in the image. The threshold value for the image segmentation is assumed to be s. Now, two independent probability distributions are defined—one for the gray levels from 0 to s and another for the pixels with intensity levels $s + 1$ to n. Let these be known as A and B, respectively. The definitions are as follows:

$$A\text{:}\ \frac{p_1}{p_s}, \frac{p_2}{p_s}, \ldots, \frac{p_n}{p_s}$$

$$B\text{:}\ \frac{p_{s+1}}{1-P_s}, \frac{p_{s+2}}{1-P_s}, \ldots, \frac{p_n}{1-P_s}$$

$$H(A) = -\sum_{i=1}^{s} \frac{p_i}{P_s} \ln \frac{p_i}{P_s}$$

$$H(B) = -\sum_{s+1}^{n} \frac{p_i}{1-P_s} \ln \frac{p_i}{1-P_s}$$

$$\psi(s) = H(A) + H(B)$$

The value of s that maximizes this entropy is considered as the optimal threshold.

3.3.3 Mean-Shift Segmentation

The next technique applied to segment the image into background and object is *mean-shift segmentation*, which is a nonparametric clustering method in which we do not need to know the number of clusters prior to segmentation and which does not constrain the shape of the clusters, that is, they can be distributed in any pattern. This is a mode-finding algorithm in which certain *neighborhoods* or groups of points are selected and the mode (center point of the regions of high concentration) of each one is calculated. Now with these modes as the center points of the neighborhoods, new modes of the new groups are calculated. This process is repeated until the modes converge and no new mode is obtained. The algorithm applied uses the following steps:

- Define a radius or a *window* size of pixel neighborhood (say, 3 × 3).
- Choose a center point for every window (or) consider every pixel, one by one, as the center of the window.
- Compute the mean/average (or) the mode (intensity value occurring with the highest frequency in the neighborhood) for each widow or pixel neighborhood.
- Now consider all the *modes* or *means* as the new center points of the window and calculate the mean or mode of the new neighborhoods (with centers at the previously obtained means or modes).
- Iterate this process until the modes converge, that is, no new values of modes or means are obtained.

Finally, a contour plot of the segmented image (on the most accurate of the results obtained from the above-mentioned three techniques) is studied to observe the contour pattern of the cancer. A contour plot is a graphic representation of three variables in two dimensions. This technique is used to map three-dimensional information onto a two-dimensional (2D) surface. Two variables x and y correspond to the axes and the third variable z corresponds to the contour levels, which is interpreted as *depth* with respect to the x–y plane.

3.3.4 Results

Raw fluorescence brain tumor images, as well as processed images using a three-step imaging algorithm, are shown in Figure 3.4a–f. It can be observed that images processed using clustering-based thresholding techniques exhibit higher contrast with respect to raw images obtained using fluorescence microscopy and the visibility of the tumor margins is enhanced significantly. Entropy thresholding also produced extremely promising results, which were superior to those by Otsu thresholding and mean-shift segmentation as they did not miss out on any object pixel. However, mean-shift segmentation did a better job of eliminating unwanted noise, although these noisy pixels could be due to the scattering of the fluorescent dye.

3.3.5 IHC and Genetics

The mutation of cells to a cancerous state is considered as a destabilization of the factors that control their natural growth or the augmentation of their receptors, which leads to autostimulation of their growth. Early clinical cancer diagnosis is almost impossible, as the common procedures are incapable of detecting cell masses/tumors of size over 1–2 cm. When cancer is evolved in the human organism, tumor cells or body tissues may produce substances that can be detected in blood, tissues, or urea via monoclonal antibodies, and are characterized as tumor markers. Sometimes, tumor biomarkers, also known as

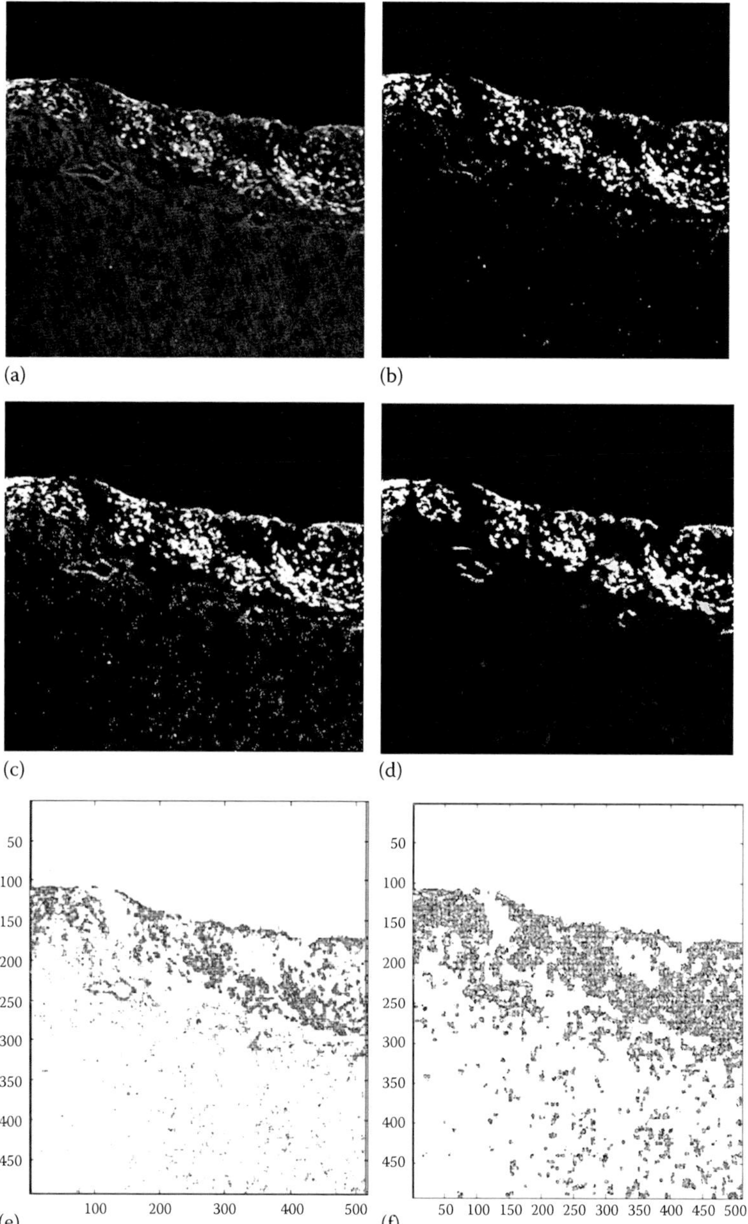

FIGURE 3.4
(a) Original raw image; (b) Otsu processed image; (c) entropy processed image; (d) mean-shift segmentation; (e) contour plot after entropy; (f) contour plot after entropy and range filtering.

oncoproteins, are present during the physiological growth of tissue. The need for detecting such markers arises in the period prior to diagnosis, when the specialist suspects the presence of cancer and during the treatment period, after diagnosing the disease, in order to attend and control its evolution. Although a wide variety of biomarkers have been tested and achieved promising outcomes, a limited number of these have been adopted in standard clinical practice over the past decade.

In human cancers, HER2 (also known as c-erB2) is activated via gene amplification, which is a genomic mutation where a small fragment at chromosome band 17q12–q21 is multiplied in a cell up to hundredfold. HER2 is colocalized, and thus most of the time coamplified with the gene *GRB7*, which is, as well, a proto-oncogene (active in breast cancer, testicular germ cell tumor, gastric cancer, and esophageal cancer). This gene amplification leads to overexpression of its protein product, which disturbs the HER receptor family signaling networks and forms heterodimers with epidermal growth factor receptor (EGFR), HER3, and HER4 upon binding of their ligands. In addition, overexpressed HER2 proteins form HER2–HER2 homodimers, the major oncogenic activation mechanism. Findings in the early 1990s revealed that antibodies to the extracellular domain of HER2 inhibit the growth of HER2-positive cell lines, which gave rise to test the most promising growth inhibitory antibody as an anticancer drug.

One of DNA's major functions is to serve as the blueprint for the manufacturing of the proteins that are used to keep cells alive. Like all proteins, the HER2 protein is the result of certain patterns of DNA. The segment of DNA that codes for HER2/neu is called an *oncogene* and it is the *HER2/neu* oncogene that produces the HER2/neu protein. All normal epithelial cells contain two copies of the *HER2* gene and express low levels of HER2 receptor on the cell surface. In some cases, during oncogenic transformation, this segment of DNA becomes damaged as the cells reproduce and the number of gene copies per cell is increased, leading to an increase in messenger RNA (mRNA) transcription and a ten- to hundredfold increase in the number of HER2 receptors on the cell's surface called overexpression.

Gene amplification/receptor overexpression has been demonstrated in breast, ovarian, bladder, gastric, and pancreatic tumors. Gene amplification is associated with aggressive cell behavior and poor prognosis. Some tumors show receptor overexpression without gene amplification and have a more favorable prognosis; yet the biologic significance of this variant is less certain. In general, the presence of HER2 amplification/overexpression appears to be a key factor in malignant transformation and is predictive of a poor prognosis in breast cancer.

The prognostic role of HER2/neu amplification or overexpression lies on the weak unfavorable prognosis in untreated breast cancer patients, although its predictive implications include resistance to hormonal therapy, resistance to chemotherapy responsiveness to doxorubicin, and, mainly, responsiveness to trastuzumab (Herceptin) therapies [10–12]. Herceptin is a monoclonal antibody that was developed specifically to target the HER2/neu protein expressed only in the cancer cells while leaving normal cells (which do not overexpress the protein) unaffected. This makes Herceptin different from chemotherapy, which kills all rapidly dividing cells, both healthy and cancerous, but sort of tamoxifen, which only work in hormone-responsive tumors.

Breast cancer was the first tumor type in which abnormalities of *HER2* gene copy number and/or expression were associated with reduced disease free and overall survival. Leaders in the field of HER2 have stressed the importance of developing a simple and accurate method of determining HER2, one that is both inexpensive and reproducible. If breast cancer is tested for HER2 status, the results will be graded as positive or negative. If the results are graded as HER2 positive, then the *HER2* genes are overproducing the HER2 protein and those cells are growing rapidly and creating the cancer. Tumors are faster growing, more aggressive, and less sensitive to chemotherapy and hormone therapy. If the results are graded as HER2 negative, then the HER2 protein is not causing the cancer. There is also a middle situation, generally considered as healthy but requiring attention every few months, where there is a controlled augmentation of the HER2 protein. The main methods for testing HER2 breast cancer are as follows [13]:

- IHC—This test measures the production of the HER2 protein by the tumor based on the intensity and the completeness of the membrane staining. The test results are ranked as 0, 1+, 2+, or 3+. If the results are 3+, the cancer is HER2 positive.
- FISH—This test uses fluorescent probes to look at the number of HER2 gene copies in a tumor cell. If there are more than two copies of the HER2 gene, then the cancer is HER2 positive.

The two techniques currently used to measure HER2 gene copies are quantitative polymerase chain reaction (PCR) and FISH. The process for PCR is fully automated and requires only minimal amounts of tumor tissue. This method will allow retrospective studies to be performed with archival tissue. FISH also only requires small tissue samples and has extreme sensitivity to detect amplification from a histologic section. However, it is not widely available in hospital laboratories. The result of this method can vary considerably if the assay is not standardized and is therefore dependent on the skill of the pathologist. A more widely used test is IHC, which measures HER2 protein expression. IHC has been specifically adapted for detection of HER2 protein using specific antibodies. The advantages of this method are that it can be used on fresh and archival tissues and that it utilizes technical and human resources readily available in pathology laboratories. Unfortunately, there are some disadvantages of this method. IHC uses different antibodies with different binding affinities and different epitope specificities, thereby creating differences in HER2 overexpression rates. In addition, HER2 overexpression scoring systems differ and often rely on subjective measures of staining intensity and pattern. When IHC staining techniques that are too sensitive are employed, it becomes problematic to differentiate between normal versus high HER2 protein levels that are associated with gene amplification. The enzyme-linked immunosorbent assay (ELISA) is another method of testing HER2 protein in serum samples. While the technology is simple and well suited to automation, it may produce significantly different results to those obtained with IHC and FISH. While IHC and FISH measure HER2 receptor protein (mostly intracellular) and gene amplification, respectively, ELISA specifically measures the levels of the extracellular HER2 receptor proteins released into the plasma from HER2-overexpressing tumors.

It is widely acknowledged that the ideal test for HER2 status is one that is simple to perform, specific, sensitive, standardized, stable over time, and allows archival tissue to be assayed. At present, the test that best meets these criteria is IHC. With standardization of laboratory testing and appropriate quality control in place, the reliability of IHC will be improved further. It is expected that FISH will become more widely used in future as well. Figure 3.5 illustrates the standardized diagnosis–treatment scheme followed by pathologists for HER2/neu evaluation [14] under the IHC assessment procedure. If no protein overexpression is detected (scores 0, +1), then breast cancer is not caused by a malfunction of the HER2/neu production mechanism, yet no specialized treatment is needed and the common protocols for cancer diagnosis and treatment are followed. However, if cancer status is assigned +3 score, implying that there exists advanced protein overexpression, then Herceptin (chemical name: trastuzumab) is adopted for blocking breast cancer cells from receiving growth signals and proliferating. In this direction, Herceptin therapy may help to slow or even stop the growth of the breast cancer and fight the disease by activating the immune system to destroy cancer cells. In an ambiguous HER2/neu status estimation (borderline situation, score +2), a retesting of the evaluation procedure must be performed in addition to a FISH testing for HER2/neu gene amplification. If the newly derived results confirm both HER2 overexpression and amplification, then Herceptin treatment is followed by the medical expert.

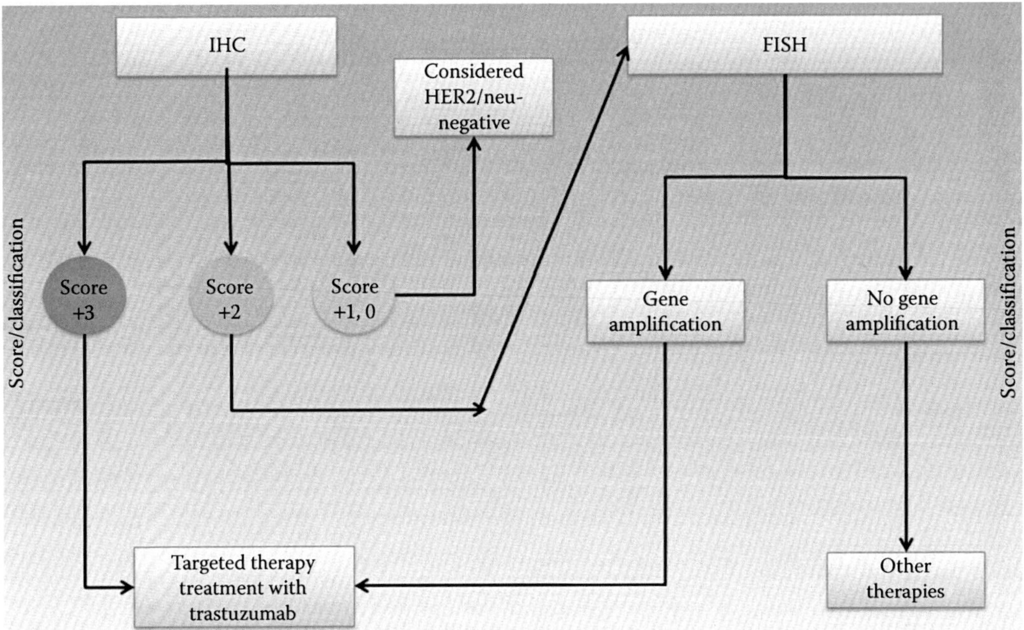

FIGURE 3.5
Standardized HER2/neu cancer diagnosis–treatment scheme.

Standard management in the treatment of many solid tumors has improved in recent years, yet many metastatic solid tumors remain incurable. Factors that limit the success of treatment include drug resistance and lack of tumor selectivity. Although progress has occurred in cytotoxic therapy for breast cancer, interest in new interventions continues. Because of the ability of the immune system to target specific responses, the area of immunotherapy has great potential in future management of cancer.

Many patients who are currently being tested for HER2 status are women with metastatic breast cancer who have been heavily pretreated. Therefore, it is important to gather information about the patient's health history, performance status, all prior treatments, emotional well-being, as well as the understanding of the disease status, the meaning of HER2 testing, and its possible implications. Nurses may collaborate with other practitioners in obtaining any or all of this information. It is also critically important to involve family members in the process of gathering accurate and thorough information.

It is also useful for oncology nurses who counsel patients to be fully aware of the advantages and disadvantages of the various testing methods and the impact that these test methods and results may have in terms of prognosis and treatment for the patient. This may require detailed communication with the pathologist about how the test sample was obtained, clarification with the laboratory about the test method employed (and its potential variability), and discussion with the oncologist to interpret the results. Knowledge of testing methods and determination of the accuracy of results are critical to appropriately identify patients that may benefit from anti-HER2 therapies.

The main contribution of this analysis covers the field of tissue evaluation. The objective is to thoroughly analyze optical microscopy as a tissue-imaging modality that enables tissue characterization from slides obtained through the procedure of IHC. The examples presented on IHC microscopy reveal the prospect for automated detection of HER2/neu protein overexpression in tissues, which can contribute to diagnosing breast cancer at its earlier stages and treating the disease with higher probability. We are interested in automatically extracting the degree of protein overexpression on the cell membranes through image processing based on the characteristic identities the diseased tissues reveal after the application of immunochemistry. The preparation and processing of tissues is a complicated biological issue and is not considered as an issue of this chapter; however, it is of substantial importance. A well-prepared tissue will secure a clear and informative image.

The Her-2/neu gene can be screened by using molecular and immunological probes that vary in their complexity, sensitivity, and specificity. In the beginning, Her-2 amplification was evaluated by Southern blotting, which was supplanted by a sensitive and rapid quantitative PCR method. FISH is a more recent technique that enables Her-2/neu amplified cells to be visualized within a tumor slice. In addition, Her-2/neu overexpression can be detected with various methods, including Northern blot and *in situ* hybridization for Her-2 RNA, Western blot, and immunoassays for Her-2 protein, but the most widely used method is IHC. FISH is a more recently developed method that can visualize the number of gene copies present in tumor cells and provide a sensitive and accurate measure of Her-2/neu gene amplification, while IHC can be easily carried out on formalin-fixed, paraffin-embedded tissues, and is more familiar, less expensive, and simpler compared to FISH. FISH measures Her-2/neu amplification in DNA level, whereas IHC measures gene overexpression in protein level and identifies cases in which the gene product is overexpressed even without being amplified. Yet the results produced cannot be fully objectively accepted by all researchers and are complex and subject to considerable variations in the hands of different teams and laboratories. The advantage of FISH versus IHC is the ability to analyze gene integrity and protein expression on two consecutive tumor sections, thus manipulating the same cells. IHC has the drawback that produces results that are conflicting due to different sensitivity and specificity of the primary antibodies used; there is variability in IHC interpretation and technical artifacts can be introduced [15].

In IHC, positive reactions with diaminobenzidine (DAB) are identified as a dark brown reaction product on the cell membrane and the specimens are graded as negative, low, medium, and high positive, based on both the percentage of positively stained cells and the staining intensity according to a scoring protocol. The specimens with high or medium IHC positivity are considered to have Her-2/neu overexpression, compatible with FDA-approved criteria for Herceptin treatment. An arbitrary scoring system needs to be assigned for Her-2/neu protein levels, which in reality cover a continuous spectrum. In fact, different scoring systems have been used. Recent pilot studies use graded values, including high, medium, and low positive and negative, whereas HercepTest uses 3+, 2+, 1+, and negative scores. Another scoring approach is mentioned in Reference 16: categorization as 3+, 2+, 1+, or 0+. When membrane staining, whether incomplete, complete, strong, or weak, is absent in less than 10% of the cells, the score is negative or 0. Staining is scored as 3+ when the surface of the tumor cell has strong intensity, 2+ when it has moderate intensity, and 1+ when the membrane has weak intensity and is incompletely stained. However, other studies used strong positive (2+), weak positive (1+), and absent (0), or a positive versus negative system, where *positive* was defined as the relative difference in cytoplasmic membrane staining between tumor cells and normal epithelial cells. Therefore, it is inevitable that some interlaboratory discrepancies may exist for at least some of the cases, especially medium or low positive ones.

In general, the HercepTest scoring is adopted. According to this, negative results are 0 and 1+ intensity (no or barely perceptible membrane staining in >10% of tumor cells) and

TABLE 3.1

Generalization of IHC Score

IHC Score	Staining Pattern
0	No slides returned
1 and 2	Considerably less invasive tumor nuclei staining than expected, compared to the test slides stained by the organizing laboratory
3	Demonstration of less invasive tumor nuclei staining than expected to stain or/and the intensity of staining is considerably weaker than expected
4 and 5	Demonstration of the proportion of nuclei of invasive tumors expected to stain, with roughly the expected staining intensity

TABLE 3.2

Standardized IHC Score Based on Practical Measurements

Score	HER2 Status	Staining Pattern
0	Negative	No staining observed, or membrane staining in less than 10% of tumor cells
+1	Negative	A faint/barely perceptible membrane staining detected in more than 10% of tumor cells. The cells are only stained in *part* of the membrane.
+2	Borderline	A weak to moderate complete membrane staining observed in more than 10% of tumor cells
+3	Positive	A strong complete membrane staining observed in more than 10% of the tumor cells

positive results are 2+ (moderate complete membrane staining in >10% of tumor cells) and 3+ (strong complete membrane staining in >10% of tumor cells). The description is summarized in Tables 3.1 and 3.2, where the adopted scoring system is described in Table 3.1 and summarized in a standardized form in Table 3.2 according to practical measurements, keeping up with the majority of bibliography.

It is not known if there is a real difference between 10% and 100% staining from a clinical point of view. Studies on HER2 do not address the percentage of positive staining in detail. In this regard, it is difficult to compare different studies because of different definitions of positivity.

IHC is a purely medical issue of deep theoretical background and the detailed explanation of its rules and procedures is not of our main interest. Our goal is to attempt to process the extracted IHC images of breast tissues and automatically determine/define the extent of HER2 overexpression on the examined breast tissues. This is not an easy part, as very few image analysis methods have been developed, and, most of all, there is not a totally and generally accepted subjective method to define the degree of malignancy. In addition, the visual processing of the prepared tissues is directly dependent on the quality of the laboratorial technique. The tissue evaluation is achieved by counting, through a standardized procedure, the total percentage of staining of the membranes of the cells forming the tissue section. If the percentage exceeds a specific value, defined according to a *scoring method*, this tissue part is considered as *affected*, so the *patient* has to proceed to the next step, which is targeting the HER2 antibody with anti-HER2 therapy.

The objective is to exploit the fact that IHC *stains* the membranes of HER2-overexpressed cells, which monitors them with a characteristic brown color [adopting the 3,3′-Diaminobenzidine (DAB) staining protocol] or red color [adopting the Papanicolaou stain (PAP) staining protocol] in the extracted microscope images, whereas the cell nuclei are highlighted with a pure blue coloring. The percentage of staining along each membrane contour, the percentage of *stained* cells with respect to all cells, and the *darkness* (the intensity of brown/red color) at each membrane are factors that reveal if a tissue is healthy or not. The difficult part is to accurately define these factors and safely interpret them according to a well-established scoring system. The outcome of this procedure is crucial. Determining a healthy tissue as *containing malignant cells* would saddle the patient with the heavy load of carrying the disease and introduce him/her to unnecessary, extremely costly, and potentially dangerous targeted treatment with Herceptin, which was found suspicious of causing cardiovascular abnormalities. Considering a defected tissue as healthy would prevent the patient from following an instantly needed therapy, thus putting his/her own life in danger!

The effect of IHC testing on the tissue sample is to stain the membranes of tumor cells, partially or completely. The percentage of tumor cells that have completely stained membranes and the intensity of that staining have been used, so far, by an experienced pathologist to derive a score. The goal is to develop an algorithm/procedure to contribute to the specialist's evaluation by segmenting the input IHC image to regions forming the tissue cells, count their number, define their contour borders associated with the cytoplasmic membrane, and deliver quantitative measurements on the distribution of membrane staining values that are relevant to deriving a score.

Generally, the membranes are assumed to be stained brown or red, according to the staining protocol implied by the utilized antibodies, whereas the nuclei are assumed to be stained blue. An important feature of the developed algorithm is to be able to logically connect membranes that are not completely stained. This is necessary in order to determine the completeness (percentage) of membrane staining for each and every tumor cell. There should be a high degree of concordance between algorithm cell boundaries and biological membranes. (It is generally accepted that no algorithm will perfectly perform this task.)

A continuous and qualitative cooperation with a medical specialist is needed in order to accurately define the parameters of the problem and insert them, step by step, into the image processing procedure. It is also important to mention the undoubtedly essential contribution of the anatomy expert to the tissue evaluation, recognizing that no automated or semiautomated algorithmic procedure can totally substitute the expert's tutoring and observation.

The objective of this chapter—the analysis of IHC images—is summarized in 3.6, which introduces the main steps of the entire procedure an individual should follow in order to fulfill a detailed and compact evaluation of a breast cancer tissue slide. In the first stage, proper and standardized actions should take place for the precise tissue preparation and preservation of its anatomical and chemical properties. This is achieved in specially equipped laboratories by pathology anatomists and staff, utilizing proper tools and protocols. In the second stage, the tissue slide is examined under the microscope, the protein overexpression is evaluated, and the diagnosis by the medical expert is performed, which will determine the further treatment and therapy of the patient. As a further step for automated tissue characterization and analysis, the corresponding images can be extracted by the camera adjusted on the microscope lens, selecting the appropriate magnification, and advanced image processing methodologies are fused and applied to the input image samples in order to algorithmically extract qualitative and quantitative results (Figure 3.7).

3.3.6 Color Models

The RGB color model is an additive color model in which red, green, and blue components are added together in various ways to reproduce a broad array of colors and is widely used in screening applications. The name of the model comes from the initials of the three additive primary colors, red, green, and blue. Pure color information is described as a linear combination

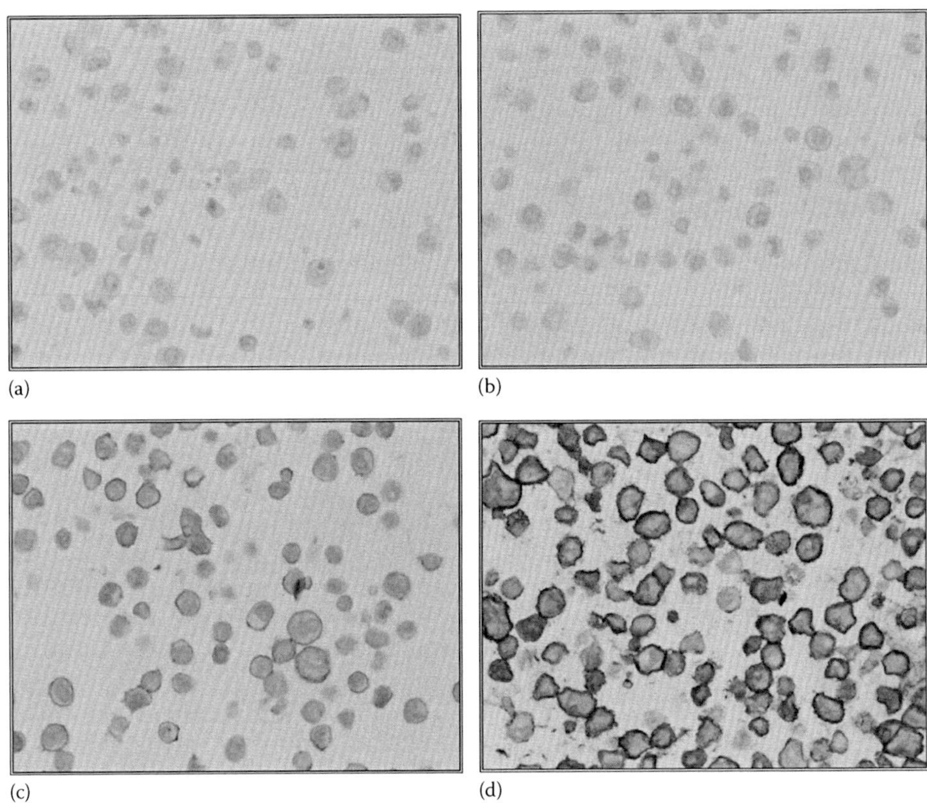

FIGURE 3.6
(See color insert.) (a) IHC score 0 image (totally normal and healthy tissue, no HER2 overexpression). (b) IHC score +1 image (healthy tissue with controlled HER2 overexpression). (c) IHC score +2 image (healthy but potentially suspicious tissue, augmented HER2 overexpression). (d) IHC score +3 image (affected tissue). (Data from UKNEQAS, *Immunocytochemistry*, 2, 2003.)

of all the three components. The term RGBA is also used to mean red, green, blue, and alpha. This is not a different color model but a representation, where alpha is an additional channel (not component) used for transparency. For an 8-bit image, each component value ranges from 0 to 255, defining the extent of the basic color it contains. It is represented as a cube in the Cartesian space with each axis representing each color component. RGB is an additive color space, modeling the way that primary color lights combine to form new colors when mixed.

In the hue, saturation, and value (HSV) model, also called hue, saturation, and brightness (HSB) or hue, saturation, and intensity (HSI), color is represented by three components: (1) hue (H), defining the pure color type (such as red and green) and ranging from 0° to 360° (with red at 0°, green at 120°, blue at 240°, etc.); (2) saturation (S) or *purity*, ranging from 0% to 100% and defining the amount of white color present (the lower the saturation of a color, the more faded the color will appear); and (3) value (V) or brightness (B) or intensity (I), ranging from 0% to 100%. It is a nonlinear transformation of the RGB color space. In order to process images aiming at extracting color information and features, it is preferable to use the HSV color model over alternative models such as RGB, because of its advantage to emulate the human color perception system. In addition, pure color features are independent of the intensity channel, which plays a key role in color segmentation as only the hue channel is processed for color determination.

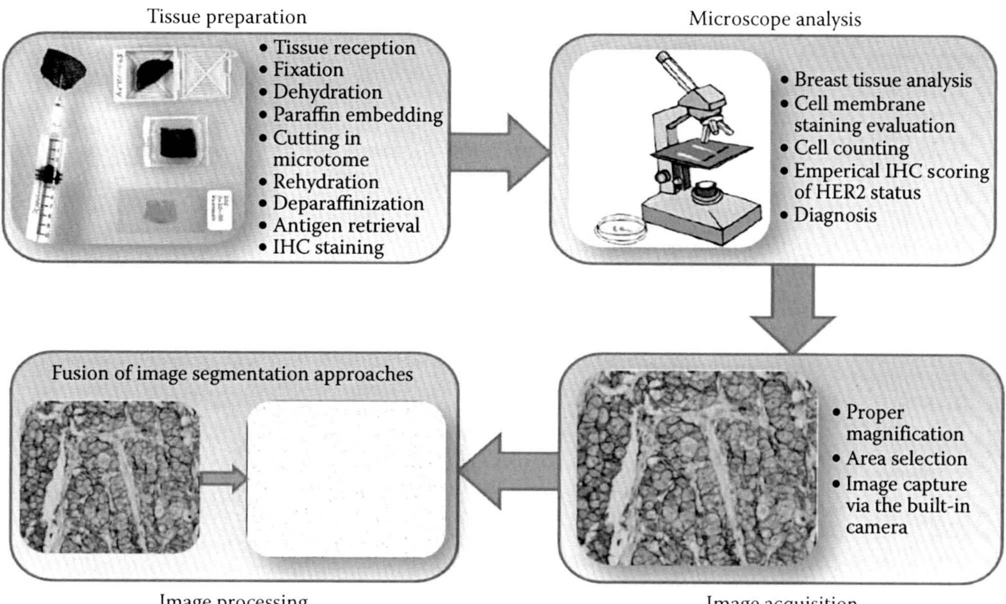

FIGURE 3.7
(See color insert.) Summary of the key steps of a unified, complete procedure for efficient analysis of an IHC image aiming at the assessment of the Her2/neu protein overexpression.

In the Lab color space, also known as International Commission on Illumination (CIE) or CIELAB, color is represented by dimension L standing for lightness and the a and b channels for the color-opponent dimensions, having the color variable independent of the color intensity. Lab color is designed to approximate human vision and its L component closely matches human perception of lightness. Thus, it can be used to make accurate color balance corrections by modifying only the a and b components, or to adjust the lightness contrast using only the L component. The $L^*a^*b^*$ color space includes all perceivable colors, exceeding the capabilities of the RGB color model. One of the most important attributes of the $L^*a^*b^*$ model is its device independence, meaning that the colors are defined independent of the nature of their creation or the device they are displayed on.

The conversion from the RGB model to the HSV is achieved using the following formula:

$$H = \frac{360}{2\pi} \times \arctan\left(\frac{\sqrt{3}}{2}(G - B),\ R - \frac{(G+B)}{2}\right) \times \frac{240}{360}$$

$$S = \sqrt{R^2 + G^2 + B^2 - RG - RB - GB} \times \frac{240}{255}$$

$$I = \frac{R + G - B}{3} \times \frac{240}{255}$$

Figure 3.8 illustrates a characteristic IHC image in the RGB color space and the distinct intensity components in the RGB, HSV, and Lab models, revealing the difference in the encoded information with respect to the various features and regions of breast tissue.

The decision significantly processed color space biomedical images affects the quality of the segmentation results. The RGB model is preferable for the representation of the images on a computer display or a monitor, whereas the other two models facilitate the

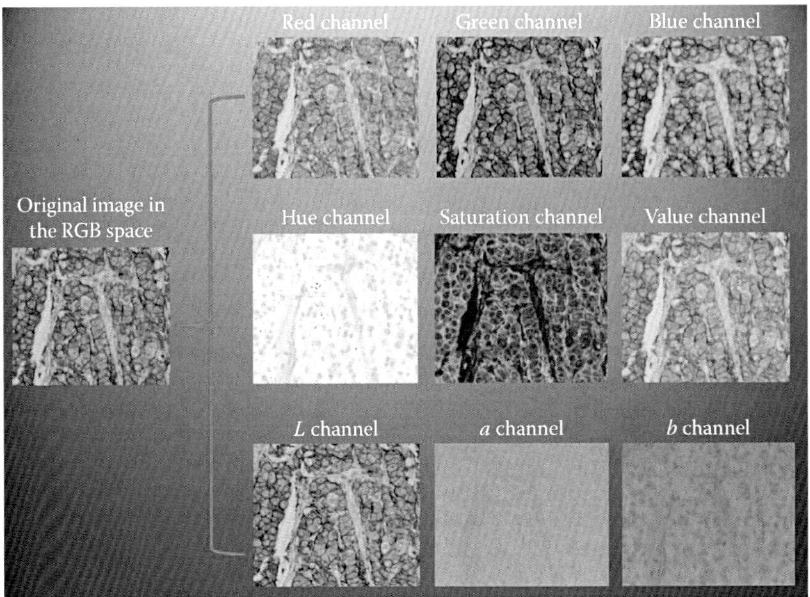

FIGURE 3.8
Original IHC image and the three intensity bands in the RGB (first row), HSV (second row), and Lab (third row) spaces.

differentiation of pure colors between the key regions of the depicted tissues, which is crucial for the detection of proteins revealed after the staining of the corresponding cell and tissue areas introduced via IHC. A representative illustration of a typical IHC image under these three basic color spaces can be viewed in Figure 3.9 before and after histogram equalization for contrast improvement. It is evident that each color model represents more efficiently distinct features of the tissue sample. Cell nuclei are more easily discriminated utilizing the HSV model, while the membrane boundary is more accurately described in the Lab color space facilitating the evaluation of the staining process. The three distinct contrast-enhanced components of each model are depicted in Figure 3.10. Notice the differentiation of the image information among the various color components or channels; each component mainly represents a particular feature/region.

3.3.7 Image Clustering

Cluster analysis or clustering is the process of grouping a set of objects into classes (clusters) according to a matching criterion (distance function) in such a way that the intercluster similarity between the elements, along with the extracluster separability, is increased. It is a common technique for statistical data analysis with numerous applications including machine learning, pattern recognition, image analysis, information retrieval, and bioinformatics.

Clustering can be achieved by various algorithms, which may differ significantly in their notion of what features constitute a cluster and of how to efficiently represent and extract classes. For these reasons, feature extraction and definition forms a key part of the clustering process. Popular representations of clusters include groups with small distances among the cluster members, dense areas of the data space, and intervals or particular statistical

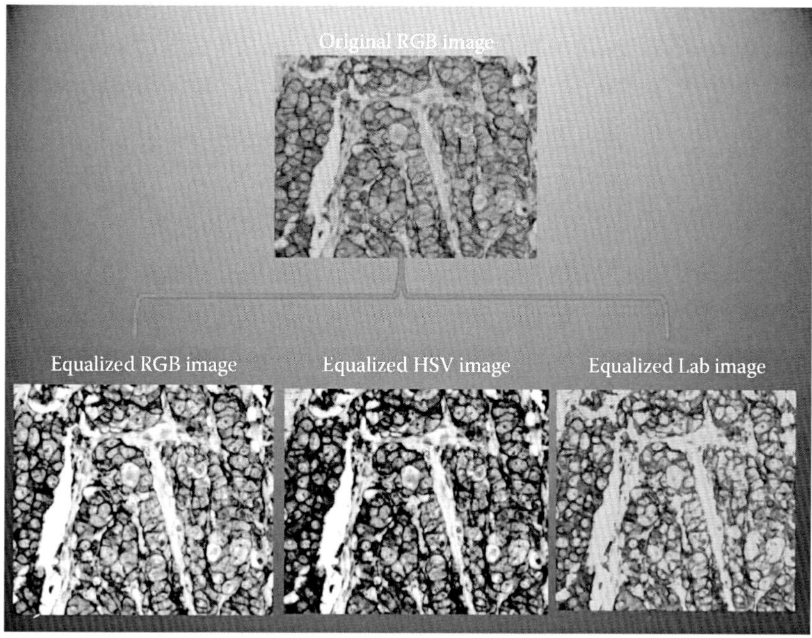

FIGURE 3.9
(See color insert.) A typical IHC image represented in the RGB, HSV, and Lab color spaces after histogram equalization for contrast improvement.

distributions. The appropriate clustering algorithm and parameter settings (including values such as the distance function to use, the density threshold, or the number of expected clusters) depend on the distinct input dataset and the expected use of the results. Cluster analysis is not an automatic task but an iterative process of information searching and extraction through an interactive optimization procedure that involves initialization, random in the majority of the cases, and trial and failure. Data preprocessing is also needed along with model parameter determination and convergence criteria in order to obtain the result with the desired properties.

Clustering is widely used in computer vision for partitioning a digital image into multiple segments (sets of pixels) with common characteristics such as color, intensity, or texture, or even locate objects and boundaries (lines and curves). The objective is to assign a label to every pixel in an image such that pixels with the same label share certain visual characteristics. The classic *K*-means algorithm and the recently reemerged and promising mean-shift algorithm are two representative methodologies for image clustering segmentation.

K-means clustering [17] is an iterative technique that is used to partition an image into *K* clusters and constitutes one of the simplest, quickest, yet representative unsupervised learning algorithms that solve classification problems. The main idea is to define *K* centroids, one for each cluster, having the limitation that the classes must be known a priori. Different locations of centers cause different results; thus, the algorithm suffers from initialization. The better choice is to position them as far from each other as possible. The next step is to take each point belonging to a given dataset and associate it with the nearest centroid, according to a well-defined criterion. Squared Euclidean distance, where each centroid is the mean of the points in that cluster, and city-block distance, where each

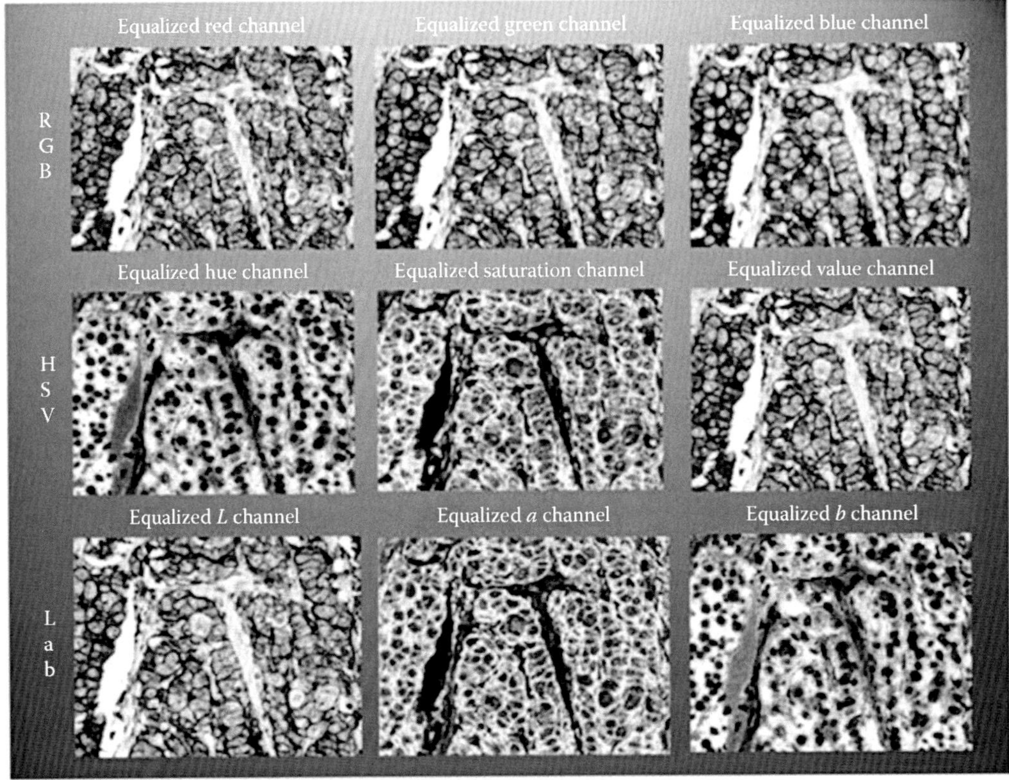

FIGURE 3.10
A typical IHC image represented in the different channels of the RGB, HSV, and Lab color spaces.

centroid is the component-wise median of the points in that cluster, are the two dominant distance metrics, defined in the following equations:

$$\text{Squared Euclidean distance } E_{ij} = \sum_{i=1}^{n} (X_i - C_j)^2$$

$$\text{city-block distance } E_{Cij} = \sum_{i=1}^{n} \left| X_i - C_j \right|$$

where:

n is the total number of points

X_i and C_j represent the candidate point i and the centroid of cluster j, respectively

When all points have been connected with a cluster center, the first step has been completed and a preliminary classification has been achieved. At this point, k new centroids as barycenters of the clusters resulting from the previous step need to be recalculated. After this update of the k centroids, a new binding has to be done between the points from the same dataset and the nearest new centroid, forming an iterative procedure. As a result of this loop, we may notice that the k centroids change their location step by step until no more changes take place. In other words, centroids do not move any more.

The basic algorithmic steps are summarized in the following scheme:

1. Pick K cluster centers, either randomly or based on some heuristic.
2. Assign each pixel in the image to the cluster that minimizes the distance between the pixel and the cluster center according to a selected distance function.
3. Recompute and update the cluster centers by averaging all of the pixels in the cluster.
4. Repeat steps 2 and 3 until convergence is attained (i.e., no change in pixel classification is observed).

In this case, the distance is the squared or absolute difference between a pixel and a cluster center. The difference is typically based on the pixel color, intensity, texture, and location, or a weighted combination of these factors. K can be selected manually, randomly, or by a heuristic. This algorithm is guaranteed to converge, but it may not return the optimal solution, as the quality of the result highly depends on the initial set of clusters and the value of K. The way to initialize the centroids is not specified and it frequently happens that suboptimal partitions are found. This poor initialization can be improved by multiple runs of the algorithm or the clustering of one or more samples first for *training* the algorithm. Another drawback is the handling of empty partitions and the production of outliers, especially in the cases of different size of the regions to be segmented. The prerequisite of the algorithm for known number of classes is particularly troublesome, since in most applications there is not such knowledge. The nature of the biomedical input image and the objective to segment cell nuclei imply the existence of three color classes within the image (yet the selection of $K = 3$): (1) the *blue* area of the cell nuclei, (2) the *brown* area of the membrane staining, and (3) the remaining tissue area. Selecting as a feature the color information derived from the S/V or a/b channels of the HSV and Lab color spaces, respectively, and the Euclidean distance as a similarity criterion, the algorithm can derive the segmented color regions of an image. A segmentation example applying the above k-means clustering scheme to an IHC image sample is illustrated in Figure 3.11.

The basic limitation of k-means clustering for a priori knowledge on the number of candidate classes and the shape of their distribution is overcome by the mean-shift clustering algorithm. Mean-shift is a robust technique that has been applied in many computer vision tasks, including image segmentation and visual tracking. It was proposed in the mid-1970s [18] but was not widely used until Cheng [19] and Comaniciu [20] applied the algorithm to computer vision and woke up the interest on it. In essence, mean-shift is an iterative mode detection algorithm in the density distribution space based on the moving to a kernel-weighted average of the observations within a smoothing window. This computation is repeated until convergence is obtained at a local density mode. The steps are as follows:

1. Estimate the density.
2. Find the modes of the density.
3. Associate each data point with one mode.

The main idea beyond mean-shift is to treat the points in the d-dimensional feature space as an empirical probability density function where dense regions in the feature space correspond to the local maxima or modes of the underlying distribution. For each data point in the examined sample, a gradient ascent procedure is performed on the local estimated density until convergence. The stationary points of this process represent the modes of the

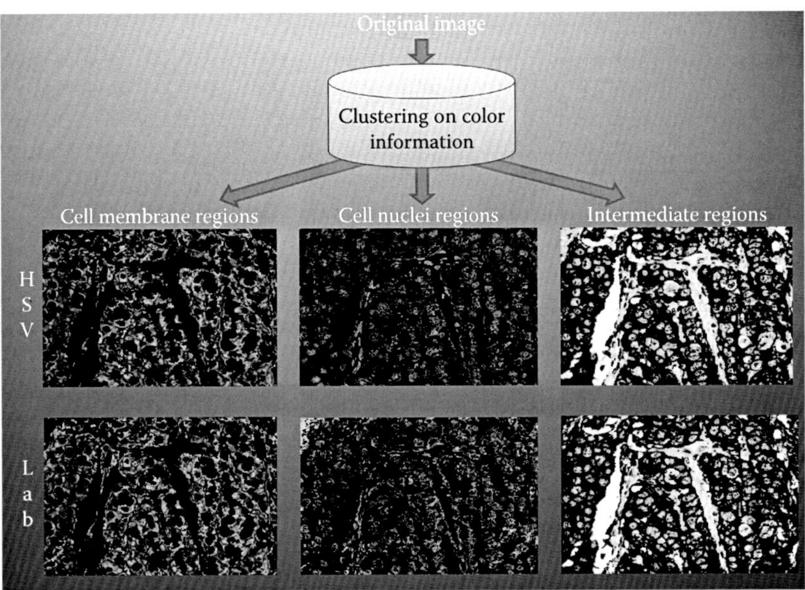

FIGURE 3.11
Example of IHC image segmentation applying the *k*-means clustering algorithm to the color information derived from the HSV and Lab color models.

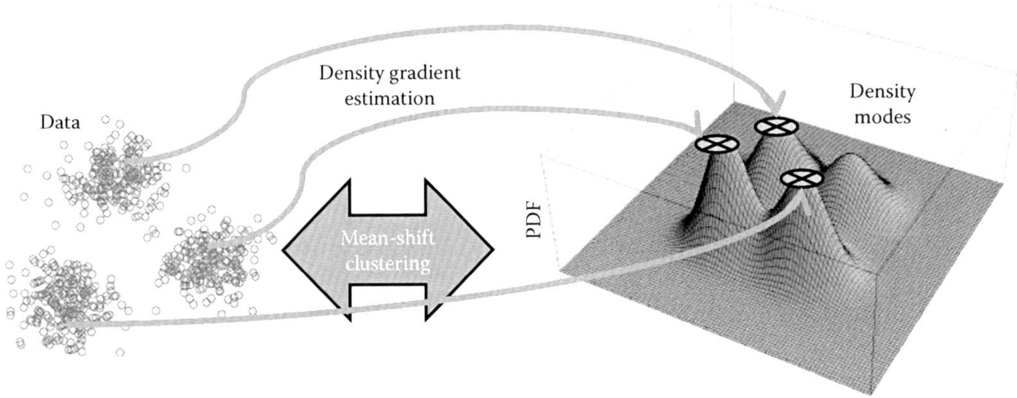

FIGURE 3.12
Schematic representation of mean-shift algorithm. PDF, probability density function.

distribution. Thus, the data points associated with the same stationary point are considered members of the same cluster. This procedure is schematically represented in Figure 3.12.

Each data point is associated with the nearby peak of the dataset's probability density function. Mean-shift defines a window around each point and computes its mean. Then it shifts the center of the window to the mean and repeats the algorithm till it converges. After each iteration, the window shifts to a denser region of the dataset.

Yet the algorithmic steps include the following:

1. Fix a window around each data point.
2. Compute the mean of data within the window.

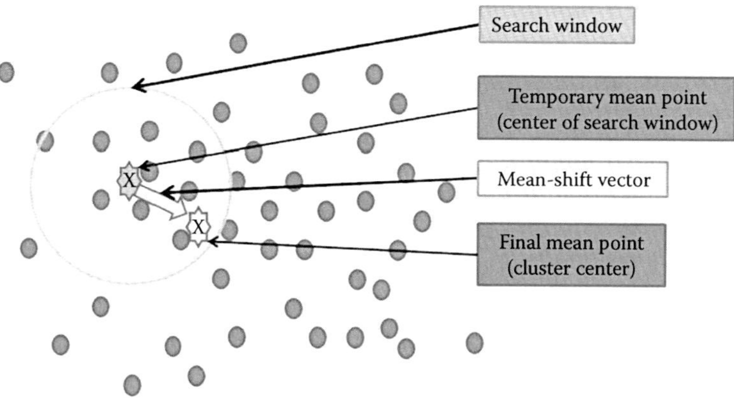

FIGURE 3.13
Schematic representation of cluster center calculation procedure.

3. Shift the window to the mean and repeat till convergence (no significant change in the mean point).
4. Repeat this procedure for all data points.

A snapshot of the clustering procedure is depicted in Figure 3.13. This algorithm assigns every data point to a cluster center (mean point) and automatically extracts the number of present classes within the input signal/image.

A summary of the basic mathematical background of the algorithm is reviewed in the following text. A kernel $\phi(x)$ is a function that satisfies the following requirements:

$$\int_{R^d} \phi(x)\mathrm{d}x = 1 \text{ and } \phi(x) \geq 0$$

Some examples of kernels include the following:

$$\text{Rectangular: } \phi(x) = \begin{cases} \alpha & a \leq x \leq b \\ 0 & \text{else} \end{cases}$$

$$\text{Normal: } \phi(x) = \alpha e^{-\frac{x^2}{2\sigma^2}}$$

$$\text{Epanechnikov: } \phi(x) = \begin{cases} \alpha(1-x^2) & |x| \leq 1 \\ 0 & \text{else} \end{cases}$$

Kernel density estimation is a nonparametric way to estimate the density function of a random variable. This is usually called as the Parzen window technique. Given a kernel K, n data points $\{x_1, x_2, \ldots, x_n\}$, and a bandwidth parameter h representing the window size, the Kernel density estimator for a given set of d-dimensional points is given as

$$\hat{f}(x) = \frac{1}{nh^d} \sum_{i=1}^{n} K\left(\frac{x - x_i}{h}\right)$$

Given a sample $S = \{s_i: s_i \in R_n\}$ and a kernel K, the sample mean using K at point x is calculated as

$$m(x) = \frac{\sum_i s_i K(s_i - x)}{\sum_i K(s_i - x)}$$

Now, *mean-shift* is given by $m(x) - x$. Let $x = m(x)$ and repeat the above procedure. This repeated movement of x is called the mean-shift algorithm. Mean-shift can be considered to be based on gradient ascent on the density contour. Applying gradient to the kernel density estimator, we obtain

$$\hat{f}(x) = \frac{1}{nh^d} \sum_{i=1}^{n} K\left(\frac{x - x_i}{h}\right) \Rightarrow \nabla \hat{f}(x) = \frac{1}{nh^d} \sum_{i=1}^{n} K'\left(\frac{x - x_i}{h}\right)$$

Using the kernel form

$$K(x - x_i) = C \cdot k\left(\left\|\frac{x - x_i}{h}\right\|^2\right)$$

where:
 C is a constant

the gradient of the density estimator becomes

$$\nabla \hat{f}(x) = \frac{2C}{nh^{d+2}} \sum_{i=1}^{n} (x - x_i) K'\left(\left\|\frac{x - x_i}{h}\right\|^2\right)$$

Setting $g(x) = -K'(x)$ as another kernel, we have

$$\begin{aligned}
\nabla \hat{f}(x) &= \frac{2C}{nh^{d+2}} \sum_{i=1}^{n} (x_i - x) g\left(\left\|\frac{x - x_i}{h}\right\|^2\right) \\
&= \frac{2C}{nh^{d+2}} \sum_{i=1}^{n} \left[x_i g\left(\left\|\frac{x - x_i}{h}\right\|^2\right) - x g\left(\left\|\frac{x - x_i}{h}\right\|^2\right) \right] \\
&= \frac{2C}{nh^{d+2}} \sum_{i=1}^{n} g\left(\left\|\frac{x - x_i}{h}\right\|^2\right) \left[\frac{\sum_{i=1}^{n} x_i g\left(\left\|\frac{x - x_i}{h}\right\|^2\right)}{\sum_{i=1}^{n} g\left(\left\|\frac{x - x_i}{h}\right\|^2\right)} - \frac{\sum_{i=1}^{n} x g\left(\left\|\frac{x - x_i}{h}\right\|^2\right)}{\sum_{i=1}^{n} g\left(\left\|\frac{x - x_i}{h}\right\|^2\right)} \right] \\
&= \frac{2C}{nh^{d+2}} \left[\sum_{i=1}^{n} g\left(\left\|\frac{x - x_i}{h}\right\|^2\right) \right] \underbrace{\left[\frac{\sum_{i=1}^{n} x_i g\left(\left\|\frac{x - x_i}{h}\right\|^2\right)}{\sum_{i=1}^{n} g\left(\left\|\frac{x - x_i}{h}\right\|^2\right)} - x \right]}_{\text{mean-shift vector}}
\end{aligned}$$

Finally, we obtain the formulas and definitions:

$$\text{KDE of } f \text{ with kernel } g\text{: } \hat{f}(x) = \frac{2C}{nh^{d+2}}\left[\sum_{i=1}^{n} g\left(\left\|\frac{x-x_i}{h}\right\|^2\right)\right] = \frac{2C_{\text{new}}}{nh^{d}}\left[\sum_{i=1}^{n} g\left(\left\|\frac{x-x_i}{h}\right\|^2\right)\right]$$

$$\text{Mean-shift vector: } m(x) = \frac{\sum_{i=1}^{n} x_i g\left(\left\|\frac{x-x_i}{h}\right\|^2\right)}{\sum_{i=1}^{n} g\left(\left\|\frac{x-x_i}{h}\right\|^2\right)} - x$$

$$\text{Gradient at } x\text{: } \nabla \hat{f}(x) = \frac{2C}{nh^{d+2}}\hat{f}(x)m(x) = \frac{2C_{\text{new}}}{nh^{d}}\hat{f}(x)m(x)$$

The mean-shift is proportional to the local gradient estimate, yet it can define a path leading to a stationary point of the estimated density, the mode of the distribution (the cluster centroid). It is also noticed that the mean-shift step is large for low-density regions corresponding to valleys and decreases as x approaches a mode (becomes zero if the point identifies with the mode).

It is remarkable that the only parameter of the algorithm is the bandwidth h. The number of classes is internally evaluated, as the mean-shift vector determines if the newly calculated center will be merged with an existing one or will be the center of a new cluster. The basic drawback of the algorithm is the low speed of convergence, but there exists a lot of research regarding implementation for speedups and improvements. In addition, a limitation of the standard mean-shift procedure is that the value of the bandwidth parameter is unspecified and application dependent [21]. As a novel scheme for improving classification accuracy, we propose the combination of clustering methods. Mean-shift clustering could be initially performed in order to calculate the number of key clusters within the image along with their centers. This information can be subsequently used for the efficient initialization of the k-means algorithm, guiding its searching mechanism to fast convergence, close to an *optimal* result. As an alternative, we can use in an opposite sequence the k-means clustering with many classes, in order to capture the intraclass details. Then, at the second stage, the mean-shift approach can be exploited in order to group together the initial, generally relaxed, clustering. An example of mean-shift clustering segmentation performed on the hue color band of an IHC image sample is illustrated in Figure 3.14.

3.3.8 Watershed Transform

Watershed transform constitutes one of the reference methodologies regarding image segmentation. The major idea beyond watershed transformation is based on the concept of topographic representation of image intensity, fused with other principal image segmentation methods including discontinuity detection, thresholding, and region processing. Because of these factors, watershed segmentation displays more effectiveness and stableness than other segmentation algorithms, producing closed contours and separating intersecting regions. Thus, in relation to the previous *clustering* approaches, the watershed transform aims directly at segmenting distinct regions with clear boundaries. Image clustering can recover compact areas based on the relatively similar intensity of internal pixels, leading to homogeneous regions. This scheme could be effectively used

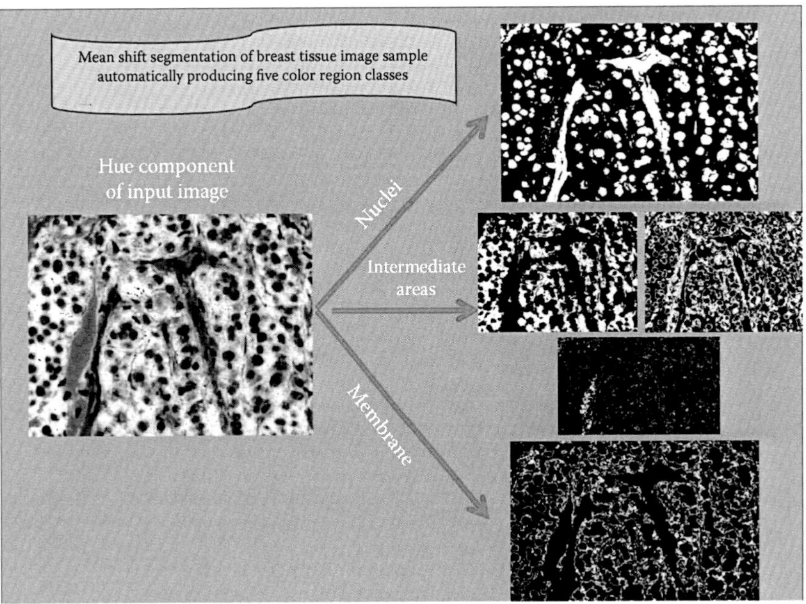

FIGURE 3.14
Example of IHC image segmentation applying the mean-shift clustering algorithm. *Left*: Equalized hue component of input image. *Right*: Output images applying mean-shift clustering to the hue component of sample image, five classes automatically calculated based on intensity differentiations. *Upper right*: Cluster representing the cell nuclei regions. *Lower right*: Cluster representing the cell membrane points. *Middle right*: Classes representing the intermediate regions within the breast tissue.

to detect the cell nuclei, which are all stained with the characteristic blue color originated from the hematoxylin–eosin coloring protocol. However, the thin and sometimes partially stained cell membrane boundaries can more efficiently represented by the lines of discontinuity defining the borders of cell regions. In this framework, it is appropriate to extract the watershed lines via the watershed transform, which is capable of connecting broken contours based on the intensity differentiation of the surrounding points from the internal area. Region-based methodologies are less efficient for this task, since neither strong nor complete membrane staining is observed in all cells forming the tissue segment.

When reporting on watershed transform, three fundamental notions arise: minima, catchment basins, and watershed lines. These definitions are illustrated in Figure 3.15. If we imagine that the bright areas of an image have *peaks* and the dark areas have *valleys*, then it might look like the topographic surface illustrated in Figure 3.15. We may observe three types of points: (1) minima, points belonging to the different minima; (2) catchment basins, points at which water would certainly fall in a single minimum; and (3) watershed lines, the highest intensity level points at which water would have equal probability to fall in more than one minimum. The goal of this segmentation scheme is to detect all of the watershed lines (the highest gray level).

The scenario underlying this method comes from geography: it is that of a landscape or topographic relief that is flooded by water, with watersheds being the separation lines of the domains of attraction of rain falling over the region. Alternatively, we may imagine the landscape being immersed in a lake, with holes pierced in local minima. Basins will form the water tanks filled up with water starting at these local minima, and at points where

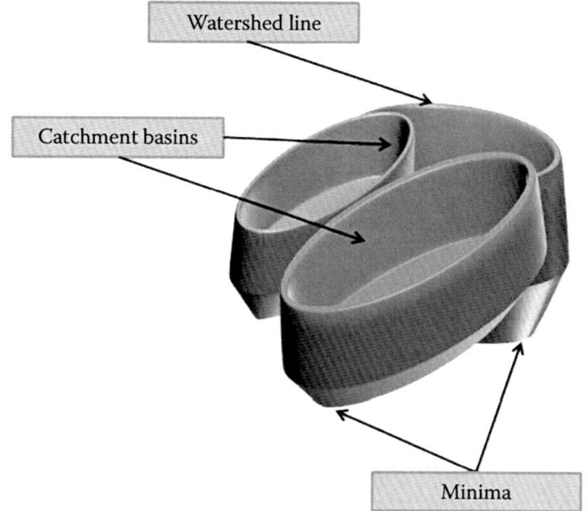

FIGURE 3.15
Fundamental definitions of the watershed transform.

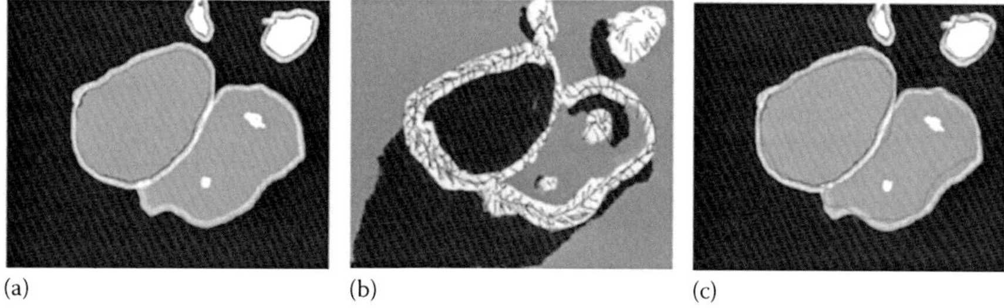

FIGURE 3.16
Initial grayscale image (a), its topographic surface (b), and the segmented watershed image (c).

water originated from different basins is crossed, dams are built. When the water level has reached the highest peak in the landscape and only the watershed lines are obvious, the process is stopped, yet the landscape has been partitioned into clearly separated water tank regions [22]. This procedure is illustrated in Figure 3.16.

In order to apply to the watershed transform for image segmentation, two approaches may be used. The first one starts from locating one basin at a time and then finds its watersheds by taking a set complement of the basin. The other performs a complete partition of the image into basins and subsequently extracts the watersheds by boundary detection. The segmentation procedure also includes an appropriate labeling of the resulting regions, implying that all points belonging to a given catchment basin have the same unique label, which is distinct from all the other assigned labels of the catchment basins. Sometimes, the watershed transform is not applied directly to the original image, but to its (morphological) gradient [23], producing watersheds at the points of gray value discontinuity (edges). The watershed algorithms can be generally divided into two classes: one based on alterations of a recursive algorithm by Vincent and Soille [22] and another based on distance functions by Meyer [23].

The Vincent and Soille algorithm includes two main steps: (1) sorting of pixels in increasing gray values, providing the capability of easily accessing pixels at a certain gray level, and (2) flooding the areas of similar gray level, successively increasing the levels starting from the image minima. The algorithm assigns a unique label to each minimum and its corresponding basin step. All points at a gray level h are initially labeled as *masked*. In the following stage, the nodes with labeled neighbors from the previous iteration are stored in a queue, forming the geodesic influence zones. Pixels that are adjacent to two or more different basins are marked as *watershed nodes*, whereas pixels that can only be reached from nodes carrying the same label are merged with the corresponding basin. Finally, the points that still remain masked form a set of new minima at level h, whose connected neighbors get a new label. In this way, the time complexity of the algorithm becomes linear, depending on the number of pixels of the input image, thus making it faster than other watershed methods. In addition, the accuracy of this algorithm is remarkable, while it can be adapted to any kind of grayscale picture and directly generalized to n-dimensional images or even to graphs. However, the Meyer method defines watershed in terms of a distance function called topographic distance. The main idea is to represent the image as a graph by linking each node with its lowest neighbors, which are pixels/areas in the same basin. This graph is successively pruned in order to keep only one lower neighbor, creating a forest, where each tree expands a distinct segment. By attaching the same label to all points originated from a common ancestor, the algorithm produces a mosaic image, where each tile represents a catchment basin of the segmented image and their boundaries form the watershed lines. Classical shortest path algorithms from the graph theory can be adopted to produce watershed algorithms, easily implementable in hardware and parallelizable among various processors. During the pixel integration procedure, each point gets its value from only one. Yet the entire procedure is translated to the problem of finding a shortest path between each pixel and a regional minimum in a weighted graph, which implies an increased algorithmic complexity.

Assume that the image f is an element of the space $C(D)$ of real twice continuously differentiable functions on a connected domain D with only isolated critical points. Then the topographical distance between points p and q in D is defined by

$$T_f(p,q) = \inf_{\gamma} \int_{\gamma} \left\| \nabla f\left(\gamma(s)\right) \right\| ds$$

where the infinum is over all paths (smooth curves) inside.

Let $f \in C(D)$ have k minima $\{m_{k \in I}\}$ (roots of its gradient producing positive second gradient) for some index set I. The catchment basin $CB(m_i)$ of a minimum m_i ($i = 1, 2, \ldots, k$) is defined as the set of points that are topographically closer to m_i than to any other regional minimum m_j:

$$CB(m_i) = \left\{ x \in D \middle| \forall j \in I \backslash \{i\} \middle| : f(m_i) + T_f(x, m_i) < f(m_j) + T_f(x, m_j) \right\}$$

The watershed of f is the set of points that do not belong to any catchment basin:

$$\text{Wshed}(f) = D \cap \left[\bigcup_{i \in I} CB(m_i) \right]$$

Therefore, the watershed transform of f assigns labels to the points of D, such that different catchment basins are uniquely labeled and a special label W is assigned to all points of the watershed of f. Thus, the Meyer's flooding algorithm is summarized as follows:

Label the Regional Minima with Different Colors

1. Select a pixel p, not colored, not watershed, adjacent to some colored pixels, and having the lowest possible gray level.
2. If p is adjacent to exactly one color, then label p with this color.
3. If p is adjacent to more than one color, then label p as watershed.
4. Repeat steps 1–3 until all pixels have been processed.

There exist a lot of open issues concerning the watershed transform. A detailed and thorough analysis and review of the existing watershed transform approaches can be found in Reference 19. First, there is the question of accuracy of watershed lines. In general, the result should be a closed contour, so the distance metrics adopted in the watershed calculation should approximate the Euclidean distance. The main drawback of the watershed method in its original form is that it produces a severe oversegmentation of the image; many small regions are extracted due to calculation of numerous local minima in the input image. The most appropriate solution is the use of markers, key subregions, or single points within each basin that significantly differ from the remaining basin region (e.g., the brightest points derived by the h-maxima transform), resulting in the so-called marked watershed transform [24]. Each initial marker has a one-to-one relationship to a specific watershed region; thus, the number of markers will equal the final number of watershed regions. The markers can be manually or automatically selected, saving computational time and improving segmentation accuracy.

An example of the application of the watershed transform is illustrated in Figure 3.17. Detailed analysis of the implementation steps will be given in Section 3.3.9. In the first application of the transform (upper image), the input image was directly the blue color channel of the original image. The selection of the specific channel was driven by the fact that the cell nuclei are stained with a dark blue color via the laboratorial procedure of IHC, yet their corresponding regions will originate from the high-intensity regions of the blue component of the RGB model. We can notice that the result is not accurate as oversegmentation has taken place. Extracting markers by applying the h-domes and using as an input the difference of markers from the blue channel of the original image, the segmentation accuracy can be significantly increased.

3.3.9 Active Contours Model

While previous techniques operate on the region similarity concept in order to perform image segmentation, there exists a complementary category of algorithms based on the dissimilarity concept at the boundary regions. Contour-based techniques are well established in segmentation bibliography, providing accurate and robust results even in a noisy environment, but having the drawbacks of initialization, local minima, and stopping criteria problems. The principle of such techniques lies on the linking of edge points extracted via an edge detection scheme, by exploiting curvilinear continuity. While some methods perform only edge linking, advanced contour-based models set specific optimization criteria in order to iteratively approximate the object borders starting from an initialized closed curve [25]. Global minimum energy searching methods have been proved generally effective in overcoming local minima problems due to the presence of artifacts introduced within the image, leading to robust convergence regarding the final contour extraction.

Chan and Vese [26] proposed a powerful and flexible methodology for active contour object detection combining curve evolution techniques, level sets, and the Mumford–Shah

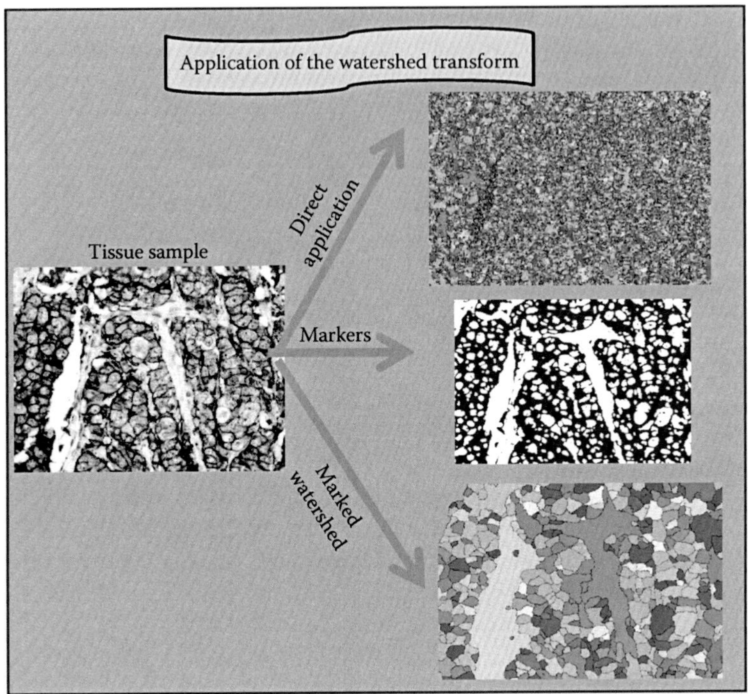

FIGURE 3.17
Application of watershed transform and the marked watershed transform to an IHC image. *Left*: Grayscale version of the equalized color image in the RGB color space, serving as the input of the transform. *Upper right*: Oversegmentation result after application of the watershed transform directly to the input image. *Middle right*: Markers representing the light regions of the input image extracted by the *h*-domes transform. *Lower right*: Improved segmentation result after the application of the watershed transform to the subtraction of marker image from the grayscale input image.

functional, accomplishing to detect corners and any topological change. The model begins with a contour in the image plane defining an initial segmentation, and then this contour gradually evolves according to a level set method until it meets the boundaries of the foreground region. According to the model, a curve C is represented via a function ϕ (the level set function) as $C = \{(x, y) | \phi(x, y) = 0\}$, where (x, y) are coordinates in the image plane, while the evolution of the curve is given by the zero-level curve at time t of function $\phi(x, y, t)$. Negative values of ϕ denote points outside the curve, whereas positive values of ϕ originate from points belonging to the internal area of the curve, as depicted in Figure 3.18.

At any given time, the level set function simultaneously defines an edge contour ($\phi = 0$) and a segment of the image ($\phi \neq 0$) and is being evolved according to the partial differential equation iteratively converging to a meaningful segmentation of the image:

$$\frac{\partial \phi}{\partial t} = |\nabla \phi| F, \quad \phi(x,y,0) = \phi_0(x,y)$$

where:

F denotes the speed of the curve evolution

$\phi_0(x, y)$ defines the initial contour the algorithm started to be generated from

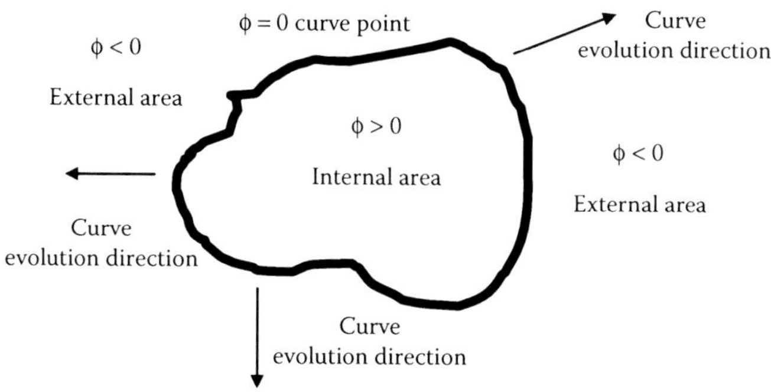

FIGURE 3.18
Description of an image region via a level set function $\phi(x)$.

A particular case is the motion by mean curvature, when F becomes the curvature of the level curve of ϕ passing through (x, y) according to the formula:

$$F = \text{div}\left[\frac{\nabla\phi(x,y)}{|\nabla\phi(x,y)|}\right]$$

The speed of the evolving curve becomes zero on the points with highest gradients; therefore, the curve stops on the desired boundary, which appears to be strong gradients.

Chan and Vese updated the classic snake model described above, introducing the energy functional term. Assuming that the image I consists of two regions of approximately constant distinct intensities I^1 and I^2 and that the object of interest is represented by the region of value I^1 (inside the curve C), the *fitting energy functional* is denoted according to equation:

$$\begin{aligned} F(c_1,c_2,C) &= F_1(c_1,c_2,C) + F_2(c_1,c_2,C) \\ &= \lambda_1 \times \int_{\text{insideC}} |I(x,y) - c_1|^2 \, \mathrm{d}x\mathrm{d}y + \lambda_2 \times \int_{\text{outsideC}} |I(x,y) - c_2|^2 \, \mathrm{d}x\mathrm{d}y \\ &\quad + \mu \times \text{Length}(C) + v \times \text{Area}(\text{inside } C) \end{aligned}$$

where:

C is any variable curve except for the object boundary C_0
constants c_1 and c_2 are the averages of I inside and outside C, respectively
μ, ν, λ_1, λ_2 are nonnegative, fixed, user-defined parameters

If $F_1(C) > 0$ and $F_2(C) \approx 0$, the curve is outside the object; if $F_1(C) \approx 0$ and $F_2(C) > 0$, the curve C is inside the object; and if $F_1(C) > 0$ and $F_2(C) > 0$, then the curve is both inside and outside the object, intersecting with its boundary. The contour of the foreground region is the solution of the minimization problem $\inf_C(F(c_1, c_2, C)) \approx 0 \approx F(c_1, c_2, C_0)$. Using the Heaviside function H and the one-dimensional (1D) Dirac measure δ_0, defined, respectively, by

$$H(z) = \begin{cases} 1, & \text{if } z \geq 0 \\ 0, & \text{if } z < 0 \end{cases}, \quad \delta_0(z) = \frac{\mathrm{d}}{\mathrm{d}z} H(z)$$

we can calculate component length, area, and the means c_1, c_2 via minimization of functional $F()$ by the following formulas:

$$\text{Length}(\phi=0)=\int_c \left|\nabla H(\phi(x,y))\right| dxdy=\int_c \delta_0(\phi(x,y))\left|\nabla(\phi(x,y))\right| dxdy$$

$$\text{Area}(\text{inside}\,\phi\geq 0)=\int_c H(\phi(x,y))dxdy$$

$$c_1(\phi)=\frac{\int_C I(x,y)H(\phi(x,y))dxdy}{\int_c H(\phi(x,y))dxdy},\quad \text{minimizing } F() \text{ with respect to } c_1$$

$$c_2(\phi)=\frac{\int_C I(x,y)\left[1-H(\phi(x,y))\right]dxdy}{\int_c [1-H(\phi(x,y))]dxdy},\quad \text{minimizing } F() \text{ with respect to } c_2$$

Keeping c_1 and c_2 constant, the goal is to minimize the energy functional with respect to the level set function $\phi(x)$ and extract the image contour. This is achieved by obtaining the Euler–Lagrange equation for $\phi(x)$ according to the conditions:

$$\begin{cases} \dfrac{\partial\phi}{\partial t}=\delta(\phi)\left[\mu\,\text{div}\left(\dfrac{\nabla\phi}{|\nabla\phi|}\right)-\nu-\lambda_1(\text{I}-c_1)^2+\lambda_2(\text{I}-c_2)^2\right] \text{ in contour } C \\ \dfrac{\delta(\phi)}{|\nabla\phi|}\times\dfrac{\partial\phi}{\partial n}=0 \quad \text{on } \partial C \end{cases}$$

A detailed description of the complete mathematical background can be found in Reference 27. Chan and Vese proposed the fixed values $\lambda_1 = \lambda_2 = 1$ and $\nu = 0$. Parameter μ defines the update weight factor for curve evolution owing to the perimeter size of the already evaluated regions, whereas λ_1, λ_2 are the update weight factors owing to the variations of the image in the external and internal areas, respectively, of the curve.

In order to reveal the efficiency of the Chan and Vese active contour scheme, an illustration of the application of our contour-based segmentation technique to an IHC image sample is provided in Figure 3.19. The algorithm is fed in the input with (1) the image sample in the Lab color space (Figure 3.19a), which provides effective differentiation between the cell membrane and cell nuclei color, and (2) the cell area contours derived from the watershed transform, which will act as an upper bound for active contour expansion. The initial contour $\phi_0(x,y)$ (Figure 3.19b) is extracted via mean-shift clustering, which produced an approximation of the areas of cell nuclei. The cell boundary curves estimated after the convergence of the algorithm and the segmentation result after 969 iterations are depicted in the second row of Figure 3.19 in its left and right parts, respectively.

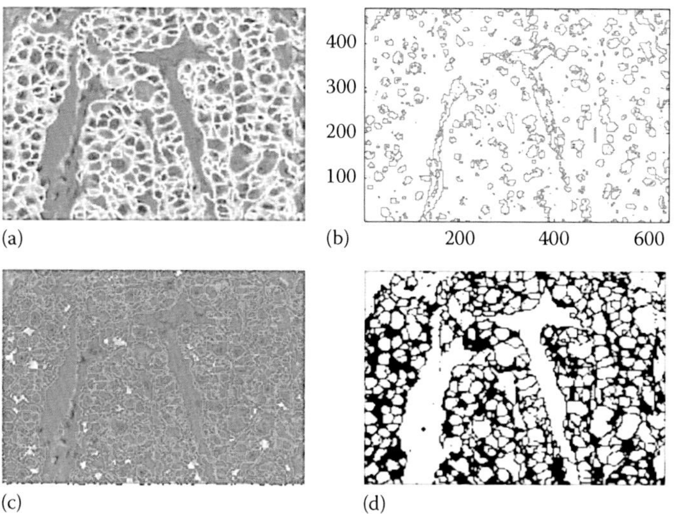

FIGURE 3.19
Application of an active contour model to an IHC image. (a) Input image derived from the pseudocolor Lab image superimposed with the watershed lines for controlling the convergence of the algorithm. (b) Initial points derived from mean-shift clustering for defining the starting points of the snake model. (c) Final contours extracted by the active contour algorithm superimposed on the pseudocolor Lab image via convergence after 969 iterations. (d) Binary image containing the segmented regions of the tissue sample.

Despite its advantages, some of the reasons due to which digital WSI is not being routinely used for diagnostic purposes are as follows:

- Sample size restrictions
- Precision of measurement between different people and between different instruments
- Accuracy of measurement—probability of error
- Measuring bias
- Calibration accuracy
- Experience of the pathologist

3.4 Genome Imaging

Genetic association is a test of a relationship between a particular phenotype and a specific allele of a gene. This approach usually begins with selecting a biological aspect of a particular condition or disease, then identifying variants in genes thought to impact on the candidate biological process, and next searching for evidence that the frequency of a particular variant (allele) is increased in populations having the disease or condition. Imaging genomics is a form of genetic association analysis, where the phenotype is not a disease phenotype but a physiological response of the brain during specific information processing. The protocol for imaging genomics involves first identifying a meaningful variation in the DNA sequence within a candidate gene. For the variant to be meaningful, it should have an impact at the molecular and the cellular level on gene or protein function (i.e., be a functional variation) and the distribution of such effects at the level of brain systems involved in specific forms of information processing should be predictable.

The quantification and visualization of specific biological processes is now possible noninvasively because of our increasing knowledge of the role that specific genes and proteins play in human health and how any specific alterations could lead to specific diseases. This knowledge is combined with the ability and technology to target these specific proteins and genes in order to better assess them, to provide detailed metabolic and molecular information. All the standard major imaging modalities such as computed tomography (CT), PET, MRI, and single-photon emission CT (SPECT) are contributing to this new area of molecular and genome imaging, and each of them has their own unique working and mechanism for producing varying levels of contrast and trade-offs in resolution (spatial and temporal) and sensitivity, for the different biological processes and mechanisms being studied. A majority of the advancements in molecular imaging are currently being carried out in animal models of various irregularities and diseases, but as the field progresses and better molecular targeting techniques are developed, translation to clinical studies for humans is inevitable and can change the future of diagnostic imaging (Figure 3.20).

The aim of performing genome or molecular imaging is to image the presence, the location, and the expression levels of certain specific genes and proteins that are said to play a vital role in the metabolism and morphology related to diseases. Specifically directed contrast agents are normally used to identify and separate the gene or protein of interest. Fluorophores and radioactive substances that bind to certain specific proteins are used as contrast agents for the purpose of labeling have been used for many years. These methods have also been implemented *in vivo*, and since the development of nuclear imaging

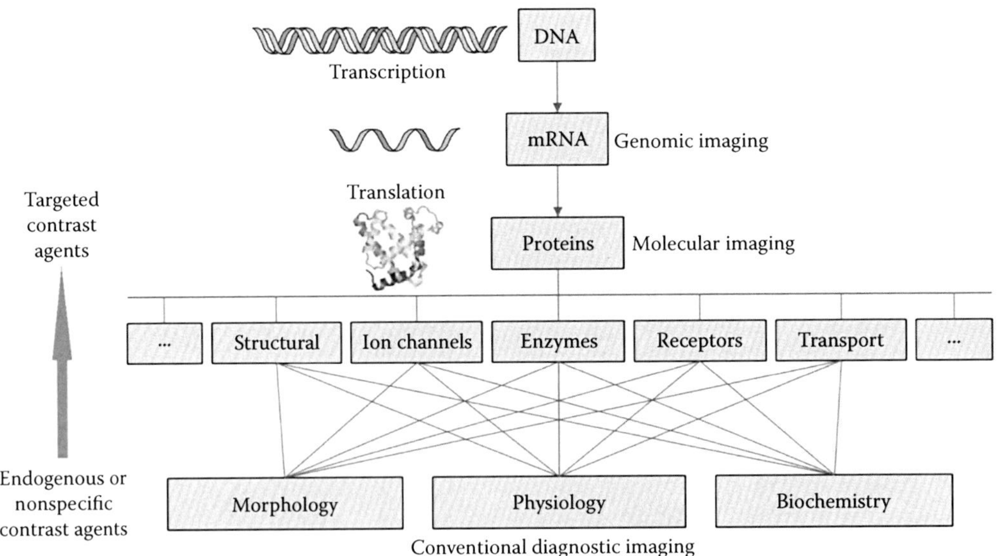

FIGURE 3.20
Schematic representation of molecular and genomic imaging. (Redrawn from Cherry, S. R., *Phys. Med. Biol.*, 49, R13, 2004.)

technology such as PET and SPECT imaging, they are being used *in vivo* for humans. When these techniques were first developed and implemented, they were used to study and map the localized neuroreceptors in the brain spatially for functional imaging.

In recent years, there has been a tremendous development in the information known concerning the precise proteins and genes, which are involved in the mechanisms and occurrence of certain diseases and also in the area of human genome sequencing. Also, since the advent of mass screening techniques such as liquid chromatography [fast protein liquid chromatography (FPLC), in particular] and other methods, it has become easy to target these proteins of interest.

A wide range of new and innovative techniques have been developed in the imaging field to generate contrast specific to genes and proteins that need to be studied. Functional neuroimaging has the capacity to view and assess the way information is processed in the brain and the metabolic changes that occur with brain function. Hence, it proves to be an extremely useful tool to characterize and study the functional genomics of the brain. The aim of genomic or molecular imaging is to understand the exact location of the gene and protein expression in the body and relate it to cognitive behavior or disease mechanism. Neuroimaging is used to scan patients with neurodevelopmental disorders and other types of diseases and to identify the location in the brain where the abnormalities are present and how and which parts are affected by it, as well as the behavior of neural pathways during routine cognitive tasks that are affected by the disease.

In cases where a complex disease is caused by mutation in a single gene or variations in a gene's associated protein (like in the case of Huntington's disease), the target gene/protein can be easily identified, but it is not necessary that diseases are caused by a single gene's mutation.

Most of the human diseases involve alterations in multiple genes and proteins and also involve environmental factors contributing to them. For example, if we talk about cancer, high-throughput screening techniques have revealed that several genes and proteins are altered. But these gene and protein alterations vary in different kinds of cancer, such that some of them can be overexpressed, whereas some of them can be underexpressed. For every different kind of cancer or other disease, changes in the gene sequence or proteins are varied. This shows us that genes and proteins play an important role in disease mechanisms.

The two main set of genes involved in cancer are *oncogenes* that regulate cell growth and *tumor suppressor* genes that are responsible for suppressing the growth of damaged cells and do not let them survive. Any mutations or changes in these genes cause the initiation and growth of different kinds of cancer when the oncogene is expressed differently than in normal cases. When there are mutations present in the expression of the tumor suppressor genes, the survival and growth of cells with a faulty DNA are altered. These genes are also involved in and responsible for the growth of normal cells; hence, the occurrence and growth of malignant cells are inextricably related to normal cell growth and division. For functional and metabolic imaging of the heterogeneity of tumors, techniques such as dynamic contrast-enhanced MRI (DCE-MRI) and Fluorodeoxyglucose PET (FDG-PET) are used for evaluating tumor response and for the diagnosis and staging of cancer. Another field that has been recently developed is *radiogenomics* in which radiographic images are related to gene expression patterns using bioinformatics. This method is used in cancer *radiomics* to understand the underlying gene expression patterns from the imaging features of the tumor. An example of this is that some features that can be viewed and studied using functional imaging such as vascular endothelial growth factor and tumor density are altered by variations in the expression of certain genes [28]. This work has been taken to the next level by predicting global gene patterns in some gliomas, glioblastoma multiforme, and hepatocellular carcinoma by studying and comparing the imaging features extracted from MRI and CT scans (Figure 3.21).

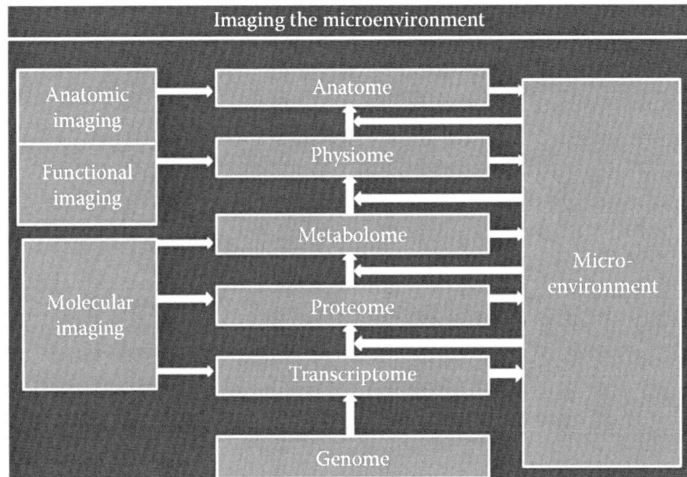

FIGURE 3.21
Functional, molecular, and anatomic imaging of cancer processes. (Data from Gillies, R. J. et al., *Clin. Radiol.*, 65, 517–521, 2010.)

As there can be a large number of genes that could possibly influence the brain, affect the behavior, and initiate diseases, candidate gene selection becomes an extremely important task. Many nongenetic factors such as age, gender, and environmental factors also have an effect along with genetic variations. Due to the advent of high-throughput screening techniques, the entire genome can be studied for polymorphisms or variations, using genome-wide association studies.

The most important step in utilizing molecular and genomic imaging is to define and identify the gene or protein (called the *target*) that is to be imaged. Thus, we need to have knowledge about the basic molecular mechanisms of the disease being studied. High-throughput screening methods are applied to identify and locate genes or proteins that have abnormal (high or low) levels of expression in specific diseases. In functional *genomics* [29], a range of technologies including DNA microarrays, laser capture microdissection, and quantitative polymerase chain reaction are being used to find genes that have a direct influence in the beginning or spread of a disease. In the study of *proteomics* [29] techniques such as 2D gel electrophoresis, FPLC, and mass spectrometry are being used to identify and separate proteins. The function and interaction between proteins can be explored using many techniques including computational methods, and the structure of these proteins also provides us information regarding the function of proteins and protein–protein interactions. These structures can be studied using methods such as electron microscopy and X-ray crystallography. New advancements in high-throughput protein screening techniques are based on microfabrication and microfluidic technologies.

3.4.1 Imaging Mass Spectrometry

Mass spectrometry is the technique used to analyze a sample of a certain material by producing and studying the spectra of the mass of the constituent atoms and molecules. Along with other functions, this method is used to study the chemical structure of certain molecules of substances of interest. Mass spectrometry is carried out by first ionizing the compounds to get charged molecules whose ratio of mass to charge will be studied. The molecules comprising the ionized sample are separated based on their mass-to-charge ratio. *Imaging mass spectroscopy* (IMS), which is also known as *mass spectroscopy imaging*, is a mass spectroscopy method that is used to image and visualize the spatial distribution of compounds such as proteins of interest, metabolites, and biomarkers. This is an important technique in the field of proteomics and metabolomics and is used to separate the proteins or metabolites of interest and visualize their spatial molecular distribution and structure simultaneously across the surface. This technique does not need chemical labeling or marking, which is the prime advantage (Figure 3.22).

There are three emerging techniques that are the frontier of IMS: secondary ion mass spectroscopy (SIMS), matrix-assisted laser desorption ionization (MALDI), and desorption electroscopy ionization (DESI). SIMS is used to analyze solid surfaces by *hitting* them with a highly focused beam of particles with high energy (called primary ion beam), which leads to the emission of secondary ions from the surface, known as the secondary ion beam that is analyzed using a time-of-flight (TOF) mass analyzer. This technique has been taken to the next level by the matrix-enhanced SIMS (ME-SIMS), which further increases the production of the secondary ions.

The other most widely used IMS technique is MALDI. When this method was developed, it was used to image and study biological proteins and peptides. Unlike SIMS, which uses a focused ion beam, MALDI utilizes a high-powered UV or laser light irradiation source in order to generate the secondary ions. The *matrix* used in this technique is basically a light-absorbing organic material, which usually has a very low molecular weight [30]. This matrix substance is used to coat the sample surface to be irradiated with the laser or UV light in order to improve the production of secondary ions. This coating greatly enhances the ionization process, and hence also reduces the laser power needed.

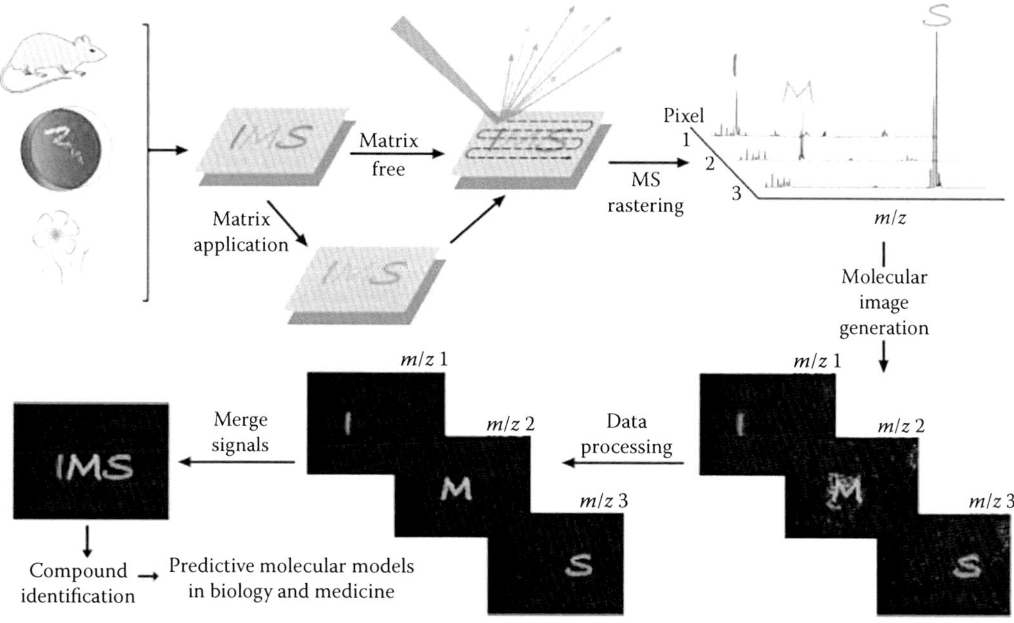

FIGURE 3.22
Overview of the general IMS workflow. (Reproduced from Watrous, J. D. et al., *J. Mass Spectrom.*, 46, 209–222, 2011. With permission.)

The last and the newest IMS technique is DESI mass spectroscopy. As the name suggests, this technique puts to use a steam or spray of charged particles as droplets in a solvent, which hit the sample surface that *wets* the surface. After this initial contact of the surface with the analyte, in all the following subsequent collisions, the particles get *desorp* secondary droplets that are analyzed, just like the secondary ions produced in the other techniques. This analysis is done using a mass spectrometer. The imaging using DESI has a lower spatial resolution and lesser sensitivity than MALDI and SIMS; it has certain advantages over them as it allows reactive imaging in which the compound of interest reacts with a chemical that can be sprayed, in order to study the interactions (Figure 3.23).

The general overview of the IMS process is explained in Figure 3.22. Samples comprising tissue sections are attached on a target surface and, if using MALDI, immediately followed by matrix application. The mass spectrometer is then made to span across the entire surface of the sample while collecting spectra at specified *sampling locations*, which are compiled into a single dataset where "the occurrence of any single mass can be visualized as a scaled false color overlay depicting the relative intensity of the ion across all sampled locations" [31]. This data can be further processed to improve its quality.

Mass spectrometry is also used for the molecular imaging of tumors to identify and locate proteins that are abnormally expressed (high or low) in tumors. In Reference 32, this technique was used for the imaging of proteins and peptides present in brain tumors, specifically glioblastomas. The mass spectra from different regions of the glioblastoma demonstrate clearly different protein expression levels, indicating that variations in protein expression levels are related to disease mechanisms, and in this case, brain tumor. In the tumor area with proliferation, some of the protein expression levels were at their highest, whereas certain proteins were localized to ischemic areas. These expression levels act as tumor markers and can be seen in specific corresponding regions of the tumor. This suggests that this technique can be used for the imaging of tumor margins where the contrast is provided by the varying levels of protein expression. This field of proteomics is also called metabolomic imaging (Figure 3.24).

One of the major strengths of IMS is that it measures and compares molecules based directly on the anatomical structural features at the cellular level. Using MALDI IMS, *molecular signatures* of diseases can be extracted [32], which comprise the expression levels of a set of proteins (5–25 kinds of proteins each). Hence, IMS and specifically MALDI have several oncology applications as well, focusing on the imaging of peptides and proteins. Protein analysis studies for diagnostic purposes (such as a comparison of the tumor region to healthy) are performed using MALDI, by extracting protein signature of different types and stages of cancer. The differences in proteins levels could also be seen in the surrounding healthy tissue

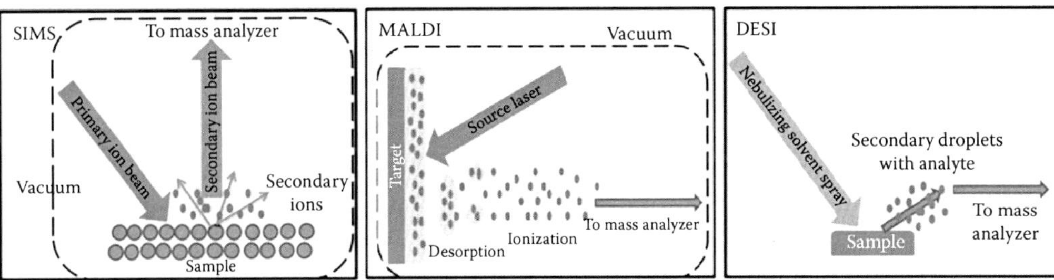

FIGURE 3.23
Ionization techniques for SIMS, MALDI, and DESI.

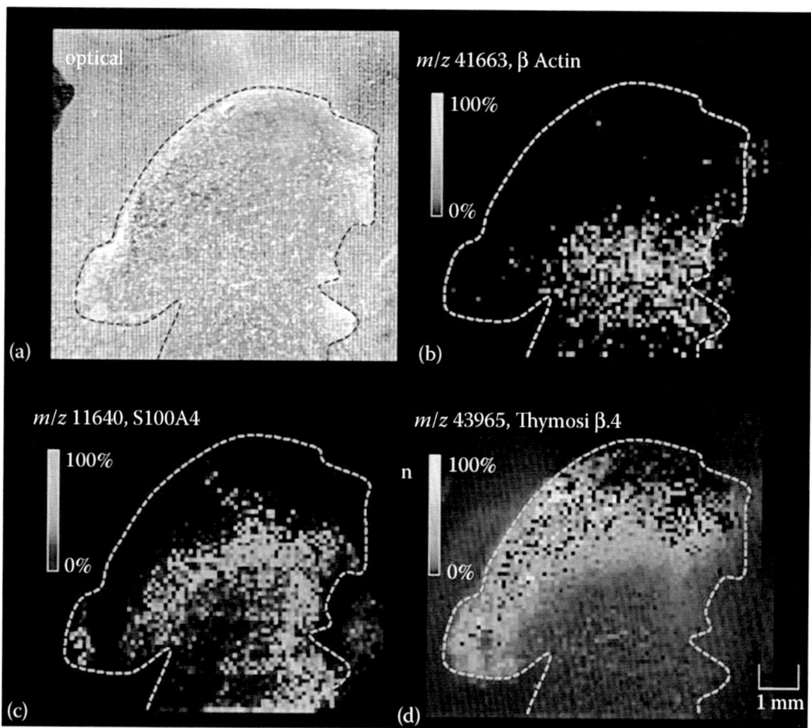

FIGURE 3.24
(a) Human glioblastoma slice mounted on a surface; (b–d) IMS images of varying protein concentration in the different regions of the tumor (higher in proliferation). (Data from Stoeckli, M. et al., *Nat. Med.*, 7, 493–496, 2001.)

near the tumor. Thus, this technique is a strong candidate for molecular assessment of tumor margins and the behavior of the area surrounding the tumor (Figure 3.25).

In Figure 3.26, a comparison between IHC-stained imaging and MALDI IMS is shown. The IHC slides on top contain lung cancer biopsy samples from two patients and the images shown are IMS results from peptide imaging.

Apart from relating to disease mechanism, another important goal of MALDI and other IMS techniques is to identify the relationship between a course of treatment (therapeutic regime) and the corresponding patient response. In such studies, environmental and other factors not related to protein expression should also be considered.

3.4.2 Liquid Chromatography—Protein Identification and Separation

Chromatography refers to the technique of separating different contents of mixtures. A fluid called the *mobile phase* carries the mixture through a structure that contains another material known as the *stationary phase*, and the different kinds of substances/contents of the mixture move at varying speeds and hence get separated. In liquid chromatography, the mobile phase is a liquid. If the stationary bed in a chromatography system is in a column or a tube, it is known as column chromatography. In certain situations, a single column is not enough to separate a few of the analytes. The part that is not separated is then transferred to a second column that has different physical and chemical properties helpful in separating the remaining contents of the mixture. This is known as 2D chromatography as the particle retention mechanism is different from the standard 1D separation [33,34].

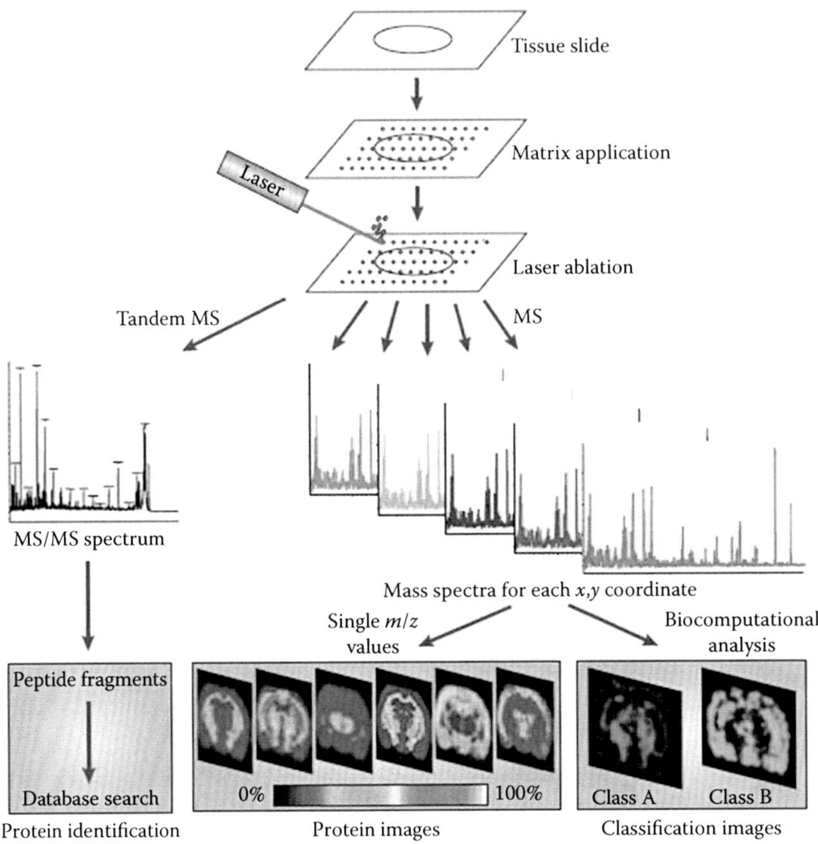

FIGURE 3.25
Workflow of MALDI. (Data from Stoeckli, M. et al., *Nat. Med.*, 7, 493–496, 2001.)

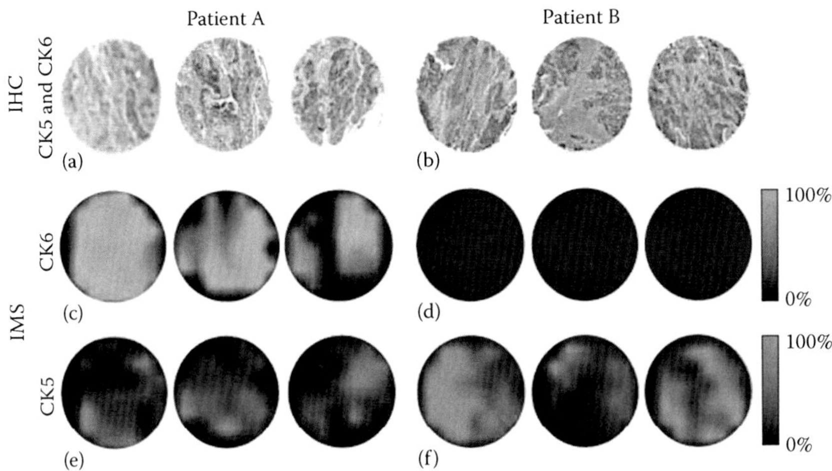

FIGURE 3.26
Comparison of IHC and MALDI IMS results. (Data from Stoeckli, M. et al., *Nat. Med.*, 7, 493–496, 2001.)

One of the recently developed techniques for the separation, analysis, and purification of proteins is FPLC. The basic concept is the same, that is, the different components of a mixture possess varying affinities for two materials (the mobile phase fluid and the stationary phase solid). In FPLC, the liquid in the mobile phase is a *buffer* solution whose flow rate is monitored at a constant value. Fluids from external storage tanks are drawn in to vary the composition of the buffer solution. The stationary phase consists of a resin with cross-linked beads (agarose) that are packed to form a column. Most commonly, these resins are *ion-exchange resins* chosen so that specific proteins bind to them due to the interaction of the ionic charges in the running buffer and then separate and return to the other buffer. When a certain mixture is composed of the proteins to be studied, it is dissolved in the running buffer solution (buffer 1) and is sent into the column. The specific proteins which we are interested in bind to the resin, and all the other components are taken out in the buffer and the proportion of the second buffer is increased steadily from 0 to maximum (100%) while maintaining the flow rate of the buffer constant. It is during this process that the proteins that are bound to the resin separate themselves from it and flow into the second buffer solution, increasing its proportion. The salt concentration of this second buffer is measured along with its concentration of the proteins. As every protein appears in the second buffer at a *peak* concentration, it is gathered and analyzed further. These proteins are further purified using chromatographic methods [34].

For biological studies, it is necessary to deliver the proteins in a pure state and in a specific quantity, and different kinds of analyses and studies require different amount of purity. Standard analysis techniques such as ELISA require that only the bulk impurities be removed, whereas other techniques such as X-ray crystallography need the proteins in pure state for structure analysis. For the purposes of protein and gene expression studies in order to relate them to disease mechanism, the proteins must be absolutely pure. After the ion-exchange chromatography performed to capture the target proteins, the purification takes multiple steps. The purification process can be used to separate the protein from the other contents of the mixture as well as to separate different kinds of proteins from each other and also to remove impurities. The separation is usually done on the basis of binding affinity, physical properties, chemical properties, and size of the protein. After the extraction step, precipitation is done along with differential solubilization followed by ultracentrifugation. In ultracentrifugation, particles are separated from each other on the basis of their mass and densities by rotating a vessel containing the mixtures at high speeds [35].

3.4.3 Functional and Metabolic Molecular Imaging

Due to all these high-throughput screening techniques, it is now becoming easy to develop therapeutic target and targeting contrast agents in order to improve functional and metabolic imaging of these gene/protein expression levels and their association with disease mechanisms. Traditionally, some fluorescent or radioactive substances are used to generate and enhance contrast. Recent developments include the use of nanoparticles for enhancing contrast in functional imaging [36]. In the recent years, many novel nanoscale systems have been developed to be used as diagnostic/therapeutic agents. Cells in the precancer stage, disease markers, fragments of viruses, and so on are some of the parameters at the molecular level, which can be detected by using nanoparticles and not by conventional diagnostic tools. The specificity and sensitivity of MRI is increased using nanoparticle-based contrast agents. In therapeutics too, nanoparticles offer controlled and targeted administration of drugs [36].

To be able to image gene expression levels, we need to develop such contrast agents that bind to the mRNA molecules, which are produced when a gene is transcribed. This technique is also known as in situ *hybridization* in which specific RNA segments that are complementary to the base sequence of mRNA of the gene to be studied are tagged with fluorescence or radioactive substances (Figure 3.27).

This technique works very well *in vitro* but was mostly unsuccessful *in vivo* due to the instability of the RNA oligonucleotides in the plasma and also due to the problems in the delivery of the contrast agent into cells. Another technique is developed, which is comparatively more successful *in vivo* and which uses reporter gene methods. Genes whose protein products (such as enzymes or even receptors) can be imaged are called reporter genes. By genetic modification, this reporter gene is switched to be under the regulation of the same promoter that controls the expression of the gene of interest. Hence, cells, tissues, or regions expressing the genes of interest also express the reporter genes. By monitoring the expression of the reporter genes, the expression of the genes of interest can also be studied (Figure 3.28).

The work of Batmanghelich et al. [37] presents a new technique that combines the two steps of selecting image features related to the disease phenotype and identifying a set of genes and proteins that are separated and analyzed to explain the previously selected image features corresponding to the disease. These steps are performed simultaneously, and then finally we assign probabilistic levels of the relevance to both the imaging and genetic factors. One of the shortcomings in genetic imaging, though, is that the sample size is too small (less subject) and there are numerous image features and genetic data. To solve this problem, only some imaging features (voxels, biomarkers) are considered. Then, a standard statistical test is used to test a set of genetic variations (such as gene expression) and a corresponding imaging biomarker. Recently, genome-wide analysis has

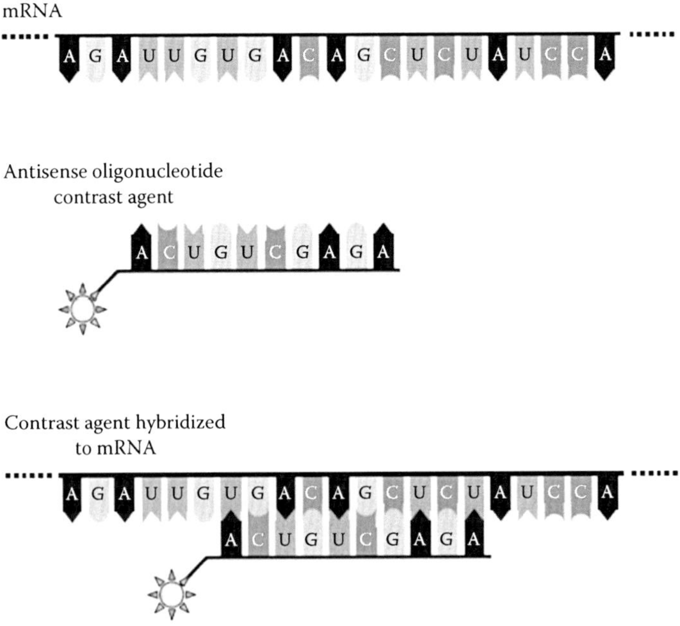

FIGURE 3.27
Top: Gene expression and transcription into mRNA. *Middle*: Complementary contrast agent. *Bottom*: Contrast agent binding (hybridizing) with the mRNA. (Data from Cherry, S. R., *Phys. Med. Biol.*, 49, R13, 2004.)

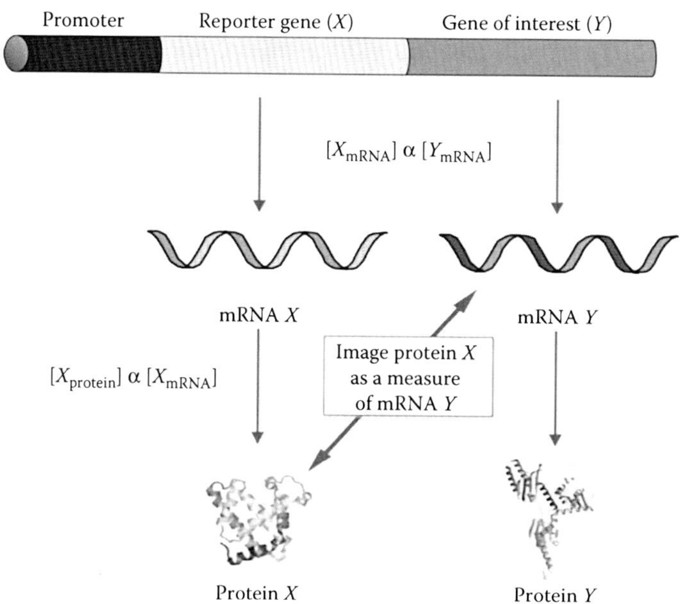

FIGURE 3.28
Imaging gene expression using reporter genes. (Redrawn from Cherry, S. R., *Phys. Med. Biol.*, 49, R13, 2004.)

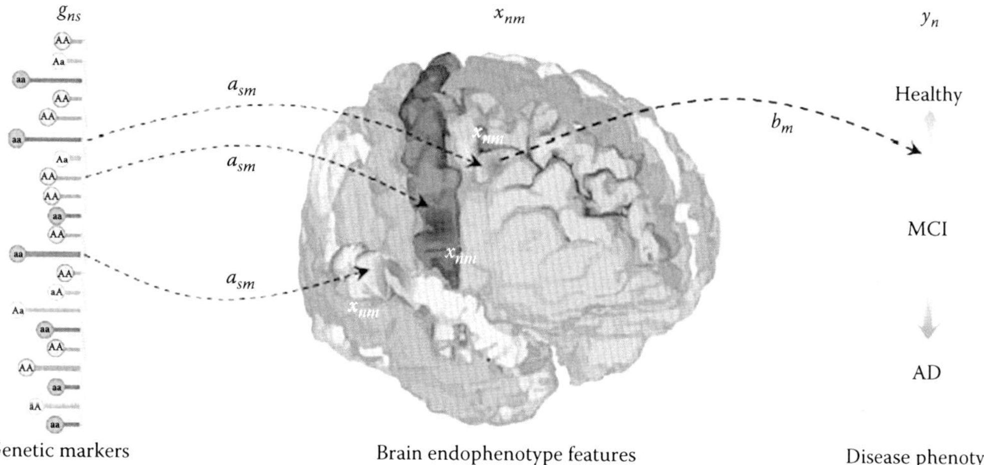

FIGURE 3.29
A schematic illustration of the relationship between genetic, imaging, and clinical measures in our model. AD, Alzheimer's disease; MCI, Mild cognitive impairment. (Data from Musa, S. M., *Nanoscale Spectroscopy with Applications*, CRC Press, Boca Raton, FL, 2013.)

been done over the set of all hundreds of thousands of voxels using univariate analysis. Multivariate techniques such as regression and regularization eliminate the shortcomings of univariate techniques. Some techniques are based on canonical correlation analysis, linear discriminant analysis, partial least squares, and so on. In those studies [37], a generative model has been presented for the relationship between genetics, imaging features, and disease mechanisms (Figure 3.29). Using this study, an inference can be drawn, and it is also possible to identify the corresponding brain locations and genetic loci. A Bayesian model based on logistic regression for predicting binary class label y_n from image features x_n is adopted, which is expressed as [37]

$$p(D,Z|G;V,\pi)=\prod_{n=1}^{N}p(y_n|\eta,x_n)\prod_{m=1}^{M}p(b_m)p(\eta_m|b_m)p(x_{nm}|b_m,a_m,g_n;v_m)\prod_{s=1}^{S}p(a_{sm}|b_m)$$

where:

$Z = \{\eta, b, A\}$ is the set of latent variables
$D = \{X, y\}$ is the set of data variables that are modeled
$\Pi = \{\sigma_\eta^2, \sigma_0^2, \alpha, \beta\}$ is the set of hyperparameters
$y = [y_1;\ldots; y_N]$ and $X = [x_1;\ldots; x_N]$

Here, a joint distribution of the hidden variables Z and the modeled variables D is constructed, given the genetic markers $G = [g_1;\ldots; g_N]$ [37].

In this modeling of the relationship between imaging features and genetic variations, the imaging features are related to the disease mechanism and can predict it, differently from the other imaging features. If m is a feature relevant for the prediction of a disease ($b_m = 1$), then the variations in the values of this feature can be explained by a subset of the genetic variations.

3.5 Conclusion

At the imaging system level, several difficulties and opportunities are present in molecular and genome imaging. In all the various functional and metabolic modalities, the imaging performance and output need to be improved to have an enhanced spatiotemporal resolution and higher sensitivity. It is very important to accurately quantify the studies done in molecular functional imaging so that diseases can be tracked and the effects of treatment can be analyzed. For this purpose, molecular imaging has been equipped with various software applications for a quantitative and statistical analysis. Even with all these developments, there are still several practical difficulties and challenges faced while performing studies with multiple modalities. This is surely a time-consuming process that needs to be managed efficiently in order to better visualize the huge volume of different kinds of data available through molecular functional imaging. The field is relatively young, which is the reason that standardization and generalized validation has not been done for the various data acquisition and analysis methods using different instruments, different types of imaging modalities, and different imaging centers.

Digital pathology and WSI provide pathologists and researchers an efficient hardware and software platform to acquire and assess images of tissue sections of interest. This approach is being adopted widely for research purposes but still needs to be standardized for diagnostic uses. These systems improve the image quality and scanning time, and allow easy storage and distribution of the slide images. Despite all the advantages, there are several aspects that need to be looked into, such as standardized guidelines, regulations, protocol, and familiarity with the devices, before these systems can be employed widely for direct patient-care and diagnostic purposes.

References

1. Ghaznavi, F., Evans, A., Madabhushi, A., and Feldman, M. (2013). Digital imaging in pathology: Whole-slide imaging and beyond. *Annual Review of Pathology: Mechanisms of Disease, 8*, 331–359.
2. Venter, J. C., Adams, M. D., Myers, E. W., Li, P. W., Mural, R. J., Sutton, G. G., Smith, H. O. et al. (2001). The sequence of the human genome. *Science, 291*(5507), 1304–1351.
3. Lander, E. S., Linton, L. M., Birren, B., Nusbaum, C., Zody, M. C., Baldwin, J., Devon, K. et al. (2001). Initial sequencing and analysis of the human genome. *Nature, 409*(6822), 860–921.
4. Hariri, A. R., and Weinberger, D. R. (2003). Imaging genomics. *British Medical Bulletin, 65*(1), 259–270.
5. Thompson, P. M., Martin, N. G., and Wright, M. J. (2010). Imaging genomics. *Current Opinion in Neurology, 23*(4), 368.
6. Ramos-Vara, J. A. (2005). Technical aspects of immunohistochemistry. *Veterinary Pathology Online, 42*(4), 405–426.
7. Onody, P., Bertrand, F., Muzeau, F., Bièche, I., and Lidereau, R. (2001). Fluorescence in situ hybridization and immunohistochemical assays for HER-2/neu status determination: Application to node-negative breast cancer. *Archives of Pathology and Laboratory Medicine, 125*(6), 746–750.
8. Aperio. (2009). *A Digital Pathology Solution for Immunohistochemistry, Digital IHC—User Guide*. Aperio, Vista, CA.
9. Giakos, G., Deshpande, A., Quang, T., Farrahi, T., Narayan, C., Shrestha, S., Zervakis, M. et al. (2013). An automated digital fluorescence imaging system of tumor margins using clustering-based image thresholding. *IEEE International Conference on Imaging Systems and Techniques*. IEEE, Beijing, People's Republic of China.
10. Paik, S., Bryant, J., Park, C., Fisher, B., Tan-Chiu, E., Hyams, D., Fisher, E. R. et al. (1998). erbB-2 and response to doxorubicin in patients with axillary lymph node-positive, hormone receptor-negative breast cancer. *Journal of the National Cancer Institute, 90*(18), 1361–1370.
11. Cobleigh, M. A., Vogel, C. L., Tripathy, D., Robert, N. J., Scholl, S., Fehrenbacher, L., Wolter, J. M. et al. (1999). Multinational study of the efficacy and safety of humanized anti-HER2 monoclonal antibody in women who have HER2-overexpressing metastatic breast cancer that has progressed after chemotherapy for metastatic disease. *Journal of Clinical Oncology, 17*(9), 2639–2648.
12. Mohsin, S. K., Weiss, H. L., Gutierrez, M. C., Chamness, G. C., Schiff, R., Digiovanna, M. P., Wang, C. X. et al. (2005). Neoadjuvant trastuzumab induces apoptosis in primary breast cancers. *Journal of Clinical Oncology, 23*(11), 2460–2468.
13. Wolff, A. C., Hammond, M. E. H., Schwartz, J. N., Hagerty, K. L., Allred, D. C., Cote, R. J., Dowsett, M. et al. (2006). American society of clinical oncology/college of American pathologists guideline recommendations for human epidermal growth factor receptor 2 testing in breast cancer. *Journal of Clinical Oncology, 25*(1), 118–145.
14. Onody, P., Bertrand, F., Muzeau, F., Bièche, I., and Lidereau, R. (2001). Fluorescence in situ hybridization and immunohistochemical assays for HER-2/neu status determination: Application to node-negative breast cancer. *Archives of Pathology and Laboratory Medicine, 125*(6), 746–750.
15. Bánkfalvi, A., Boecker, W., and Reiner, A. (2004). Comparison of automated and manual determination of HER2 status in breast cancer for diagnostic use: A comparative methodological study using the Ventana BenchMark automated staining system and manual tests. *International Journal of Oncology, 25*(4), 929–964.
16. MacQueen, J. (1967). Some methods for classification and analysis of multivariate observations. In *Proceedings of the Fifth Berkeley Symposium on Mathematical Statistics and Probability*, L. M. Le Cam, and J. Neyman, Eds. University of California Press, Berkeley, CA, Vol. 1, No. 281–297, p. 14.

17. Fukunaga, K., and Hostetler, L. (1975). The estimation of the gradient of a density function, with applications in pattern recognition. *IEEE Transactions on Information Theory, 21*(1), 32–40.
18. Cheng, Y. (1995). Mean shift, mode seeking, and clustering. *IEEE Transactions on Pattern Analysis and Machine Intelligence, 17*(8), 790–799.
19. Comaniciu, D., and Meer, P. (2002). Mean shift: A robust approach toward feature space analysis. *IEEE Transactions on Pattern Analysis and Machine Intelligence, 24*(5), 603–619.
20. Wang, J., Thiesson, B., Xu, Y., and Cohen, M. (2004). Image and video segmentation by anisotropic kernel mean shift. In *Computer Vision-ECCV 2004*. Springer, Berlin, Germany, pp. 238–249.
21. Roerdink, J. B., and Meijster, A. (2000). The watershed transform: Definitions, algorithms and parallelization strategies. *Fundamenta Informaticae, 41*(1), 187–228.
22. Meyer, F., and Beucher, S. (1990). Morphological segmentation. *Journal of Visual Communication and Image Representation, 1*(1), 21–46.
23. Vincent, L., and Soille, P. (1991). Watersheds in digital spaces: An efficient algorithm based on immersion simulations. *IEEE Transactions on Pattern Analysis and Machine Intelligence, 13*(6), 583–598.
24. Meyer, F. (1994). Topographic distance and watershed lines. *Signal Processing, 38*(1), 113–125.
25. Moga, A. N., and Gabbouj, M. (1998). Parallel marker-based image segmentation with watershed transformation. *Journal of Parallel and Distributed Computing, 51*(1), 27–45.
26. Malik, J., Belongie, S., Leung, T., and Shi, J. (2001). Contour and texture analysis for image segmentation. *International Journal of Computer Vision, 43*(1), 7–27.
27. Chan, T. F., and Vese, L. A. (2001). Active contours without edges. *IEEE Transactions on Image Processing, 10*(2), 266–277.
28. Aubert, G., and Vese, L. (1997). A variational method in image recovery. *SIAM Journal on Numerical Analysis, 34*(5), 1948–1979.
29. Gillies, R. J., Anderson, A. R., Gatenby, R. A., and Morse, D. L. (2010). The biology underlying molecular imaging in oncology: From genome to anatome and back again. *Clinical Radiology, 65*(7), 517–521.
30. Cherry, S. R. (2004). In vivo molecular and genomic imaging: New challenges for imaging physics. *Physics in Medicine and Biology, 49*(3), R13.
31. Watrous, J. D., Alexandrov, T., and Dorrestein, P. C. (2011). The evolving field of imaging mass spectrometry and its impact on future biological research. *Journal of Mass Spectrometry, 46*(2), 209–222.
32. Stoeckli, M., Chaurand, P., Hallahan, D. E., and Caprioli, R. M. (2001). Imaging mass spectrometry: A new technology for the analysis of protein expression in mammalian tissues. *Nature Medicine, 7*(4), 493–496.
33. Schwamborn, K., and Caprioli, R. M. (2010). Molecular imaging by mass spectrometry—Looking beyond classical histology. *Nature Reviews Cancer, 10*(9), 639–646.
34. Sheng, S., Skalnikova, H., Meng, A., Tra, J., Fu, Q., Everett, A., and Van Eyk, J. (2011). Intact protein separation by one- and two-dimensional liquid chromatography for the comparative proteomic separation of partitioned serum or plasma. In *Serum/Plasma Proteomics*, R. J. Simpson, and D. W. Greening, Eds. Humana Press, New York, pp. 29–46.
35. Berkowitz, S. A. (1988). Protein purification by multidimensional liquid chromatography. *Advances in Chromatography, 29*, 175–219.
36. Tribl, F., Lohaus, C., Dombert, T., Langenfeld, E., Piechura, H., Warscheid, B., Meyer, H. E. et al. (2008). Towards multidimensional liquid chromatography separation of proteins using fluorescence and isotope-coded protein labelling for quantitative proteomics. *Proteomics, 8*(6), 1204–1211.
37. Musa, S. M. (Ed.). (2013). *Nanoscale Spectroscopy with Applications*. CRC Press.
38. Batmanghelich, N. K., Dalca, A. V., Sabuncu, M. R., and Golland, P. (2013). Joint modeling of imaging and genetics. In *Information Processing in Medical Imaging*, J. C. Gee, S. Joshi, K. M. Pohl, W. M. Wells, and L. Zöllei, Eds. Springer, Berlin, Germany, pp. 766–777.

4

Developing a Comprehensive Taxonomy for Human Cell Types

Richard Conroy and Vinay Pai

CONTENTS

4.1 Introduction

Cells are the fundamental quanta of biology, the basic building blocks of life. There are about 37 trillion cells in the human body (Bianconi et al. 2013) and each cell performs a finite set of functions, depending on its development pathway and environment. The accepted wisdom is that there are between 200 and 500 distinct cell types present in an adult human body (Valentine et al. 1994; Vickaryous and Hall 2006), although a definition for what is a cell type, what are the unique characteristics of each cell type, and a systematic survey of all the cells in the human body have not been carried out. Cell types have historically been based on morphological features and qualitative measures of a small set of surface proteins, as well as knowledge of the organ from which the cells were isolated. From these studies, we know that cells *in vivo* are heterogeneous, even when they are from the same tissue and nominally of the same type, which prompts a key question: Are there a finite number of states in which human cells can exist associated with stable phenotypes? If so, are we now at the stage of being able to map these states and the relationships between them with current technologies in order to provide a more robust taxonomy system for human cells? Can we get to the stage of being able to identify an unknown cell circulating in the human body, for example, locate and characterize a tumor that potentially sheds it?

In this chapter, we review some of the recent technology and computational developments that help to measure the inherent biological heterogeneity of single cells that exist *in vivo*, and the significance of this diversity to biological and disease processes. Cells can vary significantly in size, from 1 to 500 μm, and have a wide range of number of copies of each of their biomolecules, from 1 to 10^9 copies. For mammalian cells around a volume of 1 pL, a single copy of a biomolecule corresponds to a concentration of approximately 1 pM, so sensitivity to achieve single-molecule counting is not necessarily an issue, unless there is significant dilution or the detection scheme has low efficiency. However, the dynamic range across almost 9 orders of magnitude is more challenging, and if biomolecules are being measured directly, careful thought needs to be given to avoid any bias in the results. Furthermore, there are as many as 100,000 distinct endogenous biomolecules in a cell arising from ~5,000–20,000 transcribed and translated genes, with an average cell having around 2 pg of DNA, 5–25 pg of RNA, and 20–75 pg of proteins and lipids (Cheung et al. 2013). These biomolecules exist over a wide range of timescales, from seconds for metabolites to the full lifetime of the cell in the case of extremely long-lived proteins. They can also exist in many different states due to conformational changes, interactions, variants, and modifications. Finally, cells can contain a significant number of exogenous molecules, waste materials, and dissolved gases, all of which may be indicative of their state. This complexity leads to the nontrivial question of which type and combination of measurements and what type of analysis give the most sensitive and specific phenotypic descriptions, while being fast, inexpensive, robust, and generalizable.

To begin thinking about a more systematic survey of human cell types, we need to address complexity through having high content, high-throughput measurement techniques as well as tools for analyzing the resulting multidimensional datasets and prognostic models. The current generic workflow for single-cell phenotyping involves collecting cells, sorting and organizing them and then analyzing them using imaging or 'omics technologies. Optical imaging is arguably the best suited imaging modality for studying individual cells and can be used to make measurements of morphology, molecular or functional features with minimal perturbation, as well as providing high spatial and temporal resolution. Integrating optical imaging with destructive, but more comprehensive, 'omics technologies provides the potential for extremely rich and synergistic measurements but increases the complexity of the analysis. The Visible Human Project is one example of an existing survey of the human body and how optical imaging can be applied to visualize and study the structural richness (http://www.nlm.nih.gov/research/visible/visible_human.html). Do we have the capabilities and motivation to think about extending our knowledge of the human body to the individual cell level?

Another central question of trying to assemble a taxonomy for human cells from an inventory is the significance of the environment in determining the cell type. Cell–cell communication is critical to many physiological and structural properties (Dejana 2004), and the physical–chemical properties of the environment can control differentiation (Guilak et al. 2009). Carefully established structures such as stem cell niches provide unique environments that control the function of cells (Calvi et al. 2003), and without these environments, the cells exhibit significantly different behavior. It is also estimated that fibroblast cells are present in as many as 297 distinct physical locations in the human body and that the differences in gene expression profiles are often attributable to differences in the extracellular matrix environment (Hatano et al. 2011). However, comparative and evolutionary studies have clearly determined homologous cell types conserved across species and development stages using molecular fingerprinting (Arendt 2008). There is unlikely to be a satisfactory experimental answer to this question until a more complete inventory of cells can be carried out using minimal perturbation and high-content, high-throughput screening, although it is likely that at least a taxonomy independent of the environment and common to closely related species can be assembled for cells with segregated functions.

4.2 Current Approaches to Cell Taxonomy

The current framework for cell classification in multicellular eukaryotic organisms is that there are two fundamental types: germ cells and somatic cells. Germ cells are those capable of giving rise to gametes that have the hallmarks of sexual reproduction and can be distinguished by their ability to undergo both meiosis and mitosis. Somatic cells by contrast only undergo mitosis but give rise to the majority of specialized cell types in the human body, typically estimated to be between 200 and 500 distinct cell types. These cell types are grouped into three major categories depending on the embryonic layer that gave rise to that cell type: endoderm, mesoderm, and ectoderm. The innermost layer, the endoderm, gives rise to many of the secreting cells and those involved in interior linings, digestion, and respiration. The middle layer, the mesoderm, gives rise to connective tissues, the lining of several organs, and blood cells. The outer layer, the ectoderm, forms the nervous system and organs on the surface of the body. In histology, cell types are considered under the four basic types of tissue: muscle, nervous, connective, and epithelial, providing a complementary, though distinct approach to classification.

Cell types are traditionally identified using optical and electron microscopies to qualitatively distinguish morphological and molecular features. The arrangement of cells and tissue features, for example, the positioning of proximal and distal tubules around glomeruli in the kidney, also assists in identification. The use of a panel of stains enhances the detection of a wide range of biomolecules and aids cell identification in histologic analysis. The more recent development of antibody-based techniques for labeling the presence of cell-surface proteins has greatly enhanced the identification of specific cells types. More than 320 unique clusters of differentiation (CDs), or cell-surface targets for immunophenotyping, have been identified, and the expression of these CDs can be mapped to tissues and cell types (Creighton 1999). This approach has been expanded to include antibodies against many targets. The rate at which this type of antibody-based profiling has developed is highlighted by a recent study that used antibodies to map expression of 4842 proteins in 48 human tissues and 45 human cell lines (Ponten et al. 2009).

These measurements work on the assumptions that there are strong correlations between the cell's molecular, functional, and morphological properties; cells of the same type have the greatest similarities in the chosen measurements; and there are unique mappings from these measurements to stable phenotypes. Analyzing and understanding the limitations of these underlying assumptions is a key part of developing a robust classification system. With the development of increasingly high-content and high-throughput techniques, deciphering which similarities and differences are relevant to cell typogenesis is an increasingly complicated computational challenge because of the disparate, multidimensional, and multimodal datasets.

A decade ago, the Human Cytome Project proposed a comprehensive framework for studying the complexity of cells using three levels: (1) studying the behavior of cells in their life cycle, (2) determining the molecular status of normal and diseased cell states, and (3) studying the assembly of cells into tissues (Valet et al. 2004). Their idealized workflow consisted of functional analysis of living cells, followed by multiple rounds of molecular staining of surface and intracellular structures with the data fed into pattern analysis algorithms. A more detailed submicron spatial analysis of individual cells of interest could then be carried using super-resolution microscopy techniques, or microdissected for 'omics analysis of the cell contents (Tarnok 2004). They identified a number of significant challenges, including increasing the sensitivity of 'omics approaches to enable the analysis of individual cells, the integration and automation of optical imaging systems, and the fast and efficient extraction of knowledge from multiparameter datasets (Valet and Tarnok 2004).

Another framework to a more comprehensive understanding of cell types is to analyze cell differentiation pathways. Recently, a network approach has been used to visualize differentiation pathways in human development by mining data from many sources (Galvao et al. 2010). This study found 873 distinct cell types, but noted that all cell types are not present in an adult human; for example, cell types found only in the placenta or those found during prenatal development occur only at distinct times in the life cycle. This analysis, based on graph theory, was organized in 19 distinct functional classes (C1–C19) and optimized based on similarity of parameters identified in the literature. One of the interesting aspects of this study is that the appearance and disappearance of cells types were studied where information was available. The timing of 783 distinct cell types in human embryo development was found in the literature and studied, with all 783 types utilized by 250 days post fertilization, but with fewer than half of them still present in the embryo on day 250. The most rapid proliferation of distinct cell types is during days 30–50 when surviving cell types emerge from nonsurviving progenitors, resulting in more than 400 distinct cell types being created during the first 50 days. From a network analysis perspective, the condition of critical branching occurs at around 40 days. This type of functional map derived from studying differentiation pathways highlights the modular nature of cells, and that cells with related biological functions associate to form distinct tissue types at a higher length scale.

One recent effort to develop a more complete repository of human cell information is the online CELLPEDIA/SHOGoIN database that is under development (http://shogoindb.cira.kyoto-u.ac.jp/). The researchers have been collecting gene expression profiles and images for a decade and have developed their own classification scheme based on images, gene expression, and cell differentiation data as well as physical mapping data and relevant journal articles (Hatano et al. 2011). Histological image data are included for a subset of cell types, although morphological data do not contribute to the classification; the gene expression data used come from a number of sources and cell preparations, including heterogeneous tissues, multiple unique cells, and single cells. Their scheme distinguishes stem and differentiated cells into two categories, utilizes the open Cell Ontology, and the integrative multispecies anatomy ontology, Uberon (Mungall et al. 2012). There is good mapping between the ~1500 cell types and hierarchical structures described by the Open Biological and Biomedical Ontologies and the 2260 differentiated cell types and eight-level classification hierarchy used in CELLPEDIA, after correcting for differences in the vascular system descriptors. The team is currently working on new ways to deduce differentiation pathways, and the data for the generation of induced pluripotent stem cells have become one of the major foci for the project. In their published work, the project team felt that cell location and neighborhood would become increasingly important in describing cell types and that datasets from other technologies would extend and complement the gene expression profiles. These examples of current approaches to the classification of human cell types suggest that there is greater complexity in cell types that we can resolve with widely available measurements and analytical techniques. In Sections 4.3 and 4.4, we will describe some of the recent advances in techniques, which may improve our resolution before returning to some of the computation challenges involved in analyzing the datasets from these methods.

4.3 Genomic, Transcriptomic, Proteomic, and Metabolomic Tools for Cell Classification

A range of technologies capable of comprehensively studying nearly all the constituents of one class of biomolecules from individual cells have emerged over the past decade. These technologies can provide a detailed quantification of DNA, RNA, proteins, and metabolites in a sample of cells and are used extensively to characterize development processes and diseases. The interplay between each of these layers of biomolecules within a cell is complicated, with messenger RNA (mRNA) abundance only accounting for 30%–70% of the variability observed in protein abundance (Vogel and Marcotte 2012), making the choice of which class of biomolecules to measure nontrivial. We discuss some of the recent works in single-cell genomics, transcriptomics, proteomics, and metabolomics in Sections 4.3.1 through 4.3.5 before returning to optical techniques and the informatics challenges in Sections 4.4 and 4.5.

4.3.1 Single-Cell Genomics

Although polymerase chain reaction (PCR) amplification has been applied to individual genes from individual cells for more than 25 years, DNA sequencing technology has only recently progressed to the point that it is capable of reading nearly the entire genome of individual cells. Whole-genome amplification at the single-cell level is technically challenging due to the need to amplify the entire genome a millionfold for microarray and deep sequencing, but polymerases typically used for PCR amplification have amplification bias and relatively high error rates. To address these weaknesses, two different amplification techniques—multiple strand displacement (MDA) (Raghunathan et al. 2005) and multiple annealing and looping-based amplification cycles (MALBAC) (Zong et al. 2012)—have emerged as promising techniques to provide less skewed results and better genome coverage (Blainey 2013). MDA is carried out isothermally, with a hyperbranched structure forming as primers that bind to newly formed DNA synthesized by strand displacement. MALBAC, however, uses a pool of random primers that have a common sequence and a set of variable nucleotides, which results in a quasi-linear preamplification. These techniques provide enough sequencing depth of the DNA to reconstruct >90% of the human genome, although one remaining challenge is the *de novo* reconstruction of genomes sequenced at the single-cell level.

These approaches have been used to detect single-nucleotide variants and copy number variations in individual human cells (Zong et al. 2012), to study mutation rates in sperm (Wang et al. 2012), heterogeneity of chemotherapy-resistant tumors (Hjortland et al. 2011), and pathogenic strains in biofilms (McLean et al. 2013). Sperm are a favorite cell type to study because of the genetic variations that can occur, and the application of whole-genome amplification techniques have resulted in a new estimate of 30 mutations per million base pairs (Wang et al. 2012), and the observation that 5% of sperm show aneuploidy (Zong et al. 2012). Whole-genome amplification has also been applied to single first and second polar bodies and human oocyte pronuclei to study aneuploidy and single-nucleotide variations, highlighting the potential clinical application of these techniques for more accurate and cost-effective selection of fertilized eggs for embryo transfer (Hou et al. 2013). Analysis of tumor heterogeneity to personalize therapy has also emerged as a significant area of interest because sequencing is needed to identify mutated genes and genetic instabilities. In a study of the evolution of a breast cancer, three subpopulations were identified and one was linked to a liver metastasis, suggesting a punctuated model of tumor progression with few persistent intermediate subpopulations (Navin et al. 2011). A single-cell study of acute myeloid leukemia revealed a clonal progression of multiple mutations in the hematopoietic stem cells in some patients, suggesting that preleukemic hematopoietic stem cells may harbor a range of mutations that contribute to a relapse (Jan et al. 2012).

One of the current challenges is to extend single-cell DNA analysis beyond genomics and into epigenomics and nucleosomics. Assays for analyzing methylation states of the DNA (Guo et al. 2013), histone modifications, and chromosomal conformations are rapidly moving toward single-cell sensitivity (Mensaert et al. 2014) and will help to fill in the gap in understanding between genotype and phenotype. The potential for studying the nucleosome in five dimensions in fixed or live cells is a fascinating prospect and likely to open up many new avenues of research and understanding (Mitchell et al. 2014).

4.3.2 Single-Cell Transcriptomics

Gene expression profiling has emerged over the past decade as a powerful technique for phenotyping cells. Initially, arrays were limited to large preparations of high-quality mRNA, although different techniques can now generate transcriptional profiles from many individual cell types. The abundance of each mRNA transcript varies between 1 and 5,000 copies per cell and 100,000–300,000 mRNA molecules overall, but with more than 90% of the transcriptome expressed at less than 50 copies per cell (Eberwine et al. 2014). A complete survey therefore requires methods with digital sensitivity, a wide dynamic range, and the ability to discriminate closely related variations. The general workflow to date has been to convert mRNA to complementary DNA (cDNA) with universal primers and reverse transcriptase, which is then amplified by PCR or other techniques before analysis (Wang et al. 2013). Although there are more copies and shorter sequences to amplify than for whole-genome amplification, the challenge is to avoid bias in the amplification and maintain a quantitative readout of the relative abundances of the 5,000–15,000 distinct mRNA transcripts. The use of sensitive, deep sequencing to study how often each gene is represented has now reached the stage of being able to make digital counting of the presence and prevalence of RNA transcripts in close to the complete transcriptome (Mortazavi et al. 2008). *RNA sequencing* (RNA-seq) is able to identify >75% genes than microarray techniques from mouse blastomeres, including more than 1750 previously unknown splice junctions and an estimate of 8%–19% of known transcript isoforms have at least two isoforms expressed (Tang et al. 2009).

The cost of single-cell transcriptional profiling has decreased to around $20–$40 per cell (Islam et al. 2014), while the reproducibility and throughput have increased. There has also been a proliferation of kits for DNA and RNA amplification from single cells, and the year 2013 saw the release of the first commercial, automated system for RNA-seq. Systems to conduct thousands of single-cell amplification reactions in parallel in nanowells have also been demonstrated recently (Gole et al. 2013). These changes are opening up exploration of the space between single and ensemble measurements as well as the variance in expressed genes. Eberwine et al. (2014) estimated that transcriptional profiles of tens of thousands of cells are needed to begin mapping out the degrees of freedom in a mammalian cell, although this assumes that all genes contribute equally to defining the phenotype of the cell. How the regulation or abundance of expressed genes affects the phenotype of the cells is an open question, however, particularly for low-level or variably expressed genes.

Single-cell transcriptional profiling was applied to a number of biological challenges and revealed complexity at a level not previously observed. A study amplifying all RNAs in human embryos and human embryonic stem cells found 22,687 expressed sequences, including 8,701 long noncoding RNAs (lncRNAs), with thousands of genes and lncRNAs differentially expressed at different developmental stages (Yan et al. 2013). Another analysis of the development stages of mouse and human oocytes revealed a significant portion of polymeric gene transcripts and that each development stage could be accurately delineated by a small number of functionally clustered genes (Xue et al. 2013). Single-cell transcriptional analysis also helps to reveal a degree of heterogeneity in what have been perceived to be relatively homogenous populations. Recently, investigators have studied the response of bone marrow-derived dendritic cells to lipopolysaccharides and observed an extensive, bimodal variation in mRNA abundance and splicing patterns (Shalek et al. 2013), suggesting subpopulations with different sensitivities and response time. Single-cell transcriptional profiling has also been used to study rare circulating tumor cells, shed from a tumor or metastasis, resolving hundreds of differentially expressed genes, isoforms, and single-nucleotide polymorphisms across these cells (Ramskold et al. 2012).

A number of challenges still drive technology development. Only a limited subset of cells has been profiled, with a preference for large cells and cells that are not embedded tightly in tissue. Furthermore, profiling primarily focused on cytosolic RNA, although the analysis of pre-mRNA in the nuclear envelope may also provide interesting insights. Another challenge with these approaches is that many RNA species experience post-translational modification, resulting in different variants. These challenges resulted in a number of different profiling methods emerging. Methods to detect these variants include single-cell tagged reverse transcription (Islam et al. 2011) and switching mechanism at the 5′ end of the RNA template (Ramskold et al. 2012). Another approach to transcriptional profiling is cell expression by linear amplification and sequencing (CEL-seq), which uses an *in vitro* transcription step (Hashimshony et al. 2012).

For spatial analysis of mRNA expression in tissue, a number of fluorescence *in situ* hybridization (FISH) techniques that use short fluorescent probes to detect RNA targets were developed. There are some challenges with the sensitivity, specificity, and number of parameters that can be measured by these techniques in fixed tissue, although they do have single-molecule sensitivity (Itzkovitz and van Oudenaarden 2011) and are used to measure 32 expressed genes simultaneously (Lubeck and Cai 2012). There are several proposed methods for improving the parameter space through combinatorial tagging and spatial resolution through super-resolution imaging techniques as well as extending measurements to study variants and to live cells.

4.3.3 Single-Cell Proteomics

Techniques for measuring protein translation in single cells lag behind methods for measuring transcription. The proteome is perhaps the most complex of the biomolecules to measure with around 25,000 proteins present in a cell, abundance varying over 5 orders of magnitude, and no existing technique for amplifying them. The goal for the field is to develop a high dynamic range approach that can measure copy number and ideally posttranslational modifications as well as protein–protein and protein–substrate interactions without disruptively tagging proteins before analysis. Posttranslational modification can involve structural changes such as disulfide bridges, chemical modification of amino acids such as deamidation and carbamylation, as well as the addition of functional groups such as localization signals, cofactors, or acylation and alkylation, all of which have biological significance but add several orders of magnitude more complexity to accurately capturing these modifications as part of studying the proteome. Capillary electrophoresis (CE), a wide range of chromatography and mass spectrometry (MS) techniques, immunoassays, and combinations of these techniques have been refined over many decades to study many proteins of interest, although each has its limitations in sensitivity, dynamic range, and resolution. In addition, antibody labeling approaches, such as the study mentioned earlier, have been used as a readout for known protein targets, with content limited by the quality of antibodies and the dynamic range and spectral limitations of the imaging system used.

Several microfluidic systems were developed to overcome limitations in handling the small sample size and the need for high-sensitivity readout because of the lack of amplification techniques. Microfluidic antibody chips (MACs), where a cell is trapped and mechanically lysed with a laser pulse over a large array of capture antibodies, has a detection limit of ~30 copies of a protein and a dynamic range of nearly 5 orders of magnitude (Salehi-Reyhani et al. 2011), although the challenge will be multiplexing to measure many proteins at the same time. The single-cell barcode chip uses DNA barcodes covalently coupled to capture antibodies, which are then read in a microarray reader, enabling measurement of protein copy number and protein phosphorylation for 11 intracellular proteins (Shi et al. 2012). The integration and automation of many isolation, manipulation, separation, and detection techniques into lab-on-a-chip microfluidic systems is opening up many new approaches to studying the single-cell proteome (Martini et al. 2007; Fritzsch et al. 2012; Yun et al. 2013), although many of the more advanced techniques have yet to be comprehensively validated or widely adopted into practice by the community.

Flow cytometry techniques are one of the gold standards for single-cell protein analysis, enabling up to ~15 protein levels to be measured simultaneously per cell along with other features in a high-throughput fashion (Krutzik et al. 2008). Systematically tagging green fluorescent protein (GFP) sequences to open reading frames to create a fusion protein library of strains enables about 2500 proteins, or half of the proteome, of yeast to be studied at the single-cell level (Newman et al. 2006). This study found that protein variation is normally distributed and the principal source of protein variation in many cases is due to the stochastic production and destruction of mRNA. GFP tagging is used in human cells revealing different accumulation rates for cell cycle-regulated proteins depending on mRNA lifetimes (Cohen et al. 2009); however, this approach cannot realistically be scaled to a wide range of human cells and genes because generating the libraries would be very time consuming and more technically difficult. Sequencing ribosome-protected mRNAs is an alternative approach to genome-wide investigation of protein translation and is used to study protein abundance in yeast in response to environment factors (Ingolia et al. 2009). There are some challenges with flow cytometry, primarily cells need to be fixed and permeabilized to study intracellular proteins, and it is difficult to study dynamics, interactions, and small molecules. Most flow cytometers also require a minimum number of cells and reasonable statistics to work with, making small samples or rare cells challenging to analyze. Integration of cell selection designs and multiple detection zones into microfluidic systems can provide a solution to making some of these measurements. However these devices are unlikely to displace flow cytometers as a workhorse in the laboratory and clinic in the near future.

One of the other gold standards for protein analysis is MS, often used in combination with other techniques that isolate cells and separate their contents before analysis. The advantage of MS over flow cytometry is that it can potentially be used to identify unknown proteins of interest, whereas flow cytometry relies on the availability of good-quality antibodies or peptides that bind known protein targets. In MS, the sample is bombarded by electrons, ionizing it and creating charged small biomolecules and charged fragments of large biomolecules. The resulting complex map needs to be interpreted, using a number of analytical approaches and libraries of prior results. The choice of ionization and mass selection techniques as well as the type of detector determines fragment size, sensitivity, and other characteristics. Electrospray ionization (ESI) or matrix-assisted laser desorption-ionization (MALDI) combined with time-of-flight (TOF) and Fourier transform ion cyclotron resonance can be used to study whole proteins, often referred to as a *top-down* strategy. Complementing this approach is the enzymatic digestion of proteins before MS analysis to create peptide sequences that can be analyzed using tandem mass spectrometry (MS–MS) or peptide mass fingerprinting, often referred to as a *bottom-up* approach. Different chromatography and electrophoresis techniques can be used to fractionate proteins or peptides prior to MS analysis and stable isotopes can be used to provide a degree of quantification. MS is currently capable of resolving thousands of proteins in small populations of cells over a dynamic range of 10^6, although it can only detect high-abundance peptides for single-cell samples (Altelaar and Heck 2012). However, a number of groups are working on approaches for handling small-volume samples and minimizing losses during sample preparation, fractionation, and separation that appear promising.

One significant advance in single-cell proteomic analysis is the development of mass cytometry that combines cell cytometry techniques with MS analysis (Tanner et al. 2013). In this approach, antibodies are modified to chelate isotopes typically not found in biological systems, such as the transition metals, and then used to label proteins present on the cells being studied. When single-cell droplets are ionized, cells with bound antibodies provide a distinct, quantifiable response pattern in the mass range of the isotopes. While not capable of profiling the whole proteome, requiring fixed cells and resulting in the destruction of the cell, this technique can provide analysis of up to ~100 proteins from individual cells with high sensitivity, providing a rich multidimensional protein abundance datasets at the single-cell level. It is used to examine human immune system functional states (Bjornson et al. 2013), cell cycle (Behbehani et al. 2012), and leukemia

(Amirel et al. 2013), and potentially the same labeling technique could be extended to imaging MS approaches.

Single-cell proteomics has many outstanding challenges due to the range of the proteins expressed, the variety of posttranslational modifications employed, and the significance of protein–protein and protein–substrate interactions. For example, different glycoproteins are often associated with inflammatory and cancer-related disease processes, and through their co- and posttranslation modifications, unusual glycan structures and site-specific glycoforms create an extremely rich landscape (Chandler and Goldman 2013). With an increasing interest in identifying biomarker candidates though, there are significant efforts to investigate secreted glycoproteins, which can be studied using a number of techniques including autoantibodies, MS, enzymatic methods, and arrays. The identification of protein phosphorylation is also of significant interest, because of its role in signal transduction, but many phosphoproteins are extremely labile, making consistent, quantitative measurements difficult (Bodo and Hsi 2011).

4.3.4 Single-Cell Metabolomics

The metabolome is perhaps one of the most challenging measurements at the single-cell level because of the diversity and large dynamic range of the small (<2 kDa) molecules involved. It is almost impossible to amplify or tag the molecules without drastically impacting the function and the dynamic nature of both endogenous and exogenous metabolites provides very time- and environment-dependent snapshots. Metabolites have rapid turnover rates, of the order of milliseconds to minutes, much faster than the transcriptome and proteome, which turn over in minutes to hours. Furthermore, there are many of the same metabolites present in the extracellular space (Zenobi 2013), so rapid quenching techniques and removal of the surrounding medium are required for high-fidelity measurements of cell contents.

Taking advantage of the large size of the neurons from the Californian sea slug, 50 endogenous metabolic compounds were studied and quantified across a range of different neurons using CE coupled to MS, and metabolic features were identified common to each subtype (Nemes et al. 2011). Alternatively, large plant epidermal cells were studied using laser ablation ESI (LAESI), where 35 metabolites were identified and neighboring cells with different pigmentations were observed to have different metabolic profiles (Shrestha and Vertes 2009). Working with smaller yeast cells and using silicon nanopost arrays for desorption–ionization, approximately 10% of the known >1000 yeast metabolites could be identified in samples of less than 100 cells, and the metabolic differences between stressed and control cells measured (Walker et al. 2013). Recently, with negative-ion mode MALDI, 26 metabolites have been measured in single yeast cells (Ibanez et al. 2013). MALDI-MS systems with imaging capabilities have a spatial resolution of one micron, sufficient for studying individual cells in mouse brain sections, and the potential for coregistration with optical images (Zavalin et al. 2012). Secondary ion MS and the use of stable isotopes have recently been demonstrated to provide sub-100-nm resolution and precise quantification down to 1 ppm, enabling detailed subcellular studies of metabolism (Steinhauser and Lechene 2013).

While a range of fluorescent probes were also developed to assay adenosine triphosphate, adenosine diphosphate, and adenosine monophosphate (Moro et al. 2010), and there are schemes for metabolite-responsive fluorescent protein probes (Dedecker et al. 2013), it is difficult to realize schemes for simultaneous detection of multiple metabolites. Vibrational spectroscopy is also used to identify specific metabolites, most notably beta-carotene that occurs in high concentration, and has two distinct peaks, which can be compared to MS results (Urban et al. 2011). However, the crowded spectrum from a reasonably large detection region within cells and the lack of specific bands corresponding to most metabolites make demultiplex the signal currently impossible to identify a metabolite profile, although a fingerprinting approach may be a practical intermediate step.

4.3.5 Opportunities and Challenges in Single-Cell 'omics

The proliferation of 'omics technologies with single-cell sensitivity has provoked a more detailed study of the heterogeneity observed (Elowitz et al. 2002; Sigal et al. 2006). Variability between cells can be stochastic or deterministic in nature; the stochastic component arises from both the sampling noise and the probabilistic nature of many biological processes, whereas the deterministic component is indicative of the cell responding to intrinsic (Sigal et al. 2006) and extrinsic (Raser and O'Shea 2005) signals. For example, a study of human dermal skin-derived fibroblasts found that expression patterns differed depending on lineage and environment that may have important consequences for cell transplantations and wound healing (Lorenz et al. 2008). Viral infection (Snijder et al. 2012) and aging (Bahar et al. 2006) can also perturb the transcriptional profile of a cell to different extents, compounding the challenge of building up meaningful descriptors of cell types. What is increasingly clear is that what used to be considered biological noise is a fundamental feature of cells that they use to extract information about their environment (Eldar and Elowitz 2010) and perhaps paradoxically to maintain their state without expending excessive resources on tight regulation of all pathway components.

For determining cell types, each of the different classes of biomolecules provides a different insight into the current or future state of the cell and there is a need for more correlation studies. Making quantitative measurements of two or more classes of biomolecules on the same cell is technically very difficult. One approach was described, which can assay proteins, RNA, and DNA from a single-cell lysate and found correlations of 0.37 and 0.31 between transfected plasmid DNA, transcribed mRNA, and translated protein concentrations (Stahlberg et al. 2012). This is still a long way from being able to assay the complete composition of a cell at the single-molecule level but an important step to measuring and understanding some of the features of the interactome (Sanchez et al. 1999).

4.4 Photonic Tools for Cell Classification

Our ability to control the spatial, temporal, and spectral properties of light sources, detectors, biomolecules, and nanostructures has radically changed how many biological experiments are carried out, how specimens are manipulated and increased the types of measurement possible. Optical imaging techniques can shed light on many characteristics of individual cells, including analysis of the molecular constituents of the cell, structural and morphological properties, cell function, and the interrelationship between cells. There is also a revolution in rapid prototyping, automation, and robotics, which has impacted the testing, validating, and translating new systems. Together, these technological advances push optical microscopy into an era of high-content, high-throughput imaging and drive a need for better tools to analyze and model the resulting multidimensional datasets.

Optical imaging has a number of key strengths, including it minimally perturbing the system being observed, having a large spatial–temporal dynamic range being easily multiplexed to measure multiple parameters simultaneously, and being capable of single-molecule counting with high accuracy and reproducibility. Broadly speaking, measurements are based on three different approaches: (1) static and dynamic morphological features, (2) molecular and functional measures using absorption or emission agents, and (3) spectroscopic techniques using exogenous and endogenous biomolecules. These approaches are not mutually exclusive and are often combined. Furthermore, these measurements can be made either on isolated cells, as is the case in flow cytometers, or *in situ* as part of intact tissue. In Sections 4.4.1 through 4.4.5, some recent advances in each of these areas will be described before moving on to consider some of the tools used for analyzing the resulting data.

4.4.1 Flow Cytometry

As described earlier, flow cytometry systems have become increasingly sophisticated with throughput approaching 100,000 cells/second and analysis of tens to hundreds of samples per flow channel per day (Robinson et al. 2012). However, the relatively broad emission spectra of fluorescent dyes and particles limits the number of parameters that can be routinely measured simultaneously to around 10 fluorescent channels across the visible and near-infrared (Gonzalez et al. 2008), which, in combination with scattering and other parameters, gives a set of up to 20 measurements (Lugli et al. 2010). If isotopes are used instead as labels for MS analysis, as in the recently developed TOF mass cytometry (CyTOF) technique, then 34 markers can be measured simultaneously (Bendall et al. 2011) and potentially as many as 100 are possible. An alternative approach to increasing the measurement parameter space is to carry out a tomographic reconstruction of the cell using a full range of angles or tomosynthesis using a limited set of projections. In addition to characterizing and sorting individual cells, the entire organisms such as *Caenorhabditis elegans* can also be profiled at a rate of 100/s (Dupuy et al. 2007), making the identification and isolation of well-defined phenotypes from a heterogeneous population possible. The capabilities of these systems can be enhanced by using more complex microfluidic chips and can be combined with optogenetic approaches to provide high-throughput analysis of synaptic function (Stirman et al. 2010).

A number of challenges remain in flow cytometry including how best to perform automated gating using unsupervised approaches, how to minimize perturbation of fragile cell types in high-shear environments, how to reanalyze samples with a different set of biomarkers, and how to deal with small sample sizes. A review of newly developed automated techniques as part of the FlowCAP Consortium challenges has been recently published (Aghaeepour et al. 2013), in addition to a review of computational approaches in flow cytometry (Robinson et al. 2012).

4.4.2 Morphological Imaging Techniques

Multiparameter cell phenotyping by automated imaging has emerged as a powerful method for high-throughput, high-content analysis (Failmezger et al. 2013). Morphological features can be identified using a number of label-free light microscopy techniques, resulting in measurements of abundance, area or volume, texture, density and contour length, as well as cell packing, density, and cell motility.

A number of bright-field, phase-based, and fluorescence techniques have been refined or developed over the past decade to provide the basis for high-content morphological imaging. The development of quantitative phase imaging approaches, including digital holography (Rappaz et al. 2014), spatial light interference microscopy (Ding and Popescu 2010), and ptychography (Marrison et al. 2013), has addressed some of the shortcomings of differential interference contrast microscopy. These techniques have been used to study stem cells and their microenvironment (Vega et al. 2012), and apoptosis and necrosis (Ziegler and Groscurth 2004), and for drug discovery (Evensen et al. 2010).

There is also a rapid development of tomosynthesis, tomographic, and projection techniques for reconstructing the three-dimensional (3D) morphology of cells. Optical projection tomography, achieved by rotating the specimen and using computer tomography techniques to reconstruct the 3D image of the cell, can be combined with molecular imaging to provide high-resolution, multiplexed images (Miao et al. 2012). The capture of full 3D images of zebrafish embryos with one μm resolution in tens of seconds and the measurement of 200 morphological features to identify cell phenotypes have also been recently demonstrated (Pardo-Martin et al. 2013).

The use of second- and third-harmonic generation in noncentrosymmetric biomolecules can be used to identify morphological features, in particular the arrangement of structural proteins (Campagnola et al. 2002). Multiphoton processes provide a degree of confocality, enabling deeper imaging in tissue and simplifying 3D sectioning. Higher harmonic generation has been used to study development of live zebrafish embryos, with submicron resolution and millimeter penetration depth, enabling imaging of cytoarchitectural development during embryogenesis (Sun et al. 2004). By combining second- and third-harmonic generation imaging, the lineage of cells in a zebrafish embryo up to the 1000-cell stage have also been mapped, and wave-like division cycles observed (Olivier et al. 2010).

A number of other techniques also have an impact on the identification and tracking of cell phenotypes. Selective plane illumination microscopy is used for optical sectioning of live, developing embryos with minimal photodamage (Huisken et al. 2004) as well as light sheet fluorescence microscopy (Keller 2013). Super-resolution microscopy techniques push the spatial limits enabling the study of many more subcellular features (Maglione and Sigrist 2013), whereas serial time-encoded amplified microscopy pushes the temporal resolution to frame rates of more than 10 MHz (Fard et al. 2011).

4.4.3 Imaging Agent Development

Imaging agents are crucial for many experiments because they enhance the signal-to-noise ratio, often to the level that transient single molecules can be observed. This enables very sensitive assays to be developed for both general and specific molecular targets and for very detailed studies of spatial–temporal distribution and dynamics. Imaging agents can also be used to assay function such as enzymatic activity, to measure environmental conditions such as pH and membrane potential, and to measure dynamics such as protein conformational changes. Finally, they come in many forms, from fluorescent reporter genes to synthetic dyes, quantum dots, and rare-earth doped nanoparticles, not to mention bioluminescent substrates and phosphorescent dyes. It is impossible to cover this rich field of research in any depth here, so interested readers are encouraged to read a number of recent review articles (Hellebust and Richards-Kortum 2012; Eckert et al. 2013; Klymchenko and Mely 2013; Li et al. 2013b; Yuan et al. 2013).

The development and optimization of fluorescent proteins (Tsien 1998) has significantly changed the study of phenotypes by enabling the study of the spatial–temporal distribution of specific proteins within a cell. These fluorescent DNA-encoded proteins, or similar bioluminescent proteins, can be inserted into genes and will result in tagged proteins. The field has rapidly blossomed over the past 15 years, with the generation of many new variants with different fluorescence peaks or that work on different bioluminescent substrates, split reporter designs for monitoring kinase activity or DNA repair processes, as well as a move toward fluorescent proteins optimized for *in vivo* use (Brogan et al. 2012). New classes of genetically encoded biosensors, such as ones sensitive to calcium (Chen et al. 2013) and voltage (Kralj et al. 2012), are increasing the breadth of probes for functional phenotyping.

While this chapter focuses primarily on single-cell measurement and analysis techniques, there is also an increasing toolbox of techniques for perturbation of cell functions and phenotype. Advances in DNA-encoded, light-activated proteins that perturb cells, dubbed *optogenetics*, were initially used to perturb membrane potential in neurons (Boyden et al. 2005) but are now used for many purposes including the control of transcription and epigenetic states (Konermann et al. 2013). Transcriptome-induced phenotype remodeling is used for direct cell-to-cell phenotype conversion and zinc-finger nucleases (Sul et al. 2012), transcription activator-like effector nuclease (TALEN) and clustered regularly interspaced short palindromic repeats (CRISPR) for genome engineering (Gaj et al. 2013). When combined with single-cell measurement techniques, these tools will enable better assessment of the impact of perturbations on the cell state and potentially provide insight for therapeutic techniques.

Two related challenges in imaging agent development include how to move beyond the limitations of the visible–near infrared spectral window and limitations due to the cross-reactivity and steric hindrance of antibodies, both of which limit the number of parameters that can be measured simultaneously. A number of solutions were developed based on either photodegradation or chemical degradation of immunofluorophores to enable serial rounds of antibody labeling. Using nuclear staining to enhance image registration between sequential scans, single-cell staining patterns for 61 protein antigens in a formalin-fixed, paraffin-embedded tissue slice using a chemical approach have been demonstrated with no appreciable loss of signal (Gerdes et al. 2013), providing extremely rich datasets for the analysis of tumor heterogeneity (Clarke et al. 2014). In frozen tissues, more than 90 biomolecules have been mapped using the photobleaching approach to create so-called toponomes (Schubert et al. 2006).

Creating spectral barcodes by unique combinations of fluorophores provides an alternative route to multiplexing measurements beyond the traditional —four to six distinct fluorophore emission regions. Using this approach, 11 gene transcripts were studied over 14 times and for 2199 individual cells to study serum response in normal human fibroblasts (Levsky et al. 2002). The development of super-resolution microscopy has also greatly expanded the capabilities of imaging to fill in the parameter space (spatial dimensions multiplied by number of distinct biomolecules). With a 10-nm resolution and a 15-fluorophore barcode, more than 450 targets could be localized with near-molecular accuracy within cells (Cai 2013). The use of FISH techniques is also sufficiently sensitive for counting individual transcripts, providing a route to quantitative molecular analysis at the single-cell level (Femino et al. 1998). The expression of genes in live cells has also been demonstrated in bacteria (Yu et al. 2006) and the bursting production of beta-galactosidase studied in a population of cells (Cai et al. 2006).

4.4.4 Spectroscopic Imaging Techniques

A wide variety of spectroscopic techniques can potentially be used for cell phenotyping, although many are not employed for this purpose because of low signal-to-noise, low spatial or temporal resolution, or their impracticality given the availability of other technologies. Techniques such as circular dichroism, laser-induced breakdown spectroscopy, and photoacoustic spectroscopy as well as time-resolved approaches such as dynamic light scattering can provide biologically relevant measurements; however, they are not commonly used for phenotyping individual cells.

Rotational and vibrational spectroscopy is one type of analysis that provides fingerprints helpful for phenotyping cells. The spectrum is typically measured using a Fourier transform infrared spectrometer giving a resolving power of a thousandth of a wavenumber in the infrared region of the spectrum. Vibrational spectroscopy is used to characterize multidrug resistance in sarcoma cells (Murali Krishna et al. 2005), to study the impact of carcinogens on development of Syrian hamster embryos (Walsh et al. 2009), and to differentiate transient versus persistent human papillomavirus infection in cervical cytology specimens (Kelly et al. 2010).

The spectrum arising from inelastic scattering of visible or near-infrared light from different biomolecules can also be used for identification. The Raman spectrum can be assigned to different molecular bonds with the potential for molecular fingerprinting within a small excitation volume; however, the intensity of scattering is very weak, making image collection slow. The spectrum can be interrogated in narrow band to record the presence of one species, or the full spectrum can be acquired and processed to provide more detailed chemical maps. Raman microspectroscopy has been used to characterize individual leukemia cells (Chan et al. 2008), to study the contraction cycle of single cardiomyocytes (Inya-Agha et al. 2007), to differentiate different grades of dysplasia *in vivo* (Duraipandian et al. 2013), and to combine with laser tweezers for analyzing and sorting cells (Lau et al. 2008). The low intensity of the signal can also be enhanced through a number of techniques, including coherent anti-Stokes Raman scattering (Gasecka et al. 2013) and surface-enhanced Raman scattering (Lee et al. 2014).

The autofluorescence spectrum is used for spectroscopic analysis. The second-harmonic generation spectrum of several biomolecules including collagen, retinol, nicotinamide adenine dinucleotide (NADH), and folic acid is used to identify structural and physiological characteristics of tissue (Zipfel et al. 2003). Autofluorescence analysis is also used to characterize stem cell differentiation (Santin et al. 2013) and mapping of neuroanatomy in the mouse brain (Gleave et al. 2012), and is commonly used to assess the *fundus* in age-related macular degeneration (Batioglu et al. 2013).

4.4.5 Live-Cell and Pathology Imaging Systems

Automated analysis of tissue, sometimes referred to as tissomics, provides measurements of spatial relations and cell morphology. One of the significant advantages of this approach over cytometry approaches is the capability for repeated analysis. However, there are many challenges in imaging the complexity of tissue slices, including how to achieve fast, high-fidelity 3D reconstruction, fast multispectral acquisition, how to characterize the extracellular space, and how to register different spectral and spatial slices together.

There are many examples of systems developed for digital pathology and high-content, high-throughput analysis. For example, the integration and automation of RNA interference (RNAi) screening and live-cell time-lapse microscopy was demonstrated in 2006, enabling the identification and discrimination of phenotypes resulting from the targeted genes, as well as differentiation delays in functions compared to complete arrest (Neumann et al. 2006). More recently, this technique has been used to generate multiparametric phenotypic profiles on genome-wide RNAi screen, enabling the comparison of unknown phenotypes and known phenotypes to predict the function of uncharacterized genes (Fuchs et al. 2010). The scale of current performance is illustrated by a recent RNAi screen, in which 85 morphological features were analyzed on individual cells from 1656 movies associated with 315 gene knockdowns and the results computed within a couple of days (Failmezger et al. 2013). One of the remaining technical challenges with RNAi screening is how to deal with off-target effects, although in general the choice of RNAi oligos and the use of enzyme-digested pools of RNAi can diminish these effects (Bickle 2010). Screening using a large number of quantitative parameters to describe cellular phenotypes can also improve hit rate and verification by studying all proteins participating in a process as well as potential off-pathway targets. It also provides some mechanistic insights into subpopulations with different responses. High-content systems capable of both wide-field and higher resolution confocal imaging with multiple excitation and imaging paths are increasingly common, making the use of five-dimensional imaging (three spatial dimensions, temporal and spectral) a more routine tool in analyzing phenotypes.

4.5 Bioinformatics Approaches for Automated Cell Phenotyping

There has been a long-standing interest in having fast algorithms capable of semi- or fully automated classification of cells based on quantitative measures. There are five primary reasons cited in support of automated image analysis (Carpenter et al. 2006): (1) It provides simultaneous measurement of many parameters of interest, (2) it is quantitative and reproducible, (3) it can measure individual cells across an entire population rather than scoring a select visual field, (4) it is able to detect features not readily detectable by human observers, and (5) it is much less labor intensive. Automated microscopy systems are now capable of generating more than 85,000 images per day, so the trend is to move away from observer-based scoring, while recognizing that the full interpretative ability of humans may not easily be replicated by a computer and that autonomous systems are unlikely to displace pathologists or other human clinical observers in the near future.

For computer-assisted analysis of multidimensional image sets, there are two broad approaches: (1) nonsegmented pattern recognition and counting techniques applied to whole or tiled images and (2) analysis of segmented regions of interests (ROIs) against a background. Both approaches can be classified into (1) supervised machine learning approaches where a training set is required and (2) unsupervised learning where the system is capable of finding patterns based on rule sets or other constraints. Several comprehensive books and reviews were written and interested readers are referred to them for a more complete review (Chan and Shen 2005; Russ 2006; Bovik 2010; Li et al. 2013a; Sommer and Gerlich 2013).

4.5.1 Nonsegmented Approaches

Pattern recognition and counting techniques approach the problem of image analysis from a different perspective to segmentation algorithms; segmentation approaches try to identify and characterize an ROI using measureable features, whereas pattern recognition tries to distinguish the characteristics of an image or arbitrary tiles of an image, often in comparison with control images. The advantages of pattern recognition techniques are that they are not limited by the accuracy of segmentation models, making them more widely applicable to different tissue and cell growth preparations, and they are fast and easy to implement (Sommer and Gerlich 2013). Some of the disadvantages include increasing computation times for multidimensional large datasets and increased sensitivity to background artifacts.

A number of nonsegmented approaches were reported including color-spatial schemes that build features such as color histograms or spatial frequency distributions (Huang et al. 1999) and shape feature vectors (Mahmoudi et al. 2003). The PhenoRipper software package uses an unsupervised, nonsegmentation approach that utilizes a color-spatial analysis of a tiled image followed by clustering (Rajaram et al. 2012). Using this approach, the researchers reported analyzing datasets that could not automatically be segmented and more than a 20-fold reduction in computation time for analyzing a 100,000-image dataset. With the advent of whole-slide imaging and increasing computation power available, pixel-level, object-level, and semantic-level patterns are increasingly being used for analyzing pathology slides in addition to segmented data (Kothari et al. 2013).

4.5.2 Segmentation Approaches

The goal with image segmentation is to identify ROIs within the image that can then be analyzed based on a number of features and given an overall classification. These features can include size, morphology, and texture descriptors, and may result in tens to hundreds of measurements being extracted per ROI. Typically, there are a number of preprocessing steps to ensure image quality and consistency prior to analysis, which may include registration, removal of artifacts, noise reduction, other types of filtering, and drift correction.

A number of general-purpose segmentation algorithms are used in other areas of machine vision and are also used for cell segmentation including thresholding, histogram analysis, and edge detection, each of which is based on the statistical analysis of the pixels present in the image. These approaches are sufficient when there are strong signals and sparse cells, although in complex specimens relevant morphological features can be missed if cells are not well segmented (Hill et al. 2007). To address these concerns, other theoretical approaches were developed, including watershed segmentation, variational methods, and level set algorithms (Rittscher 2010). Model-based approaches were also developed, including Markov random fields, deformable models, atlas-guided approaches, shape fitting, and artificial neural networks that provide alternative approaches (Pham et al. 2000). For object tracking in movies, there are also many techniques, each with its own advantages and disadvantages (Uchida 2013). A great variety of techniques underscore how difficult reliable image segmentation is, particularly for unlabeled, complex, 3D tissue. Labeling cell nuclei, organelles, and membranes with contrast agents greatly simplifies the segmentation process but increases the complexity and time required for sample preparation.

4.5.3 Data Reduction

One of the hurdles to wider use of image-based phenotype screens is the extraction of biological meaning from the measured parameters. Typically, a three-step process is employed: the first step is to make measurements using either segmented or nonsegmented approach, then to reduce the dimensions of the dataset, and finally to cluster similar images or regions together or use a classifier to assign a phenotype. Three general data reduction techniques are used for the second step: (1) factor analysis, (2) feature selection schemes, and (3) recursive feature extraction using support vector machines (SVM). Factor analysis is used to identify groups of parameters that correlate with each other in a multiparameter dataset, for example, reducing 36 measured parameters to 6 orthogonal factors (Young et al. 2008). A comparison of eight feature selection schemes found that stepwise discrimination analysis (SDA) and the genetic algorithm provided the highest accuracy, with SDA reducing a 84-dimensional data to 39 features (Huang et al. 2003). Using an SVM algorithm to find optimal hyperplanes distinguishing treated from untreated cells, a dose-clustering algorithm was capable of reducing 300 parameters to 20 informative features, and corresponded with little loss of classification accuracy (Loo et al. 2007).

4.5.4 Unsupervised Learning Approaches

Unsupervised techniques do not assume any prior knowledge about the image or ROIs being analyzed. The most common unsupervised approach used in microscopy is clustering, which seeks to group objects based on feature similarity. Many clustering algorithms were developed, including *K*-means, spectral, weighted gap statistics (Yan and Ye 2007), internal indices (Tibshirani et al. 2001), and jump methods (Sugar and James 2003). One of the significant advantages of unsupervised approaches is that they can be used to identify unknown phenotypes.

One integrated clustering approach that is applied to single-cell immune system data generated by CyTOF is the spanning tree progress of density-normalized events (Qiu et al. 2011). By applying density-dependent downsampling, agglomerative clustering, construction of a minimum spanning tree, and then upsampling, a tree of approximately 150 clusters can be put together describing the immune system. This approach is successfully used to study the impact of small-molecule regulators on human peripheral blood mononuclear cell signaling dynamics and cell-to-cell communication (Bodenmiller et al. 2012).

4.5.5 Supervised Learning Approaches

The goal of supervised learning approaches is to identify the optimal object-to-class mapping by identifying a set of attributes, or features, that distinguish a dataset from the control data and that can then be validated and tuned as it predicts the class membership of a new image or ROI. The classes are predefined by the supervisor, and one of the significant components of the process is to train the classifier with a dataset describing the features associated with each class.

On the face of it, feature extraction is relatively easy. Algorithms to automatically measure lengths, areas, volumes, contour lengths, and textures are common; however, the real challenge is in making them sensitive to real differences between cells and insensitive to the conditions under which the cell is imaged, which turns out to be a nontrivial task. This has resulted in many other features being developed, including ones based on histogram analysis, frequency analysis, structural representations, and deformation features providing a huge feature space that can be measured (Uchida 2013).

Many classifiers exist and they can be broken down into three broad categories: nearest neighbor, discriminant functions, and ensemble methods. Commonly used classifiers include *k*-nearest neighbor, decision trees, neural networks, and SVMs. Once trained, these classifiers can be very efficient. In one study, a nearest neighbor approach was used to classify nearly 80 million images into 75,000 classes (Torralba et al. 2008). One significant challenge, however, with all these approaches is that they are dependent on having a sufficiently large dataset to work on; otherwise, if the parameter space of the classifier is larger than the number of images, overfitting can occur. There also needs to be a large enough training set per class, with coverage of all features being extracted, a curse of the complexity being employed. Using data dimensionality, reduction techniques can address this issue to some extent (Van der Maaten et al. 2009), although significant planning and control experiments are essential in developing a robust, supervised classification system.

The performance of extracted features can be analyzed in a filtering process, again with many different algorithms available (Candia et al. 2014). Many features are not orthogonal, allowing the dimensionality of the dataset to be reduced using statistical techniques. An interesting question is whether the analysis of a sufficient number of spatial imaging features can compensate for the spectral limitations in analyzing molecular biomarkers. Recently, an analysis of 19 million cell divisions using 200 quantitative features and an SVM classifier was able to study the role of nearly 1000 genes in cell division using a small interfering RNA library, a feat that would have required several hundred well-characterized molecular probes (Neumann et al. 2010).

4.5.6 Image Analysis Software

In addition to several commercially available image analysis packages, a number of open-source software packages have been developed over the past 15 years with the growing use of high-content screening in scientific research and the pharmaceutical industry (Eliceiri et al. 2012). One of the first open-source systems designed for researchers is CellProfiler, which has a modular design and uses primary and secondary clumped object analyses for image segmentation, yielding a diverse range of measures of cellular and subcellular features, including size, shape, texture, and area (Carpenter et al. 2006). The GCellIQ package that runs in MATLAB® uses an iterative cluster phenotype model with improved gap statistics (Yin et al. 2008), whereas Fuji is a flexible platform for rapid algorithm testing based on the ImageJ platform (Schindelin et al. 2012). These approaches can require high-computational overheads and require familiarity with and tuning of parameters for each condition being imaged. PhenoRipper tries to reduce these barriers to analysis by using an unsupervised, nonsegmentation approach, simplified parameters, and intuitive user interfaces (Rajaram et al. 2012). A number of archives exist for open-source bioinformatics techniques and datasets, including Bioconductor (http://www.bioconductor.org) that brings together packages based on the statistical program language R.

4.5.7 Datasets, Interoperability, and Standards

One of the biggest challenges in high-content, high throughput analysis is having quantitative and standardized techniques for collecting and analyzing data that are well characterized, can be broadly applied, and are robust. There is also a growing need for interoperability, standards, and libraries to facilitate the exchange of information, access to datasets for testing and comparison of algorithms, and archiving of significant results, and to support publications. The Open Microscopy Environment Consortium is a broad-based group that aims to bring together a number of open-source packages in a move toward developing a more cohesive platform for accessing, processing, analyzing, sharing, and publishing image data (http://www.openmicroscopy.org/site). A number of repositories have also emerged to serve the microscopy community, distinct from the other clinical imaging systems. In addition to CELLPEDIA mentioned earlier, a number of image archives have also been developed, including The Cell: An Image Library which is a public repository of reviewed and annotated images and videos (https://www.cellimagelibrary.org/), the Allen Brain Atlas which is a multimodal, multiresolution neuroanatomical atlas containing extensive gene expression data (http://www.brain-map.org/), and CellFinder which maps validated gene and protein expression, phenotypes, and images to related cell types (http://cellfinder.de/).

There is still a need for better dissemination of imaging datasets in standardized formats along with analytical tools, associated metadata, and 'omics information. Metadata describing sample preparation, experimental parameters, and calibration will help improve reliability and reproducibility of results. PhenomicDB is a multiorganism phenotype–genotype database that aggregates information on many common model systems to enable comparative phenomics studies (http://www.phenomicdb.de/). For flow cytometry data, Cytobank (http://www.cytobank.org/) provides a platform for managing, analyzing, and sharing information, including software, metadata, and workflows.

One of the other big informatics challenges, particularly in comparative studies between organisms, is the lack of a universal structured format to describe phenotypic data. The annotation of rich disease states and mutant model systems is underdescribed by the lack of well-structured, computable representations for complex phenotypic datasets at the cell and tissue level, such as high-content imaging and gene expression profiles. A number of vocabularies to describe cell phenotypes were developed by the major cell line repositories, including the American Type Cell Collection and the European Collection of Cell Cultures, although these vocabularies are often species specific, based on the tissue of origin, or are not updated to have a full range of classifiers. The comprehensive and unbiased analysis of phenotypic data to develop classifications for well-defined cell types that are species independent, while including details on a wide range of macroscopic properties, is conceptually and computationally challenging. The Cell Ontology (http://cellontology.org/) is one example of a species-independent conceptual framework that tries to fill this gap between the biomolecular ontologies of the OBO Foundry and the previous vocabularies not based on phenotypic data that were developed, complemented by efforts to build specific ontologies for cell lines, beta cells, and neurons.

4.6 Challenges and Opportunities

Biophotonics plays a major role in reshaping how we define cell phenotypes as the primary technique for nondestructive analysis because of its sensitivity, scalability, and capacity to perform different types of measurements. The challenge of collecting and interpreting a multidimensional dataset favors semiautomated cytometry approaches using molecular probes against specific targets. However with increasing computational power there are an increasing number of approaches using endogenous image-based approaches to study spatial–temporal relationships with parameters that can be reproducibly measured but that have fuzzy biologic interpretation. The use of exogenous contrast agents will always simplify the process, however, and is likely to remain the preferred approach when appropriate agents exist and there is not a bottleneck in sample preparation.

In addition to optical imaging and 'omics methods, a number of other techniques are used for phenotyping cells, including mechanical stretching (Gossett et al. 2012) and dry mass analysis (Cheung et al. 2013) as well as characterization by subcellular features such as mitochondria. An increasing range of technologies are also being developed for studying the mass composition of cells (Popescu et al. 2014), with one study showing that single-cell measurements of nucleic and protein mass show significant variation across and within cell types (Cheung et al. 2013). There has also been very active development of phenotyping methods in other areas of the life sciences, including plant biotechnology (Sozzani and Benfey 2011). Finally, other parts of the electromagnetic spectrum are also used to image cells, most notably the soft X-ray region (McDermott et al. 2012).

There are many opportunities to integrate and automate different forms of analysis. The gap between high-content approaches where the throughput is normally limited by time and cost considerations and high-throughput approaches that are not as sensitive and measure fewer parameters is continuing to decrease, spurring the development of new technologies and computational techniques. As mentioned earlier, single-cell analysis techniques measuring the genome, transcriptome, proteome, and metabolome are having a significant impact across a range of fields, including cancer biology, immunology, stem cell research, neuroscience, and comparative biology. Other emerging areas of opportunity include susceptibility and persistence of pathogens in cells, validation of techniques in tissue engineering and regenerative medicine, and better monitoring of stem cell differentiation pathways. Comprehensive mapping of the interaction between the different biomolecules present in the cell is another area of opportunity, for example, mapping microRNA–mRNA and DNA–protein interactions, as well as studying the spatial and temporal dynamics and interactions within the nucleosome. The validation of many techniques in regenerative medicine also requires an understanding of whether engineered cells and tissues can safely and effectively supplement or replace tissues in the human body (Tsuji et al. 2010). It is clear that these techniques are increasingly finding applications in the clinic beyond more traditional uses of flow cytometry, karyotyping, and fertility treatments. They also provide a complementary form of analysis to the biophotonic tools and there are significant opportunities for utilizing the power of imaging and 'omics tools to drive biomedical discovery.

A broad goal in the field is to link genotype and phenotype through the combined analysis of phenomic and genomic data. The rapid proliferation of sequenced genomes, the growing array of collections with systematically mutated strains, and the development of high-throughput technologies have rapidly opened up the field of functional genomics in the past decade (Hunt-Newbury et al. 2007). Initially, phenotypic models were based on endpoint observations such as viability, but the automation of imaging has enabled researchers to begin linking dynamical processes (Neumann et al. 2010) and systematic analysis of complex pathways *in vivo* (Hamilton et al. 2005) to high-throughput systems to develop more complex phenotypic descriptions. For example, the DevStaR machine learning system can systematically and quantitatively characterize *C. elegans* in a functional genomic assay using a hierarchical object recognition approach to provide an automated identification of development stages in a mixed worm population (White et al. 2013).

A number of other short- to medium-term trends are emerging that spur technology development in different directions. The desire to minimize perturbation of cells and patients has led some researchers to develop virtual biopsies, wherein sufficient molecular and structural biomarkers can be extracted from nondestructive, relatively quick imaging assays. These may be fluid biopsies utilizing optical techniques to analyze blood or cerebral spinal fluid, or multiscale, multimodal techniques to analyze solid, deep tissues, but in either case the trend is toward a more mechanistic-based understanding of complex diseases and the level of heterogeneity that needs to be measured in order to effectively personalize care. Another trend is toward multifactorial, high-content, high-throughput studies where integration, automation, and reductions in cost and time enable longitudinal studies across multiple cell populations and combinations of environmental variables.

This multiplexing adds an additional level of complexity to automated analysis of datasets and other issues associated with big data.

An area of particular relevance for cell phenotyping but beyond the scope of this chapter is the identification, isolation, preparation, and preservation of samples. After a cell of interest is identified, it needs to be isolated for analysis, a process that is relatively simple in a fluid but typically requires enzymatic digestion for solid tissues. This digestion has two downsides: (1) the spatial information of arrangement of the tissue is lost and (2) some fragile biomarkers can be disrupted. Laser capture microdissection addresses these downsides by selectively cutting around the cells of interest with submicron precision to release cell and maintain the tissue morphology without contaminating the surrounding tissue (Decarlo et al. 2011), although it does face a number of efficiency challenges. The choice of preservation technique can have a significant and detrimental impact at the single-cell level, as indicated by a recent study that has found measureable changes in methylation levels caused by the freeze–thaw cycle (Riesco and Robles 2013).

In addition to the development of tools, there are also many areas of opportunity for computational methods. With the collection and analysis of huge amounts of data, the development of computational models for gene regulation at the single-cell level (Rosenfeld et al. 2005) is also an area of considerable interest. The increasing amount of verified data from single-cell metabolomic studies will also provide the needed rate information for mathematical models of cellular metabolism (Jol et al. 2012). However, some of the limitations of using molecular, physiological, and structural characterization have been highlighted by recent comparative genomics studies. In fast-evolving species, cell type-specific markers associated with slower evolving species were often found to be missing or modified, complicating comparison (Kortschak et al. 2003).

Eberwine et al. (2014) drew an analogy with the microbiome and raised the possibility that the organizing principle of tissues may be thought of as a functionally coherent assembly of an ecology of cells, whose interactions characterize the system-level phenotype. This ecosystem model of evolving communities provides a new viewpoint to frame biological processes and the emergence of disease and underscores the need to study phenotypes at the single-cell level. Although it is unclear whether there is a finite and measureable range of phenotypes or if they can be arranged in a robust taxonomy, it is clear that optical measurement and manipulation techniques and multidimensional analytical techniques will play a central role in helping to understand the biological significance of heterogeneity and will lead to a more comprehensive approach for the classification of cell phenotypes.

References

Aghaeepour, N., G. Finak et al. (2013). "Critical assessment of automated flow cytometry data analysis techniques." *Nature Methods* **10**(3): 228–238.

Altelaar, A. F. and A. J. Heck (2012). "Trends in ultrasensitive proteomics." *Current Opinion in Chemical Biology* **16**(1/2): 206–213.

Amirel, A. D., K. L. Davis et al. (2013). "viSNE enables visualization of high dimensional single-cell data and reveals phenotypic heterogeneity of leukemia." *Nature Biotechnology* **31**(6): 545–552.

Arendt, D. (2008). "The evolution of cell types in animals: Emerging principles from molecular studies." *Nature Reviews Genetics* **9**(11): 868–882.

Bahar, R., C. H. Hartmann et al. (2006). "Increased cell-to-cell variation in gene expression in ageing mouse heart." *Nature* **441**(7096): 1011–1014.

Batioglu, F., S. Demirel et al. (2013). "Fundus autofluorescence imaging in age-related macular degeneration." *Seminars in Ophthalmology*. http://www.ncbi.nlm.nih.gov/pubmed/23952079.

Behbehani, G. K., S. C. Bendall et al. (2012). "Single-cell mass cytometry adapted to measurements of the cell cycle." *Cytometry A* **81**(7): 552–566.

Bendall, S. C., E. F. Simonds et al. (2011). "Single-cell mass cytometry of differential immune and drug responses across a human hematopoietic continuum." *Science* **332**(6030): 687–696.

Bianconi, E., A. Piovesan et al. (2013). "An estimation of the number of cells in the human body." *Annals of Human Biology* **40**(6): 463–471.

Bickle, M. (2010). "The beautiful cell: High-content screening in drug discovery." *Analytical and Bioanalytical Chemistry* **398**(1): 219–226.

Bjornson, Z. B., G. P. Nolan et al. (2013). "Single-cell mass cytometry for analysis of immune system functional states." *Current Opinion in Immunology* **25**(4): 484–494.

Blainey, P. C. (2013). "The future is now: Single-cell genomics of bacteria and archaea." *FEMS Microbiology Reviews* **37**(3): 407–427.

Bodenmiller, B., E. R. Zunder et al. (2012). "Multiplexed mass cytometry profiling of cellular states perturbed by small-molecule regulators." *Nature Biotechnology* **30**(9): 858–867.

Bodo, J. and E. D. Hsi (2011). "Phosphoproteins and the dawn of functional phenotyping." *Pathobiology* **78**(2): 115–121.

Bovik, A. C. (2010). *Handbook of Image and Video Processing*. Elsevier Academic Press, Boston, MA.

Boyden, E. S., F. Zhang et al. (2005). "Millisecond-timescale, genetically targeted optical control of neural activity." *Nature Neuroscience* **8**(9): 1263–1268.

Brogan, J., F. Li et al. (2012). "Imaging molecular pathways: Reporter genes." *Radiation Research* **177**(4): 508–513.

Cai, L. (2013). "Turning single cells into microarrays by super-resolution barcoding." *Briefings in Functional Genomics* **12**(2): 75–80.

Cai, L., N. Friedman et al. (2006). "Stochastic protein expression in individual cells at the single molecule level." *Nature* **440**(7082): 358–362.

Calvi, L. M., G. B. Adams et al. (2003). "Osteoblastic cells regulate the haematopoietic stem cell niche." *Nature* **425**(6960): 841–846.

Campagnola, P. J., A. C. Millard et al. (2002). "Three-dimensional high-resolution second-harmonic generation imaging of endogenous structural proteins in biological tissues." *Biophysical Journal* **82**(1 Pt 1): 493–508.

Candia, J., J. R. Banavar et al. (2014). "Understanding health and disease with multidimensional single-cell methods." *Journal of Physics Condensed Matter* **26**(7): 073102.

Carpenter, A. E., T. R. Jones et al. (2006). "CellProfiler: Image analysis software for identifying and quantifying cell phenotypes." *Genome Biology* **7**(10): R100.

Chan, T. and J. Shen (2005). *Image Processing and Analysis: Variational, PDE, Wavelet, and Stochastic Methods*. SIAM, Philadelphia, PA.

Chan, J. W., D. S. Taylor et al. (2008). "Nondestructive identification of individual leukemia cells by laser trapping Raman spectroscopy." *Analytical Chemistry* **80**(6): 2180–2187.

Chandler, K. and R. Goldman (2013). "Glycoprotein disease markers and single protein-omics." *Molecular & Cellular Proteomics* **12**(4): 836–845.

Chen, T. W., T. J. Wardill et al. (2013). "Ultrasensitive fluorescent proteins for imaging neuronal activity." *Nature* **499**(7458): 295–300.

Cheung, M. C., R. LaCroix et al. (2013). "Intracellular protein and nucleic acid measured in eight cell types using deep-ultraviolet mass mapping." *Cytometry A* **83**(6): 540–551.

Clarke, G. M., J. T. Zubovits et al. (2014). "A novel, automated technology for multiplex biomarker imaging and application to breast cancer." *Histopathology* **64**(2): 242–255.

Cohen, A. A., T. Kalisky et al. (2009). "Protein dynamics in individual human cells: Experiment and theory." *PloS One* **4**(4): e4901.

Creighton, T. E. (1999). *Encyclopedia of Molecular Biology*. John Wiley, New York.

Decarlo, K., A. Emley et al. (2011). "Laser capture microdissection: Methods and applications." *Methods in Molecular Biology* **755**: 1–15.

Dedecker, P., F. C. De Schryver et al. (2013). "Fluorescent proteins: Shine on, you crazy diamond." *Journal of the American Chemical Society* **135**(7): 2387–2402.

Dejana, E. (2004). "Endothelial cell-cell junctions: Happy together." *Nature Reviews Molecular Cell Biology* **5**(4): 261–270.

Ding, H. and G. Popescu (2010). "Instantaneous spatial light interference microscopy." *Optics Express* **18**(2): 1569–1575.

Dupuy, D., N. Bertin et al. (2007). "Genome-scale analysis of in vivo spatiotemporal promoter activity in *Caenorhabditis elegans*." *Nature Biotechnology* **25**(6): 663–668.

Duraipandian, S., W. Zheng et al. (2013). "Near-infrared-excited confocal Raman spectroscopy advances in vivo diagnosis of cervical precancer." *Journal of Biomedical Optics* **18**(6): 067007.

Eberwine, J., J. Y. Sul et al. (2014). "The promise of single-cell sequencing." *Nature Methods* **11**(1): 25–27.

Eckert, M. A., P. Q. Vu et al. (2013). "Novel molecular and nanosensors for in vivo sensing." *Theranostics* **3**(8): 583–594.

Eldar, A. and M. B. Elowitz (2010). "Functional roles for noise in genetic circuits." *Nature* **467**(7312): 167–173.

Eliceiri, K. W., M. R. Berthold et al. (2012). "Biological imaging software tools." *Nature Methods* **9**(7): 697–710.

Elowitz, M. B., A. J. Levine et al. (2002). "Stochastic gene expression in a single cell." *Science* **297**(5584): 1183–1186.

Evensen, L., W. Link et al. (2010). "Imaged-based high-throughput screening for anti-angiogenic drug discovery." *Current Pharmaceutical Design* **16**(35): 3958–3963.

Failmezger, H., H. Frohlich et al. (2013). "Unsupervised automated high throughput phenotyping of RNAi time-lapse movies." *BMC Bioinformatics* **14**: 292.

Fard, A. M., A. Mahjoubfar et al. (2011). "Nomarski serial time-encoded amplified microscopy for high-speed contrast-enhanced imaging of transparent media." *Biomedical Optics Express* **2**(12): 3387–3392.

Femino, A. M., F. S. Fay et al. (1998). "Visualization of single RNA transcripts in situ." *Science* **280**(5363): 585–590.

Fritzsch, F. S., C. Dusny et al. (2012). "Single-cell analysis in biotechnology, systems biology, and biocatalysis." *Annual Review of Chemical and Biomolecular Engineering* **3**: 129–155.

Fuchs, F., G. Pau et al. (2010). "Clustering phenotype populations by genome-wide RNAi and multiparametric imaging." *Molecular Systems Biology* **6**: 370.

Gaj, T., C. A. Gersbach et al. (2013). "ZFN, TALEN, and CRISPR/Cas-based methods for genome engineering." *Trends in Biotechnology* **31**(7): 397–405.

Galvao, V., J. G. Miranda et al. (2010). "Modularity map of the network of human cell differentiation." *Proceedings of the National Academy of Sciences of the United States of America* **107**(13): 5750–5755.

Gasecka, A., A. Daradich et al. (2013). "Resolution and contrast enhancement in coherent anti-Stokes Raman-scattering microscopy." *Optics Letters* **38**(21): 4510–4513.

Gerdes, M. J., C. J. Sevinsky et al. (2013). "Highly multiplexed single-cell analysis of formalin-fixed, paraffin-embedded cancer tissue." *Proceedings of the National Academy of Sciences of the United States of America* **110**(29): 11982–11987.

Gleave, J. A., M. D. Wong et al. (2012). "Neuroanatomical phenotyping of the mouse brain with three-dimensional autofluorescence imaging." *Physiological Genomics* **44**(15): 778–785.

Gole, J., A. Gore et al. (2013). "Massively parallel polymerase cloning and genome sequencing of single cells using nanoliter microwells." *Nature Biotechnology* **31**(12): 1126–1132.

Gonzalez, V. D., N. K. Bjorkstrom et al. (2008). "Application of nine-color flow cytometry for detailed studies of the phenotypic complexity and functional heterogeneity of human lymphocyte subsets." *Journal of Immunological Methods* **330**(1/2): 64–74.

Gossett, D. R., H. T. Tse et al. (2012). "Hydrodynamic stretching of single cells for large population mechanical phenotyping." *Proceedings of the National Academy of Sciences of the United States of America* **109**(20): 7630–7635.

Guilak, F., D. M. Cohen et al. (2009). "Control of stem cell fate by physical interactions with the extracellular matrix." *Cell Stem Cell* **5**(1): 17–26.

Guo, H. S., P. Zhu et al. (2013). "Single-cell methylome landscapes of mouse embryonic stem cells and early embryos analyzed using reduced representation bisulfite sequencing." *Genome Research* **23**(12): 2126–2135.

Hamilton, B., Y. Dong et al. (2005). "A systematic RNAi screen for longevity genes in *C. elegans*." *Genes & Development* **19**(13): 1544–1555.

Hashimshony, T., F. Wagner et al. (2012). "CEL-Seq: Single-cell RNA-Seq by multiplexed linear amplification." *Cell Reports* **2**(3): 666–673.

Hatano, A., H. Chiba et al. (2011). "CELLPEDIA: A repository for human cell information for cell studies and differentiation analyses." *Database* **2011**: bar046.

Hellebust, A. and R. Richards-Kortum (2012). "Advances in molecular imaging: Targeted optical contrast agents for cancer diagnostics." *Nanomedicine* **7**(3): 429–445.

Hill, A. A., P. Lapan et al. (2007). "Impact of image segmentation on high-content screening data quality for SK-BR-3 cells." *BMC Bioinformatics* **8**: 340. doi:10.1186/1471-2105-8-340.

Hjortland, G. O., L. A. Meza-Zepeda et al. (2011). "Genome wide single cell analysis of chemotherapy resistant metastatic cells in a case of gastroesophageal adenocarcinoma." *BMC Cancer* **11**: 455.

Hou, Y., W. Fan et al. (2013). "Genome analyses of single human oocytes." *Cell* **155**(7): 1492–1506.

Huang, J., S. R. Kumar et al. (1999). "Spatial color indexing and applications." *International Journal of Computer Vision* **35**(3): 245–268.

Huang, K., M. Velliste et al. (2003). "Feature reduction for improved recognition of subcellular location patterns in fluorescence microscope images." *Proceedings of SPIE*. http://proceedings.spiedigitallibrary.org/volume.aspx?volumeid=3565.

Huisken, J., J. Swoger et al. (2004). "Optical sectioning deep inside live embryos by selective plane illumination microscopy." *Science* **305**(5686): 1007–1009.

Hunt-Newbury, R., R. Viveiros et al. (2007). "High-throughput in vivo analysis of gene expression in *Caenorhabditis elegans*." *PLoS Biology* **5**(9): e237.

Ibanez, A. J., S. R. Fagerer et al. (2013). "Mass spectrometry-based metabolomics of single yeast cells." *Proceedings of the National Academy of Sciences of the United States of America* **110**(22): 8790–8794.

Ingolia, N. T., S. Ghaemmaghami et al. (2009). "Genome-wide analysis in vivo of translation with nucleotide resolution using ribosome profiling." *Science* **324**(5924): 218–223.

Inya-Agha, O., N. Klauke et al. (2007). "Spectroscopic probing of dynamic changes during stimulation and cell remodeling in the single cardiac myocyte." *Analytical Chemistry* **79**(12): 4581–4587.

Islam, S., U. Kjallquist et al. (2011). "Characterization of the single-cell transcriptional landscape by highly multiplex RNA-seq." *Genome Research* **21**(7): 1160–1167.

Islam, S., A. Zeisel et al. (2014). "Quantitative single-cell RNA-seq with unique molecular identifiers." *Nature Methods* **11**(2): 163–166.

Itzkovitz, S. and A. van Oudenaarden (2011). "Validating transcripts with probes and imaging technology." *Nature Methods* **8**(4): S12–S19.

Jan, M., T. M. Snyder et al. (2012). "Clonal evolution of preleukemic hematopoietic stem cells precedes human acute myeloid leukemia." *Science Translational Medicine* **4**(149): 149ra118.

Jol, S. J., A. Kummel et al. (2012). "System-level insights into yeast metabolism by thermodynamic analysis of elementary flux modes." *PLoS Computational Biology* **8**(3): e1002415.

Keller, P. J. (2013). "In vivo imaging of zebrafish embryogenesis." *Methods* **62**(3): 268–278.

Kelly, J. G., K. T. Cheung et al. (2010). "A spectral phenotype of oncogenic human papillomavirus-infected exfoliative cervical cytology distinguishes women based on age." *Clinica Chimica Acta* **411**(15/16): 1027–1033.

Klymchenko, A. S. and Y. Mely (2013). "Fluorescent environment-sensitive dyes as reporters of biomolecular interactions." *Progress in Molecular Biology and Translational Science* **113**: 35–58.

Konermann, S., M. D. Brigham et al. (2013). "Optical control of mammalian endogenous transcription and epigenetic states." *Nature* **500**(7463): 472–476.

Kortschak, R. D., G. Samuel et al. (2003). "EST analysis of the cnidarian *Acropora millepora* reveals extensive gene loss and rapid sequence divergence in the model invertebrates." *Current Biology* **13**(24): 2190–2195.

Kothari, S., J. H. Phan et al. (2013). "Pathology imaging informatics for quantitative analysis of whole-slide images." *Journal of the American Medical Informatics Association* **20**(6): 1099–1108.

Kralj, J. M., A. D. Douglass et al. (2012). "Optical recording of action potentials in mammalian neurons using a microbial rhodopsin." *Nature Methods* **9**(1): 90–95.

Krutzik, P. O., J. M. Crane et al. (2008). "High-content single-cell drug screening with phosphospecific flow cytometry." *Nature Chemical Biology* **4**(2): 132–142.

Lau, A. Y., L. P. Lee et al. (2008). "An integrated optofluidic platform for Raman-activated cell sorting." *Lab on a Chip* **8**(7): 1116–1120.

Lee, S., H. Chon et al. (2014). "Rapid and sensitive phenotypic marker detection on breast cancer cells using surface-enhanced Raman scattering (SERS) imaging." *Biosensors & Bioelectronics* **51**: 238–243.

Levsky, J. M., S. M. Shenoy et al. (2002). "Single-cell gene expression profiling." *Science* **297**(5582): 836–840.

Li, J., L. Chen et al. (2013b). "Cage the firefly luciferin!—A strategy for developing bioluminescent probes." *Chemical Society Reviews* **42**(2): 662–676.

Li, F., Z. Yin et al. (2013a). "Bioimage informatics for systems pharmacology." *PLoS Computational Biology* **9**(4): e1003043.

Loo, L. H., L. F. Wu et al. (2007). "Image-based multivariate profiling of drug responses from single cells." *Nature Methods* **4**(5): 445–453.

Lorenz, K., M. Sicker et al. (2008). "Multilineage differentiation potential of human dermal skin-derived fibroblasts." *Experimental Dermatology* **17**(11): 925–932.

Lubeck, E. and L. Cai (2012). "Single-cell systems biology by super-resolution imaging and combinatorial labeling." *Nature Methods* **9**(7): 743–748.

Lugli, E., M. Roederer et al. (2010). "Data analysis in flow cytometry: The future just started." *Cytometry A* **77**(7): 705–713.

Maglione, M. and S. J. Sigrist (2013). "Seeing the forest tree by tree: Super-resolution light microscopy meets the neurosciences." *Nature Neuroscience* **16**(7): 790–797.

Mahmoudi, F., J. Shanbehzadeh et al. (2003). "Image retrieval based on shape similarity by edge orientation autocorrelogram." *Pattern Recognition* **36**(8): 1725–1736.

Marrison, J., L. Raty et al. (2013). "Ptychography—A label free, high-contrast imaging technique for live cells using quantitative phase information." *Scientific Reports* **3**: 2369.

Martini, J., W. Hellmich et al. (2007). "Systems nanobiology: From quantitative single molecule biophysics to microfluidic-based single cell analysis." *Sub-Cellular Biochemistry* **43**: 301–321.

McDermott, G., D. M. Fox et al. (2012). "Visualizing and quantifying cell phenotype using soft X-ray tomography." *BioEssays* **34**(4): 320–327.

McLean, J. S., M. J. Lombardo et al. (2013). "Genome of the pathogen *Porphyromonas gingivalis* recovered from a biofilm in a hospital sink using a high-throughput single-cell genomics platform." *Genome Research* **23**(5): 867–877.

Mensaert, K., S. Denil et al. (2014). "Next-generation technologies and data analytical approaches for epigenomics." *Environmental and Molecular Mutagenesis* **55**: 155–170. doi:10.1002/em.21841.

Miao, Q., A. P. Reeves et al. (2012). "Multimodal 3D imaging of cells and tissue, bridging the gap between clinical and research microscopy." *Annals of Biomedical Engineering* **40**(2): 263–276.

Mitchell, A. C., R. Bharadwaj et al. (2014). "The genome in three dimensions: A new frontier in human brain research." *Biological Psychiatry* **75**(12): 961–969.

Moro, A. J., P. J. Cywinski et al. (2010). "An ATP fluorescent chemosensor based on a Zn(II)-complexed dipicolylamine receptor coupled with a naphthalimide chromophore." *Chemical Communications* **46**(7): 1085–1087.

Mortazavi, A., B. A. Williams et al. (2008). "Mapping and quantifying mammalian transcriptomes by RNA-Seq." *Nature Methods* **5**(7): 621–628.

Mungall, C. J., C. Torniai et al. (2012). "Uberon, an integrative multi-species anatomy ontology." *Genome Biology* **13**(1): R5.

Murali Krishna, C., G. Kegelaer et al. (2005). "Characterisation of uterine sarcoma cell lines exhibiting MDR phenotype by vibrational spectroscopy." *Biochimica et Biophysica Acta* **1726**(2): 160–167.

Navin, N., J. Kendall et al. (2011). "Tumour evolution inferred by single-cell sequencing." *Nature* **472**(7341): 90–94.

Nemes, P., A. M. Knolhoff et al. (2011). "Metabolic differentiation of neuronal phenotypes by single-cell capillary electrophoresis-electrospray ionization-mass spectrometry." *Analytical Chemistry* **83**(17): 6810–6817.

Neumann, B., M. Held et al. (2006). "High-throughput RNAi screening by time-lapse imaging of live human cells." *Nature Methods* **3**(5): 385–390.

Neumann, B., T. Walter et al. (2010). "Phenotypic profiling of the human genome by time-lapse microscopy reveals cell division genes." *Nature* **464**(7289): 721–727.

Newman, J. R., S. Ghaemmaghami et al. (2006). "Single-cell proteomic analysis of *S. cerevisiae* reveals the architecture of biological noise." *Nature* **441**(7095): 840–846.

Olivier, N., M. A. Luengo-Oroz et al. (2010). "Cell lineage reconstruction of early zebrafish embryos using label-free nonlinear microscopy." *Science* **329**(5994): 967–971.

Pardo-Martin, C., A. Allalou et al. (2013). "High-throughput hyperdimensional vertebrate phenotyping." *Nature Communications* **4**: 1467.

Pham, D. L., C. Y. Xu et al. (2000). "Current methods in medical image segmentation." *Annual Review of Biomedical Engineering* **2**: 315–337.

Ponten, F., M. Gry et al. (2009). "A global view of protein expression in human cells, tissues, and organs." *Molecular Systems Biology* **5**: 337.

Popescu, G., K. Park et al. (2014). "New technologies for measuring single cell mass." *Lab on a Chip* **14**(4): 646–652.

Qiu, P., E. F. Simonds et al. (2011). "Extracting a cellular hierarchy from high-dimensional cytometry data with SPADE." *Nature Biotechnology* **29**(10): 886–891.

Raghunathan, A., H. R. Ferguson et al. (2005). "Genomic DNA amplification from a single bacterium." *Applied and Environmental Microbiology* **71**(6): 3342–3347.

Rajaram, S., B. Pavie et al. (2012). "PhenoRipper: Software for rapidly profiling microscopy images." *Nature Methods* **9**(7): 635–637.

Ramskold, D., S. Luo et al. (2012). "Full-length mRNA-Seq from single-cell levels of RNA and individual circulating tumor cells." *Nature Biotechnology* **30**(8): 777–782.

Rappaz, B., B. Breton et al. (2014). "Digital holographic microscopy: A quantitative label-free microscopy technique for phenotypic screening." *Combinatorial Chemistry & High Throughput Screening* **17**(1): 80–88.

Raser, J. M. and E. K. O'Shea (2005). "Noise in gene expression: Origins, consequences, and control." *Science* **309**(5743): 2010–2013.

Riesco, M. F. and V. Robles (2013). "Cryopreservation causes genetic and epigenetic changes in zebrafish genital ridges." *PloS One* **8**(6): e67614.

Rittscher, J. (2010). "Characterization of biological processes through automated image analysis." *Annual Review of Biomedical Engineering* **12**: 315–344.

Robinson, J. P., B. Rajwa et al. (2012). "Computational analysis of high-throughput flow cytometry data." *Expert Opinion on Drug Discovery* **7**(8): 679–693.

Rosenfeld, N., J. W. Young et al. (2005). "Gene regulation at the single-cell level." *Science* **307**(5717): 1962–1965.

Russ, J. C. (2006). *The Image Processing Handbook*. CRC Press, Boca Raton, FL.

Salehi-Reyhani, A., J. Kaplinsky et al. (2011). "A first step towards practical single cell proteomics: A microfluidic antibody capture chip with TIRF detection." *Lab on a Chip* **11**(7): 1256–1261.

Sanchez, C., C. Lachaize et al. (1999). "Grasping at molecular interactions and genetic networks in *Drosophila melanogaster* using FlyNets, an Internet database." *Nucleic Acids Research* **27**(1): 89–94.

Santin, G., M. Paulis et al. (2013). "Autofluorescence properties of murine embryonic stem cells during spontaneous differentiation phases." *Lasers in Surgery and Medicine* **45**(9): 597–607.

Schindelin, J., I. Arganda-Carreras et al. (2012). "Fiji: An open-source platform for biological-image analysis." *Nature Methods* **9**(7): 676–682.

Schubert, W., B. Bonnekoh et al. (2006). "Analyzing proteome topology and function by automated multidimensional fluorescence microscopy." *Nature Biotechnology* **24**(10): 1270–1278.
Shalek, A. K., R. Satija et al. (2013). "Single-cell transcriptomics reveals bimodality in expression and splicing in immune cells." *Nature* **498**(7453): 236–240.
Shi, Q., L. Qin et al. (2012). "Single-cell proteomic chip for profiling intracellular signaling pathways in single tumor cells." *Proceedings of the National Academy of Sciences of the United States of America* **109**(2): 419–424.
Shrestha, B. and A. Vertes (2009). "In situ metabolic profiling of single cells by laser ablation electrospray ionization mass spectrometry." *Analytical Chemistry* **81**(20): 8265–8271.
Sigal, A., R. Milo et al. (2006). "Variability and memory of protein levels in human cells." *Nature* **444**(7119): 643–646.
Snijder, B., R. Sacher et al. (2012). "Single-cell analysis of population context advances RNAi screening at multiple levels." *Molecular Systems Biology* **8**: 579.
Sommer, C. and D. W. Gerlich (2013). "Machine learning in cell biology—Teaching computers to recognize phenotypes." *Journal of Cell Science* **126**(Pt 24): 5529–5539.
Sozzani, R. and P. N. Benfey (2011). "High-throughput phenotyping of multicellular organisms: Finding the link between genotype and phenotype." *Genome Biology* **12**(3): 219.
Stahlberg, A., C. Thomsen et al. (2012). "Quantitative PCR analysis of DNA, RNAs, and proteins in the same single cell." *Clinical Chemistry* **58**(12): 1682–1691.
Steinhauser, M. L. and C. P. Lechene (2013). "Quantitative imaging of subcellular metabolism with stable isotopes and multi-isotope imaging mass spectrometry." *Seminars in Cell & Developmental Biology* **24**(8/9): 661–667.
Stirman, J. N., M. Brauner et al. (2010). "High-throughput study of synaptic transmission at the neuromuscular junction enabled by optogenetics and microfluidics." *Journal of Neuroscience Methods* **191**(1): 90–93.
Sugar, C. A. and G. M. James (2003). "Finding the number of clusters in a dataset: An information-theoretic approach." *Journal of the American Statistical Association* **98**(463): 750–763.
Sul, J. Y., T. K. Kim et al. (2012). "Perspectives on cell reprogramming with RNA." *Trends in Biotechnology* **30**(5): 243–249.
Sun, C. K., S. W. Chu et al. (2004). "Higher harmonic generation microscopy for developmental biology." *Journal of Structural Biology* **147**(1): 19–30.
Tang, F. C., C. Barbacioru et al. (2009). "mRNA-Seq whole-transcriptome analysis of a single cell." *Nature Methods* **6**(5): 377–382.
Tanner, S. D., V. I. Baranov et al. (2013). "An introduction to mass cytometry: Fundamentals and applications." *Cancer Immunology, Immunotherapy* **62**(5): 955–965.
Tarnok, A. (2004). "New technologies for the human cytome project." *Journal of Biological Regulators and Homeostatic Agents* **18**(2): 92–95.
Tibshirani, R., G. Walther et al. (2001). "Estimating the number of clusters in a data set via the gap statistic." *Journal of the Royal Statistical Society Series B Statistical Methodology* **63**: 411–423.
Torralba, A., R. Fergus et al. (2008). "80 million tiny images: A large data set for nonparametric object and scene recognition." *IEEE Transactions on Pattern Analysis and Machine Intelligence* **30**(11): 1958–1970.
Tsien, R. Y. (1998). "The green fluorescent protein." *Annual Review of Biochemistry* **67**: 509–544.
Tsuji, O., K. Miura et al. (2010). "Therapeutic potential of appropriately evaluated safe-induced pluripotent stem cells for spinal cord injury." *Proceedings of the National Academy of Sciences of the United States of America* **107**(28): 12704–12709.
Uchida, S. (2013). "Image processing and recognition for biological images." *Development, Growth & Differentiation* **55**(4): 523–549.
Urban, P. L., T. Schmid et al. (2011). "Multidimensional analysis of single algal cells by integrating microspectroscopy with mass spectrometry." *Analytical Chemistry* **83**(5): 1843–1849.
Valentine, J. W., A. G. Collins et al. (1994). "Morphological complexity increase in metazoans." *Paleobiology* **20**(2): 131–142.
Valet, G., J. F. Leary et al. (2004). "Cytomics—New technologies: Towards a human cytome project." *Cytometry A* **59**(2): 167–171.

Valet, G. and A. Tarnok (2004). "Potential and challenges of a human cytome project." *Journal of Biological Regulators and Homeostatic Agents* **18**(2): 87–91.

Van der Maaten, L., E. Postma et al. (2009). "Dimensionality reduction: A comparative review." *Journal of Machine Learning Research* **10**: 1–41.

Vega, S. L., E. Liu et al. (2012). "High-content imaging-based screening of microenvironment-induced changes to stem cells." *Journal of Biomolecular Screening* **17**(9): 1151–1162.

Vickaryous, M. K. and B. K. Hall (2006). "Human cell type diversity, evolution, development, and classification with special reference to cells derived from the neural crest." *Biological Reviews of the Cambridge Philosophical Society* **81**(3): 425–455.

Vogel, C. and E. M. Marcotte (2012). "Insights into the regulation of protein abundance from proteomic and transcriptomic analyses." *Nature Reviews Genetics* **13**(4): 227–232.

Walker, B. N., C. Antonakos et al. (2013). "Metabolic differences in microbial cell populations revealed by nanophotonic ionization." *Angewandte Chemie* **52**(13): 3650–3653.

Walsh, M. J., S. W. Bruce et al. (2009). "Discrimination of a transformation phenotype in Syrian golden hamster embryo (SHE) cells using ATR-FTIR spectroscopy." *Toxicology* **258**(1): 33–38.

Wang, J., H. C. Fan et al. (2012). "Genome-wide single-cell analysis of recombination activity and de novo mutation rates in human sperm." *Cell* **150**(2): 402–412.

Wang, Q., X. Zhu et al. (2013). "Single-cell genomics: An overview." *Frontiers in Biology* **8**(6): 569–576.

White, A. G., B. Lees et al. (2013). "DevStaR: High-throughput quantification of *C. elegans* developmental stages." *IEEE Transactions on Medical Imaging* **32**(10): 1791–1803.

Xue, Z. G., K. Huang et al. (2013). "Genetic programs in human and mouse early embryos revealed by single-cell RNA sequencing." *Nature* **500**(7464): 593–597.

Yan, L. Y., M. Y. Yang et al. (2013). "Single-cell RNA-Seq profiling of human preimplantation embryos and embryonic stem cells." *Nature Structural & Molecular Biology* **20**(9): 1131–1139.

Yan, M. and K. Ye (2007). "Determining the number of clusters using the weighted gap statistic." *Biometrics* **63**(4): 1031–1037.

Yin, Z., X. B. Zhou et al. (2008). "Using iterative cluster merging with improved gap statistics to perform online phenotype discovery in the context of high-throughput RNAi screens." *BMC Bioinformatics* **9**: 264. doi:10.1186/1471-2105-9-264.

Young, D. W., A. Bender et al. (2008). "Integrating high-content screening and ligand-target prediction to identify mechanism of action." *Nature Chemical Biology* **4**(1): 59–68.

Yu, J., J. Xiao et al. (2006). "Probing gene expression in live cells, one protein molecule at a time." *Science* **311**(5767): 1600–1603.

Yuan, H., J. K. Register et al. (2013). "Plasmonic nanoprobes for intracellular sensing and imaging." *Analytical and Bioanalytical Chemistry* **405**(19): 6165–6180.

Yun, H., K. Kim et al. (2013). "Cell manipulation in microfluidics." *Biofabrication* **5**(2): 022001.

Zavalin, A., E. M. Todd et al. (2012). "Direct imaging of single cells and tissue at sub-cellular spatial resolution using transmission geometry MALDI MS." *Journal of Mass Spectrometry* **47**(11): i.

Zenobi, R. (2013). "Single-cell metabolomics: Analytical and biological perspectives." *Science* **342**(6163): 1243259.

Ziegler, U. and P. Groscurth (2004). "Morphological features of cell death." *News in Physiological Sciences* **19**: 124–128.

Zipfel, W. R., R. M. Williams et al. (2003). "Live tissue intrinsic emission microscopy using multiphoton-excited native fluorescence and second harmonic generation." *Proceedings of the National Academy of Sciences of the United States of America* **100**(12): 7075–7080.

Zong, C., S. Lu et al. (2012). "Genome-wide detection of single-nucleotide and copy-number variations of a single human cell." *Science* **338**(6114): 1622–1626.

5

Functional Near-Infrared Spectroscopy and Its Applications in Neurosciences

Fengmei Lu, Xiaohong Lin, Jing Li, and Zhen Yuan

CONTENTS

5.1 Introduction

Since the mid-1970s, functional near-infrared spectroscopy (fNIRS) has been developing a noninvasive technique to investigate brain cerebral hemodynamic levels associated with brain activity under different stimuli by measuring the absorption coefficient of the near-infrared (NIR) light between 650 and 950 nm [1–8]. Compared to other functional imaging modalities, such as functional magnetic resonance imaging (fMRI) and positron emission tomography (PET), fNIRS has the advantages of portable, convenience, and low cost, and more importantly, it offers unsurpassed high temporal resolution and quantitative information for both oxyhemoglobin and deoxyhemoglobin, which is essential for revealing rapid changes of dynamic patterns of brain activities including changes of blood oxygen, blood volume, and blood flow.

As a neuroimaging method, fNIRS enables continuous and noninvasively monitoring of changes in blood oxygenation and blood volume related to human brain function.

fNIRS can be implemented in the form of a wearable and noninvasive or minimally intrusive device, and it has the capacity to monitor brain activity under real-life conditions and in everyday environments. During neural stimulus processing, there will be local increase in blood flow, blood volume, and blood oxygenation in a stereotyped hemodynamic response. Recently, advances in the understanding of light propagation in diffusive media (also known as photon migration) and technical developments in optoelectronics components have made it possible to extract valuable optical/hemodynamic information from human brain. Different fNIRS instruments including commercial systems or laboratory prototypes were developed and used effectively in preclinical and clinical studies. Obtaining measurements of the hemodynamic response of localized regions of the brain allows for inferences to be made regarding local neural activity. In this chapter, first the history and basic theory of fNIRS will be discussed. Then the representative instrumentations for fNIRS will be introduced. Finally, the applications of fNIRS in neurological disorders and cognitive neurosciences will be discussed.

5.2 Basic Principles of fNIRS

Frans Jöbsis, the founder of *in vivo* fNIRS, reported the first real-time noninvasive detection of hemoglobin oxygenation using transillumination spectroscopy in 1977 [4]. Jöbsis and Chances also used fNIRS to study cerebral oxygenation in human subjects after concerning the applications of this technique in laboratory animals [9–11]. Later Ferrari investigated the effects of carotid artery compression on regional cerebral blood oxygenation and blood volume of cerebrovascular patients together with the data on newborn brain measurements by utilizing prototype fNIRS instruments [12,13]. Importantly, Delpy performed the first quantitative measurement of various oxygenation and hemodynamic parameters in newborn infants including changes in oxygenated (HbO_2), deoxygenated (HbR), and total hemoglobin (HbT) concentrations; cerebral blood volume; and cerebral blood flow [14,15].

fNIRS is based on the physical and physiological mechanisms that human tissues are relatively transparent to light in the NIR spectral window and the relatively high attenuation of NIR light in tissue is due to the main chromophore hemoglobin (the oxygen transport red blood cell protein) located in small vessels of the microcirculation, such as capillary and arteriolar and venular beds. It is weakly sensitive to blood vessels >1 mm because they completely absorb the light. Given the fact that arterial blood volume fraction is approximately 30% in human brain [16,17], the fNIRS technique offers the possibility to obtain information mainly concerning oxygenation and blood volume changes occurring within the venous compartments.

fNIRS is a noninvasive and safe neuroimaging technique that utilizes laser diode and/or light-emitting diode light sources spanning the optical window and flexible fiber optics to carry the NIR light from (source) and to (detector) tissues. Fiber optics are very suitable for any head position and posture. fNIRS measurements can be performed in natural environments without the need for restraint or sedation. Adequate depth of NIR light penetration (almost one-half of the source–detector distance) can be achieved using a source–detector distance around 3 cm. The selection of the optimal source–detector distance depends on NIR light intensity and wavelength, as well as the age of the subject and the head region measured. As a consequence of the complex light scattering effect by different tissue layers, the length of the NIR light path through tissue is longer than the physical distance between the source and the detector.

5.3 Theory Model for fNIRS

According to Beer's law [6], the wavelength-dependent tissue optical density changes can be written in terms of the changes of the chromophores including HbO_2 and HbR at time t with wavelength λ:

$$\begin{bmatrix} \Delta \mathrm{OD}(r,t)\big|_{\lambda_1} \\ \Delta \mathrm{OD}(r,t)\big|_{\lambda_2} \end{bmatrix} = \mathrm{DPF}(r)l(r) \begin{bmatrix} \varepsilon_1(\lambda_1) & \varepsilon_2(\lambda_1) \\ \varepsilon_1(\lambda_2) & \varepsilon_2(\lambda_2) \end{bmatrix} \begin{bmatrix} \Delta \mathrm{HbO_2}(r,t) \\ \Delta \mathrm{HbR}(r,t) \end{bmatrix} \tag{5.1}$$

where:

OD is the optical density as determined from the negative log ratio of the detected intensity of light with respect to the incident intensity of light using continuous-wave (CW) measurements

ΔOD is the optical density change (unitless quantity) at the position r

DPF(r) is the unitless differential path length factor

$l(r)$ (mm) is the distance between the source and the detector

$\varepsilon_i(\lambda)$ is the extinction coefficient of the ith chromophore at wavelength λ of laser source

ΔHbO_2 and ΔHbR (μM) are the chromophore concentration changes for oxy- and deoxy-hemoglobin, respectively

After multiplying the inverse matrix of the extinction coefficients for both sides of Equation 5.1, the time series matrix for the changes of HbO_2 and HbR is written as

$$\begin{bmatrix} \Delta \mathrm{HbO_2}(r,t) \\ \Delta \mathrm{HbR}(r,t) \end{bmatrix} = \frac{\begin{bmatrix} Q_{\mathrm{HbO_2}}(r,t) \\ Q_{\mathrm{HbR}}(r,t) \end{bmatrix}}{\mathrm{DPF}(r)l(r)} \tag{5.2}$$

where:

$Q(r,t)$ vectors are the product of the inversion matrix of the extinction coefficients and the optical density change vectors

Similar operational procedures could be extended to the nth wavelength case based on regularization methods:

$$\begin{bmatrix} \Delta \mathrm{HbO_2} \\ \Delta \mathrm{HbR} \end{bmatrix} = (E^T R^{-1} E)^{-1} E^T R^{-1} \frac{\begin{bmatrix} \Delta \mathrm{OD}\big|_{\lambda_1} \\ \Delta \mathrm{OD}\big|_{\lambda_2} \\ \cdots \\ \Delta \mathrm{OD}\big|_{\lambda_n} \end{bmatrix}}{\mathrm{DPF} \times l}; \quad E = \begin{bmatrix} \varepsilon_1(\lambda_1) & \varepsilon_2(\lambda_1) \\ \cdots & \cdots \\ \varepsilon_1(\lambda_n) & \varepsilon_2(\lambda_n) \end{bmatrix} \tag{5.3}$$

where:

the matrix $\boldsymbol{E}$ is the extinction coefficient matrix

$\boldsymbol{R}$ is defined as the *a priori* estimate of the covariance of the measurement error

The change of total hemoglobin concentration ΔHbT (μM) is defined as the sum of ΔHbO_2 and ΔHbR.

5.4 fNIRS Instrumentations

Different fNIRS instruments with related key features, advantages and disadvantages, and parameters measurable by using different fNIRS techniques were developed and are listed in Table 5.1 [18]. Briefly, three typical signal measurement techniques using NIR

TABLE 5.1

Three Typical fNIRS Measurement Systems

	fNIRS Technique-Based Instrumentation		
Main Characteristics	**CW**	**FD**	**TD**
Sampling rate (Hz)	≤100	≤50	≤10
Spatial resolution (cm)	≤1	≤1	≤1
Penetration depth with a 4 cm source–detector distance	Low	Deep	Deep
Discrimination between cerebral and extracerebral tissue (scalp, skull, CSF)	n.a.	Feasible	Feasible
Possibility to measure deep brain structures	Feasible on newborns	Feasible on newborns	Feasible on newborns
Instrument size	Some bulky, some small	Bulky	Bulky
Instrument stabilization	n.r.	n.r.	Required
Transportability	Some easy, some feasible	Feasible	Feasible
Instrument cost	Some low, some high	Very high	Very high
Telemetry	Available	Difficult	Not easy
Measurable parameters			
$[HbO_2]$, [HbR], [HbT]	Yes, changes	Yes, absolute value	Yes, absolute value
Scattering and absorption coefficient and path length measurement	No	Yes	Yes
Tissue HbO_2 saturation measurement (%)	No	Yes	Yes

Source: Marco, F. and Valentina Q., *NeuroImage* 3, 49, 2012.

CSF, cerebrospinal fluid; HbO_2, oxyhemoglobin; HbT, HbO_2 + HbR; HbR, deoxyhemoglobin; n.a., not available; n.r., not required.

light are currently being used for optical tissue imaging/fNIRS: CW, time-domain (TD), and frequency-domain (FD) methods. CW fNIRS systems directly measure the intensity of light transmitted or reflected through the tissue. The light source used in CW systems generally has a constant intensity or is modulated at a low frequency (a few kilohertz). TD systems use short laser pulses, with temporal spread below a nanosecond, and detect the increased spread of the pulse after passing through the tissue. FD systems use an amplitude-modulated source at a high frequency (a few hundred megahertz) and measure the attenuation of amplitude and phase shift of the transmitted signal. Typically, in this approach, a radiofrequency oscillator drives a laser diode and provides a reference signal for phase measurement. Among the three methods, the CW approach is relatively cheap and easy to implement. So far, CW setups have also been the most used optical neuroimaging/spectroscopy systems.

The development of fNIRS instrumentation started in 1992 with a single-channel system with low temporal resolution and poor sensitivity. In 1995, the multichannel systems (the first 10-channel system) were reported. The present high temporal resolution multichannel systems, using the three different fNIRS techniques and complex data analysis systems, provide simultaneous multiple measurements and display the results in the form of a map or image over a specific cortical area or the whole brain. The potential that exists for fNIRS more than for any other neuroimaging modality is represented by the realization of multichannel wearable and/or wireless systems that allow fNIRS measurements even in normal daily activities (Figure 5.1).

The main commercially available transportable fNIRS systems are provided in Table 5.2, which utilize fiber optic bundles [18], and the typical instrumentations are also provided in Figure 5.2 [19]. The disadvantage of using fiber optic bundles is that the fibers are often heavy and of limited flexibility, perhaps provoking discomfort (especially in patients). In addition, these fNIRS systems require that the subject's head position not move beyond the length of the fiber optic bundles. Multichannel wearable and/or wireless systems could make fNIRS measurements more comfortable. These advanced fNIRS systems would represent a useful tool for evaluating brain activation related to cognitive tasks performed in normal daily activities. The main commercially available wearable/wireless fNIRS systems are provided in Table 5.3.

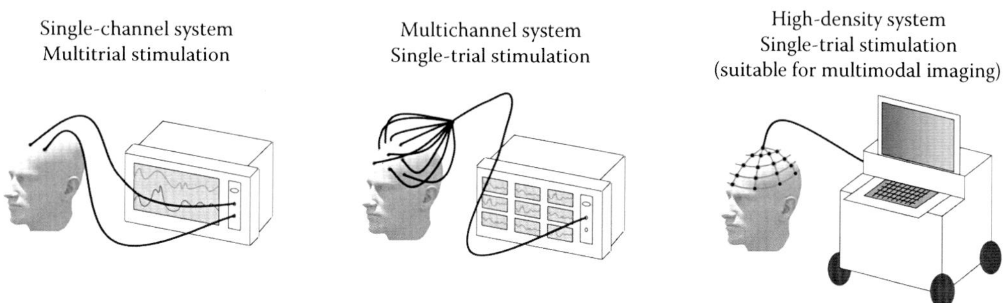

FIGURE 5.1
Sketch of the development of fNIRS instrumentation from single channel with low temporal resolution and poor sensitivity up to the multichannel systems.

TABLE 5.2

The Commercially Available Transportable fNIRS Systems

Instrument	Technique	Year of Release	Number of Channels	Company	Website
Dynot Compact	CW	2004	288–2049	NIRx, Los Angeles, CA	http://www.nirx.net
ETG-4000	CW	2003	52	Hitachi, Japan; Twinsburg, OH	http://www.hitachimed.com
ETG-7100	CW	2007	72–120	Hitachi, Japan; Twinsburg, OH	http://www.hitachimed.com
OXYMON MkIII	CW	2003	Up to 96	Artinis, Elst, The Netherlands	http://www.artinis.com
NIRO-200	CW	2008	10	Hamamatsu, Japan	http://www.hamamatsu.com
NIRS2 CE	CW	2007	16	TechEn, Inc., Milford, MA	http://www.nirsoptix.com
CW6	CW	2009	20–1024	TechEn, Inc., Milford, MA	http://www.nirsoptix.com
FOIRE-3000	CW	2007	52	Shimadzu, Kyoto, Japan	http://www.med.shimadzu.co.jp
NIRScout	CW	2008	128–1536	NIRx, Los Angeles, CA	http://www.nirx.net
HD-NI	CW	2009	Over 200	Cephalogics, Cambridge, MA	http://www.alliedminds.com
Imagent	FD	2001	Up to 128	ISS, Champaign, IL	http://www.iss.com

Source: Marco, F. and Valentina Q., *NeuroImage* 3, 49, 2012.

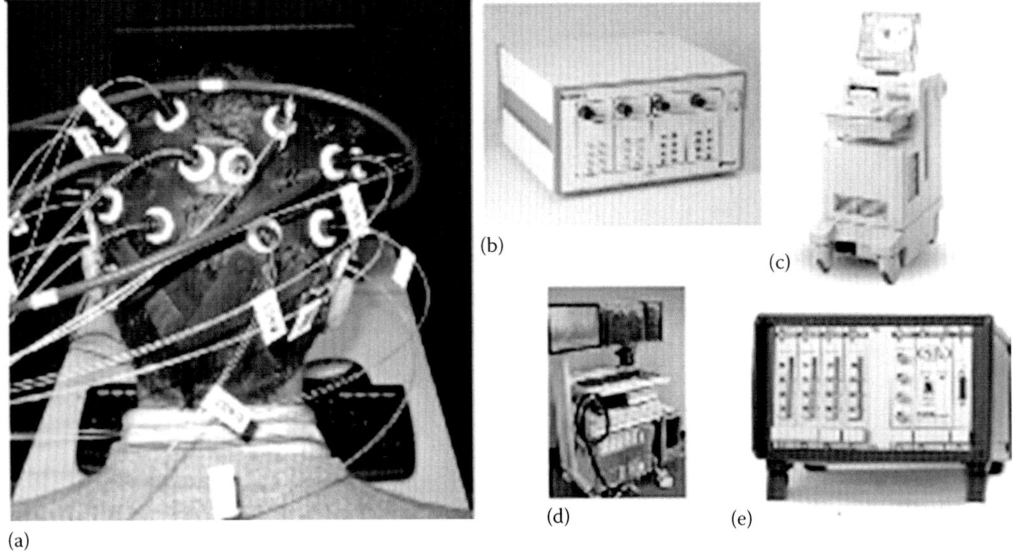

FIGURE 5.2
(a) A typical example of an fNIRS probe holder placed on the head of a participant (bilateral parietal lobe). In this setup, thin optical fibers (diameter 0.4 mm) convey NIR light to the participant's head (note that each location comprises two optical fibers, one for each wavelength), whereas optical fiber bundles (diameter 3 mm) capture the light that is scattered through the brain tissue. (b) The ISS Imagent (http://www.iss.com/biomedical/instruments/imagent.html). (c) The Hitachi ETG 4000 (http://www.hitachi-medical-systems.eu/products-and-services/optical-topography/etg-4000.html). (d) The Nirsoptix Brain Monitor (http://www.nirsoptix.com/CW6.php). (e) The NIRScout (http://www.nirx.net/imagers/nirscout-xtended). (From Cutini, S. et al. *J. Near Infrared Spectrosc.*, 20, 75–92, 2012. With permission from IM Publications).

TABLE 5.3

The Main Commercially Available Wearable/Wireless fNIRS Systems

Instrument	Year of Release	Wireless	Number of Channels	Company	Website
fNIR 1100	2009	No	16	fNIR Devices, Potomac, MD	http://www.fnirdevices.com
fNIR 1100w[a]	2011	Yes	2 or 4	fNIR Devices, Potomac, MD	http://www.fnirdevices.com
HOT 121B	2011	Yes	2	Hitachi, Japan; Twinsburg, OH	http://www.hitachimed.com
NIRSport	2011	Yes	Up to 256	NIRx, Los Angeles, CA	http://www.nirx.net
OEG-16	2009	No	16	Spectratech, Yokohama, Japan	http://www.spectratech.co.jp
OEG-SpO_2[b]	2011	Yes	16	Spectratech, Yokohama, Japan	http://www.spectratech.co.jp
PocketNIRS Duo	2010	Yes	2	DynaSense, Shizuoka, Japan	http://www.dynasense.co.jp
PortaLite[c]	2011	Yes	1	Artinis, Elst, The Netherlands	http://www.artinis.com
WOT	2009	Yes	22	Hitachi, Japan; Twinsburg, OH	http://www.hitachimed.com

Source: Marco, F. and Valentina Q., *NeuroImage* 3, 49, 2012.
Note: Channel (or measurement point) is the midpoint between the source and the detector.
[a] Children probe available.
[b] Measure of arterial SpO_2 saturation.
[c] Brain oximeter.

5.5 Applications of fNIRS in Neurosciences

Since the mid-1990s, most of the research work done in neurosciences using fNIRS has been focused on quantitative analysis and imaging of human and small animal brain functions. These two techniques were used to localize or monitor the cerebral responses under different stimuli, including visual [20–22], auditory [23], somatosensory [24], motor [25–27], and even language [28]. Further, the researchers also investigated the neurological disorders using different measurement instrumentations, which addressed the neurovascular and neurometabolic coupling mechanisms for different diseases, such as seizure and epilepsy [29–32], depression [33–35], Alzheimer's disease (AD) [36–38], stroke rehabilitation [39–42], Parkinson's disease (PD), and psychiatry diseases.

These initial studies served to demonstrate the potential assets offered by fNIRS and provided an initial framework for future studies, which to date have covered a variety of perspectives in neurosciences from basic research, neurological disorders to neuropsychology (Table 5.4).

5.5.1 Neurology

5.5.1.1 Epilepsy

Epilepsy is a common chronic neurological disease involving recurrent, unprovoked seizures that affects approximately 3% of the human population during their lifetime.

TABLE 5.4
Main Fields of fNIRS Applications in Neurosciences

Neurology	**Alzheimer's disease**
	Dementia
	Depression
	Epilepsy
	Parkinson's disease
	Post-neurosurgery dysfunctions
	Rehabilitation
	Stroke recovery
Psychiatry	Anxiety disorders
	Childhood disorders
	Eating disorders
	Mood disorders
	Personality disorders
	Substance-related disorders
	Schizophrenic disorders
Psychology/education	Attention
	Body representation
	Comprehension
	Developmental disorders
	Developmental psychology
	Emotion
	Functional connectivity
	Gender differences
	Language
	Memory
	Perception
	Reasoning
	Social brain
Basic research	Brain computer interface
	Fusion
	Neuroergonomics
	Pain research
	Sleep research
	Sports sciences research

Seizures are caused by the synchronous, rhythmic firing of a population of neurons, lasting from seconds to minutes. Approximately 60%–70% of patients experienced focal or partial seizures, whereas another 30%–40% of patients have generalized seizures. Epilepsy in approximately 70% of patients could be controlled by medication, whereas for medically refractory seizures, resection of the epileptogenic zone may be considered, and thus a series of presurgical evaluations will be taken to assess brain structure abnormality and clinical feature of seizures. For patient with new onset seizures, neuroimaging will help determine whether the seizure is acute provoked or unprovoked and if any immediate treatment need to be taken. Determination of whether there is an underlying brain lesion in one of the primary steps in evaluating new-onset seizure disorders. Common causes of acute seizures are brain tumor, perinatal hypoxic or hypoxemic events, malformations of cortical development for young children, head trauma for young adults, and stroke for elderly people. Potentially epileptogenic lesion detected by neuroimaging would affirm focal seizure disorders and differentiate them from primary generalized seizure disorders, which are based on either genetic or idiopathic rather than a focal epileptogenic lesion.

fNIRS measurements were implemented to monitor the effects of spontaneous and chemically induced seizures in patients with epilepsy. Initial studies found that the blood volume increased rapidly for about 30–60 seconds after the seizure onset on the focus side, and fNIRS could be utilized to monitor the blood volume continuously during seizures [29–32]. Epileptic patients were also examined during spontaneously occurring complex partial seizures [43], in which extremely large increases in blood volume and oxygenated hemoglobin concentration were identified, stressing the importance of fNIRS as a useful and simple bedside tool to assess the brain function. Further work was performed [44] to better understanding of the pathophysiology of seizures in childhood epilepsy. In particular, fNIRS also showed its potential value in the presurgical investigation of patients with refractory epilepsy [30]. Importantly, simultaneous electrophysiology [electroencephalogram (EEG)] and fNIRS recordings were conducted to examine the localization of the ictal onset zone and assess language lateralization in a young epileptic patient as part of the presurgical evaluation. The patient underwent a prolonged EEG-fNIRS recording, while electroclinical and electrical seizures were recorded [30]. Results were compared to those obtained with other presurgical techniques [single-photon emission computerized tomography (SPECT), FDG-PET, EEG-fMRI, and EEG-magnetoencephalography] and showed good concordance for ictal onset zone localization. This illustrated that multichannel EEG-fNIRS has the potential to contribute favorably to presurgical investigation in young patients (Figure 5.3).

5.5.1.2 Alzheimer's Disease

fNIRS was employed to investigate the AD. For example, Hock et al. [38] measured the hemodynamic changes in the frontal cortex using the NIRO 500 systems, while patients with probable AD were performing a verbal fluency task (VFT). Whereas elderly healthy subjects showed increases in the local concentration changes of HbO_2 and HbT, Alzheimer's patients showed significant decreases in the concentration changes shortly after stimuli onset. In particular, the changes of hemodynamic parameters were more pronounced in the parietal cortex than in the frontal cortex. Hock et al. suggested that their findings of a regional reduction in oxygen supply during the activation of brain function might be of relevance to the development and time course of neurodegeneration. They further suggested that degenerating brain areas might show a reduction in hemoglobin oxygenation during cognitive tasks in favor of other healthier brain regions due to alterations in functional brain organization. Another investigation also employed a VFT to investigate changes in prefrontal activation in patients with AD [45], in which they found that good performance by healthy controls on the VFT was associated with a predominance of left hemispheric activation, whereas in patients with AD, a low number of correct responses were associated with the loss of this asymmetric activation pattern.

Additionally, the pathway for the olfactory response in early stages of AD was also explored [46], in which control subjects show a clearly definable response with increased HbO_2 and decreased HbR bilaterally to the smell of vanilla. By contrast, response amplitudes

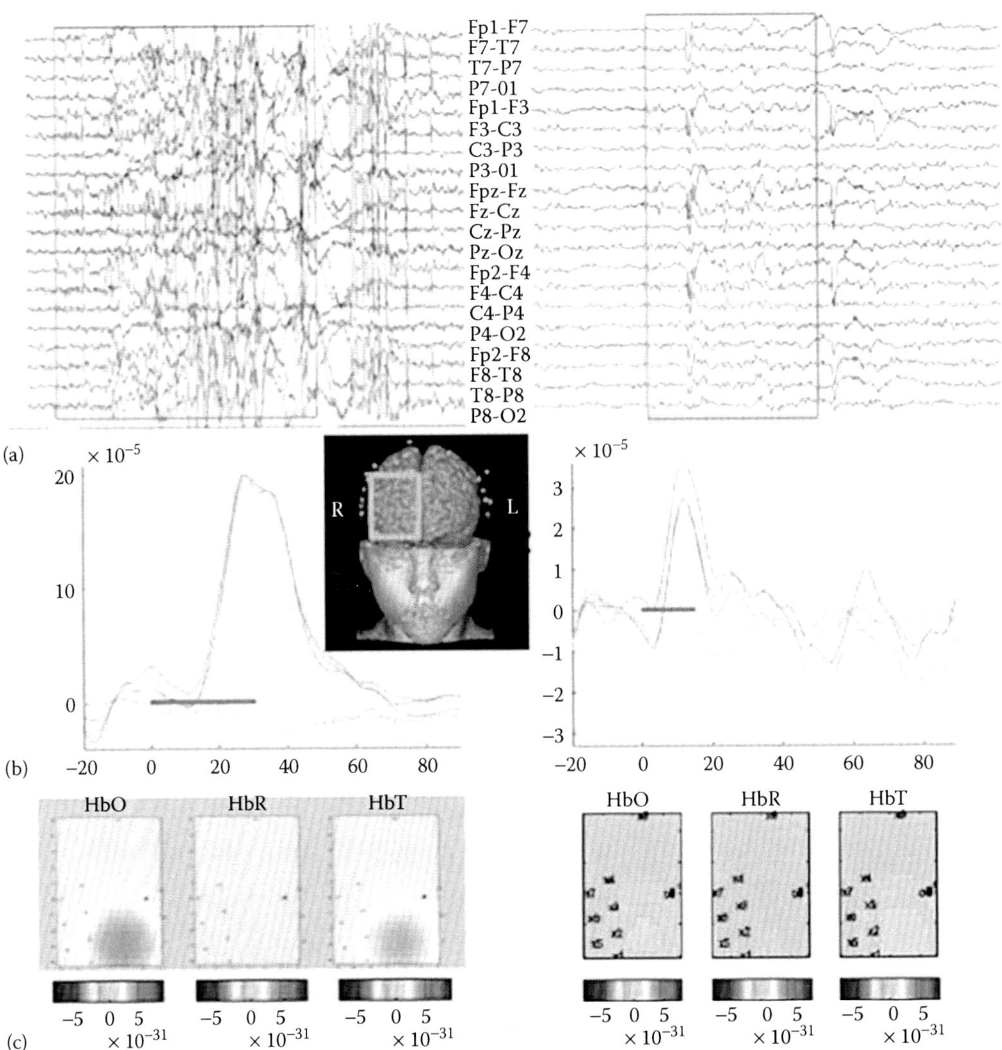

FIGURE 5.3
fNIRS-EEG results during seizures. (a) EEG data during electroclinical (left) and electrical (right) seizures are shown on the top. Simultaneous fNIRS data during seizures are shown by graphs (b) and hemoglobin map images (c). (b) Graphs show fNIRS cerebral activation in five fNIRS channels covering right frontopolar region. Cerebral activation is characterized by an enhanced HbO (solid lines) as well as a small and late decrease of HbR (dotted lines), happening a few seconds after the beginning of the seizure. The horizontal black bar shows the seizure duration. A typical initial dip (short reduction of HbO) is also obtained before the activation. This hemodynamic signal was measured on the right frontal area (see 3D MRI reconstruction in the middle), which is indicated by the gray rectangle. It is important to note that the scale between both graphs is greatly different, which shows that cerebral activation is much more prominent during the electroclinical seizure compared to the electrical seizure. The importance of this amplitude difference between both hemodynamic responses did not allow using the same scale. (c) Enhancement of HbT and HbO is shown by hemoglobin map images, where the rectangle is the same as the right frontal rectangle seen on the 3D MRI reconstruction. This activation corresponds temporally to the peak of *HbO* enhancement. Here, HbO represents HbO_2. (Reprinted from *Seizure*, 17, Gallagher, A. et al., Non-invasive pre-surgical investigation of a 10 year old epileptic boy using simultaneous EEG-NIR, 576–582, Copyright 2008, with permission from Elsevier).

to vanilla in AD patients were found to be in the range of those to sham stimuli, suggesting early deterioration of olfactory responses in AD.

5.5.1.3 Parkinson's Disease

fNIRS also holds meaningful promise for investigations of neurovascular activity in PD because it can afford against motion artifacts [19]. So far, two pilot fNIRS studies have been completed to study the basic hemodynamic mechanism of PD [47,48]. For example, Murata et al. [47] studied the cerebral blood oxygenation changes in the frontal lobe induced by direct stimulation of the thalamus or globus pallidus in patients with PD. Under conditions of neural activation of the frontal lobe, increased HbO_2 and HbT concentrations and decreased HbR concentrations during stimulus processing were observed, which demonstrated that fNIRS is able to identify neural activation patterns of cerebral cortex, especially in blood oxygen. Based on a similar paradigm, Sakatani et al. [48] found that the electrical stimulation of the same brain regions caused various cerebral blood oxygenation changes in the frontal lobe, which were similar to those found during cognitive tasks. Sakatani et al. suggested that the multiplicity of the cerebral oxygenation changes in the frontal lobe might be corrected with complex neural networks in the frontal lobe, which has many neuronal connections to other cortical areas or the basal ganglia.

5.5.1.4 Stroke

To study the brain activation and reorganization during motor recovery and rehabilitation effect in stroke, fNIRS system is more suitable, which allows evaluation of brain activities under motor tasks with larger head and body motions [49–51]. In addition, the optical techniques can be conducted in a quieter condition, which makes experiments interaction with auditory stimuli much easier and reduces the stress of agitated patients post stroke. Therefore, fNIRS system provides several advantages over PET and fMRI because it does not have to limit the rehabilitation tasks to fine motor or constrained movements or limited interaction with environment.

An fNIRS gait study of stroke patients was implemented, in which cortical activities were measured during hemiparetic gait activities on a treadmill both before and after 2 months of inpatient rehabilitation [50]. Initial asymmetries in activation in the medial primary sensorimotor cortex (SMC), premotor cortex (PMC), and supplementary motor area (SMA) were identified with increased oxygenated hemoglobin levels in the unaffected hemisphere, compared to the affected hemisphere during gait activities. Improvement in asymmetrical SMC activation was well correlated with improvement on gait tasks, which demonstrated that the recovery of locomotion after stroke may be associated with improved asymmetry in SMC activation and enhanced PMC activation in the affected hemisphere.

A preliminary study of cortical activation during cycling was performed to demonstrate the feasibility of fNIRS in stroke rehabilitation. The utilization of cycling for stroke rehabilitation is because it has similar reciprocal movement pattern with ambulation and may share the same neural mechanism of gait.

The cortical activation patterns from bilateral SMC, SMA, and PMC were measured by 20 channels of fNIRS signal resulting from 4 detectors and 14 pairs of source (each source location comprising two optical fibers for two wavelengths, 690 and 830 nm, respectively) with a sampling rate of 19.8 Hz, as shown in Figure 5.4. A single-distance method with an interoptode distance of 3.0 cm was used for probe design. The cortical mappings based on the task-related change of HbO_2 are given in Figure 5.5. The brain mappings show the

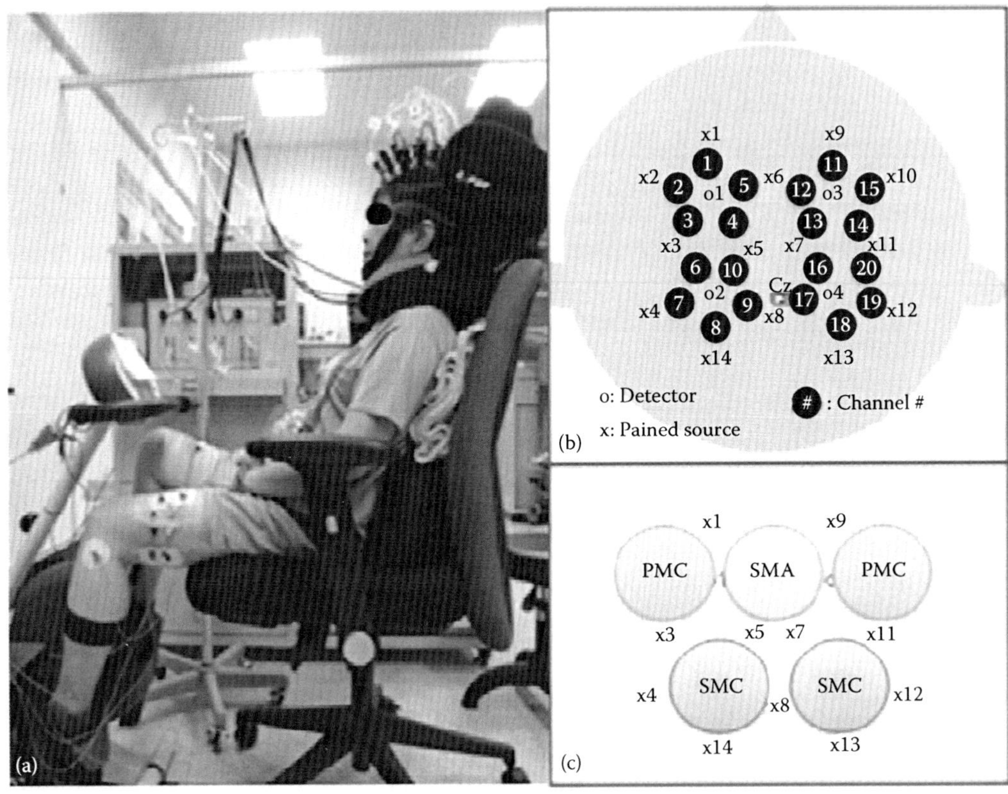

FIGURE 5.4
The experimental setup. (a) A stroke subject with optical cap mounted on the head for pedaled cycling studies. (b) Schematic diagram showing the optode locations. A total of four detectors (marked as o's) and 14 pairs of sources (marked as x's) were arranged on the scalp to enable 20-channel measurement. The center of o2 and o4 was located in the Cz position from which the cortical areas could be covered by the channels shown in (c). (From Lin, P. et al., *J. Med. Biol. Eng.*, 29, 210–221, 2009. With permission.)

significantly different cortical activation patterns in the stroke patient and the controls under different conditions. The symmetrical activation of SMC was prominently observed during active cycling in normal control, whereas the cortical maps in the stroke patient show asymmetrical activation patterns during active cycling compared with those of the normal control. In addition, the activation of SMC and PMC of the affected hemisphere appeared to be more enhanced during active condition than that during passive cycling. Further, the less asymmetrical patterns of SMC and PMC could be observed during passive compared to active cycling, which implies the involvement of SMC and PMC in the active component of cycling movement.

5.5.2 Psychiatry

5.5.2.1 Anxiety/Mood Disorders

fNIRS was employed for the investigations of anxiety disorders such as panic disorder. Early studies were conducted to reveal the frontal activation in patients with panic disorder without depression, a group characterized as having both negative emotions and

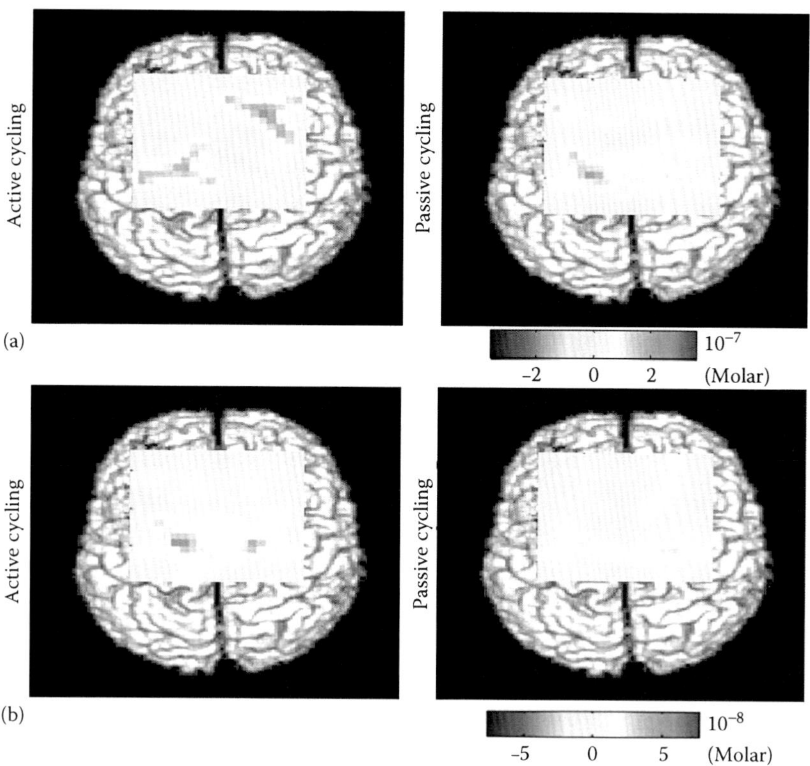

FIGURE 5.5
The cortical mapping during cycling conditions of passive cycling and active cycling for a left hemiparesis stroke patient (a) and a control (b). (From Lin, P. et al., *J. Med. Biol. Eng.*, 29, 210–221, 2009. With permission.)

avoidance-withdrawal behavior [52]. fNIRS measures were implemented for both patients and controls with the following conditions: confrontation at rest with neutral stimuli (e.g., mushroom), anxiety-relevant stimuli (e.g., spider and snake), or anxiety-irrelevant but emotionally relevant stimuli (e.g., erotic picture). The fNIRS findings indicated that HbO_2 concentrations in patients were significantly lower than in control subjects when confronted with anxiety-relevant or anxiety-irrelevant but not emotionally relevant stimuli. However, no significant evidence of frontal brain asymmetry was identified when patients or control subjects observed any of the stimuli. Akiyoshi et al. [52] suggested that patients with panic disorder might be characterized as having a greater decrease in the activation of a left frontal avoidance-withdrawal system in situations with a negative valence, thus providing biological evidence for disturbed cortical processing in panic disorder.

fNIRS was also applied to study the mood disorders such as depression and bipolar disorder. To date efforts in mood disorders have been focused on investigating hemispheric activation patterns [35]. Okada et al. found that nearly half of patients with major depression show nondominant hemisphere response patterns compared to healthy controls. Importantly, increased HbT reached to a markedly lesser degree in the supposedly dominant (left) hemisphere compared to that in the nondominant (right) hemisphere, based on handedness. By contrast, the other half of patients with depression showed a bilateral response pattern where, in response to the mirror-drawing task (MDT), HbT, HbO_2, and HbR were comparable in both hemispheres. Shifts in the response pattern based on the course of depression led Okada et al. to suggest that there may be some correlations between the course of depression and the response patterns. They also found a sex difference, with women more likely to show the bilateral response pattern and men a more dominant hemisphere response pattern. As such, the prolonged and continuous monitoring of cerebral activation with fNIRS not only helped to elucidate course-related changes in response patterns in depression but also highlighted the role of sex.

5.5.2.2 Schizophrenic Disorders

Among the neuropsychiatric conditions, schizophrenia is the most liberally utilized for fNIRS investigations. This may be because schizophrenia is a complex and enigmatic disorder or group of disorders, and despite decades of investigation has yet to be fully understood. fNIRS investigations were implemented to capture the disturbances in interhemispheric integration of brain oxygen metabolism and hemodynamics [53]. Okada et al. used an MDT stimulus and found that control groups showed distinct and well-integrated patterns of changes in HbO_2, HbR, and HbT during the MDT. By contrast, half of the patients with schizophrenia were identified with *dysregulated patterns* in the frontal regions between hemispheres, such that increases in HbO_2 were not paralleled by decreases in HbR, which demonstrated that certain symptoms of schizophrenia might be related to problems in interhemispheric integration and brain network.

Similar work was done to examine the relationship between lateralized frontal activation patterns during the execution of a continuous performance test [54]. Interestingly, no overall or hemispheric activation effects were identified in their cohort. However, a lack of lateralized activation did exist for the schizophrenia group. In particular, a trend toward higher left relative to right HbO_2 and HbR ratios at rest and during activation was observed in subjects with schizophrenia. This suggested that there may be a reduced specific lateralized frontal reactivity, possibly based on a left hemisphere functional deficit in schizophrenia.

Additionally, a recent investigation has utilized frontally based tasks such as random number generation (RNG), ruler-catching (RC), and sequential finger-to-thumb (SFT) tasks to show that there are task-dependent profile of functional abnormalities observed in schizophrenic frontal brain metabolism [55]. Specifically, during the RNG task, HbT and HbO_2 concentrations increased and HbR decreased, but the responses were significantly smaller in schizophrenic patients. During the RC task, HbO_2 in patients with schizophrenia tended to decrease, in contrast to the mostly increasing response in control subjects. No group difference was observed during the SFT task.

5.5.3 Neuropsychology

5.5.3.1 Memory and Attention

A typical pattern characterized by an increase in HbO_2 and a decrease in HbR was identified during a word memory task using fNIRS measurements [56]. In particular, a very unique neural stimulus task was implemented to study the hemodynamic activity during the execution of a visual n-back task [57]. Memory and selective attention are both involved in this paradigm: the participant, watching a sequence of stimuli, has to identify whether the current stimulus matches the one appeared from n steps earlier in the sequences. The task difficulty can be controlled by changing the value of n. And the activity of the prefrontal cortex (PFC) in participants was investigated while performing the *n*-back task, which successively showed task-relevant and task-irrelevant faces. Participants were required to respond as key press action when they identify a presented face of the relevant category (i.e., a white-bordered face) matching the previously presented face of the same category (pointed out by the black arrow above one picture). The left postcentral cortex and the middle frontal/precentral cortices bilaterally were activated by relevant stimuli, which probably maintain the features of the relevant stimuli in respect of a verbal rehearsal strategy. By contrast, superior, middle, and inferior parts of the right PFC were activated by irrelevant stimuli, consistently with the selective inhibition needed to properly perform the task. The results above manifested that the prefrontal activity during working memory tasks reflected processes of selection, inhibition, and maintenance of information, as well as attentional monitoring. The study of information encoding and retrieval, which are known to involve partially overlapping neural circuits, also focused on the activity of PFC [58]. A recent investigation on taste encoding and retrieval with fNIRS indicated that the hemodynamic activity observed during taste retrieval was significantly stronger in the bilateral frontopolar and dorsolateral prefrontal regions, particularly in the right hemisphere than that observed during encoding [59], which are generally in accordance with the hemispheric encoding/retrieval asymmetry theory [60].

As the extensive study of memory, attention was widely investigated with fNIRS in the meantime. For example, an optical study [61] investigated the hemodynamic changes of the PFC during a continuous performance test, in which participants were instructed to respond to infrequent letters appearing on a screen but not to frequent ones. An increase in HbO_2 and HbR was observed in the dorsolateral PFC (DLPFC), but the in-depth explanation of the results was limited because of the lack of a suitable control task.

In addition, the neural substrates of visuospatial attention were extensively studied, although it is not clear whether the attentional resources required for each visual hemifield are functionally separated between the cerebral hemispheres. The results of an fNIRS investigation that recorded the brain activity during a visuospatial task indicated that an increase in attentional load resulted in a greater increase in the brain activity in the left visual hemifield than that from the right visual hemifield [62]. This asymmetry was observed in all the examined brain areas, including the bilateral occipital and parietal cortices; the strongest activations were shown in posterior parietal cortex, including the intraparietal sulcus. The asymmetry of inhibitory interactions between the hemispheres was construed as the reason of the results mentioned above.

Interestingly, a recent study investigated the neural correlates of visual short-term memory with fNIRS [63]. During the research, a spatially cued variant of the change-detection task was adopted to record hemodynamic activities that responded to unilaterally encoded objects. Participants were shown an arrow head pointing to the left or right side of the screen. Following the offset of the arrow, participants were shown a memory array, which was made up of either four or eight colored squares, averagely distributed to the left and right sides of fixation. Participants were required to focus their eyes on fixation and memorize the color of the squares on the side of the memory array cued by the arrow head. The offset of the memory array was followed by another array of colored squares in the same positions as occupied by the squares of the memory array. Participants had to press the relevant button to indicate whether a change in color had happened or not. This task provides very divergent results for EEG and fMRI: EEG data [64] indicate that a contralateral negative activity is elicited by the maintenance of unilaterally encoded objects, as the intensity is proportional to the number of objects retained in visual short-term memory, whereas fMRI results show a modulation of the blood oxygenation level dependent (BOLD) response in reference to the number of memorized elements in the parietal lobes of both hemispheres [65]. Similar to fMRI results, fNIRS investigation revealed a memory-related increase in HbO_2 in the posterior parietal cortex bilaterally, even though objects had to be encoded unilaterally in the absence of eye movements. The high similarity of such results with those obtained with fMRI [65] suggests that fMRI/fNIRS and EEG techniques may reveal distinct neural signatures of the mechanisms supporting visual short-term memory.

5.5.3.2 Language and Comprehension

fNIRS studies on language was focused on the neural correlates of language during the generation of words beginning with a certain letter, in the letter version, or including the same category, in the semantic version. The effects of age and gender on brain activation during a letter VFT were investigated using fNIRS [66]. The results revealed that young participants showed more activity in the left DLPFC than the old and there was no lateralization in all the participants. Interestingly, there was no effect of gender, although a previous study [67] showed a lower frontal and temporal activity in females than in males.

fNIRS study also evaluated the short- and long-term reliability of the brain activity using a test–retest phonological VFT [68]. Participants were asked to repeat the same task after 3 and 53 weeks from the first and second data collection, respectively. Increases in HbO_2 and decreases in HbR were found in the inferior DLPFC and in part of the temporal cortex in both sessions, whereas a smaller hemodynamic response was observed in the second session. A study [69] investigated the lateralization and the functional connectivity of the frontal cortex in response to a VFT and baseline random jaw movement with fNIRS. An increase in HbO_2 and a decrease in HbR were observed during VFT in the PFC; nevertheless, the hemodynamic response in the anterior frontal during VFT was not significantly different from that during random jaw movement. Furthermore, a bilateral activation and a symmetrical connectivity were found in the PFC in both tasks, whereas a left cortical dominance and asymmetry connectivity were shown during the VFT.

In addition, fNIRS was applied in overt picture naming experiment on bilinguals [70] to investigate the brain mechanisms of bilinguals' ability to use two languages at the same time. Kovelman et al. [70] compared the use of signed and spoken language in rapid alternation from a bilingual to a monolingual mode during an overt picture naming task. Although bilinguals showed an accurate performance in both conditions, the results revealed that bilinguals, when bilinguals executed the task in the bilingual mode, showed a greater signal activation within posterior temporal regions than in the monolingual mode, indicating a role of these areas in the bilingual language switching ability. Effects of the use of a second language were evaluated in an fNIRS study [71] on translation and language switching in native Dutch short participants who were proficient in English. The participants were instructed to translate visually presented sentences aloud, from English in their native language or vice versa or alternating. The involvement of the left frontal cortex, which includes the Broca's area, was evidenced by a consistent and incremental increase in HbO_2 and a smaller decrease in HbR in all conditions, although no differences of the direction of the translation.

Further, fNIRS was employed for investigation of language representation to study the brain activity during social interactions [72], in which an ordinary face-to-face conversation was monitored with healthy subjects in the sitting position. The results revealed a negative correlation between the magnitude of frontopolar activity and the cooperativeness score of the subjects assessed using the temperament and character inventory, but no correlation was found between the PFC and the cooperativeness score.

Interestingly, language processing was implemented using fNIRS [72] to investigate the semantic and syntactic decision tasks. Participants were instructed to judge whether the presented sentences are semantically or syntactically right. Similar performances or reaction times were performed for different cases and it was found that the left inferior gyrus was more activated in the syntactic processing than in the semantic processing.

Finally, lots of fNIRS studies investigated language comprehension through the discrimination of phonemes, words, or speeches. A recent work [73] aimed at investigating the implicit processing of phonotactic cues by simultaneously applying event-related potentials (ERPs) and fNIRS. Pseudo-words, either phonotactically legal (when the first consonant was equal to the onset of a German word) or illegal (when the first consonant was not with respect to German), were presented to native German speakers. The hypothesis that was connected with the task was that the lexical activation processes in the auditory speech might be modulated by the prelexical cue processing in the spoken word recognition. The fNIRS results revealed a significant HbR decrease in the left frontotemporal hemisphere for phonotactically legal pseudo-words with respect to that observed for illegal pseudo-words. These findings were confirmed by a larger N400 effect for phonotactically legal pseudo-words in ERPs, providing evidence that processing of legal phonotactic components contributes to language comprehension.

5.5.3.3 Emotion

The neural correlates of emotion in humans have been intensely investigated in recent years. As we all know that amygdala is a subcortical structure that is crucially involved in emotion processing, several studies were conducted to figure out whether cortical activity is also related to emotion processing. The fMRI study manifested that the PFC is involved in emotional regulation [74], with a broad hemispherical asymmetry between the left and right PFCs for both positive and negative emotions [75]. fNIRS was also utilized to investigate the activity of the left and right medial PFCs [76], when participants were instructed to react to a set of pictures in two conditions in which the self-monitoring requirement was manipulated: fewer self-monitoring (i.e., just look at the emotional pictures presented) or higher self-monitoring (i.e., try to feel like the emotion expressed by the stimulus presentation). It was observed that a significant increase in HbO_2 concentrations was observed in the frontal left hemisphere when the task required higher self-monitoring processes, particularly when highly emotional pictures were viewed. This result suggests that self-monitoring processes are tightly involved in certain aspects of emotional induction and may contribute to frontal activation. And other investigations also revealed that the frontal cortex is tightly related to the maintenance of attentional demands of a task [77] and action monitoring [78]. The relationship between emotion processing and prefrontal activation was investigated in a recent fNIRS experiment [79], which indicated that prefrontal activation exists in both emotion regulation and emotion induction. It is worth mentioning that the effect was located in the left PFC with only HbR difference.

In addition, fNIRS was also applied for the investigation of examining the effect of emotional content on the occipital cortex during visual stimulation [79]. Neutral stimuli produced a smaller decrease of HbR in specific areas of the occipital cortex in comparison with the decrease induced by emotionally positive and negative stimuli. A recent study with fNIRS [80] showed that the emotional valence of the stimuli modulated the hemodynamic response peak latency of the visual cortex. Particularly, the hemodynamic response peak latency elicited by processing of positive pictures was lower than that caused by negative pictures. Besides the theoretical implications for all the results above, it is noteworthy that fNIRS has a higher temporal resolution with respect to fMRI, confirming that fNIRS can be effectively used under the right circumstances and with the appropriate objectives. fNIRS has also been employed to measuring the modulation of auditory cortex by emotional auditory stimuli recently, with the results that pleasant and unpleasant sounds produced stronger auditory cortex activity compared to neutral sounds; the results suggest that human auditory cortex activity is modulated by complex emotional sounds.

References

1. Tak, S., Yoon, S. J., Jang, J., Yoo, K., Jeong, Y., Ye, J. C. (2011). Quantitative analysis of hemodynamic and metabolic changes in subcortical vascular dementia using simultaneous near-infrared spectroscopy and fMRI measurements. *NeuroImage* 55, 176–184.
2. Yuan, Z., Zhang, Q., Sobel, E., Jiang, H. (2010). Image-guided optical spectroscopy in diagnosis of osteoarthritis: A clinical study. *Biomed. Opt. Express* 1, 74–86.
3. Yodh, A., Chance, B. (1995). Spectroscopy and imaging with diffusing light. *Phys. Today* 48, 34–40.
4. Jobsis, F. F. (1977). Noninvasive, infrared monitoring of cerebral and myocardial oxygen sufficiency and circulatory parameters. *Science* 198, 1264–1267.
5. Huppert, T., Diamond, S., Franceschini, M., Boas, D. (2009). Homer: A review of time-series analysis methods for near-infrared spectroscopy of the brain. *Appl. Opt.* 48, 280–298.
6. Yuan, Z. (2013). A spatiotemporal and time-frequency analysis of functional near infrared brain signals using ICA method. *J. Biomed. Opt.* 18(10), 106011.
7. Yuan, Z., Ye. J. (2013). Fusion of fNIRS and fMRI data: Identifying when and where hemodynamic signal are changing in human brains. *Front. Hum. Neurosci.* doi:10.3389/fnhum.2013.00676.
8. Yuan, Z. (2013). Combing ICA and Granger causality to capture brain network dynamics with fNIRS measurements. *Biomed. Opt. Express* 4(11), 2629–2643.
9. Brazy, J. E., Lewis, D. V., Mitnick, M. H., Jöbsis vander Vliet, F. F. (1985). Noninvasive monitoring of cerebral oxygenation in preterm infants: Preliminary observations. *Pediatrics* 75, 217–225.
10. Chance, B. (1991). Optical method. *Annu. Rev. Biophys. Biophys. Chem.* 20, 1–28.
11. Jöbsis, F. F. (1999). Discovery of the near-infrared window into the body and the early development of near-infrared spectroscopy. *J. Biomed. Opt.* 4, 392–396.
12. Ferrari, M., Zanette, E., Giannini, I., Sideri, G., Fieschi, C., Carpi, A. (1986). Effects of carotid artery compression test on regional cerebral blood volume, hemoglobin oxygen saturation and cytochrome-c-oxidase redox level in cerebrovascular patients. *Adv. Exp. Med. Biol.* 200, 213–221.
13. Ferrari, M., De Marchis, C., Giannini, I., Di Nicola, A., Agostino, R., Nodari, S., Bucci, G. (1986). Cerebral blood volume and hemoglobin oxygen saturation monitoring in neonatal brain by near IR spectroscopy. *Adv. Exp. Med. Biol.* 200, 203–211.
14. Wyatt, J. S., Cope, M., Delpy, D. T., Wray, S., Reynolds, E. O. (1986). Quantification of cerebral oxygenation and haemodynamics in sick newborn infants by near infrared spectrophotometry. *Lancet* 2, 1063–1066.
15. Reynolds, E. O., Wyatt, J. S., Azzopardi, D., Delpy, D. T., Cady, E. B., Cope, M., Wray, S. (1988). New non-invasive methods for assessing brain oxygenation and haemodynamics. *Br. Med. Bull.* 44, 1052–1075.
16. Delpy, D. T., Cope, M. (1997). Quantification in tissue near-infrared spectroscopy. *Philos. Trans. R. Soc. Lond., B Biol. Sci.* 352, 649–659.
17. Ito, H., Kanno, I., Fukuda, H. (2005). Human cerebral circulation: Positron emission tomography studies. *Ann. Nucl. Med.* 19, 65–74.
18. Marco, F., Valentina Q. (2012). A brief review on the history of human functional near-infrared spectroscopy (fNIRS) development and fields of application. *NeuroImage* 3, 49.

19. Cutini, S., Moro, S., Bisconti, S. (2012). Functional near infrared optical imaging in cognitive neuroscience: An introductory review. *J. Near Infrared Spectrosc.* 20, 75–92.
20. Heekeren, H. R., Obrig, H., Wenzel, R., Eberle, K., Ruben, J., Villringer, K., Kurth, R., Villringer, A. (1997). Cerebral haemoglobin oxygenation during sustained visual stimulation—a near-infrared spectroscopy study. *Philos. Trans. R. Soc. Lond., B Biol. Sci.* 352, 743–750.
21. Meek, J. H., Elwell, C. E., Khan, M. J., Romaya, J., Wyatt, J. S., Delpy, D. T., Zeki, S. (1995). Regional changes in cerebral hemodynamics as a result of a visual stimulus measured by near infrared spectroscopy. *Proc. R. Soc. Lond., B* 261, 351–356.
22. Ruben, J., Wenzel, R., Obrig, H., Villringer, K., Bernarding, J., Hirth, C., Heekeren, H., Dirnagl, U., Villringer, A. (1997). Haemoglobin oxygenation changes during visual stimulation in the occipital cortex. *Adv. Exp. Med. Biol.* 428, 181–187.
23. Sakatani, K., Chen, S., Lichty, W., Zuo, H., Wang, Y. P. (1999). Cerebral blood oxygenation changes induced by auditory stimulation in newborn infants measured by near infrared spectroscopy. *Early Hum. Dev.* 55, 229–236.
24. Franceschini, M. A., Fantini, S., Thompson, J. H., Culver, J. P., Boas, D. A. (2003). Hemodynamic evoked response of the sensorimotor cortex measured non-invasively with near infrared optical imaging. *Psychophysiology* 40, 548–560.
25. Colier, W. N., Quaresima, V., Oeseburg, B., Ferrari, M. (1999). Human motor-cortex oxygenation changes induced by cyclic coupled movements of hand and foot. *Exp. Brain Res.* 129, 457–461.
26. Hirth, C., Obrig, H., Villringer, K., Thiel, A., Bernarding, J., Muhlnickel, W., Flor, H., Dirnagl, U., Villringer, A. (1996). Non-invasive functional mapping of the human motor cortex using near-infrared spectroscopy. *Neuroreport* 7, 1977–1981.
27. Kleinschmidt, A., Obrig, H., Requardt, M., Merboldt, K. D., Dirnagl, U., Villringer, A., Frahm, J. (1996). Simultaneous recording of cerebral blood oxygenation changes during human brain activation by magnetic resonance imaging and near-infrared spectroscopy. *J. Cereb. Blood Flow Metab.* 16, 817–826.
28. Sato, H., Takeuchi, T., Sakai, K. L. (1999). Temporal cortex activation during speech recognition: An optical topography study. *Cognition* 73, B55–B66.
29. Adelson, P. D., Nemoto, E., Scheuer, M., Painter, M., Morgan, J., Yonas, H. (1999). Noninvasive continuous monitoring of cerebral oxygenation periictally using near-infrared spectroscopy: A preliminary report. *Epilepsia* 40, 1484–1489.
30. Gallagher, A., Lassonde, M., Bastien, D., Vannasing, P., Lesage, F., Grova, C. et al. (2008). Non-invasive pre-surgical investigation of a 10 year old epileptic boy using simultaneous EEG-NIR. *Seizure* 17, 576–582.
31. Steinhoff, B. J., Herrendorf, G., Kurth, C. (1996). Ictal near infrared spectroscopy in temporal lobe epilepsy: A pilot study. *Seizure* 5, 97–101.
32. Watanabe, E., Maki, A., Kawaguchi, F., Yamashita, Y., Koizumi, H., Mayanagi, Y. (2000). Noninvasive cerebral blood volume measurement during seizures using multichannel near infrared spectroscopic topography. *J. Biomed. Opt.* 5, 287–290.
33. Eschweiler, G. W., Wegerer, C., Schlotter, W., Spandl, C., Stevens, A., Bartels, M. (2000). Left prefrontal activation predicts therapeutic effects of repetitive transcranial magnetic stimulation (rTMS) in major depression. *Psychiatr. Res.* 99, 161–172.
34. Matsuo, K., Kato, T., Fukuda, M., Kato, N. (2000). Alteration of hemoglobin oxygenation in the frontal region in elderly depressed patients as measured by near-infrared spectroscopy. *J. Neuropsychiatr. Clin. Neurosci.* 12, 465–471.
35. Okada, F., Takahashi, N., Tokumitsu, Y. (1996). Dominance of the "nondominant" hemisphere in depression. *J. Affect. Disord.* 37, 13–21.
36. Frostig, R. D., Lieke, E. E., Tso, D. Y., Grinvald, A. (1990). Cortical functional architecture and local coupling between neuronal activity and the microcirculation revealed by in vivo high-resolution optical imaging of intrinsic signals. *Proc. Natl. Acad. Sci. U.S.A.* 87, 6082–6086.
37. Hanlon, E. B., Itzkan, I., Dasari, R. R., Feld, M. S., Ferrante, R. J., McKee, A. C., Lathi, D., Kowall, N. W. (1999). Near-infrared fluorescence spectroscopy detects Alzheimer's disease in vitro. *Photochem. Photobiol.* 70, 236–242.

38. Hock, C., Villringer, K., Muller-Spahn, F., Hofmann, M., Schuh-Hofer, S., Heekeren, H., Wenzel, R., Dirnagl, U., Villringer, A. (1996). Near infrared spectroscopy in the diagnosis of Alzheimer's disease. *Ann. N.Y. Acad. Sci.* 777, 22–29.
39. Chen, W. G., Li, P. C., Luo, Q. M., Zeng, S. Q., Hu, B. (2000). Hemodynamic assessment of ischemic stroke with near-infrared spectroscopy. *Space Med. Med. Eng.* 13, 84–89.
40. Nemoto, E. M., Yonas, H., Kassam, A. (2000). Clinical experience with cerebral oximetry in stroke and cardiac arrest. *Crit. Care Med.* 28, 1052–1054.
41. Saitou, H., Yanagi, H., Hara, S., Tsuchiya, S., Tomura, S. (2000). Cerebral blood volume and oxygenation among post stroke hemiplegic patients: Effects of 13 rehabilitation tasks measured by near-infrared spectroscopy. *Arch. Phys. Med. Rehabil.* 81, 1348–1356.
42. Vernieri, F., Rosato, N., Pauri, F., Tibuzzi, F., Passarelli, F., Rossini, P. M. (1999). Near infrared spectroscopy and transcranial Doppler in monohemispheric stroke. *Eur. Neurol.* 41, 159–162.
43. Villringer, A., Planck, J., Stodieck, S., Botzel, K., Schleinkofer, L., Dirnagl, U. (1994). Noninvasive assessment of cerebral hemodynamics and tissue oxygenation during activation of brain cell function in human adults using near infrared spectroscopy. *Adv. Exp. Med. Biol.* 345, 559–565.
44. Haginoya, K., Munakata, M., Kato, R., Yokoyama, H., Ishizuka, M., Iinuma, K. (2002). Ictal cerebral hemodynamics of childhood epilepsy measured with near-infrared spectrophotometry. *Brain* 125(Pt 9), 1960–1971.
45. Fallgatter, A. J., Roesler, M., Sitzmann, L., Heidrich, A., Mueller, T. J., Strik, W. K. (1997). Loss of functional hemispheric asymmetry in Alzheimer's dementia assessed with near-infrared spectroscopy. *Brain Res. Cogn. Brain Res.* 6(1), 67–72.
46. Fladby, T., Bryhn, G., Halvorsen, O., Rose, I., Wahlund, M., Wiig, P. et al. (2004). Olfactory response in the temporal cortex of the elderly measured with near-infrared spectroscopy: A preliminary feasibility study. *J. Cereb. Blood Flow Metab.* 24(6), 677–680.
47. Murata, Y., Katayama, Y., Oshima, H., Kawamata, T., Yamamoto, T., Sakatani, K. et al. (2000). Changes in cerebral blood oxygenation induced by deep brain stimulation: Study by near-infrared spectroscopy (NIRS). *Keio J. Med.* 49(Suppl. 1), A61–A63.
48. Sakatani, K., Katayama, Y., Yamamoto, T., Suzuki, S. (1999). Changes in cerebral blood oxygenation of the frontal lobe induced by direct electrical stimulation of thalamus and globus pallidus: A near infrared spectroscopy study. *J. Neurol. Neurosurg. Psychiatr.* 67(6), 769–773.
49. Mayai, I., Yagura, H., Oda, I., Konishi, I., Eda, H., Suzuki, T. et al. (2002). Premotor cortex is involved in restoration of gait in stroke. *Ann. Neurol.* 52(2), 188–194.
50. Mayai, I., Yagura, H., Hatakenaka, M., Oda, I., Konishi, I., Kubota, K. (2003). Longitudinal optical imaging study for locomotor recovery after stroke. *Stroke* 34(12), 2866–2870.
51. Lin, P., Lin, S., Penny, T., Chen, J. (2009). Application of near infrared spectroscopy and imaging for motor rehabilitation in stroke patients. *J. Med. Biol. Eng.* 29(5), 210–221.
52. Akiyoshi, J., Hieda, K., Aoki, Y., Nagayama, H. (2003). Frontal brain hypoactivity as a biological substrate of anxiety in patients with panic disorders. *Neuropsychobiology* 47(3), 165–170.
53. Okada, F., Tokumitsu, Y., Hoshi, Y., Tamura, M. (1994). Impaired interhemispheric integration in brain oxygenation and hemodynamics in schizophrenia. *Eur. Arch. Psychiatr. Clin. Neurosci.* 244(1), 17–25.
54. Fallgatter, A. J., Strik, W. K. (2000). Reduced frontal functional asymmetry in schizophrenia during a cued continuous performance test assessed with near-infrared spectroscopy. *Schizophr. Bull.* 26(4), 913–919.
55. Shinba, T., Nagano, M., Kariya, N., Ogawa, K., Shinozaki, T., Shimosato, S. et al. (2004). Near-infrared spectroscopy analysis of frontal lobe dysfunction in schizophrenia. *Biol. Psychiatr.* 55(2), 154–164.
56. Matsui, M., Tanaka, K., Yonezawa, M., Kurachi, M. (2007). Activation of the prefrontal cortex during memory learning: Near-infrared spectroscopy study. *Psychiatr. Clin. Neurosci.* 61, 31.
57. Schreppel, T., Egetemeir, J., Schecklmann, M., Plichta, M. M., Pauli, P., Ellgring, H., Fallgatter, A. J., Herrmann, M. J. (2008). Activation of the prefrontal cortex in working memory and interference resolution processes assessed with near-infrared spectroscopy. *Neuropsychobiology* 57, 188.

58. Rugg, M. D., Johnson, J. D., Park, H., Uncapher, M. R. (2008). Encoding-retrieval overlap in human episodic memory: A functional neuroimaging perspective. *Prog. Brain Res.* 169, 339.
59. Okamoto, M., Wada, Y., Yamaguchi, Y., Kyutoku, Y., Clowney, L., Singh, A. K., Dan, I. (2011). Process-specific prefrontal contributions to episodic encoding and retrieval of tastes: A functional NIRS study. *NeuroImage* 54, 1578.
60. Tulving, E., Kapur, S., Craik, F. I., Moscovitch, M., Houle, S. (1994). Hemispheric encoding/retrieval asymmetry in episodic memory: Positron emission tomography findings. *Proc. Natl. Acad. Sci. U.S.A.* 91, 2016.
61. Toichi, M., Findling, R. L., Kubota, Y., Calabrese, J. R., Wiznitzer, M., McNamara, N. K., Yamamoto, K. (2004). Hemodynamic differences in the activation of the prefrontal cortex: Attention vs. higher cognitive processing. *Neuropsychologia* 42, 698.
62. Harasawa, M., Shioiri, S. (2011). Asymmetrical brain activity induced by voluntary spatial attention depends on the visual hemifield: A functional near-infrared spectroscopy study. *Brain Cogn.* 75, 292.
63. Cutini, S., Scarpa, F., Scatturin, P., Jolicœur, P., Pluchino, P., Zorzi, M., Dell'acqua R. (2011). A hemodynamic correlate of lateralized visual short-term memories. *Neuropsychologia* 49, 1611.
64. Vogel, E. K., Machizawa, M. G. (2004). Neural activity predicts individual differences in visual working memory capacity. *Nature* 428, 748.
65. Robitaille, N., Marois, R., Todd, J., Grimault, S., Cheyne D., Jolicoeur, P. (2010). Distinguishing between lateralized and nonlateralized brain activity associated with visual short-term memory: fMRI, MEG, and EEG evidence from the same observers. *NeuroImage* 53, 1334.
66. Herrmann, M. J., Walter, A., Ehlis A. C., Fallgatter, A. J. (2006). Cerebral oxygenation changes in the prefrontal cortex: Effects of age and gender. *Neurobiol. Aging* 27, 888.
67. Kameyama, M., Fukuda, M., Uehara T., Mikuni, M. (2004). Sex and age dependencies of cerebral blood volume changes during cognitive activation: A multichannel near-infrared spectroscopy study. *NeuroImage* 22, 1715.
68. Schecklmann, M., Ehlis, A. C., Plichta M. M., Fallgatter, A. J. (2008). Functional near-infrared spectroscopy: A long-term reliable tool for measuring brain activity during verbal fluency. *NeuroImage* 43, 147.
69. Chaudhary, U., Hall, M., DeCerce, J., Rey, G., Godavarty, A. (2011). Frontal activation and connectivity using near-infrared spectroscopy: Verbal fluency language study. *Brain Res. Bull.* 84, 197.
70. Kovelman, I., Shalinsky, M. H., White, K. S., Schmitt, S. N., Berens, M. S., Paymer, N., Petitto, L. A. (2008). Dual language use in sign-speech bimodal bilinguals: fNIRS brain-imaging evidence. *Brain Lang.* 109, 112.
71. Quaresima, V., Ferrari, M., van der Sluijs, M. C. P., Menssen, J., Coller, W. N. J. M. (2002). Lateral frontal cortex oxygenation changes during translation and language switching revealed by non-invasive near-infrared multi-point measurements. *Brain Res. Bull.* 59, 235.
72. Suda, M., Takei, Y., Aoyama, Y., Narita, K., Sato, T., Fukuda M., Mikuni, M. (2010). Frontopolar activation during face-to-face conversation: An in situ study using near-infrared spectroscopy. *Neuropsychologia* 48, 441.
73. Noguchi, Y., Takeuchi, T., Sakai, K. L. (2002). Lateralized activation in the inferior frontal cortex during syntactic processing: Event-related optical topography study. *Hum. Brain Map.* 17, 89.
74. Davidson, R. J., Irwin, W. (1999). The functional neuroanatomy of emotion and affective style. *Trends Cogn. Sci.* 3, 11.
75. Dolcos, F., Iordan A. D., Dolcos, S. (2011). Neural correlates of emotion-cognition interactions: A review of evidence from brain imaging investigations. *J. Cogn. Psychol.* 23, 669.
76. Herrmann, M. J., Ehlis A. C., Fallgatter, A. J. (2003). Prefrontal activation through task requirements of emotional induction measured with NIRS. *Biol. Psychol.* 64, 255.
77. MacDonald, A. W., Cohen, J. D., Stenger V. A., Carter, C. S. (2000). Dissociating the role of the dorsolateral prefrontal and anterior cingulate cortex in cognitive control. *Science* 288, 1835.
78. Luu, P., Flaisch, T., Tucker, D. M. (2000). Medial frontal cortex in action monitoring. *J. Neurosci.* 20, 464.

79. Glotzbach, E., Mühlberger, A., Gschwendtner, K., Fallgatter, A. J., Pauli P., Herrmann, M. J. (2011). Prefrontal brain activation during emotional processing: A functional near infrared spectroscopy study (fNIRS). *Open Neuroimag. J.* 5, 33.
80. Herrmann, M. J., Huter, T., Plichta, M. M., Ehlis, A. C., Alpers, G. W., Mühlberger, A., Fallgatter, A. J. (2008). Enhancement of activity of the primary visual cortex during processing of emotional stimuli as measured with event-related functional near-infrared spectroscopy and event-related potentials. *Hum. Brain Map.* 29, 28.

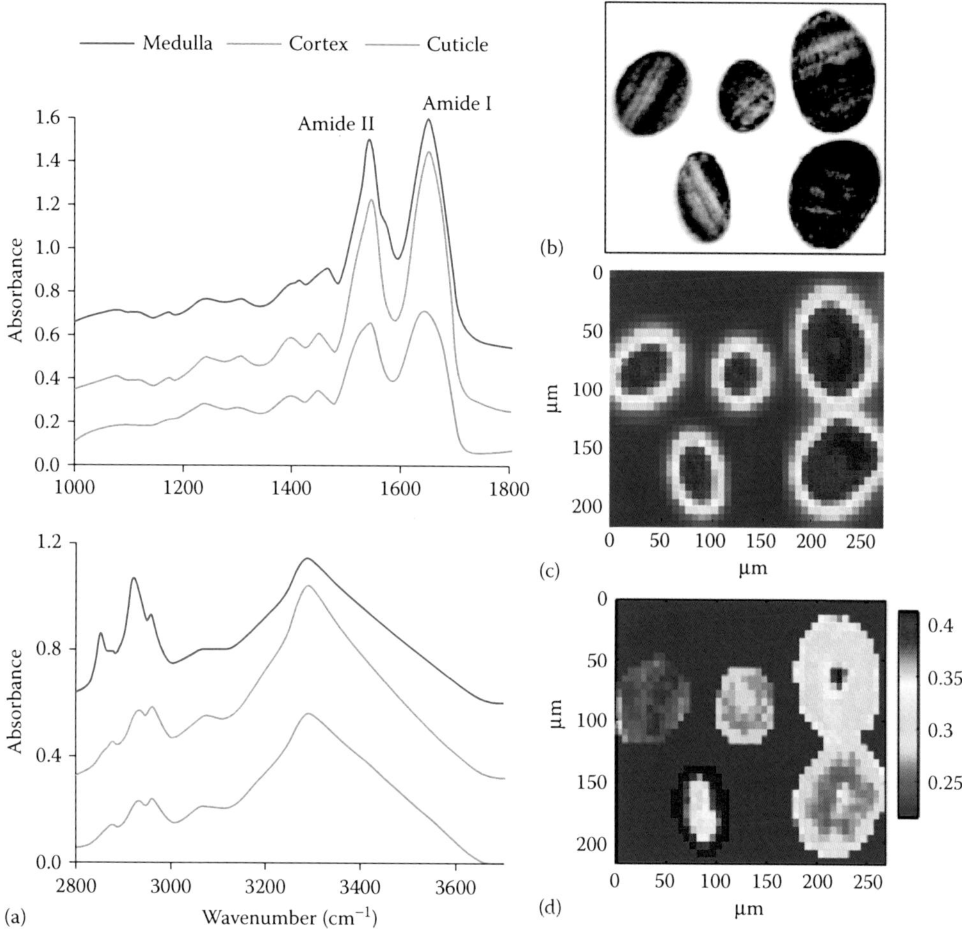

FIGURE 1.2
Optical and IR imaging of hair cross sections. (a) Representative IR spectra obtained from the medulla, cuticle, and cortex in the spectral regions of 1000–1800 and 2800–3700 cm^{-1}. Each spectrum was obtained from a single pixel with dimensions of 6.25 μm × 6.25 μm. (b) Optical image of microtomed hair cross sections. (c) IR image of protein distribution. (d) IR image of lipid distribution. Image was obtained from integration peak area ratios of 2850 cm^{-1} (C–H stretching of CH_2 groups in lipid chains) to 2960 cm^{-1} (C–H stretching of CH_3 terminal groups in proteins).

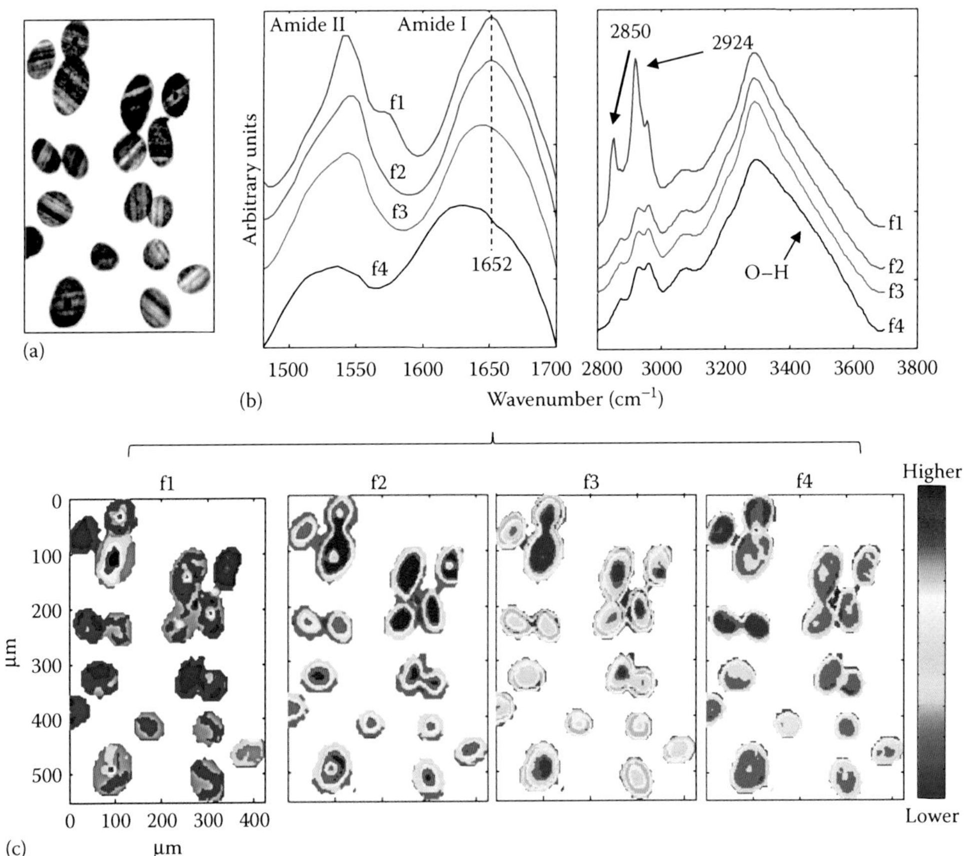

FIGURE 1.4
(a) An optical micrograph of microtomed (5 μm thick) hair cross sections. Factor analysis was conducted on an IR image acquired from hair cross sections. (b) The four distinct factor loadings in the regions of 1480–1700 and 2830–3700 cm^{-1} generated by the ISys score segregation algorithm are offset and labeled as f1–f4. (c) The spatial distribution of factor scores for each of the loadings in the spectral region of 1480–1700 cm^{-1} as marked. Dark blue indicates the lowest score, whereas green, yellow, orange, and red indicate progressively higher scores. Factor loadings and score images have been assigned to different micro-regions in hair as described in the text.

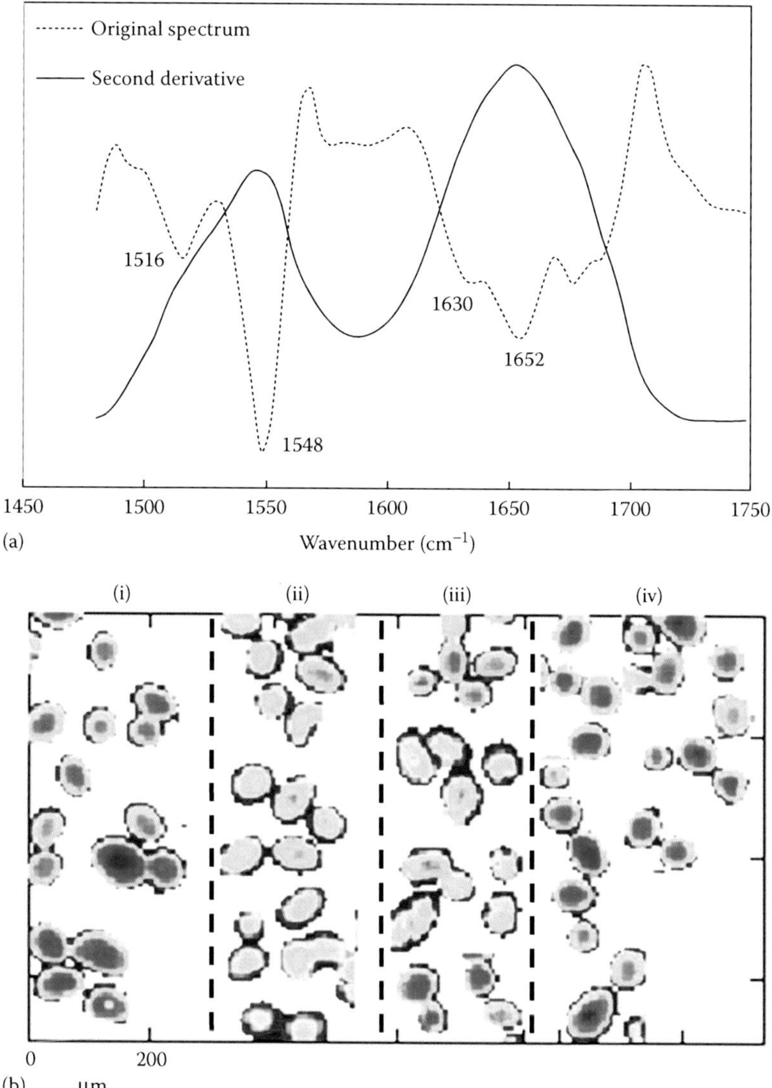

FIGURE 1.5
(a) IR spectra and the second derivative curve in the amide region of 1480–1700 cm^{-1}. The spectrum was obtained from the cortical region of undamaged European dark brown hair. The assignments of marked peaks are described in Table 1.1. (b) β-Sheet distributions in hair cross sections. The color coding is red > orange > yellow > green > blue. The maps are constructed from the ratio of peak intensities at 1516 cm^{-1} (attributed to β-sheet formation) and 1548 cm^{-1} (α-helix band). Hair samples include European dark brown hair (i), thermally treated hair without polymer pretreatment (ii), thermally treated hair with pretreatment of 0.5% hydroxyethylcellulose (iii), and thermally treated hair with pretreatment of 1% poly(vinylpyrrolidone-*co*-acrylic acid-*co*-laurylmethacrylate) + 0.5% hydroxyethylcellulose (iv).

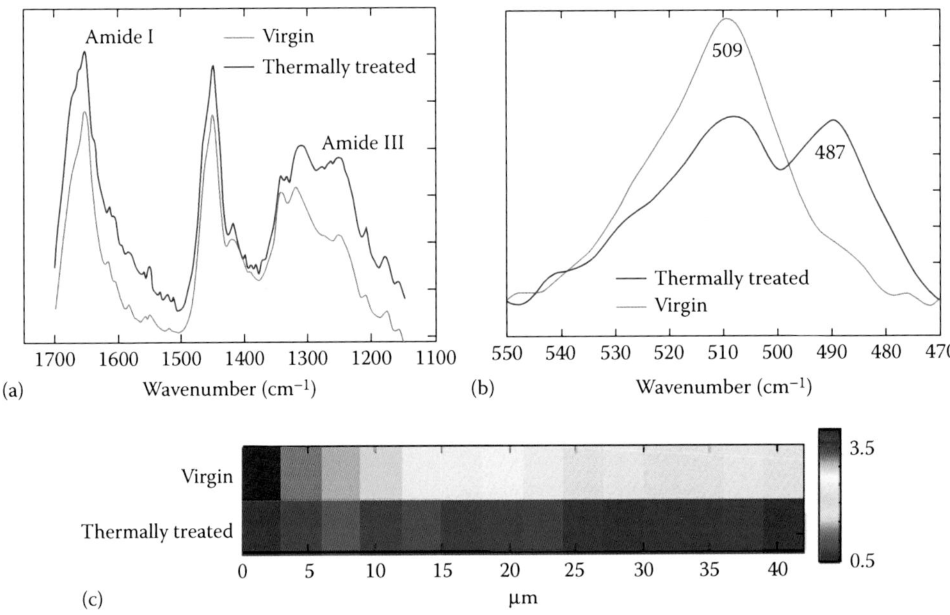

FIGURE 1.6
Representative Raman spectra in the spectral region of (a) 1150–1700 cm^{-1} and (b) 470–550 cm^{-1}. The Raman spectra were obtained from the cortical region of both thermally treated and virgin hair fibers. (c) Raman image of disulfide bond distribution along a confocal line. The confocal line was acquired from the hair surface to 42 μm deep into the fiber structure. The schematic diagram of confocal Raman acquisition is explained in Figure 1.3b. The map was created by the intensity ratio of the band at 509 cm^{-1}, arising from the S–S stretching mode, to the band at 1004 cm^{-1} (phenylalanine in keratin).

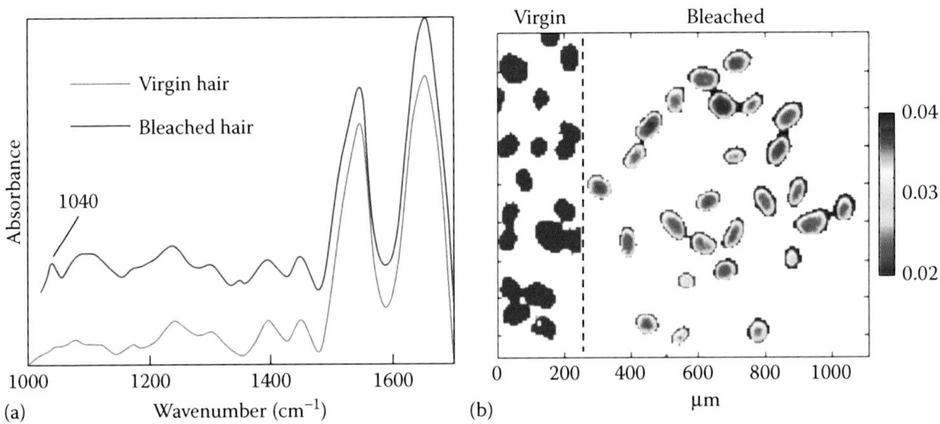

FIGURE 1.7
(a) IR spectra acquired from the cortical regions of virgin and bleached hair in the spectral regions of 1000–1700 cm^{-1}. The marked band at 1040 cm^{-1} in the spectrum of bleached hair arises from S=O stretching mode of the sulfonate groups. (b) IR images of sulfonate distribution in hair cross sections (5 μm thick) microtomed from virgin and bleached hair fibers.

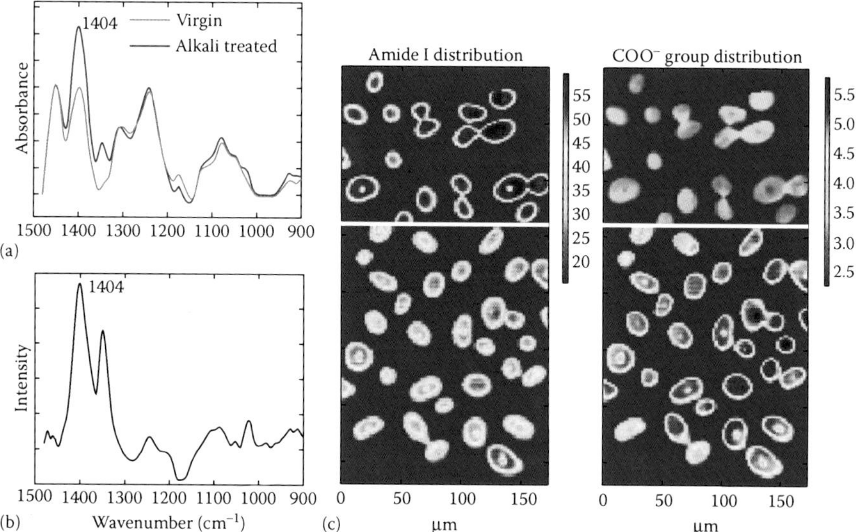

FIGURE 1.8
(a) Mean IR spectra acquired from 20 hair cross sections microtomed from virgin and alkali-treated hair in the spectral region of 900–1500 cm^{-1}. (b) The difference spectrum obtained by subtracting the mean spectrum of virgin from the mean spectrum of alkali-treated hair. (c) IR images of Amide I and COO^- group distributions. *Upper*: virgin hair. *Lower*: alkali-treated hair.

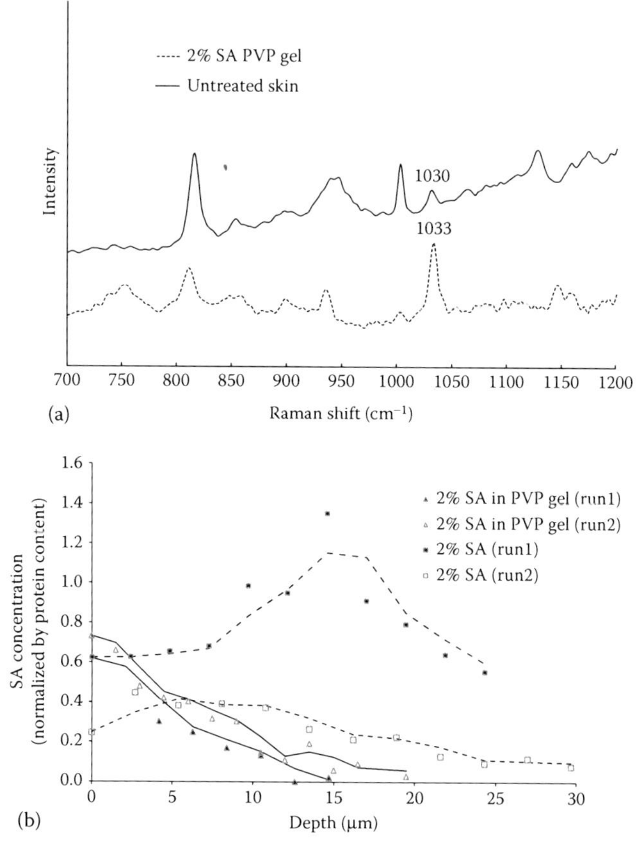

FIGURE 1.16
(a) Representative Raman spectrum of untreated pig skin acquired at a depth of ~6 μm beneath the SC surface, along with the spectrum of 2% salicylic acid (SA) in PVP gel. (b) The relative concentration of SA change in skin with depth after 2% SA was applied on the skin surface for 1 hour with different vehicles.

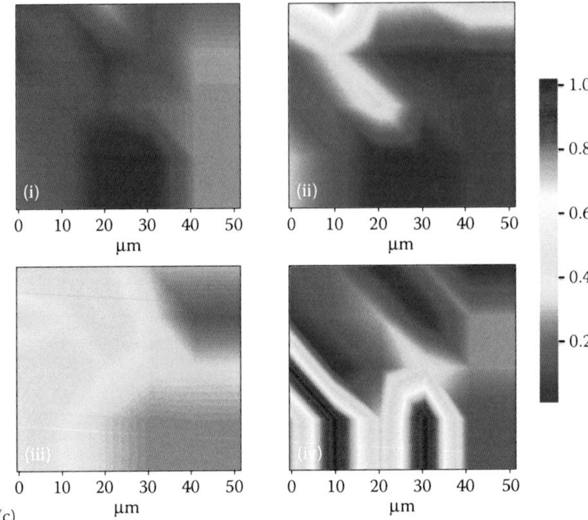

FIGURE 1.16
Continued (c) SA distributions inside the stratum corneum. Each Raman map was acquired laterally at ~6–8 μm below the skin surface. The skin was treated with 2% SA in a PVP gel for 1 hour (i), 2% SA for 1 hour (ii), 2% SA in a PVP gel for 2.5 hours (iii), and 2% SA for 2.5 hours (iv).

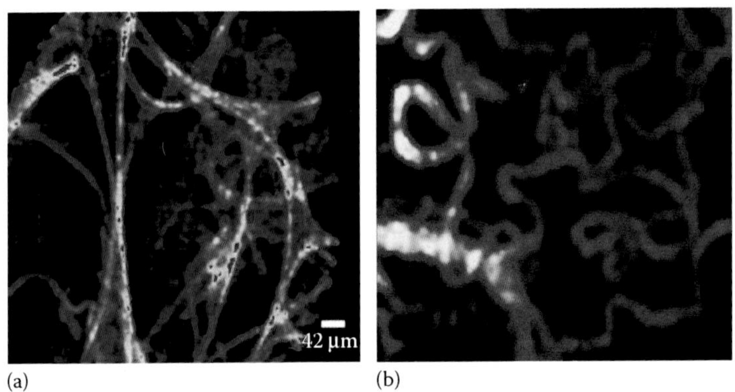

FIGURE 2.5
Confocal images obtained *in vivo* with a flexible fiber probe of 650 μm. (a) Normal human alveoli: Visualization of normal distal lung, with distinct alveolar microarchitecture. The signal shown is tissue autofluorescence; no dye was applied in this case. (b) *In vivo* angiogenesis imaging: Visualization of tumoral vessels in a mouse prostate after FITC-dextran (500 kDa) injection in the tail. The site was accessed through a microincision in the skin at the site of the tumor. Field of view is 400 × 280 μm. (From Ntziachristos, V., *Annu. Rev. Biomed. Eng.*, 8, 1–33, 2006. With permission.)

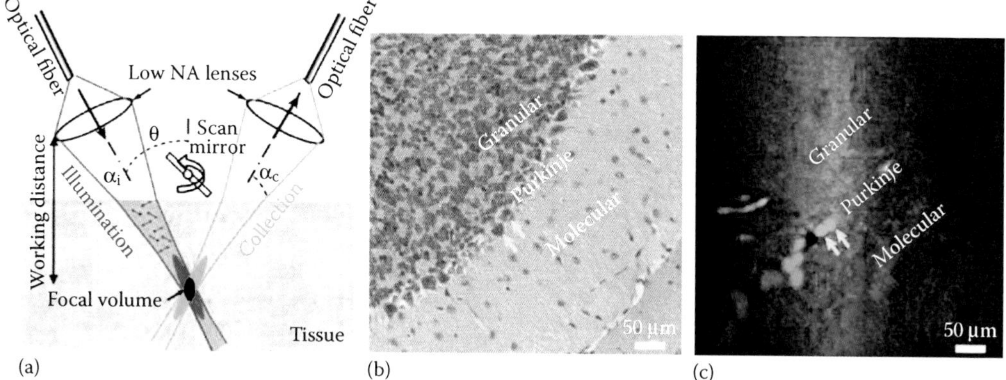

FIGURE 2.6
Dual-axis confocal microscope for *in vivo* imaging. (a) Schematic of architecture: Two low NA objectives are oriented with the illumination and collection beams crossed at an angle so that the focal volume is defined at the intersection of the photon beams, offering a significant reduction of the axial resolution, long working distances, large dynamic range, and rejection of light scattered along the illumination path. (b) Fluorescence images from the cerebellum of a transgenic mouse that expresses GFP driven by a β-actin-CMV (Cytomegalovirus) promoter. The image was collected at an axial depth of 30 μm and the scale bar is 50 μm. (c) Corresponding histology showing the Purkinje cell bodies, marked by the arrows, aligned side by side in a row, which separate the molecular from the internal granular layer. (From Ntziachristos, V., *Annu. Rev. Biomed. Eng.*, 8, 1–33, 2006. With permission.)

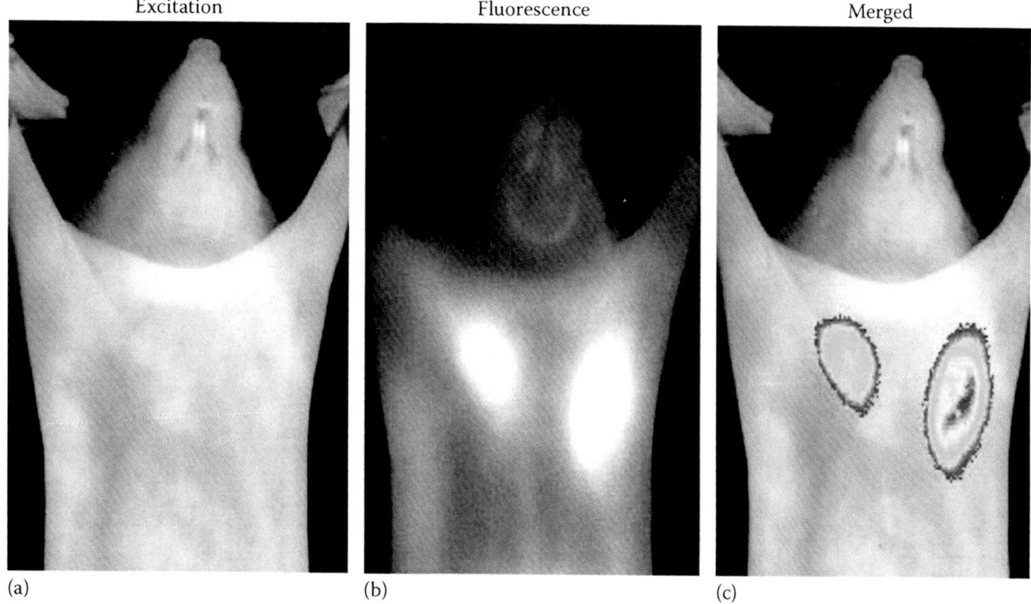

FIGURE 2.7
In vivo epi-illumination imaging of cathepsin activity from a nude female mouse with two HT1080 tumors implanted subcutaneously. (a) Image obtained at the emission wavelength; (b) fluorescence image; (c) merged image, that is, superposition of the fluorescence image shown in color on the excitation image. A threshold was applied on the fluorescence image to remove low-intensity background signals and allow for the simultaneous visualization of (a) and (b). (From Ntziachristos, V., *Annu. Rev. Biomed. Eng.*, 8, 1–33, 2006. With permission; Courtesy of Stephen Windsor; the cathepsin-sensitive probe was kindly provided by Dr. Ching Tung, both with the Center for Molecular Imaging Research, Massachusetts General Hospital and Harvard Medical School, Boston, Massachusetts.)

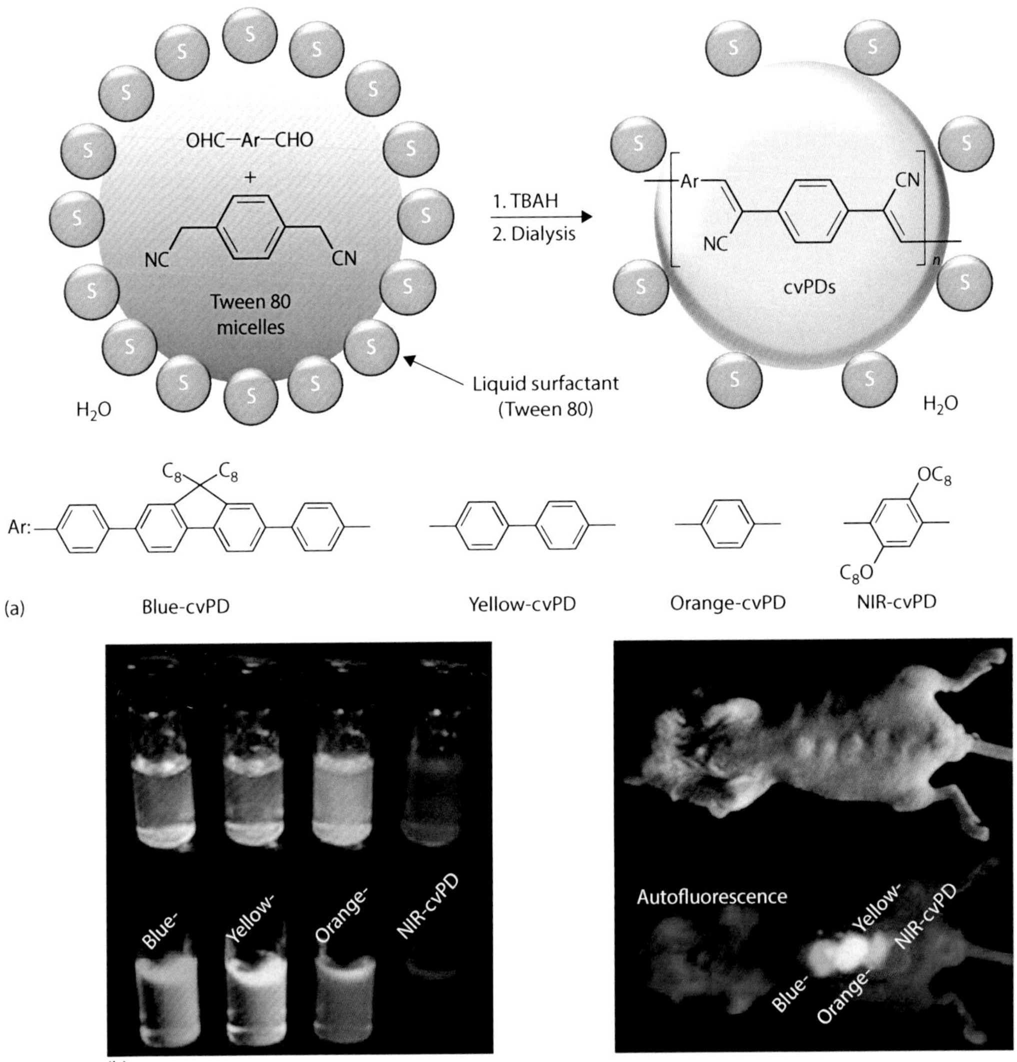

FIGURE 2.11
(a) Schematic diagram depicting colloidal synthesis of cvPDs through tetrabutylammonium hydroxide (TBAH)-catalyzed Knoevenagel condensation in the hydrophobic core of solvent-free aqueous micelles. (b) True-color photographs of water-dispersed cvPDs (left) and a cvPD-injected live mouse (right) under room light (top) and UV excitation at 365 nm for fluorescence (bottom). (Kim, S. et al., *Chem. Commun.*, 46, 1617–1619, 2010. Reproduced by permission of The Royal Society of Chemistry.)

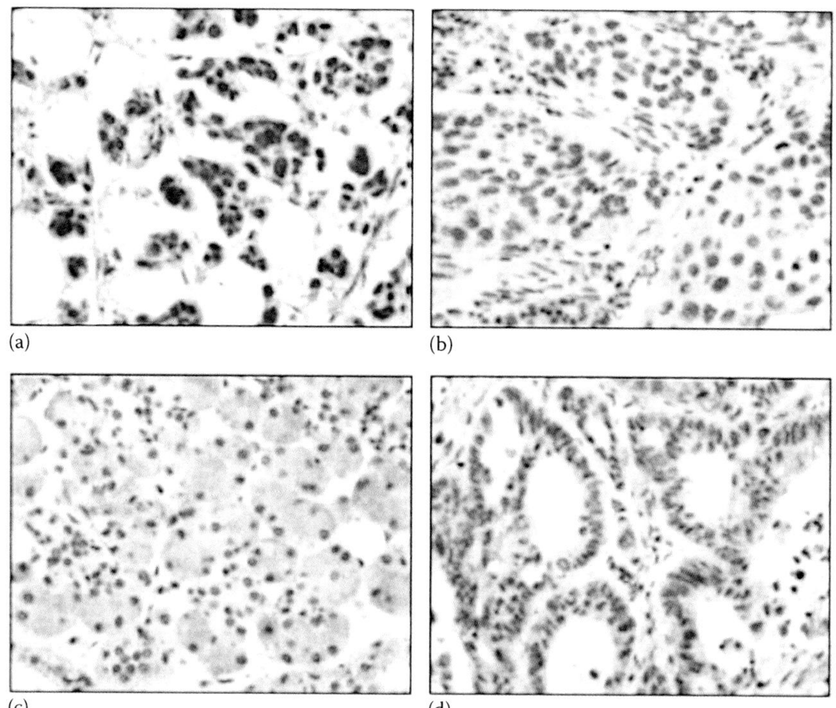

FIGURE 3.1
A typical digital IHC image: (a) Breast carcinoma; (b) lung carcinoma; (c) pancreas tissue; (d) colon carcinoma.

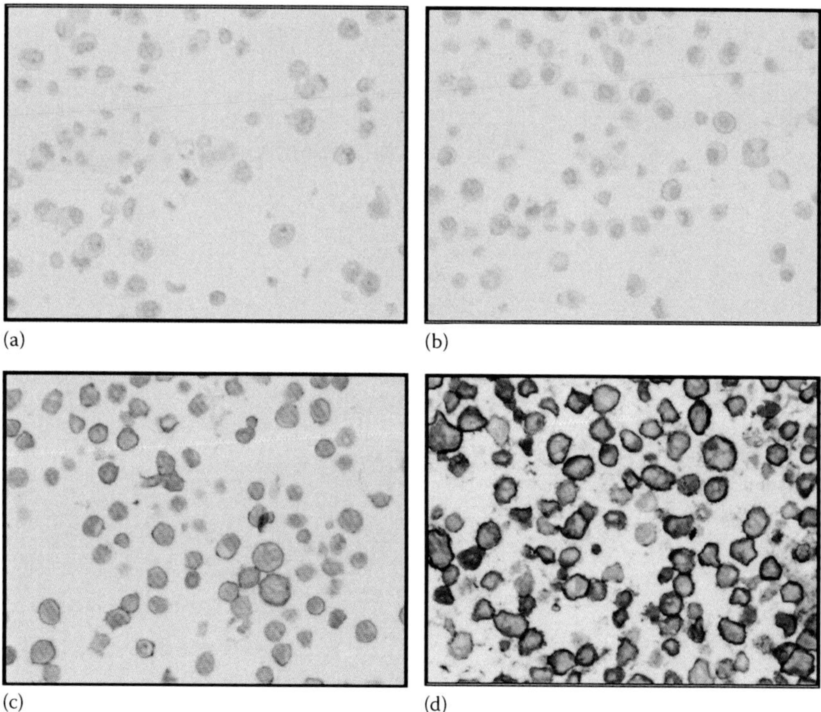

FIGURE 3.6
(a) IHC score 0 image (totally normal and healthy tissue, no HER2 overexpression). (b) IHC score +1 image (healthy tissue with controlled HER2 overexpression). (c) IHC score +2 image (healthy but potentially suspicious tissue, augmented HER2 overexpression). (d) IHC score +3 image (affected tissue). (Data from UKNEQAS, *Immunocytochemistry*, 2, 2003.)

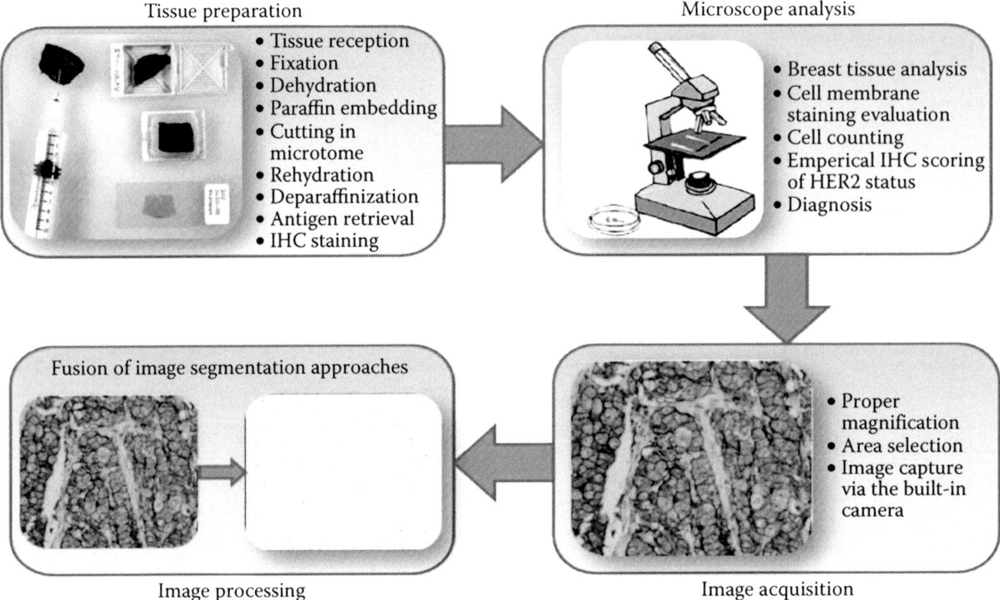

FIGURE 3.7
Summary of the key steps of a unified, complete procedure for efficient analysis of an IHC image aiming at the assessment of the Her2/neu protein overexpression.

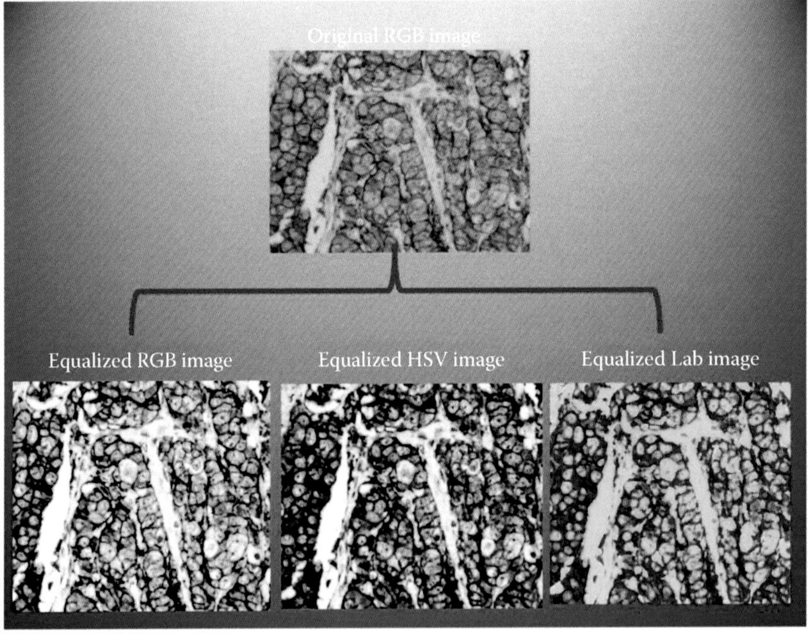

FIGURE 3.9
A typical IHC image represented in the RGB, HSV, and Lab color spaces after histogram equalization for contrast improvement.

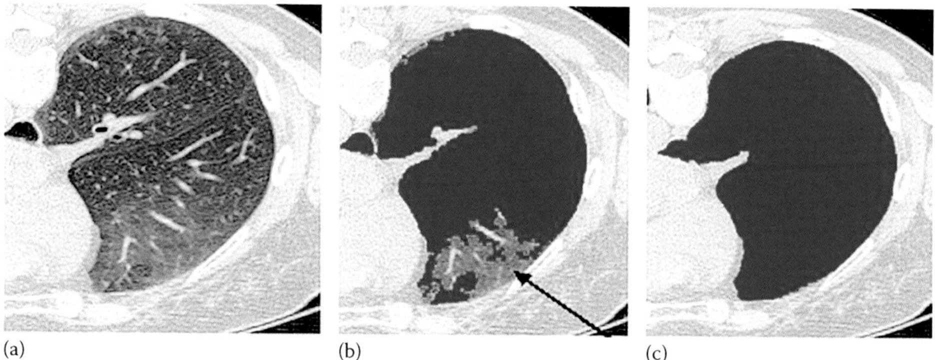

FIGURE 6.7
Lung segmentation results on scleroderma subjects. (a) Original scleroderma image scanned at residual volume. (b) Results of thresholding algorithm at 400 HU with post-processing. (c) Results from the rib curvature technique. The regions indicated by the arrow are the high attenuation regions not segmented by the fixed thresholding algorithm. (Data from Prasad, M.N. et al., *Acad. Radiol.*, 15, 1173–1180, 2008.)

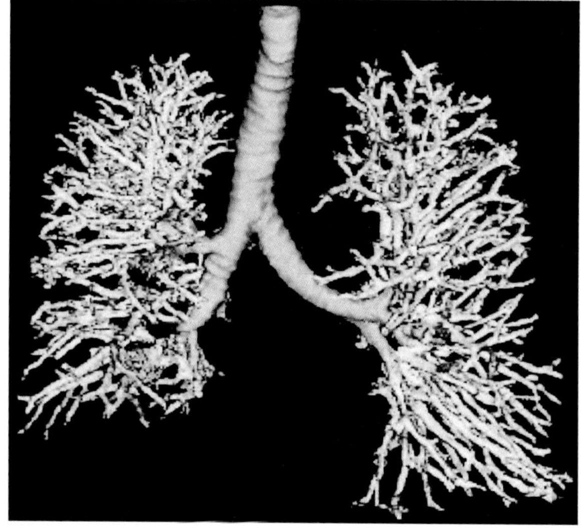

FIGURE 6.8
Bronchovascular structure. The bronchial is indicated in pink and the vascular tree in yellow. (Data from Zavaletta, V.A. et al., *Acad. Radiol.*, 14, 772–787, 2007.)

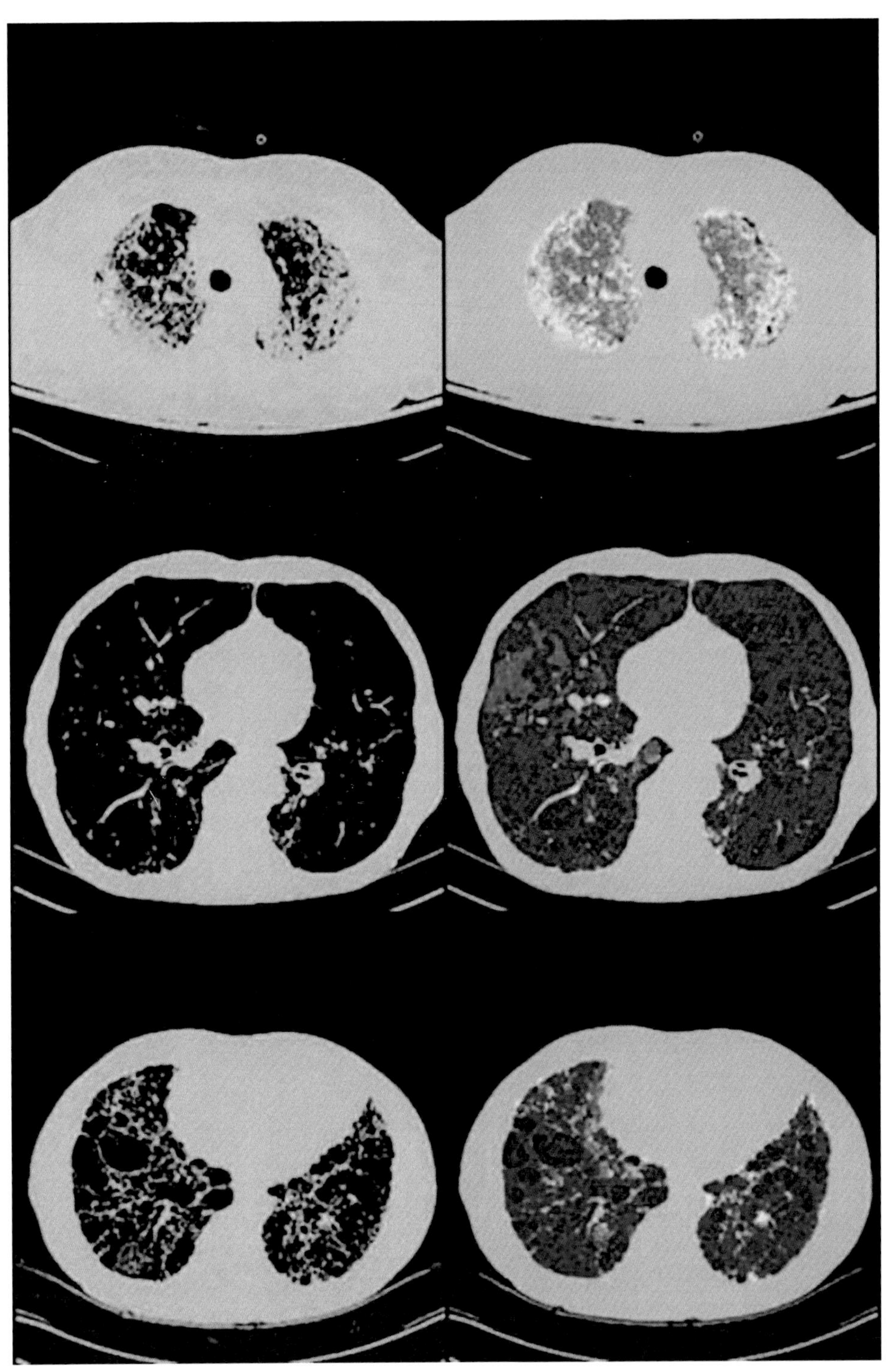

FIGURE 6.11
Quantification results depicted by color-coded overlay in several cases. *Left*: Original CT image. *Right*: Corresponding result (normal, green; ground-glass opacity, yellow; reticular opacity, cyan; honeycombing, blue; emphysema, red; consolidation, pink). (Data from Park, S.O. et al., *Korean J. Radiol.*, 10, 455–463, 2009.)

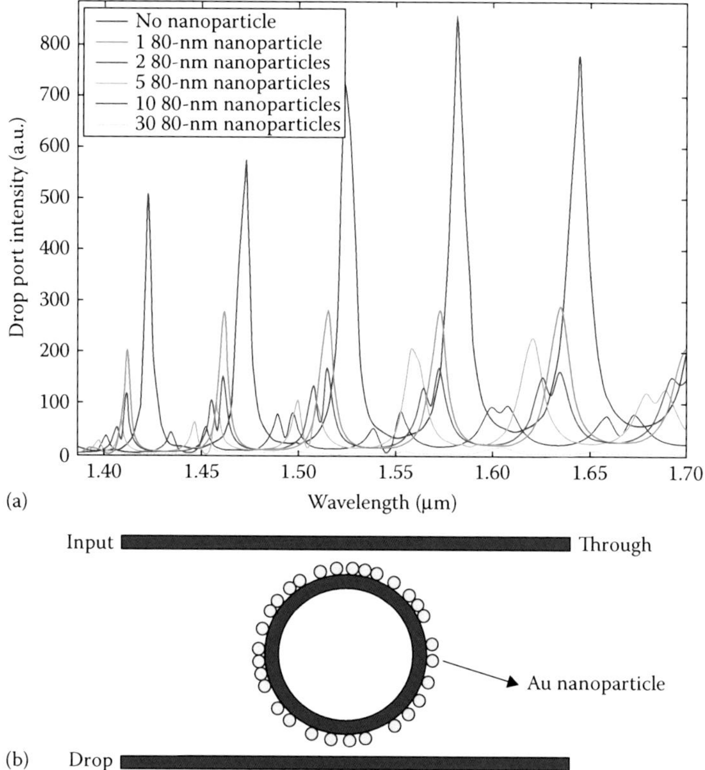

FIGURE 8.2
(a) The Drop port intensity with different numbers of 80 nm-sized Au nanoparticles adsorbed on the microring resonator. (b) The example of multiple Au nanoparticles randomly distributed on the microring resonator.

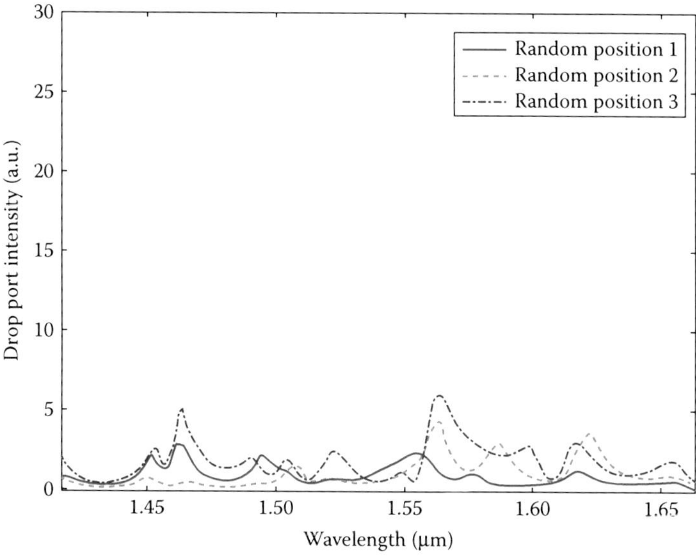

FIGURE 8.3
The Drop port intensity with three random positions of 30 Au nanoparticles of 80 nm in size.

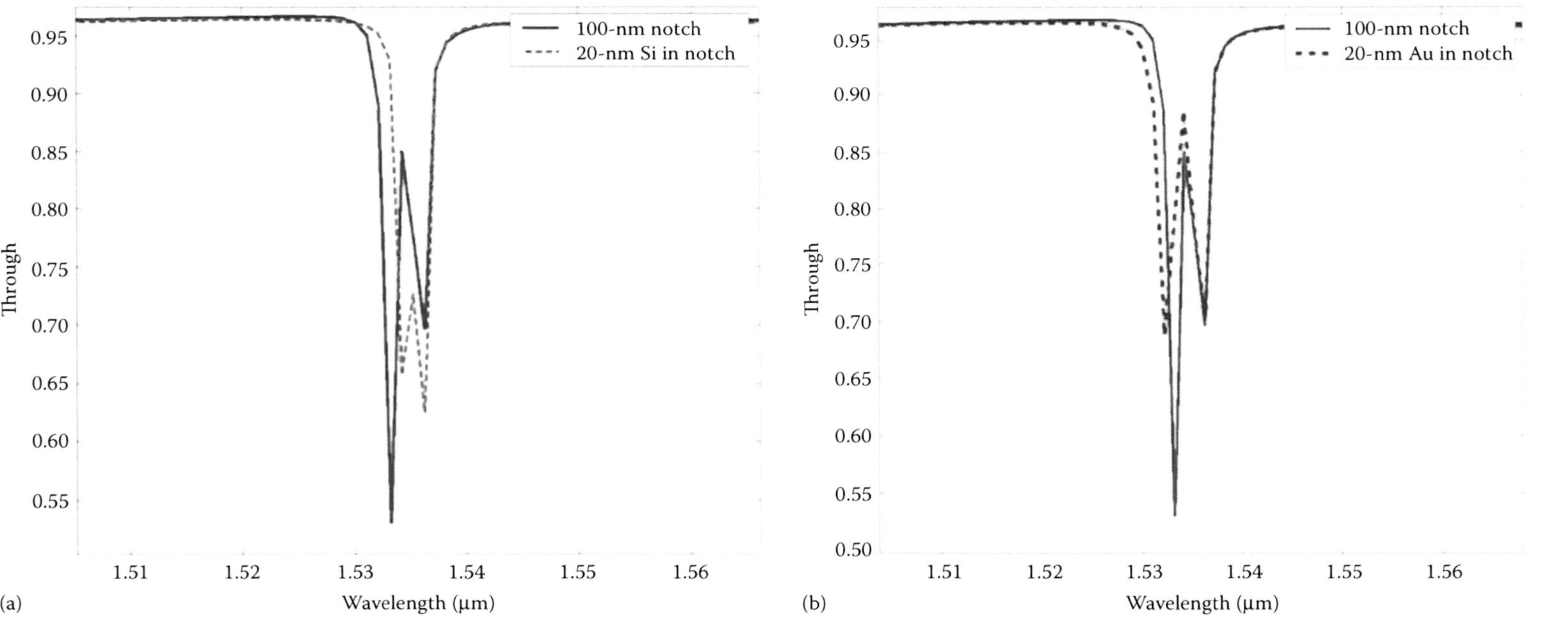

FIGURE 8.9
(a) Si nanoparticle tip of 20 nm in diameter is put inside the 100 nm-sized notch of the microring resonator. The Through port from the coupled waveguide is shown. It is observed that the bonding mode wavelength (resonance at shorter wavelength) is *red* shifted with 1.1 nm, whereas the wavelength of the antibonding mode (resonance at longer wavelength) is almost unchanged; the splitting bandwidth is smaller due to the red shift of the bonding mode. (b) Au nanoparticle tip of 20 nm in diameter is put inside the notch size of the microring resonator. We see the dramatic different shift from the Au nanoparticle for the bonding state (resonance at shorter wavelength). The 1.0 nm *blue* shift is observed instead and the antibonding state (resonance at longer wavelength) is still unchanged; the splitting bandwidth is larger than the original splitting due to the notch with air.

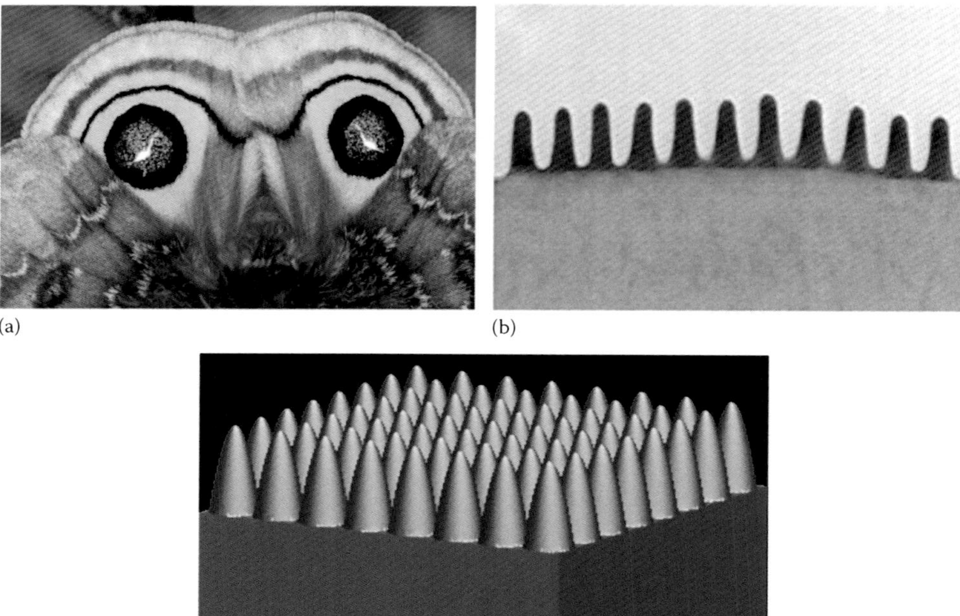

FIGURE 8.10
Illustration of moth's eye structures. (a) The image of the moth's eye. (b) The nanoscale features under SEM. The typical size is as follows: height 250 nm and pitch 300 nm. (c) The 3D view of the moth's eye structure.

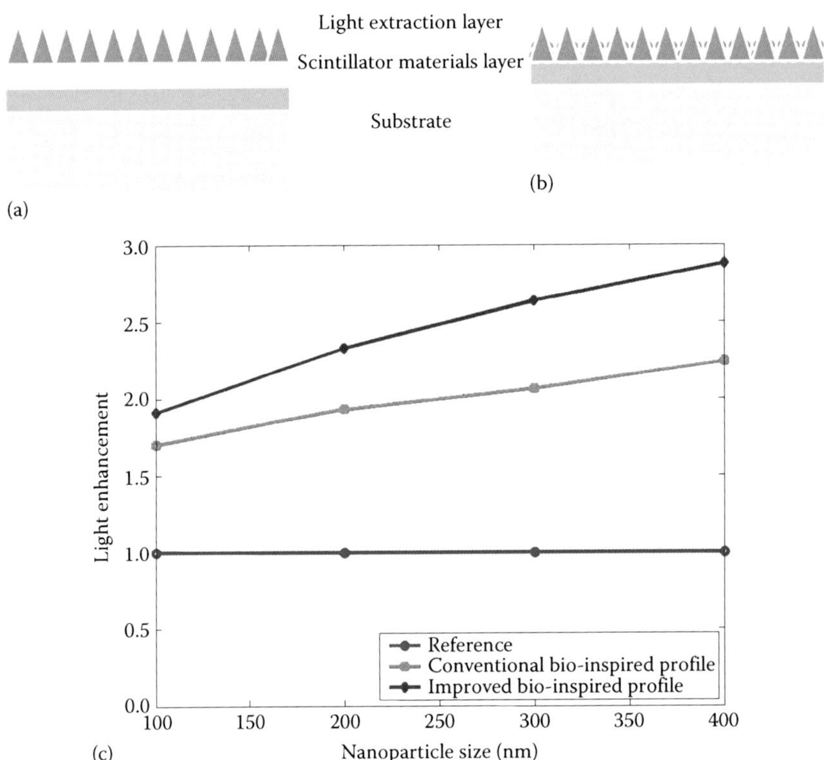

FIGURE 8.11
The simulation device structure for two different bio-inspired photonic structures. (a) The conventional moth's eye structure with flat sidewall. (b) The improved bio-inspired moth's eye structure with intentionally designed roughness on the sidewall. (c) The FDTD simulation results, which show an enhancement for both bio-inspired light extraction structure, whereas the improved bio-inspired moth's eye structures show more light output enhancement.

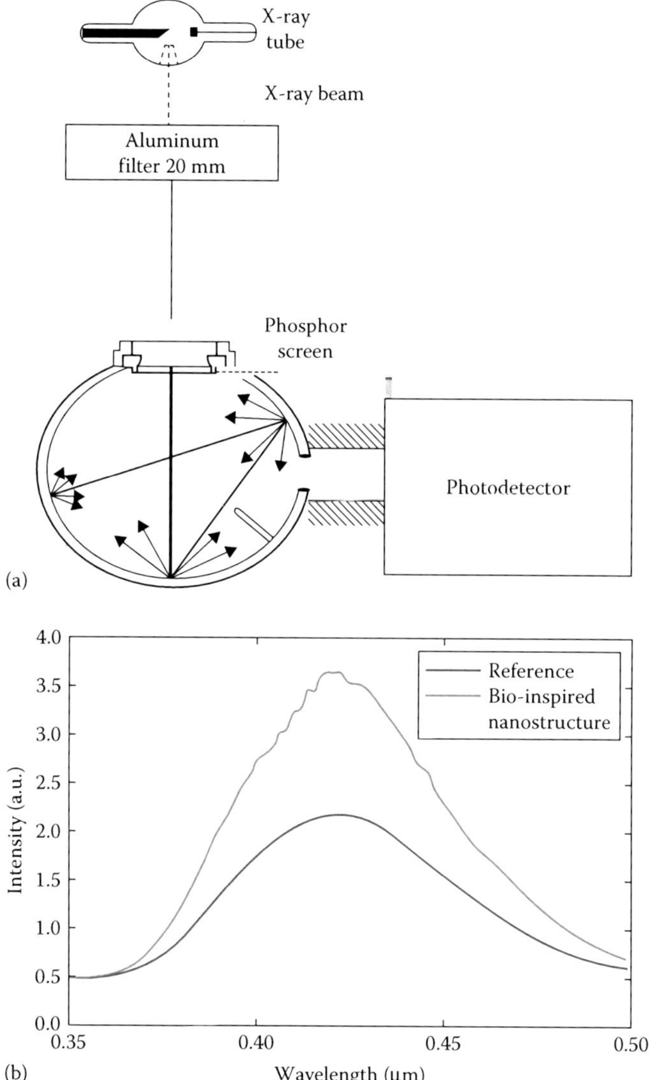

FIGURE 8.13
The X-ray mammographic unit was employed to demonstrate the light enhancement of the Lu_2SiO_5:Ce thin film with bio-inspired moth's eye-like nano photonic structures. (a) A General Electric Senographe DMR and X-ray mammographic unit with molybdenum anode target and molybdenum filter. (b) The comparison between the two Lu_2SiO_5:Ce thin films. The blue curve is the light output from the referenced Lu_2SiO_5:Ce thin film without any light extraction structures and the green curve is the light output from the Lu_2SiO_5:Ce thin film with improved bio-inspired moth's eye-like photonic structures.

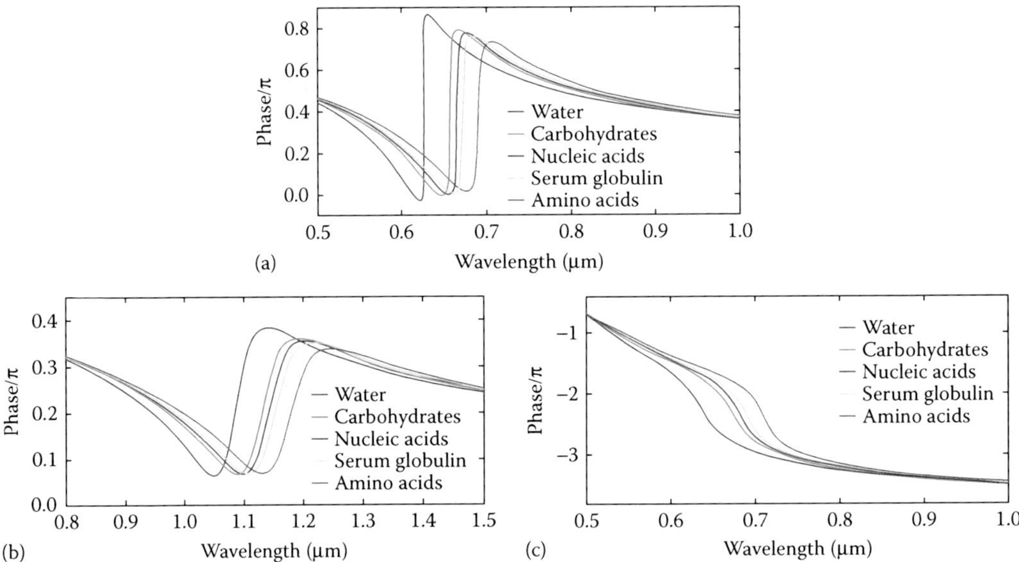

FIGURE 9.15
Phase plots for (a) PMA, (b) PMSA, and (c) PSMA structures in spectral interrogation with different biosamples.

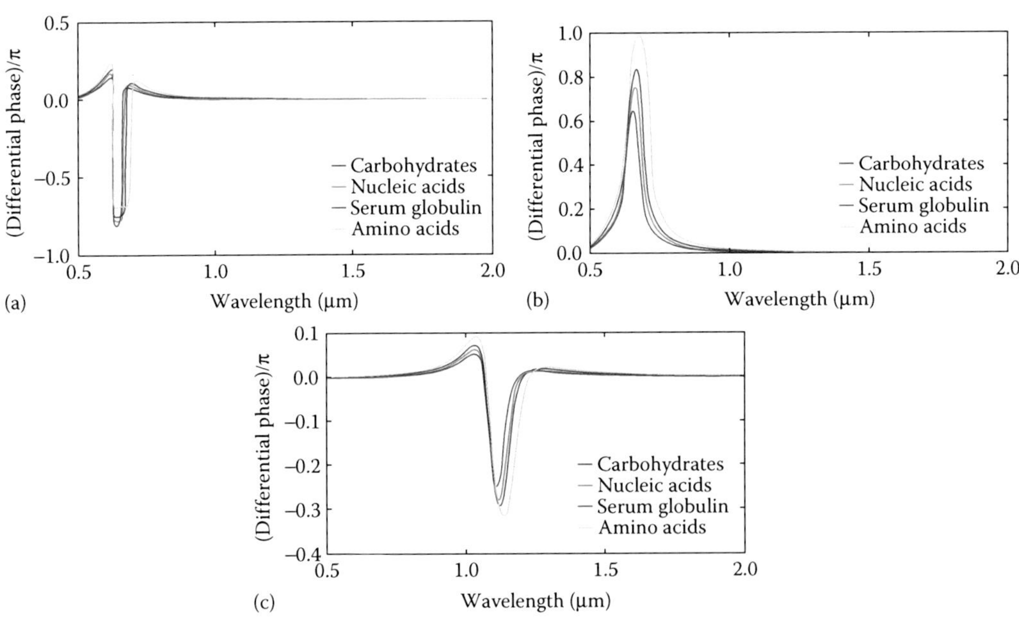

FIGURE 9.16
Differential phase plots for (a) PMA, (b) PSMA, and (c) PMSA structures in spectral interrogation for different biosamples.

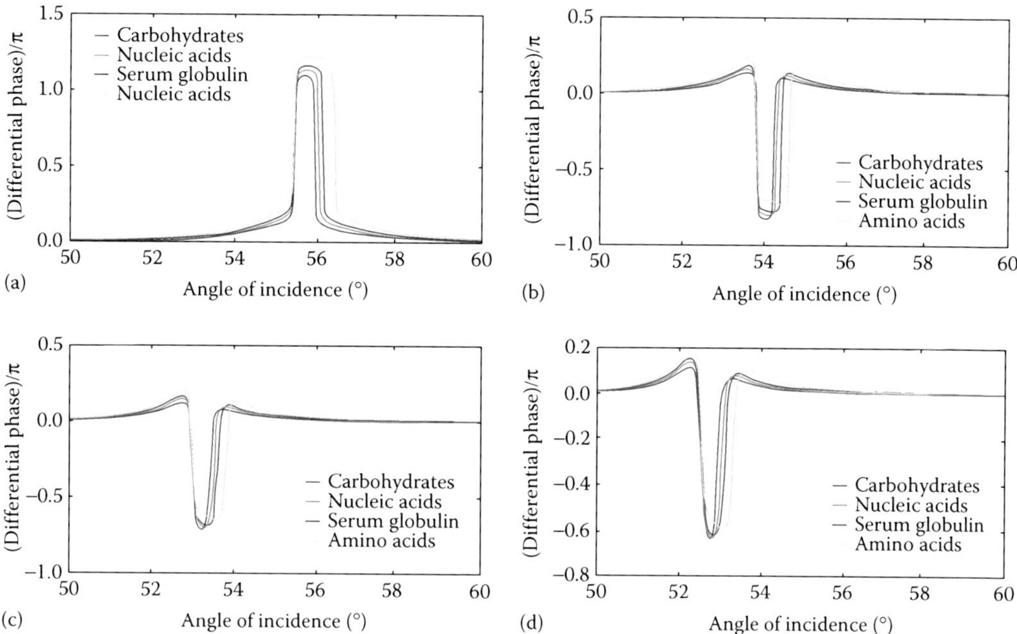

FIGURE 9.17
Differential phase plots for the PMA structure for specific wavelengths of (a) 550 nm, (b) 633 nm, (c) 700 nm, and (d) 750 nm in angular interrogation mode for different biosamples.

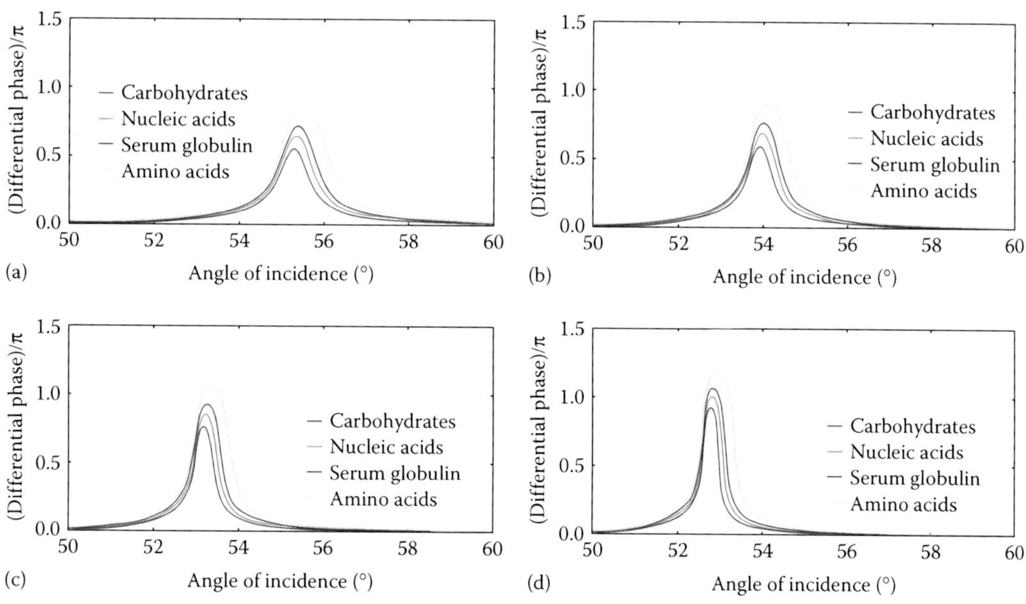

FIGURE 9.18
Differential phase plots for the PSMA structure for specific wavelengths of (a) 550 nm, (b) 633 nm, (c) 700 nm, and (d) 750 nm in angular interrogation mode for different biosamples.

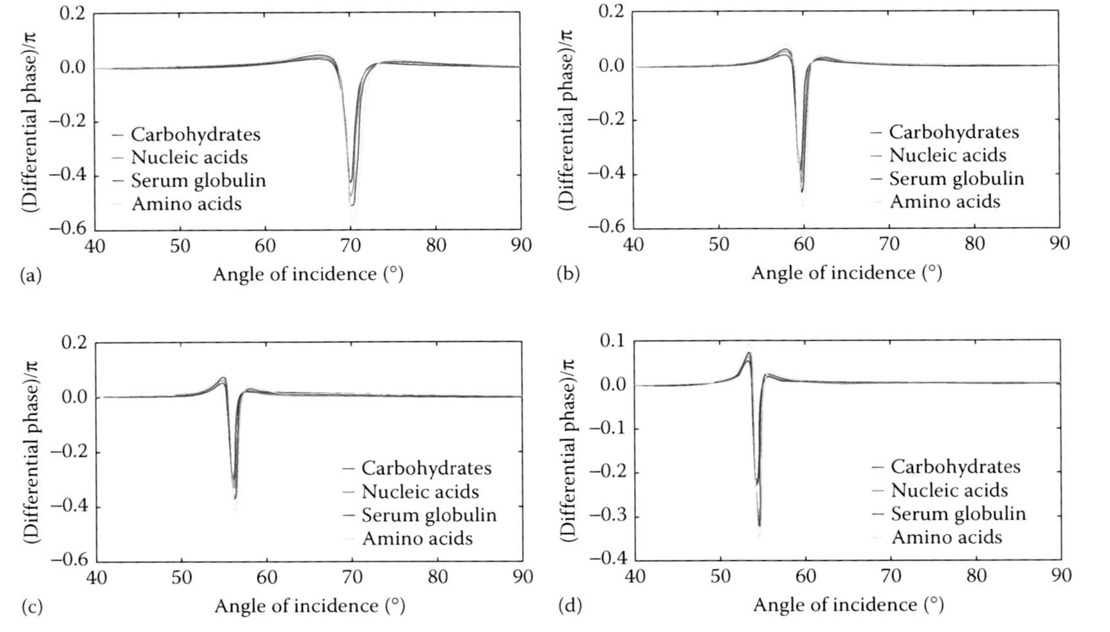

FIGURE 9.19
Differential phase plots for the PMSA structure for specific wavelengths of (a) 750 nm, (b) 850 nm, (c) 950 nm, and (d) 1050 nm in angular interrogation mode for different biosamples.

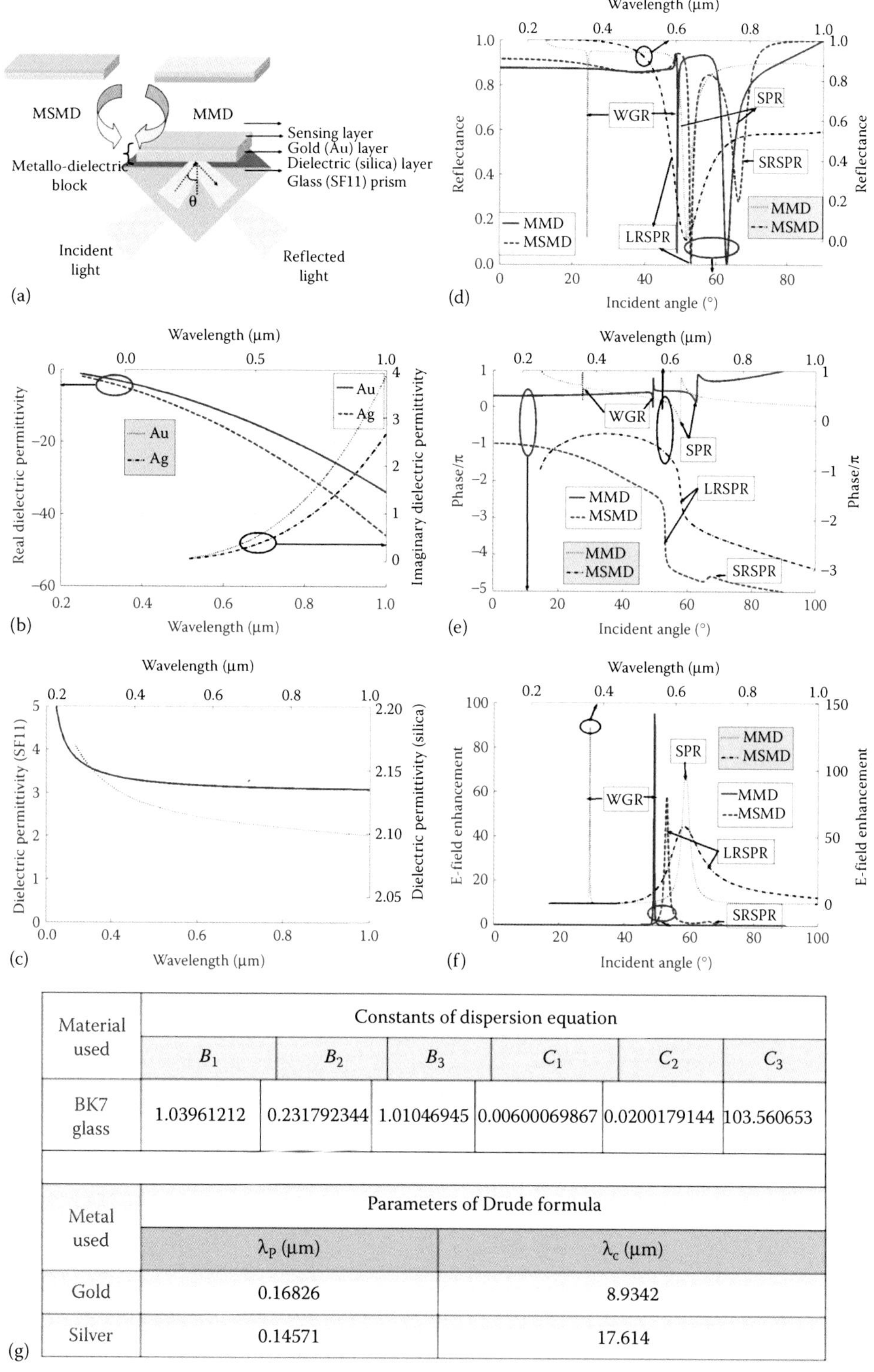

Material used	Constants of dispersion equation					
	B_1	B_2	B_3	C_1	C_2	C_3
BK7 glass	1.03961212	0.231792344	1.01046945	0.00600069867	0.0200179144	103.560653

Metal used	Parameters of Drude formula	
	λ_p (μm)	λ_c (μm)
Gold	0.16826	8.9342
Silver	0.14571	17.614

FIGURE 9.20
(a) Schematic diagram of MMD and MSMD nanoplasmonic structures. Spectral dependence of dielectric permittivity for (b) Ag and Au layer (Drude formula) and (c) SF11 glass prism and silica dielectric layer (Sellmeier dispersion relation) along with the 2D angular and spectral interrogation of (d) reflectance, (e) phase and (f) E-field enhancement, and (g) constants of the Sellmeier dispersion and Drude model. (With kind permission from Springer Science + Business Media: *Plasmonics*, Circular phase response-based analysis for swapped multilayer metallo-dielectric plasmonic structures, 9, 2014, 237–249, Bera, M. and Ray, M.)

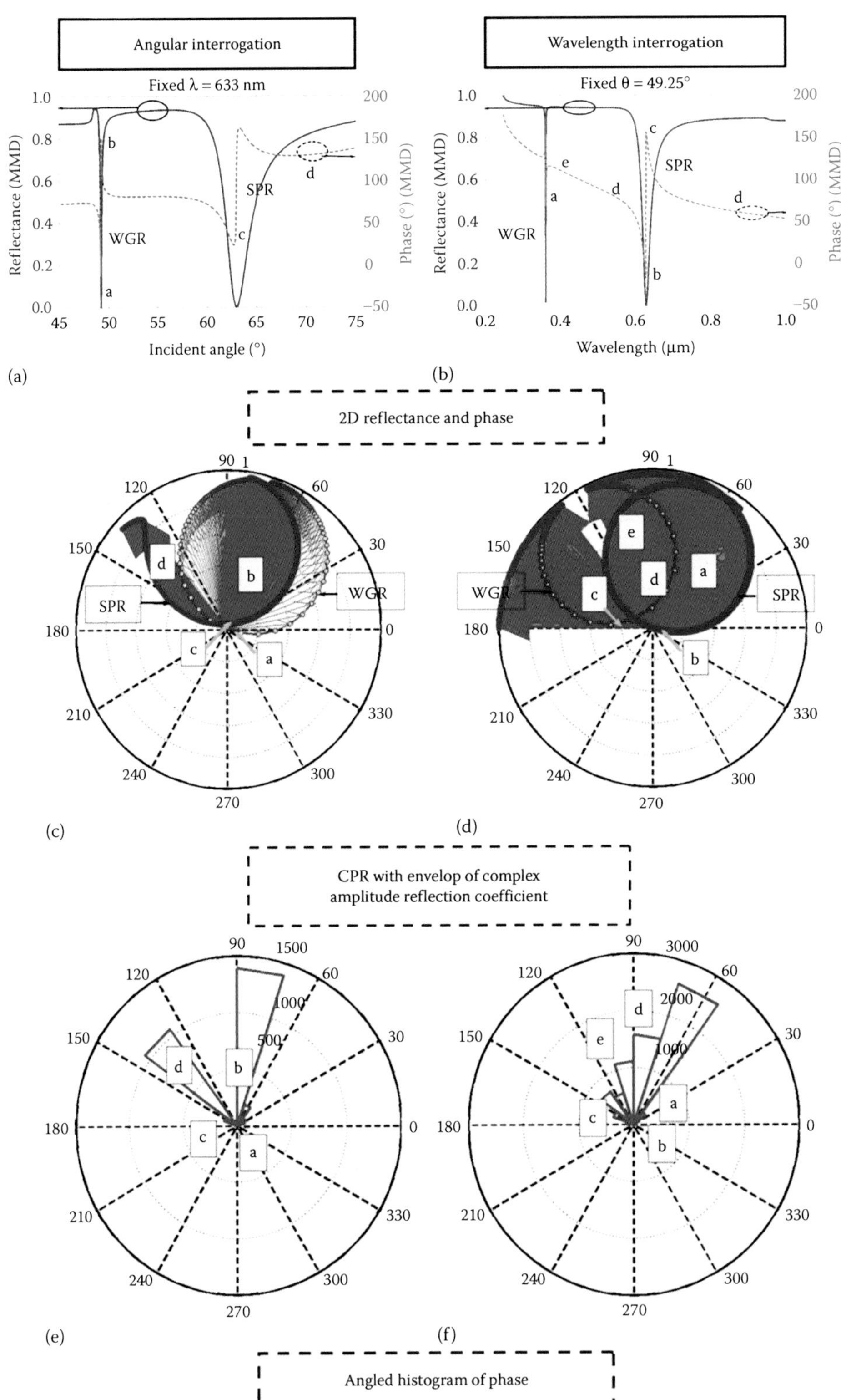

FIGURE 9.21
Simulated graphical representation for angular (left) and spectral (right) interrogation showing 2D phase and reflectance (a,b), CPR (c,d), and angled histogram (e,f) plots for MMD nanoplasmonic structure. (With kind permission from Springer Science + Business Media: *Plasmonics*, Circular phase response-based analysis for swapped multilayer metallo-dielectric plasmonic structures, 9, 2014, 237–249, Bera, M. and Ray, M.)

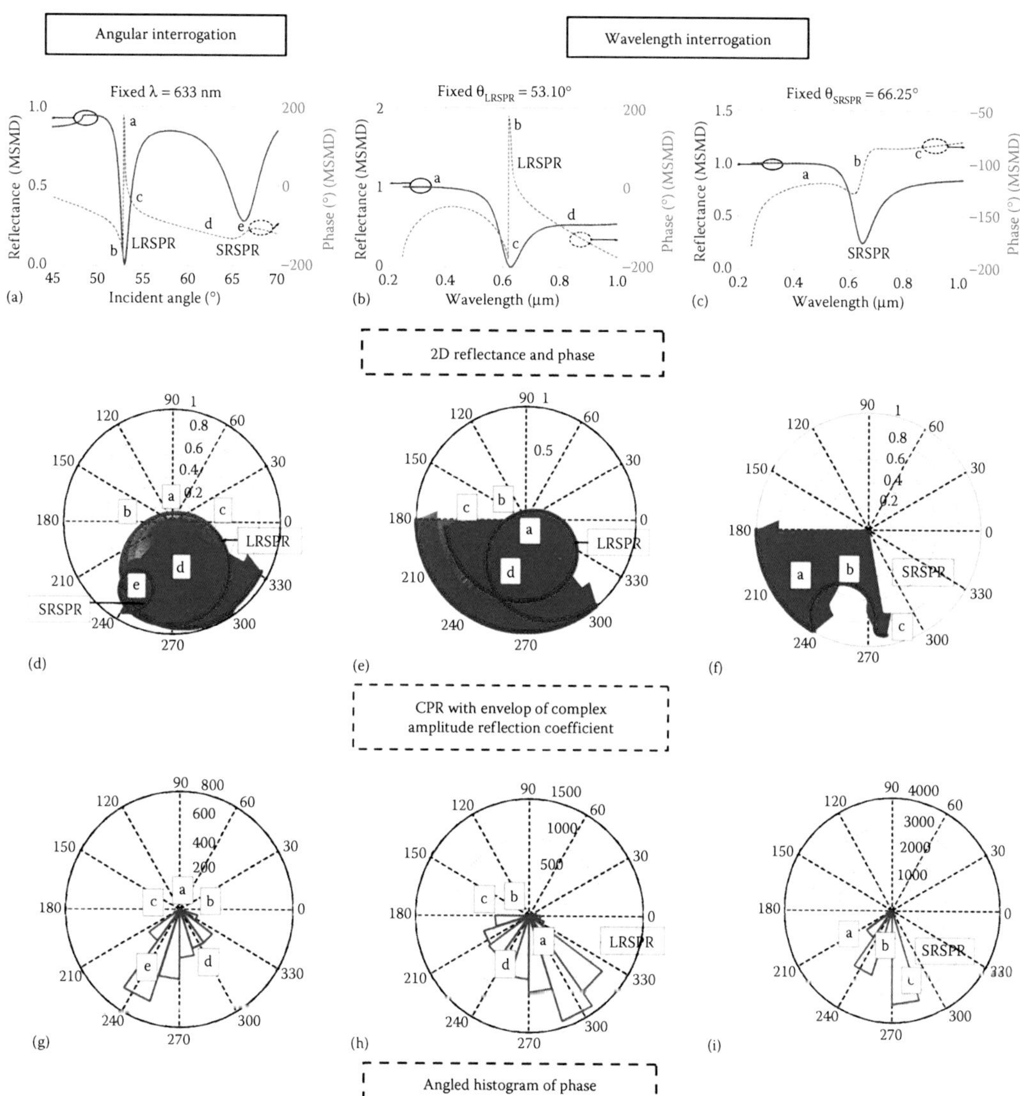

FIGURE 9.22
Simulated graphical representation for angular (left) and spectral (middle: LRSPR and right: SRSPR) interrogation showing 2D phase and reflectance (a–c), CPR (d–f), and angled histogram (g–i) plots for MSMD nanoplasmonic structure. (With kind permission from Springer Science + Business Media: *Plasmonics*, Circular phase response-based analysis for swapped multilayer metallo-dielectric plasmonic structures, 9, 2014, 237–249, Bera, M. and Ray, M.)

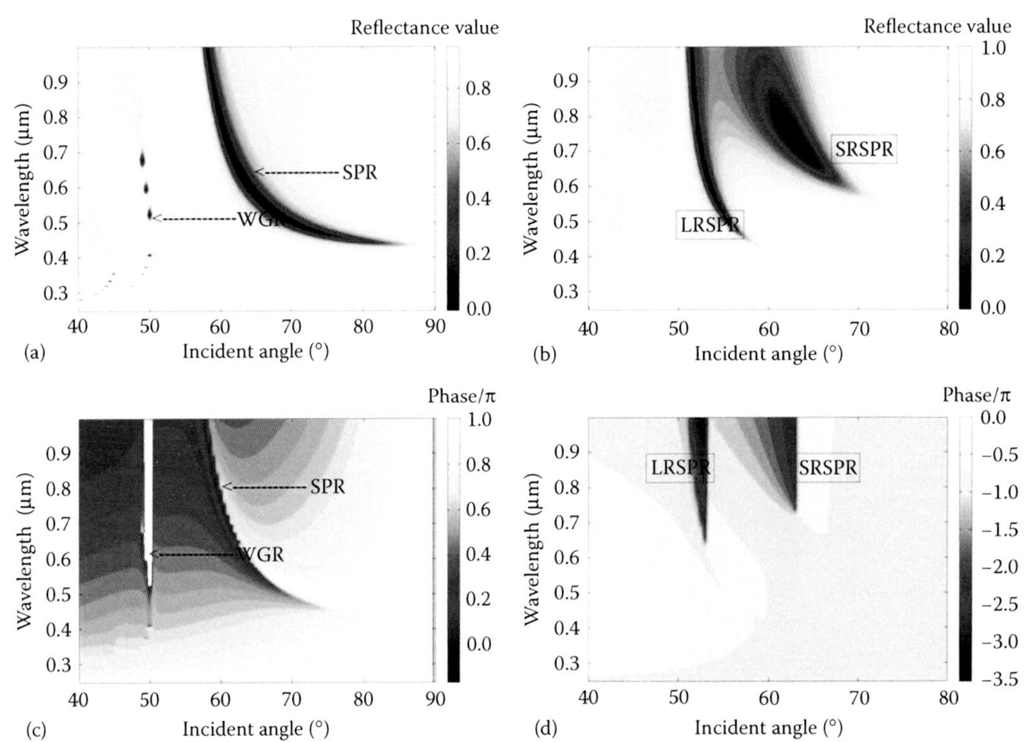

FIGURE 9.23
Simultaneous angular and spectral interrogation: (a and b) reflectance contour plot and (c and d) phase contour plot for MMD (a and c) and MSMD (b and d) structures, respectively. (With kind permission from Springer Science + Business Media: *Plasmonics*, Circular phase response-based analysis for swapped multilayer metallo-dielectric plasmonic structures, 9, 2014, 237–249, Bera, M. and Ray, M.)

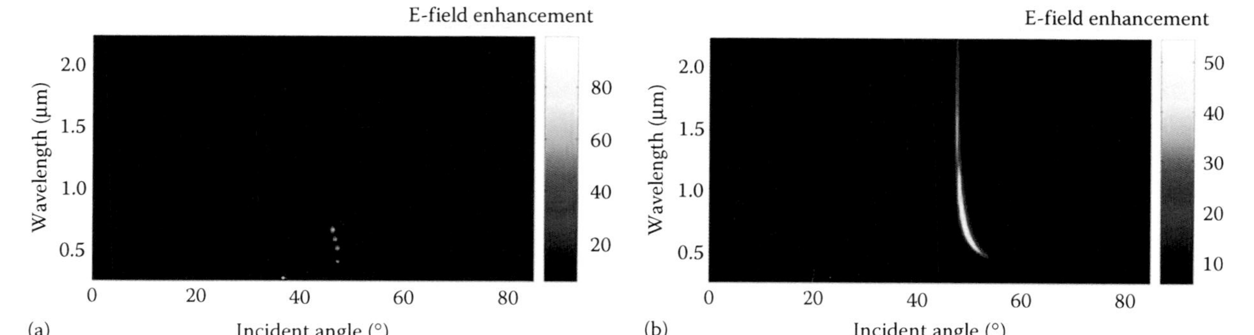

FIGURE 9.24
Simultaneous angular and spectral interrogation for E-field enhancement of (a) MMD and (b) MSMD structures. (With kind permission from Springer Science + Business Media: *Plasmonics*, Circular phase response-based analysis for swapped multilayer metallo-dielectric plasmonic structures, 9, 2014, 237–249, Bera, M. and Ray, M.)

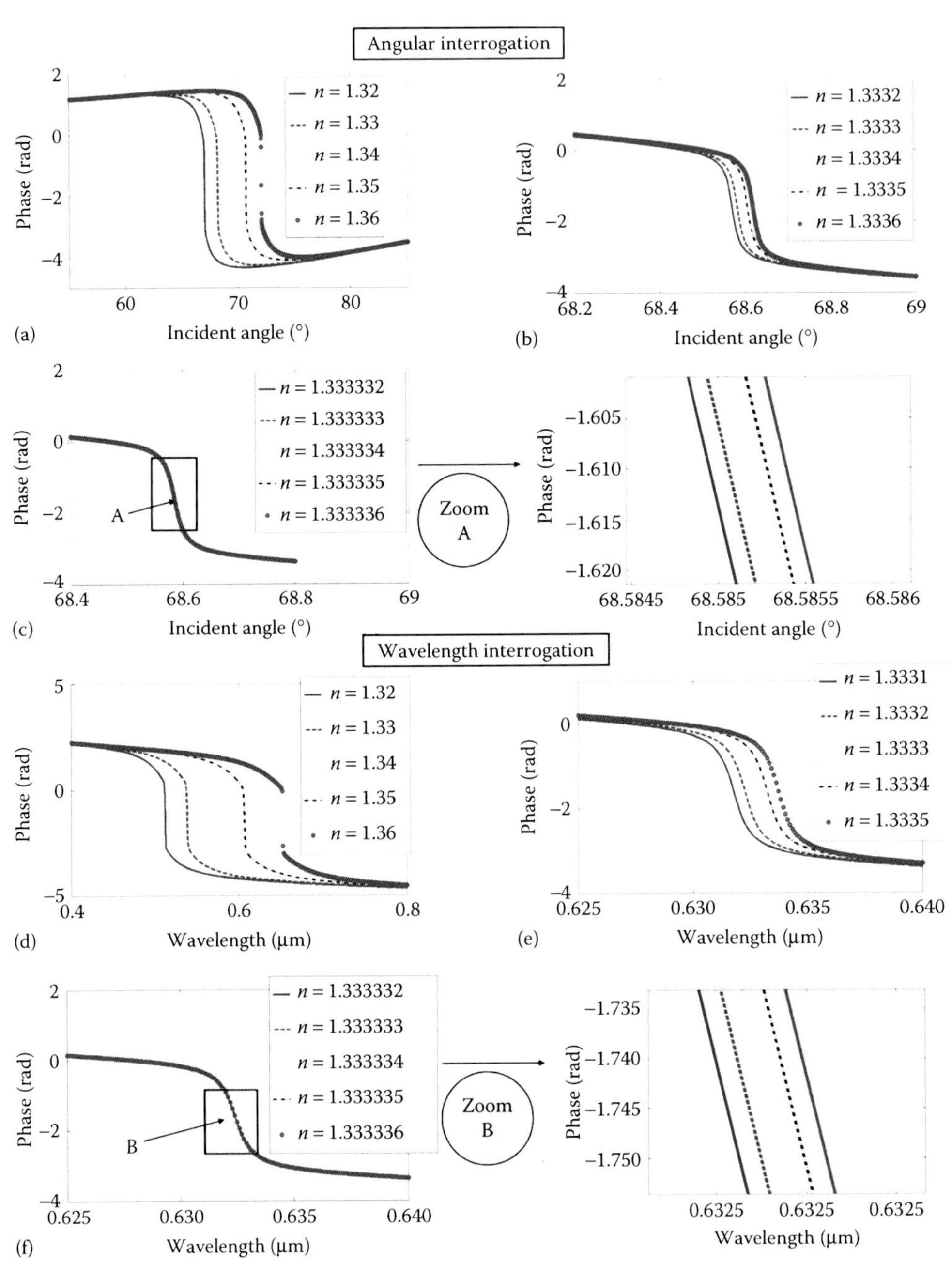

FIGURE 9.28
Phase jump shifts due to the change of the RI of the sensing medium of the order of (a and d) 10^{-2}, (b and e) 10^{-4}, and (c and f) 10^{-6} in angular and wavelength interrogation. (With kind permission from Taylor & Francis: *J. Mod. Opt.*, Resonance parameters based analysis for metallic thickness optimization of a bimetallic plasmonic structure, 61, 2014, 182–196, Bera, M., Banerjee, J., and Ray, M.)

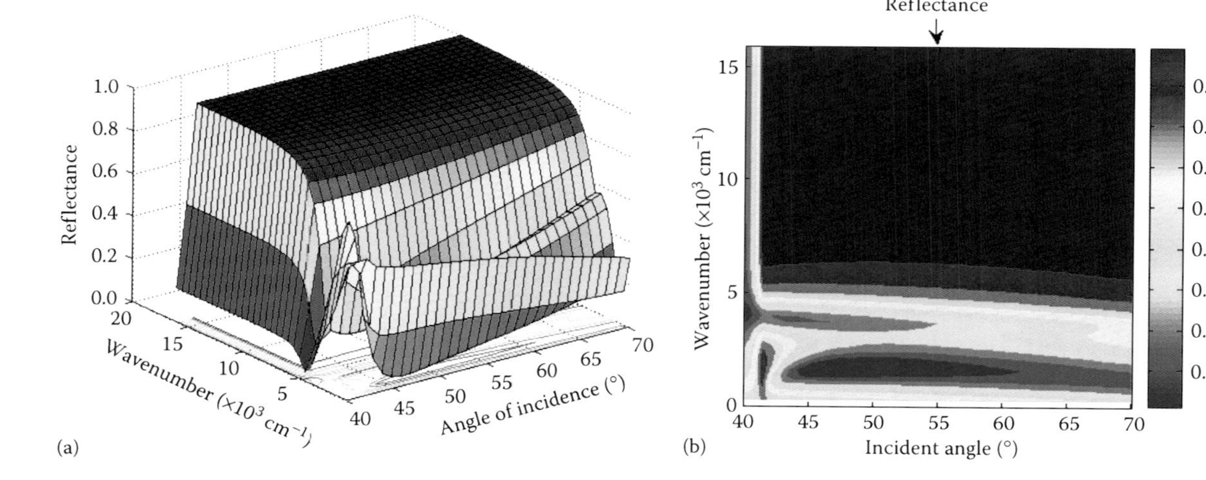

FIGURE 9.48
(a) 3D SPR spectra and (b) contour plot for ZnO:Al film showing variation with wavenumber and angle of incidence.

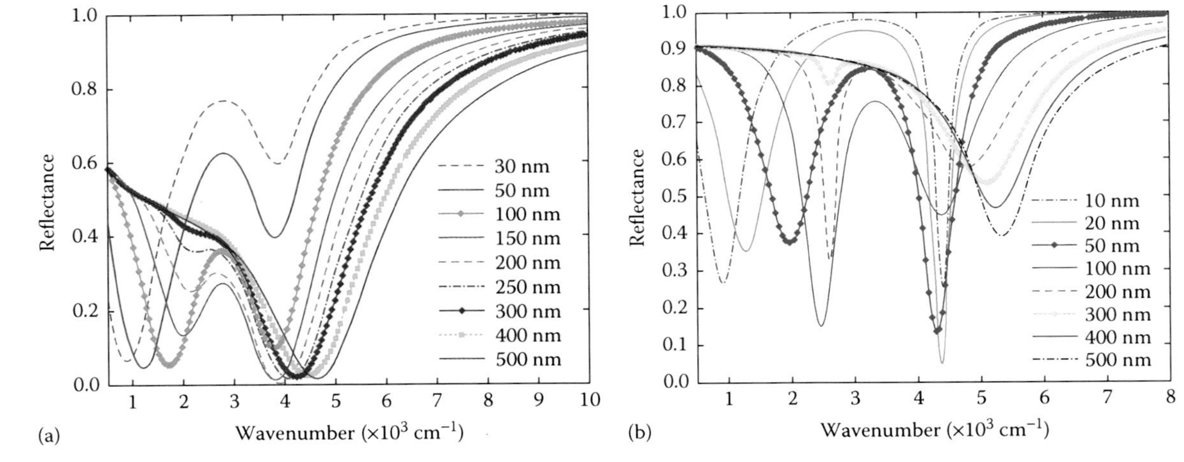

FIGURE 9.49
Variation of reflectance profile with different increasing thicknesses of (a) ZnO:Al (from 30 to 500 nm) and (b) ZnO:Ga (from 10 to 500 nm) at 45° incident angle.

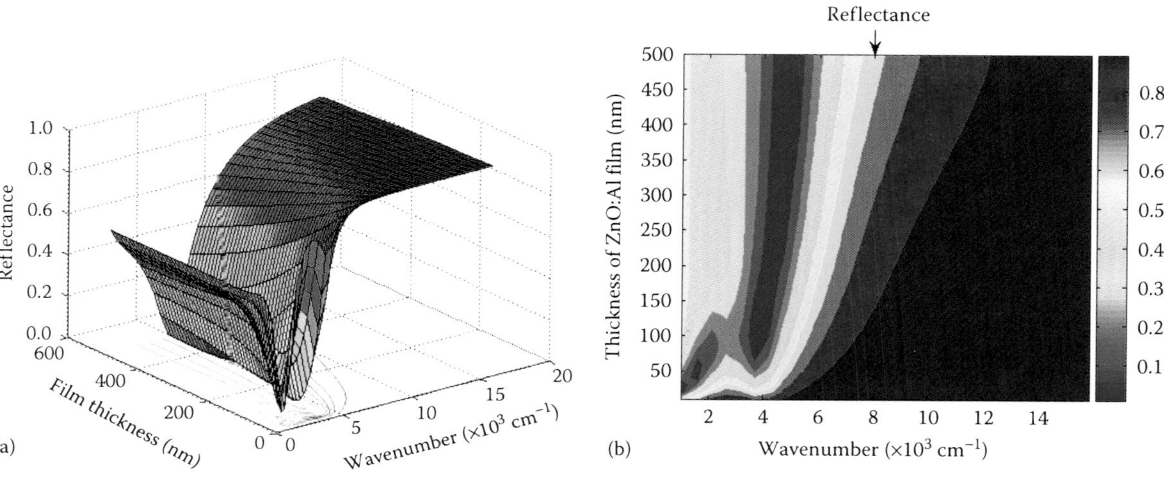

FIGURE 9.50
(a) 3D SPR spectra and (b) contour plot for ZnO:Al film showing variation with wavenumber and thickness of the film.

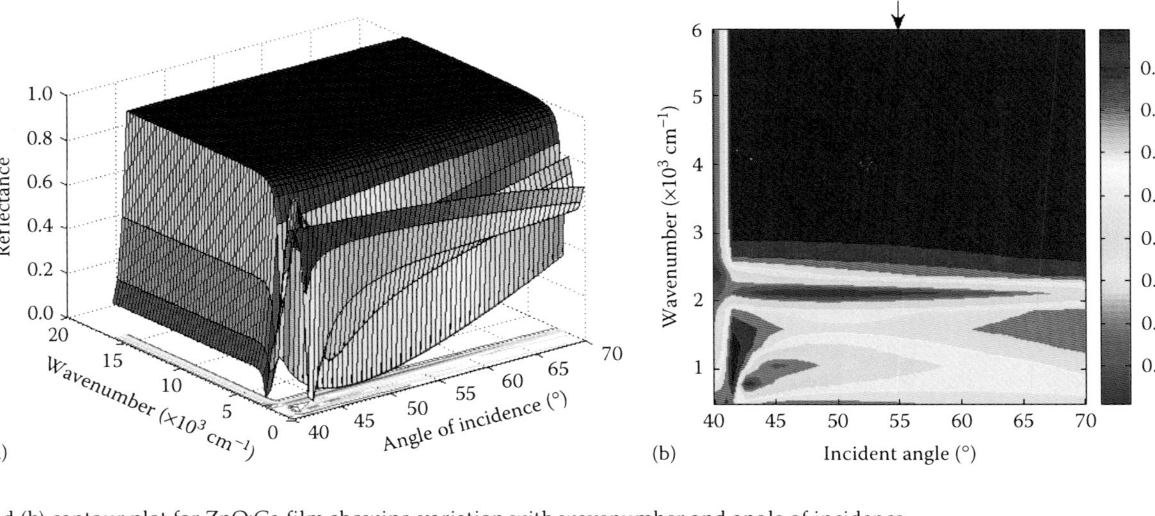

FIGURE 9.51
(a) 3D SPR spectra and (b) contour plot for ZnO:Ga film showing variation with wavenumber and angle of incidence.

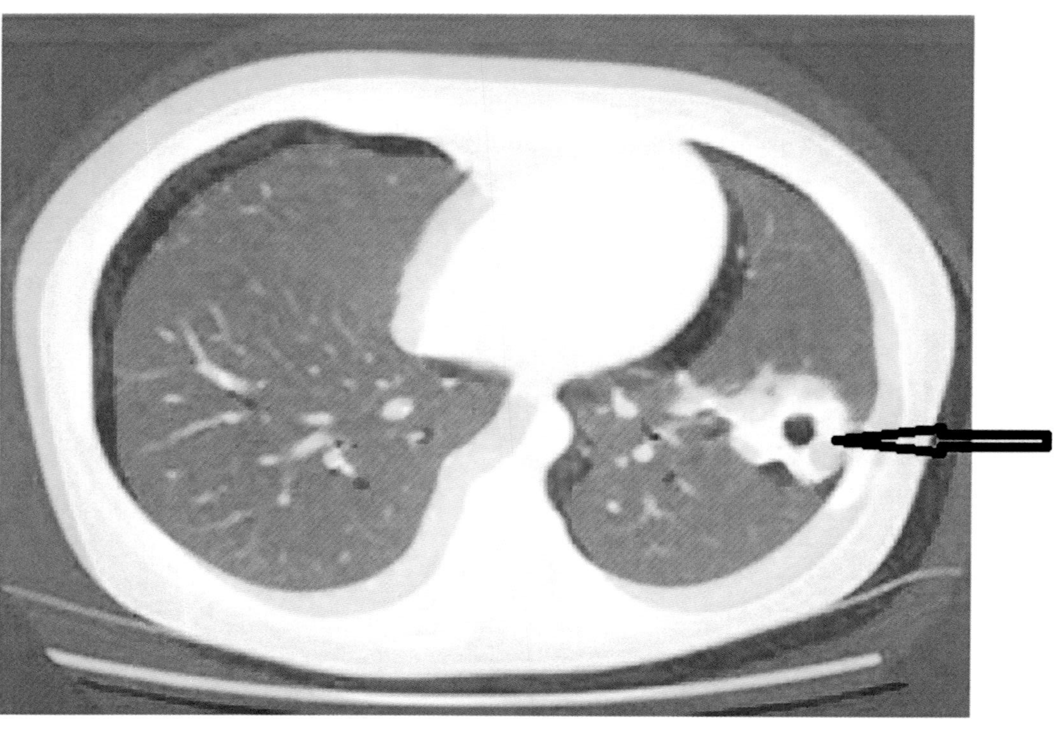

FIGURE 11.2
Simple reconstruction of 3D image from simple reconstruction from 2D CT lung. The lung cancer can be seen in the figure (black arrow).

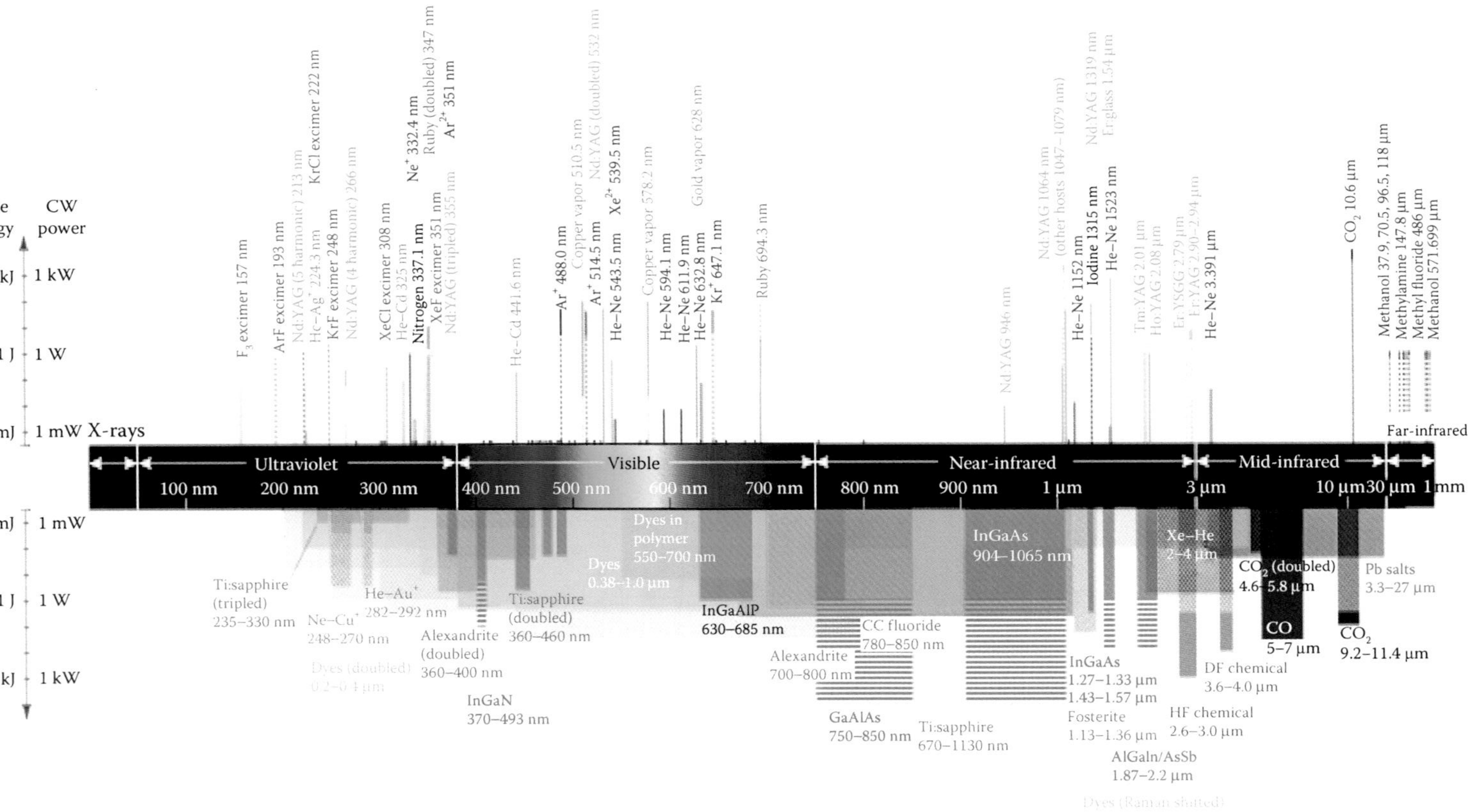

FIGURE B.2
Wavelengths of commercially available lasers. (Data from Weber, M.J., *Handbook of Laser Wavelengths*, CRC Press, Boca Raton, FL, 1999.)

6

Computer-Aided Diagnosis of Interstitial Lung Diseases Based on Computed Tomography Image Analysis

M. Anthimopoulos, S. Christodoulidis, A. Christe, and S. Mougiakakou

CONTENTS

6.1 Introduction

The term *interstitial lung disease* (ILD) refers to more than 200 chronic lung disorders that are characterized by scarring (i.e., fibrosis) and/or inflammation of the interstitium, the tissue between the air sacs of the lungs (alveoli). However, many of the disorders classified as ILD also involve other lung compartments so the more general term *diffuse parenchymal lung disease* (DPLD) is often used. Although ILD is a heterogeneous group of histologically different diseases, most of them have rather similar clinical presentations. ILDs usually cause lung tissue to permanently lose the ability to breathe and carry oxygen into the bloodstream. The most common symptoms are exertional dyspnea, persistent nonproductive cough, dry squeaks, crackles, finger clubbing, wheezing, loss of appetite, and fatigue. These symptoms may refer to different ILDs and even resemble other lung conditions or medical problems. ILD accounts for 15% of all cases seen by pulmonologists [1] and can be caused by (1) long-term exposure to hazardous materials (e.g., asbestos, fumes, and gases), (2) certain drugs or medications, (3) infections, (4) genetic abnormalities, and (5) autoimmune diseases, such as rheumatoid arthritis. In most cases, however, the causes remain unknown and the lung manifestations are described as idiopathic interstitial pneumonia (IIP). In 2002, an international multidisciplinary consensus, including the American Thoracic Society and the European Respiratory Society, proposed a classification for IIPs [2]

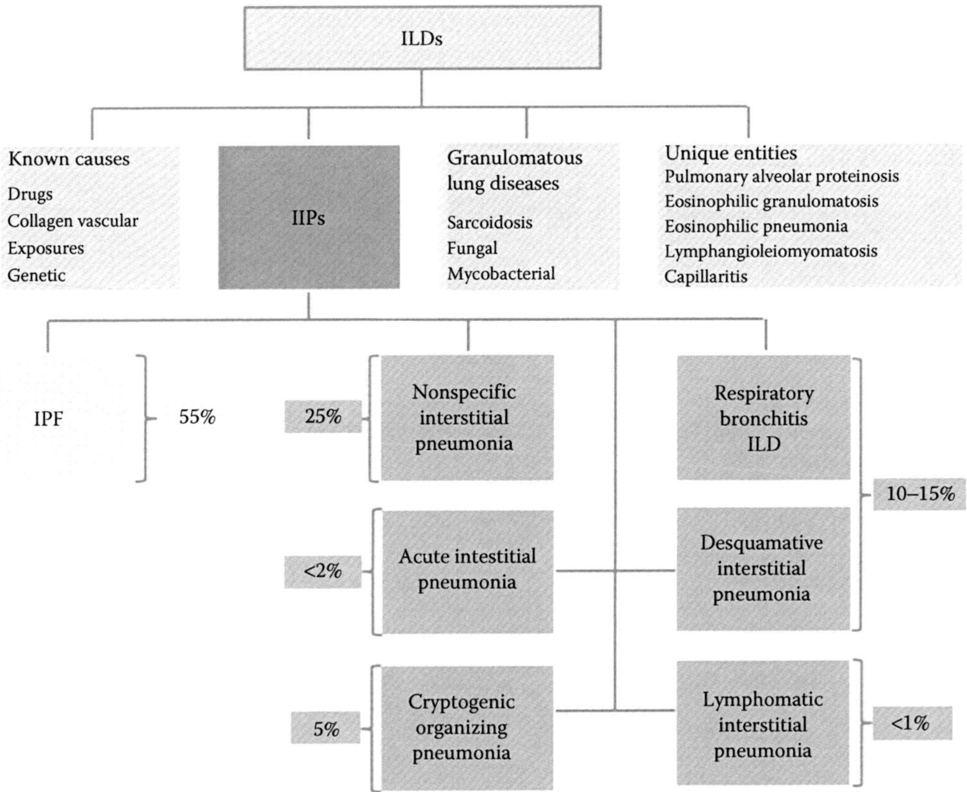

FIGURE 6.1
Classification of interstitial lung disease. (Data from Wells, A.U., *Eur. Respir. Rev.*, 22, 158–162, 2013.)

in order to establish a uniform set of definitions and criteria for their diagnosis. Figure 6.1 depicts the aforementioned classification as reproduced and modified by Wells et al. [3].

A patient with suspected ILD is questioned about his/her clinical history (first symptoms, environmental exposures, family history) and undergoes thorough physical examination (presence of crackles, finger clubbing, joint swelling, or tight skin), chest radiography, and pulmonary function testing. Moreover, the association with systemic diseases or causative agents is investigated and usually computed tomography (CT) is performed and assessed. Figure 6.2 demonstrates a chest radiograph and two CT slices of a typical idiopathic pulmonary fibrosis (IPF) case. An IPF case like the one shown in Figure 6.2 can be confirmed radiologically from the CT scan; however, in many cases of ILD, additional invasive procedures are required for the final diagnosis such as bronchoalveolar lavage and histological confirmation. Although diagnosis can often be made with transbronchial biopsies, some patients even require surgical biopsy. Performing a surgical biopsy exposes the patient to the potential risks of general anesthesia and requires at least a short hospital stay, which thus increases common health-care costs. Moreover, all histological biopsies have a potential risk for bleeding and are not always diagnostic.

In order to avoid the dangerous and costly histological biopsies while maintaining a high diagnostic accuracy, a lot of research has been done on the optimal imaging techniques and their interpretation. ILD diagnosis is currently done by assessing the distribution of various ILD textural patterns in the lung CT images. Typical ILD patterns in CT

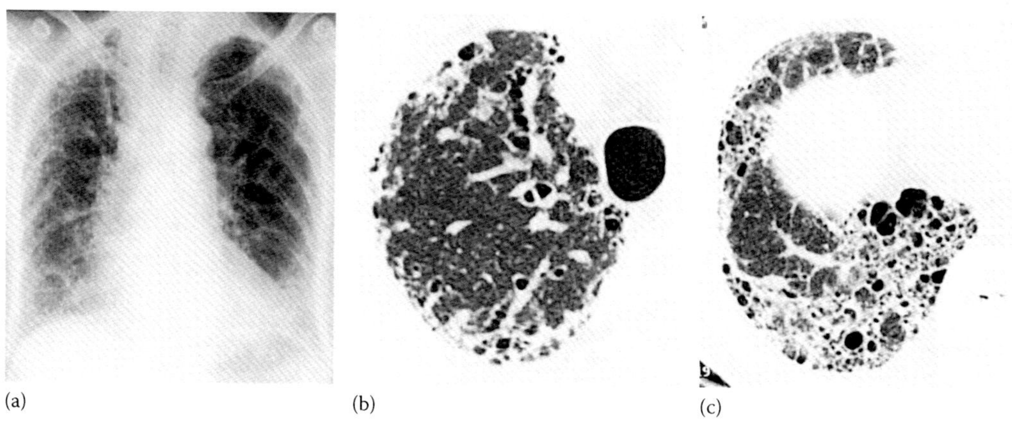

FIGURE 6.2
Idiopathic pulmonary fibrosis. (a) Chest radiograph shows typical peripheral reticular opacity, most marked at the bases, with honeycombing. Lower lobe volume loss (not present in this case) is also common. Chest radiographs may occasionally be normal in patients with IPF. (b and c) CT images show basal predominant, peripheral predominant reticular abnormality with traction bronchiectasis and honeycombing, typical of IPF. (Data from American Thoracic Society; European Respiratory Society. ATS/ERS international multidisciplinary consensus classification of the idiopathic interstitial pneumonias, *Am. J. Respir. Crit. Care Med.*, 165, 277–304, 2002.)

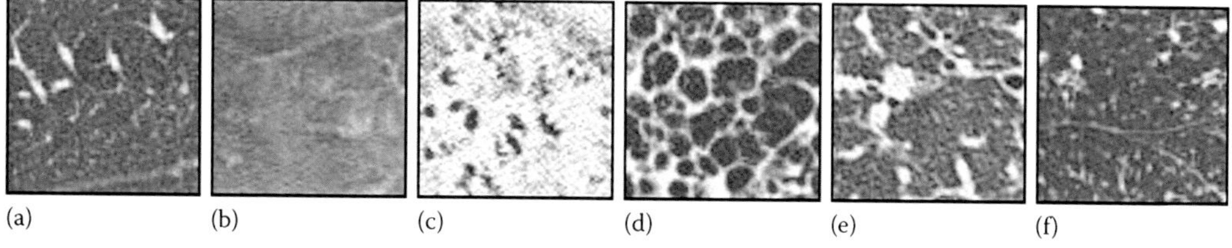

FIGURE 6.3
Examples of typical patterns found in HRCT images of ILD cases: (a) Healthy; (b) ground-glass opacity; (c) consolidation; (d) honeycombing; (e) reticulation; (f) micronodules.

images are ground-glass opacity, consolidation, honeycombing, reticulation, and pulmonary nodules (Figure 6.3). Ground-glass opacity is a nonspecific CT finding characterized by a slight increase in lung density, by partial filling of air spaces by liquids. Consolidation is a similar but more extreme condition of alveolar space filled with liquid, presenting very high attenuation values. Honeycombing is characterized by the presence of small fibrotic cysts with irregularly thickened walls, whereas a reticular pattern is a collection of small linear opacities that produce an appearance resembling a net. Pulmonary nodules are usually roundish growths on the lung and may vary in size. Apart from the types of the existing patterns, their position and extent also play an important role in the diagnosis. For example, IPF is characterized by subpleural and basal honeycombing, whereas nonspecific interstitial pneumonia is characterized by concentric peripheral ground-glass opacity without honeycombing.

High-resolution CT (HRCT) is generally considered an appropriate protocol for the lungs due to the specific radiation attenuation properties of the lung tissue. It provides thin slices (0.625–2 mm) with sufficient resolution (pixel size: 0.625–1 mm), revealing the valuable morphological detail of the pulmonary patterns. Recently, multidetector CT (MDCT) tends to replace HRCT, since it enables a better visualization of the pulmonary patterns from much denser (up to 320 slices per scan) and thinner (slice width: 0.5 mm) slices, which allow the study of the actual three-dimensional (3D) textural patterns by providing almost isotropic data. However, due to the huge amount of radiological data, the lack of strict clinical guidelines, and the resemblance between the different ILD findings, the problem of radiological ILD diagnosis is time consuming and requires an extensive experience, a fact that reduces significantly the diagnostic accuracy of the physicians and increases the inter- and intraobserver variability up to 50% [4].

Computer-aided diagnosis (CAD) systems are software tools designed to provide diagnostic assistance to physicians, based on different medical imaging techniques, advanced image analysis, and artificial intelligence. The use of computers in medical image analysis has been exploited since the 1960s; however, the recent advances in image processing and machine learning along with the enhanced capabilities of modern computers provided a new boost to this research area. A CAD system cannot be considered as a diagnostic substitute of a physician but rather as a *second opinion* provided in an automatic or semiautomatic manner in order to improve the diagnostic performance of the physician. The use of a CAD scheme can often improve the management of a disease by assisting in the early detection and recognition of lesions, and the quantitative analysis of disease severity, or even by suggesting appropriate treatment. In the remainder of this chapter, we will review the literature of the proposed CAD systems for ILDs, based on HRCT and MDCT imaging protocols, and discuss the potentials of the existing technology, the remaining challenges, and the future trends.

6.2 CAD for ILDs

Various CAD systems were proposed for the diagnosis of lung diseases based on CT imaging. Some of them addressed ILD pattern quantification problem of automatic detection and recognition of ILD abnormalities, whereas a limited number of them attempted a final differential diagnosis (DDx). A typical CAD system for ILD consists of two major stages (Figure 6.4): (1) lung segmentation and (2) ILD pattern quantification. The first stage of lung segmentation can be further split into two substeps: (1) lung field segmentation and (2) bronchovascular tree segmentation. In the second stage, the existing ILD pathologies are identified by classifying local regions of the segmented lung parenchyma according to the considered ILD patterns. To this end, a variety of feature sets were employed, describing usually texture characteristics of the local region. In most cases, a feature pool is created by accumulating features of different types, while a selection procedure follows. Noisy features are discarded, different combinations of features are tested, and finally a small subset of the initial pool is kept providing the highest discrimination among the classes with the lowest possible dimensionality.

Then, supervised classification techniques are usually employed in order to recognize the already described area by assigning it to one of the considered ILD patterns. Although most of the proposed ILD CAD systems aim to the quantification of the ILD findings in the lung, a CAD system should ideally go one step further by providing the DDx. Something like that would require all available information such as the types, the extent, the spatial distribution, and the combination of the detected patterns as well as other clinical parameters, taken from the medical history of the patients or from other laboratory tests. By the

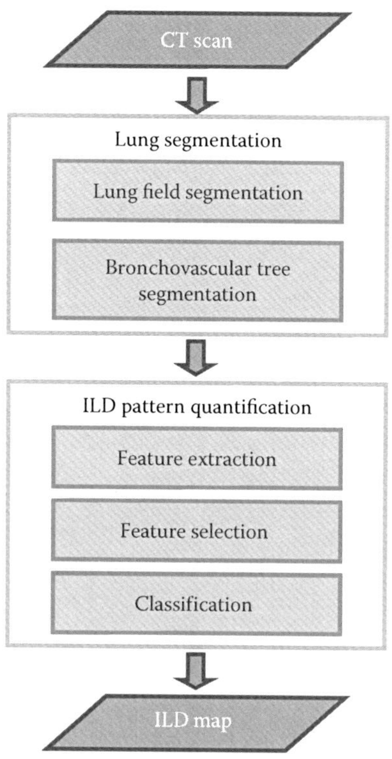

FIGURE 6.4
Generic flowchart of a CAD system for ILDs.

end of this section and after presenting the proposed approaches for the segmentation and quantification of the ILD lung, we will refer to any attempts that have been done so far toward a DDx.

6.2.1 Lung Segmentation in Presence of ILD

Image segmentation is the process of dividing an image into multiple meaningful parts usually belonging to different objects. In the context of medical image analysis, segmentation refers to the separation of different organs, tissues, or pathologies in order to be studied individually. Lung tissue segmentation constitutes a preprocessing step that plays a crucial role in the accurate lung CT image analysis in CAD systems. It mainly involves the border identification of the lung tissue with the surrounding tissue, the separation of the left and right lung volumes, and the localization of the lung lobes. In some cases, segmentation of the bronchovascular trees is also needed in order to isolate the lung parenchyma and facilitate its quantitative analysis. Many of the early CAD systems proposed the use of manual or semiautomatic segmentation, a task though that requires human effort and time, especially for the new imaging technologies such as the MDCT protocol that produces vast amounts of data per patient. This fact raised the need for accurate and automatic lung segmentation methods as a first step for any CAD system addressing lung pathologies. In this section, we will not refer to the entire literature of lung segmentation, but instead we will focus on the methods proposed for the segmentation of the lung fields and the bronchovascular tree in CT images with mild or severe ILD.

6.2.1.1 Lung Field Segmentation

Lung tissue normally presents substantially lower density than its surrounding tissues, resulting in a large contrast in Hounsfield units (HUs) within thorax CT images. Therefore, many of the conventional lung segmentation methods rely on simple intensity thresholding techniques followed by morphological operations and connected component analysis for the refinement of the results. However, these simple intensity methods become unreliable in cases containing pathologies with high density such as ILDs. The high density of the ILD patterns that corresponds to high attenuation values along with their often peripheral and basal manifestations can cause severe undersegmentation problems propagating the error to all subsequent steps of the CAD system. Uchiyama et al. [5] employed morphological opening followed by grayscale thresholding. The opening operation smoothes the CT image by removing small bright structures such as vessels while maintaining the overall gray levels and larger structures. A gray-level histogram is constructed from the smoothed image and the optimal threshold is defined as the value maximizing the separation between the two peaks of the image's histogram. After binarizing the image with the adaptively computed threshold, the final segmentation result is produced. However, in cases where consolidation patterns were present, the method failed so the segmentation was performed manually. In order to deal with the problem of erroneously discarded lung regions due to peripheral ILD patterns, some intensity-based methods proposed the use of a refinement stage that will restore the lost regions.

Kim et al. [6] adopted a segmentation method originally proposed in a system for automated lung nodule detection by Goo et al. [7]. The method is based on a gray-level thresholding step followed by connected component analysis for eliminating the background regions. Then, an edge detection technique forms the contour of the lung fields, whereas its inner area is filled and selected as the lung region. However, the thresholding will also eliminate nodules or vessels in contact with the chest wall so a bridging algorithm is employed, which calculates the convexity of each point on the contour and places an imaginary bridge tangential to each contour point. The image pixels newly encompassed by the contours are then included within the lung segmentation regions. Zavaletta et al. [8] employed the method of Hu et al. [9], which involves three main steps. In the first step, the lung region is extracted after using a threshold that is computed iteratively by averaging the two mean gray-level values of the body and nonbody voxels. The initial threshold is selected based on the CT number for pure air (–1000 HU) and the CT number for voxels within the chest. Three-dimensional connected component analysis and region growing follow, in order to discard small disconnected regions. In the second step, dynamic programming is used for the separation between the left and right lung fields. The anterior and posterior junction lines are detected by calculating the maximum cost path through a graph with weight proportional to pixel gray levels. In the final step, a sequence of morphological operations is used to smooth the irregular lung boundary shape. Sluimer et al. [10] used gray-level thresholding combined with connected component labeling and a *rolling ball* operation to correct undersegmentation. The rolling bar filter was originally proposed by Armato et al. [11] to avoid loss of juxtapleural nodules prior to their pulmonary nodule detection method. A two-dimensional ball is successively placed tangential to each contour and an indentation is identified when the ball filter touches the contour at more than one point. Then, the endpoints of the indentation are linearly connected, extending the contour and including the area of the indentation into the lung field. Figure 6.5 provides an example of the rolling ball operation for the refinement of lung segmentation. Although the method seems to perform satisfactorily for restoring lost juxtapleural nodules during

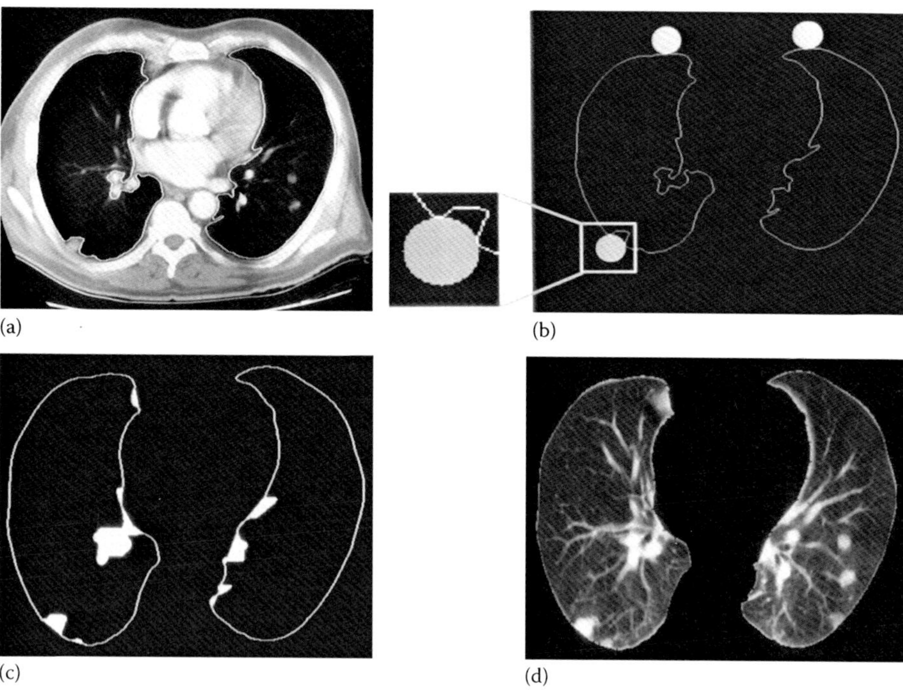

FIGURE 6.5
Example of the *rolling ball* operation proposed by Armato et al. to correct undersegmentation. (a) Original section image with lung segmentation contours superimposed. A juxtapleural nodule and hilar vessels are excluded from the right lung contour. (b) Initial placement of the rolling ball filter for each lung segmentation contour. The inset shows the filter spanning a contour indentation. (c) Lung segmentation contours after interpolation is used to bridge indentations. White areas indicate regions that are included within the new lung segmentation regions. (d) Segmented lung regions after implementation of the rolling ball algorithm. (Data from Armato, S.G. et al., *Radiographics*, 19, 1303–1311, 1999.)

lung segmentation, it could not deal cases with severe consolidation; hence, these cases were manually corrected, as reported by Armato et al. [11].

Since intensity methods failed to segment successfully lungs with severe ILD pathologies, some researchers attempted to incorporate texture information based on the a priori knowledge about the ILD texture characteristics. Korfiatis et al. [12] proposed a two-step coarse to fine method for the segmentation of pathological lung cases with ILD findings. First, an unsupervised segmentation scheme based on *k*-means clustering is used to partition the thorax region into four clusters corresponding to lung field tissue, fat, muscle, and bone, based on pixel gray-level values. The first step typically results in undersegmentation due to pathological lung areas erroneously assigned to fat, muscle, or bone clusters. To this end, an iterative neighborhood labeling of border pixels, corresponding to the outermost pixels of each distinct region of the initial estimate, is performed using a support vector machine (SVM) classifier based on local texture analysis. Both gray-level and wavelet coefficient statistics features are extracted from a 9×9 pixel analyzing region of interest (ROI) centered at every border pixel of the initial lung border. Wang et al. [13] proposed a method for segmenting lungs with severe ILD. First, the airways are segmented and removed from the CT images to prevent interference in lung segmentation. A CT value thresholding follows to obtain an initial lung estimate including normal and mild

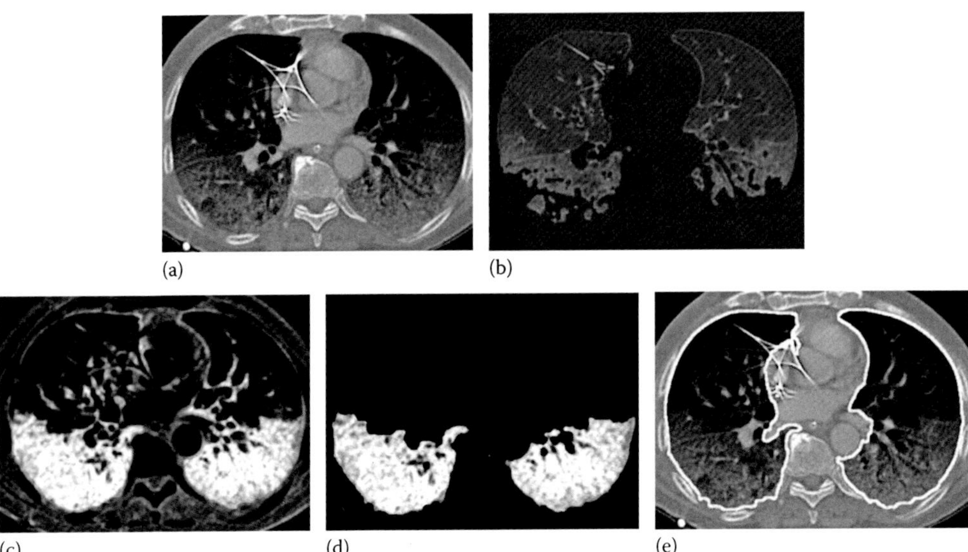

FIGURE 6.6
Example of the lung segmentation method proposed by Wang et al. (a) Original CT image of lungs with severe interstitial lung disease; (b) initial estimate of the lungs; (c) combined texture feature image; (d) the identified lung regions with severe interstitial lung disease; (e) the final lung segmentation result obtained by combining the initial estimate and the lung regions with severe ILD, followed by a postprocessing to fill the holes in the segmented lung regions. (Data from Wang, J. et al., *Med. Phys.*, 36, 4592–4599, 2009.)

ILD lung parenchymas. Then, texture features, based on the co-occurrence matrix, are heuristically combined and thresholded to further identify abnormal lung regions with severe ILD. Finally, the identified abnormal lung regions are fused with the initial lungs to generate the lung segmentation result. Figure 6.6 demonstrates an example of the aforementioned method. Despite the enhanced performance of the texture-based approaches compared to intensity methods, they heavily depend on the assumption that the texture characteristics of the existing pathologies are known a priori, which is not always true since segmentation constitutes the first stage of an ILD CAD system.

In order to exploit all available information while avoiding the strong dependency from lung's intensity and special texture characteristics, additional thorax anatomical features were proposed in some works. Sluimer et al. [14] proposed a pathological lung segmentation approach based on a segmentation-by-registration scheme in which a normal lung scan is guiding the segmentation after being elastically registered to a scan containing severe ILD. The same elastic transformation is applied to the segmentation map of the healthy lung, producing the result. As a refinement the use of a probabilistic atlas was proposed instead of a single binary mask. The atlas is produced by registering 15 normal lung images and averaging their masks. In this way, a band of voxels is created within the 3D lung mask having values between 0 and 1. When the registration is successful, this band contains the actual lung border that is identified in a second phase by a kth nearest neighbor (k-NN) voxel classification using a set of nine features; the image intensity and its derivatives are computed in three different scales. Prasad et al. [15] proposed an adaptive thresholding technique that exploits the fact that the curvature of the ribs and the curvature of the lung boundary are closely matched. The algorithm is as follows: first, the ribs are segmented by applying thresholding followed by morphological filtering and a polynomial is fitted to represent the ribs' curvature. Then, the threshold value is varied to find

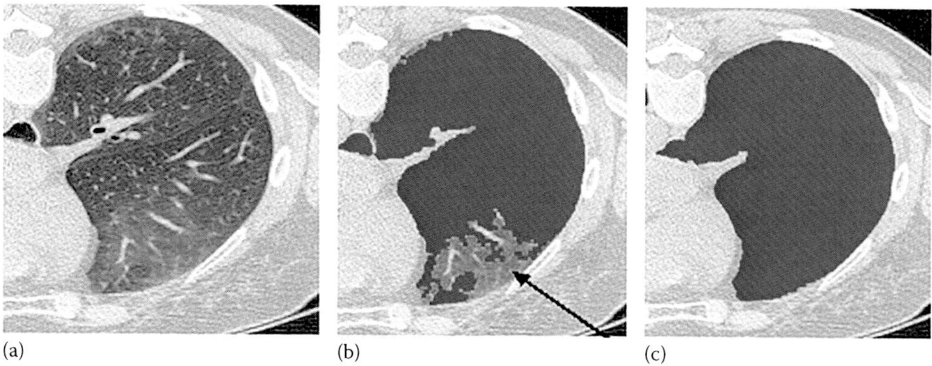

FIGURE 6.7
(See color insert.) Lung segmentation results on scleroderma subjects. (a) Original scleroderma image scanned at residual volume. (b) Results of thresholding algorithm at 400 HU with post-processing. (c) Results from the rib curvature technique. The regions indicated by the arrow are the high attenuation regions not segmented by the fixed thresholding algorithm. (Data from Prasad, M.N. et al., *Acad. Radiol.*, 15, 1173–1180, 2008.)

a lung boundary sufficiently close to the curve of the ribs based on a naïve Bayes classifier. An example of the method application is given in Figure 6.7. Hua et al. [16] converted the segmentation problem into a maximum flow problem that combines anatomical information, image intensity, and image gradient. Initially, a pre-segmentation step is performed based on the method of Hu et al. [9] and then a graph is constructed in a narrow band around the pre-segmented lung surface. A *k*-NN classifier is used in order to calculate for each voxel the probability of being on the lung boundary based on six features: the intensity, the gradient, and four features related to the position of the voxel as well as its distance from the rib convex hull region and the spinal cord. Finally, a maximum flow algorithm is employed to compute the optimal cut in the constructed graph that minimizes the total cost of the nodes. The nodes' cost is inversely related to the probability assigned to them earlier and the constructed graph corresponds to the lung boundary.

6.2.1.2 Bronchovascular Segmentation

The term *bronchovascular structure* or *tree* refers to the networks of bronchi and the pulmonary blood vessels (Figure 6.8). The bronchi constitute roughly cylindrical airways of decreasing the radius starting from the trachea and bifurcating (or trifurcating) iteratively into the lungs, whereas the vessels encompass the pulmonary arteries and veins that carry the blood from and to the heart. The presence of bronchovascular tissue in the CT slices can often obstruct the detection and recognition of ILD findings because the representation of the vessels and airway walls in HU is higher than the surroundings and really close to high attenuation pathologies such as ILD. However, the inner part of the bronchi often mimics pathologies such as emphysema due to the low attenuation values of the contained air. Furthermore, the shape and texture characteristics of the CT slices of the tree often resemble common ILD patterns (e.g., honeycombing, reticulation) producing undesirable false alarms (Figure 6.9). In order to deal with this challenge, the early proposed CAD systems [17,18] considered the bronchovascular cross sections in the CT slices as an additional texture pattern to be detected, because the low number of CT slices per case did not provide sufficient information about the 3D shape of the tree. Therefore, airways and vessels were conceived by the systems as characteristic two-dimensional (2D) textures cut by the scan plane. The 2D texture though of these 3D tubular structures strongly depends

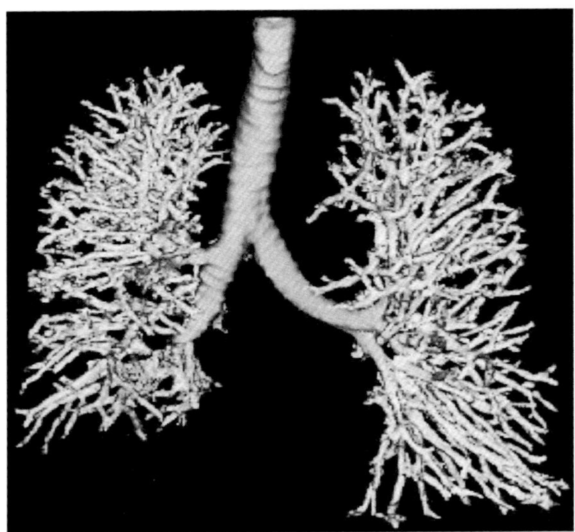

FIGURE 6.8
(See color insert.) Bronchovascular structure. The bronchial is indicated in pink and the vascular tree in yellow. (Data from Zavaletta, V.A. et al., *Acad. Radiol.*, 14, 772–787, 2007.)

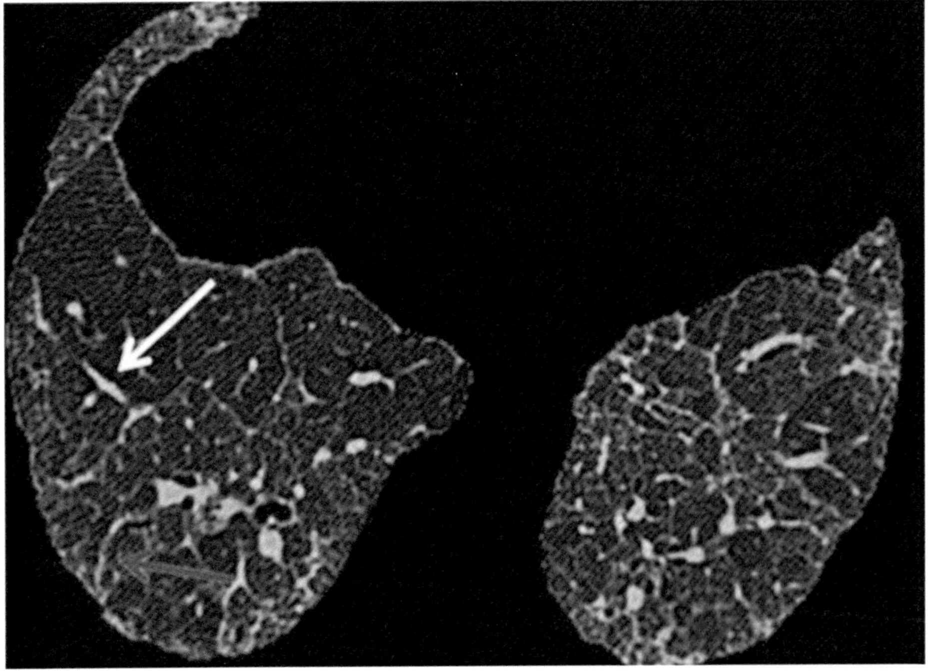

FIGURE 6.9
Example depicting similarity in radiologic appearance of vessel segments (bright arrow) to reticular patterns (dark arrow). (Data from Korfiatis, P. et al., *IEEE Trans. Inf. Technol. Biomed.*, 15, 214–220, 2011.)

on their orientation to the scan plane (longitudinal when parallel, oval when oblique, and rounded when perpendicular) producing undesirable intraclass visual diversity. More recently, the enhanced capabilities of the MDCT protocol for an almost 3D representation of the lung volumes have enabled the development of methods performing segmentation and exclusion of the bronchovascular tree as a preprocessing step. By exploiting the 3D connectivity, various systems attempted to trace the airway and vessel branches and remove them before proceeding to the ILD quantification stage achieving a considerable improvement in the accuracy for the latter.

The methods proposed for segmenting lung airways can be roughly categorized into knowledge-based segmentation, region growing, centerline extraction, and mathematical morphology, or a combination of the above [4]. Zavaletta et al. [8] performed segmentation of the trachea and its central bronchial branches with an iterative process of six- or eight-neighbor region growing algorithm thresholded at different levels to optimally extract as much of the tracheobronchial tree as possible, but prevent inclusion of erroneous low-attenuating pathologies (such as emphysema) by limiting the number of connected components from the seed point by a method similar to that of Aykac et al. [19]. Xu et al. [20] demonstrated the improvement in classification of emphysema after bronchovascular exclusion by using the method of Tschirren et al. [21]. The method employed a fuzzy connectivity system, based on small cylindrical adaptive ROIs used to follow airway branches and avoid leakage (Figure 6.10). The basic idea was that voxels from the image

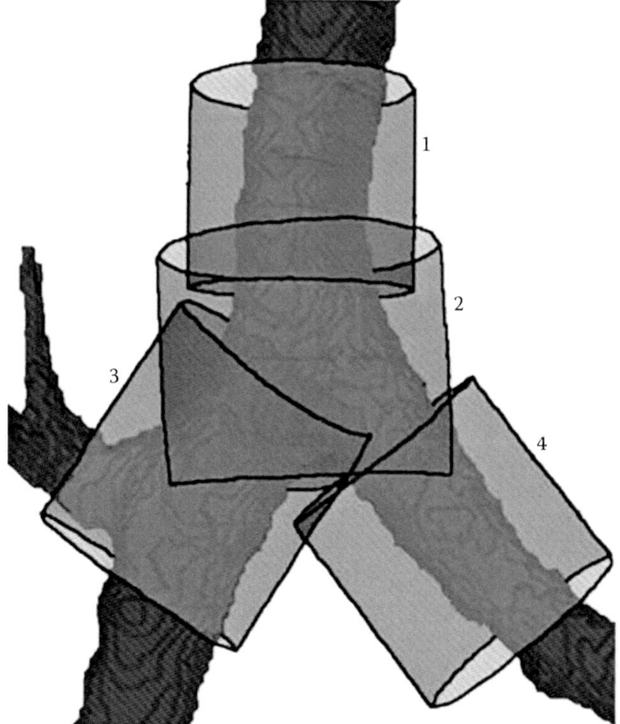

FIGURE 6.10
Basic concept of airway tree segmentation. Adaptive cylindrical ROIs follow airway tree branches as the segmentation proceeds. Segmentation is performed in a small area only, which keeps the computing time down. Possible problems (leaks) can be detected early and addressed. (Data from Tschirren, J. et al., *IEEE Trans. Med. Imaging*, 24, 1529–1539, 2005.)

were compared with a seed voxel and the similarity was expressed as a fuzzy membership value. As a similarity measure between two voxels, an affinity value based on gray values was adopted and defined for adjacent voxel only. Voxels with a similarity over a threshold were assigned in a region that is following the airway tree branches. After the region growing stage, the resulting tree is skeletonized to identify the centerlines and the branch point locations that will be used for the refinement of the airway edges. Wang et al. [13] proposed the segmentation and exclusion of the airways from the CT images in order to prevent interference in lung segmentation. To this end, the trachea is initially localized as the biggest air component of the first 30 slices by simple thresholding and 3D region growing. The rest of the airways is identified by connected component labeling, followed by morphological opening.

Segmentation of the pulmonary blood vessels is a challenging problem that is still under investigation. Although some methods were proposed for vessels segmentation in healthy lungs [4], only very few attempts were done in lungs with high-attenuating pathologies such as ILD. The majority of these systems employ vessel enhancement filters based on Hessian matrix analysis. Zavaletta et al. [8] used a Hessian-based 3D line enhancement filter followed by manual thresholding to include as much vasculature as possible with the least amount of high attenuation pathology. Xu et al. [20] adopted the method of Shikata et al. [22] to extract the pulmonary vascular tree from MDCT images. The method integrates a line filter that enhances the 3D tube-like structures and a vessel tracker attains connectivity of the vessel segments. Korfiatis et al. [23] employed a vessel tree segmentation algorithm, proposed for vessel tree segmentation in CT angiography images [24]. Specifically, a multiscale line enhancement filter, designed to enhance vessels and vessel bifurcation points, was applied on the segmented lung volume, based on the analysis of eigenvalues of the Hessian matrix at multiple scales. An expectation maximization segmentation algorithm was used to segment vessel tree volume by segmenting high-response voxels at each level. Finally, reconstruction of segmented vessel components was performed. In a more recent work, Korfiatis et al. [25] proposed a new two-stage method for the segmentation of the vessel tree in presence of ILD. The first stage of the method implements a technique proposed by Zhou et al. [24], utilizing a 3D multiscale vessel enhancement filter based on the eigenvalue analysis of the Hessian matrix and unsupervised segmentation. The second stage of the method employs a machine learning voxel classification refinement, using 3D texture features and an SVM classifier.

6.2.2 ILD Pattern Quantification

The term *quantification* includes the detection and recognition of various ILD patterns as well as the identification of their extent in the lung. Since ILDs are generally manifested as texture alterations of the lung parenchyma, most of the proposed systems employ texture classification schemes on local ROIs or volumes of interest (VOIs) depending on the 2D or 3D capabilities of the used CT imaging technology. By sliding the local classifier over the already segmented lung fields, a map of the entire lung is produced, representing the ILD quantification that can be used for attempting a final DDx either by the physicians or by automated computerized systems (Figure 6.11). The basic characteristics of a texture classification system like the one described above are the chosen feature set, the classification method, the size and shape of the chosen ROI/VOI, the considered ILD patterns, and the dataset used for training and testing the system. Table 6.1 summarizes the basic characteristics of the state-of-the-art methods for ILD quantification that will be described in the following text.

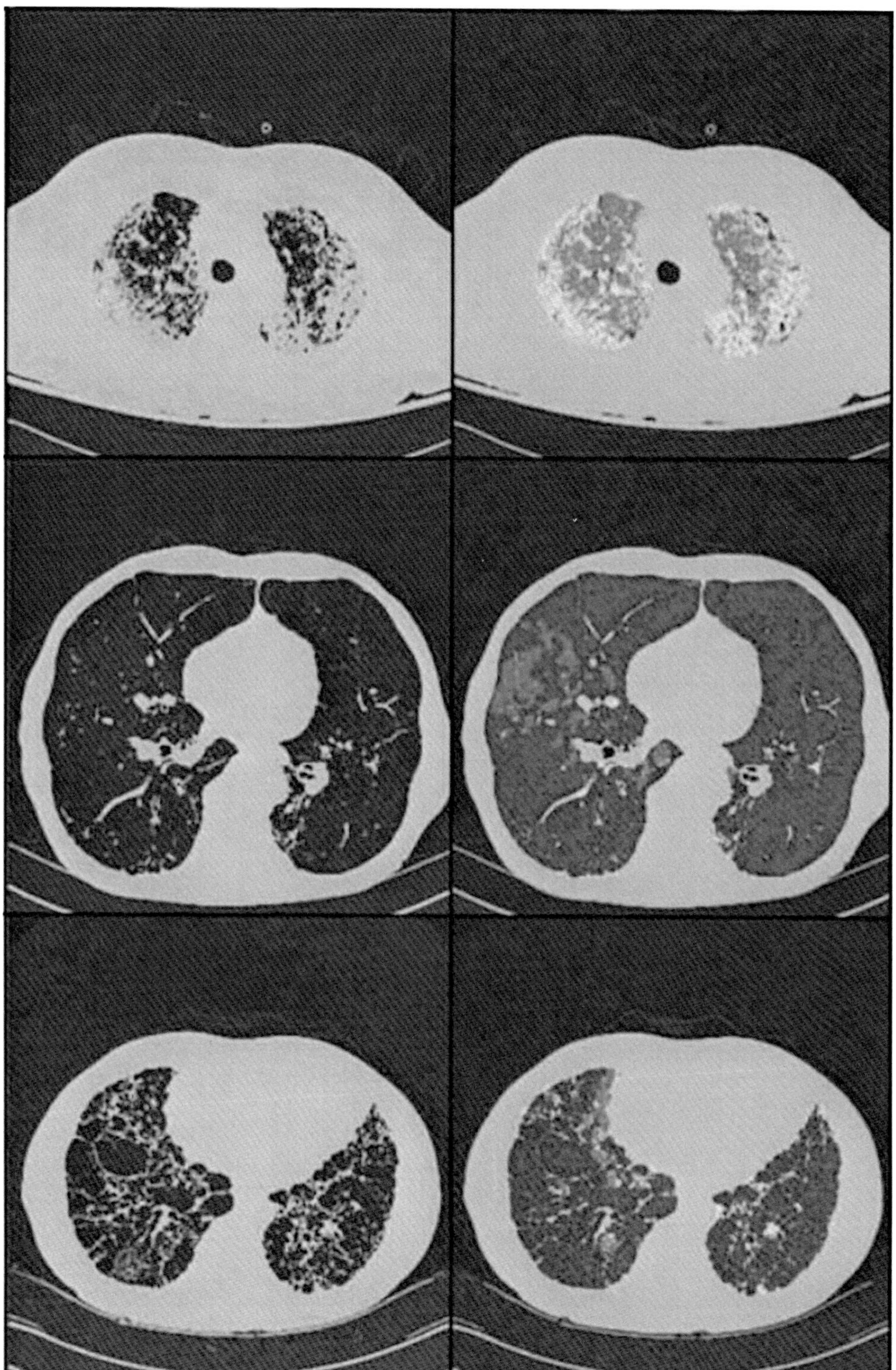

FIGURE 6.11
(See color insert.) Quantification results depicted by color-coded overlay in several cases. *Left*: Original CT image. *Right*: Corresponding result (normal, green; ground-glass opacity, yellow; reticular opacity, cyan; honeycombing, blue; emphysema, red; consolidation, pink). (Data from Park, S.O. et al., *Korean J. Radiol.*, 10, 455–463, 2009.)

TABLE 6.1
Basic Characteristics of the Proposed Studies for ILD Quantification

Study	Patterns	Features	Classification	Data	Performance
Heitmann et al. [26]	Ground glass	First-order statistics	SOM	HRCT 120 scans 20 subjects	Accuracy: 76%
		GLCM		ROI: 3 × 3 – 20 × 20	
		Edge/contrast functions			
Delorme et al. [17]	Emphysema	First-order statistics	LD	HRCT 10 subjects	Accuracy: 70.7%
	Ground glass	GLCM		ROI: 5 × 5	
	Interlobular	RLM			
	Fibrosis				
	Vessels				
	Bronchus				
Uppaluri et al. [18]	Emphysema	First-order statistics	Bayesian	HRCT 72 subjects	Accuracy: 93.5%
	Ground glass	GLCM		ROI: 31 × 31	
	Honeycombing	RLM			
	Nodular	Fractal analysis			
	Bronchovascular				
Chabat et al. [27]	Centrilobular	First-order statistics	Bayesian	HRCT 33 subjects	Sensitivity: 60.3%
	Emphysema	GLCM		ROI: circular radius 22	Specificity: 86.7%
	Panlobular	RLM			
	Emphysema				
	Constrictive bronchiolitis				
Sluimer et al. [28]	Abnormal	First-order statistics after filter bank	LD	HRCT 116 subjects	Area under the ROC curve: Az = 0.86
		Gaussian	QD	ROI: circular radius 40	
		Laplacian	SVM		
		First-order derivative	*k*-NN		
		Second-order derivative (multiscale)			
Uchiyama et al. [5]	Ground glass	First-order statistics	ANN	HRCT 315 images 105 subjects	Sensitivity: 97.4%
	Reticular and linear	Air percentage		ROI: 32 × 32, 96 × 96	Specificity: 88.0%
	Nodular	Geometric measures (white top-hat, black top-hat)			
	Honeycombing				
	Emphysema				
	Consolidation				

Az, Area under the ROC curve; ROC, Receiver operating characteristic.

(*Continued*)

TABLE 6.1

(Continued) Basic Characteristics of the Proposed Studies for ILD Quantification

Study	Patterns	Features	Classification	Data	Performance
Xu et al. [39]	Emphysema	First-order statistics	SVM	MDCT	Accuracy:
	Ground glass	3D RLM	Bayesian	20 subjects	Bayesian: 86.2%
	Honeycombing	3D GLCM		ROI: 21 × 21 × 21	SVM: 83.8%
	Normal nonsmokers	3D fractal dim			
	Normal smokers				
Zavaletta et al. [8]	Ground glass	Signatures of adaptively binned histograms	EMD from canonical signatures of considered classes	MDCT 18 scans	Average sensitivity: 71.4%
	Reticular			ROI: 15 × 15 × 15	Specificity: 93%
	Honeycombing				
	Emphysema				
Park et al. [30]	Ground glass	First-order statistics of	SVM	HRCT images	Accuracy: 89%
	Consolidation	Gray level		ROI: circular radius 32	
	Reticular	Gradient			
	Honeycomb	Top-hat			
	Emphysema	RLM			
		GLCM			
		Low attenuation area geometry			
Wang et al. [43]	Abnormal	3D RLM	QD	MDCT 68 scans	Sensitivity: 86%
		3D GLCM		ROI: 64 × 64 × 64	Specificity: 90%
Vo et al. [32]	Emphysema	Wavelet	MKL	HRCT 89 slices	Average sensitivity: 94.16%
	Honeycombing	Contourlet		38 subjects ROI: 32 × 32	Specificity: 98.68%
	Ground glass				
Korfiatis et al. [34]	Reticular	3D GLCM	*k*-NN	MDCT 13 scans	True/false positive rate ground glass: 0.638/0.361 reticular 0.942/0.147
	Ground glass			ROI: 21 × 21 × 21	
Mariolis et al. [40]	Ground glass	3D RLE	PNN	MDCT 30 scans	Accuracy: 99.8%
	Reticular	3D Laws	*k*-NN Bayesian MLR		
	Honeycombing				

EMD, Earth mover's distance; MLR, Multinomial logistic regression.

(Continued)

TABLE 6.1

(Continued) Basic Characteristics of the Proposed Studies for ILD Quantification

Study	Patterns	Features	Classification	Data	Performance
Song et al. [29]		Gabor-LBP	Minimum discrepancy criterion	HRCT 95 scans	Accuracy: 83%
	Emphysema	Multi-coordinate histogram of oriented gradients		ROI: 31 × 31	
	Ground glass	Gray-level histogram			
	Fibrosis				
	Micronodules				

The first CAD systems addressing ILDs were limited to the analysis of 2D texture structures because the available, at the time HRCT protocol could provide just a few slices per case with large intervals. Most of these systems proposed the use of already established texture description methods derived from the first-order statistics (statistics of the gray-level distribution), the gray-level co-occurrence matrix (GLCM) [35], the run-length matrix (RLM) [36], and the fractal analysis [37], and fed to various classifiers for the classification of 2D ROIs. Heitmann et al. [26] used three single self-organizing maps (SOMs), the actual CT intensity values (HU), some textural features, and simple expert rules, in order to classify ground-glass opacities (GGOs). The first-order statistics and GLCM were used by SOMs on detecting the areas of high contrast (usually tissue close to pleura, bronchus, or blood vessels) and the CT intensity values (HU) to train the third one on detecting the GGO. A simple rule was used to combine all three SOM results into one. The hybrid network correctly classified 91 of 120 scans from 20 patients. Delorme et al. [17] proposed a pixel-based texture classification scheme considering six lung patterns: normal, emphysema, GGO, interlobular fibrosis, vessels, and bronchus. The first-order statistics, GLCM, and RLM features were used in a linear discriminant (LD) classification framework achieving an accuracy of 70.7% on a dataset with 5 × 5 ROIs from 10 patients. Chabat et al. [27] used similar texture features and a Bayesian classifier to classify 880 circular ROIs with 22-pixel radius into four lung patterns: centrilobular emphysema, panlobular emphysema, constrictive obliterative bronchiolitis, and normal. The achieved sensitivity and specificity were 73.6% and 91.2%, respectively. Uppaluri et al. [18] proposed the adaptive multiple feature method (AMFM) with 22 texture features for the classification of 31 × 31 rectangular ROIs into six tissue patterns: honeycombing, GGO, bronchovascular, nodular, emphysema, and normal. Table 6.2 provides all the considered features as proposed in Reference 18. The features were again derived from the first-order statistics, GLCM, RLM, and fractal analysis. However, besides the popular geometric fractal dimension (GFD), Uppaluri et al. [18] proposed the computation of the local stochastic fractal dimension (SFD) from which the first-order statistics were computed and added to the feature set. A feature selection procedure follows, where the features with low discrimination ability or high correlation with the rest were discarded. The classification was done by a Bayesian classifier, which achieved an overall accuracy of 93.5% on a dataset of 72 patients.

Although the AMFM was generally accepted as a reliable method for the quantification of ILDs, new systems were proposed experimenting with modern classification techniques and feature sets providing a new perspective to the problem. Filter banks,

TABLE 6.2

The 22 Features as Proposed by Uppaluri et al. within the AMFM System

Feature Category	Description	Features
Gray-level distribution	Occurrence frequency of gray levels	Mean, variance, skewness, kurtosis, gray-level entropy
Run-length	Heterogeneity and tonal distribution of the gray levels	Short-run emphasis, long-run emphasis, gray-level nonuniformity, run-length nonuniformity, run percentage
Co-occurrence matrix	Spatial relationships between gray tones	Angular second moment, correlation, contrast, entropy, inertia, inverse difference moment
Fractal	Roughness or smoothness of the texture	GFD, SFD mean, SFD variance, SFD skewness, SFD kurtosis, SFD entropy

Source: Uppaluri, R. et al., *Am. J. Respir. Crit. Care Med.*, 160, 648–654, 1999.

morphological operations, and spectral analysis were involved in the feature extraction procedure, whereas *k*-NNs, artificial neural networks (ANNs), SVMs, and multiple kernel learning (MKL) were introduced for the classification problem. Sluimer et al. [28] proposed the convolution of the CT images with a filter bank and used the first-order statistics of the result as texture descriptors. The filter bank was composed of the symmetrical Gaussian and Laplacian filters together with the first and second partial image derivatives in six different angles and four different scales (Figure 6.12). The chosen ROIs were circular with a radius of 40 pixels and the considered classes were normal and abnormal. After the feature extraction, a sequential feature search was employed to select the most informative features. For the classification, four classifiers were compared: LD, quadratic classifier (QD), SVM, and *k*-NN. The study concludes that *k*-NN and SVM achieved the best performance with an accuracy of 86.2% on a dataset of 116 patients. Uchiyama et al. [5] introduced the use of geometric features based on the black and white top-hat morphological transformations along with gray-level first-order features forming a total number of six features. The first-order features include the mean, the standard deviation, and the fraction of the area of air density components. For the geometric features, the black and white top-hat transformations were applied to the ROIs and the nodular, line, and multilocular components were isolated with empirical rules. The average pixel values of the resulting images were used as measures for the fraction of the different components in each ROI. For the classification, a three-layered feed-forward ANN was used to classify 32 × 32 ROIs into one of six classes (GGO, reticular and linear opacities, nodular opacities, honeycombing, emphysematous change, and consolidation), achieving a sensitivity of 90.1% and a specificity of 83.7% on a dataset of 106 patients. Park et al. [30] used some of the popular texture features such as gray-level and gradient statistics, RLM and GLCM features combined with shape features computed on the top-hat transformations, and the low attenuation clusters. An SVM was employed to classify circular ROIs (32-pixel radius) from 92 cases into six patterns (normal, GGO, nodular, reticular, honeycombing, and emphysema) with an accuracy equal to 89%. Vo et al. [32] used wavelet- and contourlet-based features and MKL to classify 32 × 32 ROIs into four classes (normal, emphysema, honeycombing, and GGO) with average sensitivity and specificity equal to 94.16% and 98.68%, respectively. The exact number of features where 18 of 66 features were from the Haar wavelet transform, 24 features were from the contourlet transform [38], and 24 more features from the directional Gaussian derivatives. Song et al. [29] proposed two texture descriptors: the rotation-invariant Gabor-local binary pattern (RGLBP) texture descriptor and multi-coordinate histogram of oriented gradient

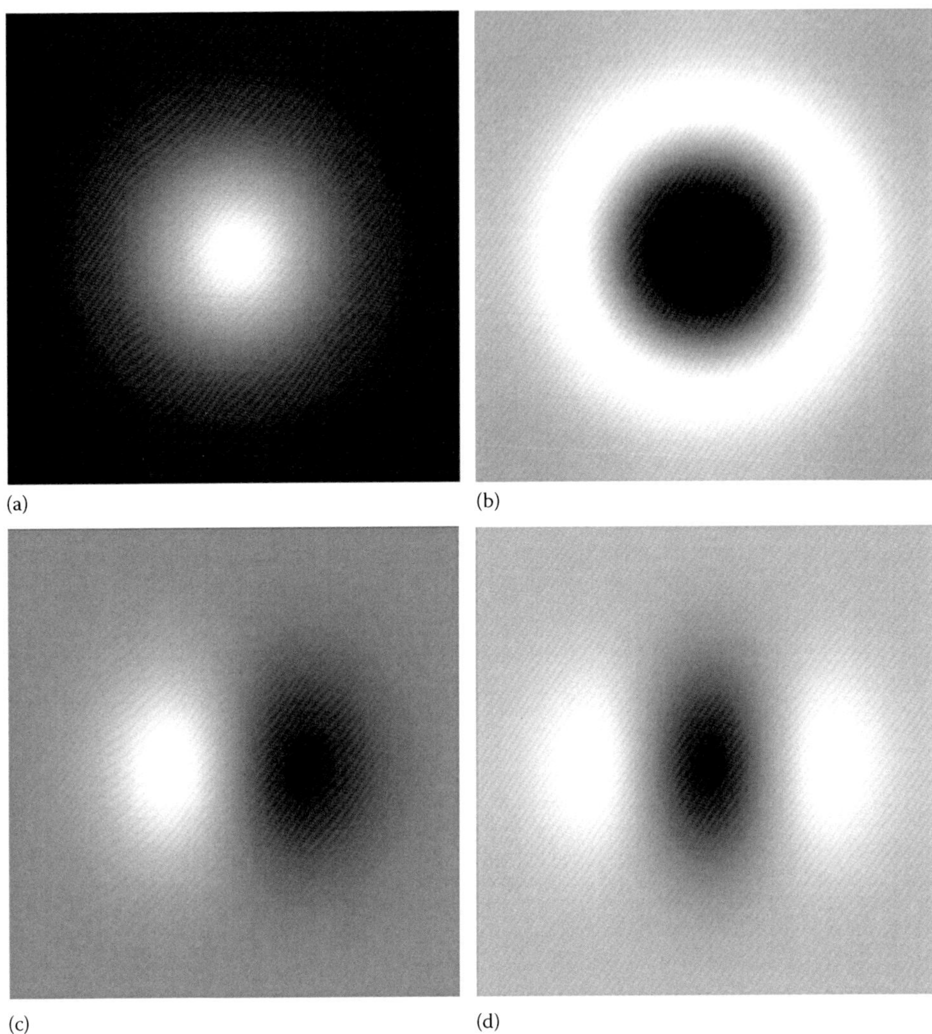

FIGURE 6.12
The four basic filters used by Sluimer et al. (a) Gaussian. (b) Laplacian. (c) Gaussian, first-order derivative. (d) Gaussian, second-order derivative. (Data from Sluimer, I. et al., *Med. Phys.*, 30, 3081–3090, 2003.)

(MCHOG) gradient descriptor. The features computed by the aforementioned descriptors together with a simple gray-level histogram were used for the classification into five categories of lung tissues (normal, emphysema, ground glass, fibrosis, and micronodules). The classification is done with the so-called patch-adaptive sparse approximation (PASA) method, which is based on the minimum discrepancy criterion.

Recently, proposed systems have exploited the new possibilities of MDCT scanners that are able to achieve almost isotropic 3D submillimeter-resolution acquisition. Hence, some of the already proposed 2D texture feature sets were expanded and computed on the 3D lung parenchyma. Xu et al. [39] proposed the use of a 3D AMFM for the classification of $21 \times 21 \times 21$ VOIs from 20 cases into five classes: emphysema, GGO, honeycombing, normal nonsmokers, and normal smokers. From each VOI, the first-order statistics, RLM, GLCM, and fractal features were extracted. An SVM and a Bayesian classifier were used, and the best result

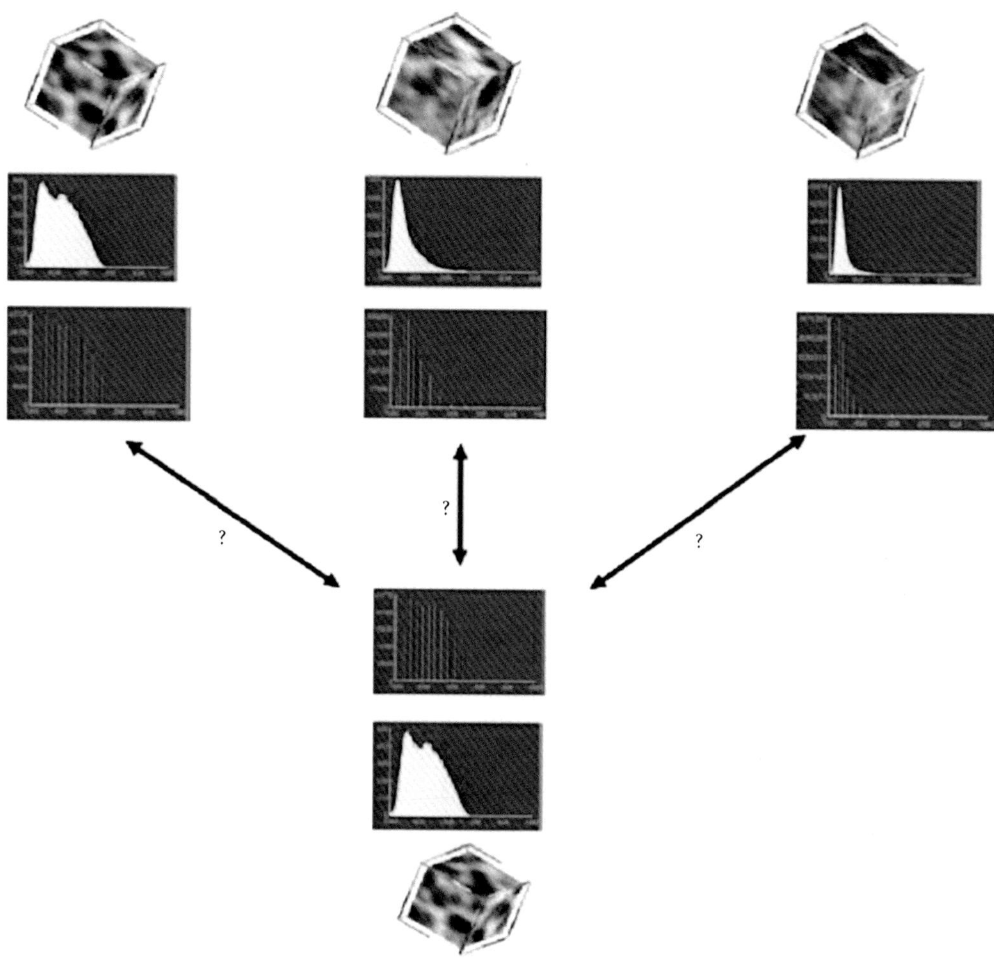

FIGURE 6.13
The main point of the algorithm. The top three rows show the cubes of data that are known along with their histograms and canonical signatures. The bottom three rows show correspond to an unknown cube of data. The idea of the algorithm is to compare the unknown signature with the known signatures using the earth mover's distance as the metric. The unknown cube of data is assigned to the known cube's class for which the signatures are most similar. The symbol "?" expresses the query "which are the distances of the unknown signature at the bottom from the known signatures on the top and which one is the minimum?" (Data from Zavaletta, V.A. et al., *Acad. Radiol.*, 14, 772–787, 2007.)

was achieved by the latter with an accuracy of 86.2%. However, Xu et al. [39] conclude that both classifiers are comparable when it comes to classifying patterns associated with ILDs. Zavaletta et al. [8] used signatures from adaptively binned histograms for the description of the 15 × 15 × 15 VOIs and the earth mover's distances [33] from the canonical class signatures for the classification (Figure 6.13). The adaptive binning of the histograms is performed by clustering the values with k-means and the signature is defined as a set of paired values (μ_i, w_i) where μ_i is the centroid of the cluster and w_i is the number of voxels in it. The canonical signature for a class is computed by combining the signatures for each of the training VOIs and reclustering the distribution into K clusters. The creation of a canonical signature allows for a more computationally efficient way to match signatures instead of computing the distance between all training signatures and all test signatures. The overall classification accuracy after the removal of the bronchovascular structure was 72.7% for classifying into five classes (normal, GGO, reticular, honeycombing, and emphysema) originated from 18 cases. Korfiatis et al. [34] employed 3D GLCM features and a *k*-NN classifier for classifying 21 × 21 × 21 VOIs from 14 patients into three classes (reticular, GGO, and normal). Thirteen different GLCM features were computed for every one of the 13 used directions. The mean and range of each GLCM feature over the 13 directions comprised a total of 26 GLCM-based features for each distance. By considering five different distances (d = 1–5), the 130 values of the final feature set were computed and a stepwise discriminant analysis followed to reduce the dimensionality. Mariolis et al. [40] combined 3D RLM features with 3D Laws feature set in order to describe VOIs with various sizes from 30 patients. Four methods were used [probabilistic neural networks (PNNs), *k*-NN, Bayesian and multinomial logistic regression] for the classification into four classes (normal, GGO, reticular, and honeycombing), with PNN and *k*-NN achieving the best results on the order of 99.8%. However, Mariolis et al. [40] do not report the chosen values for various parameters such as the size of the VOI.

6.2.3 Differential Diagnosis

The final stage of a CAD system is responsible for combining the results of the previous stages in order to provide a global DDx, which is the procedure of identifying the presence of a disease out of a set of possible alternatives. Despite the large number of publications addressing the problems of lung segmentation and ILD pattern quantification, just very few approaches attempted to proceed with a final DDx for the entire lung. Van Ginneken et al. [41] proposed an automatic method for the segmentation of lung fields into 42 regions (Figure 6.14) followed by a classification step that assigns to each region a confidence value for being abnormal. The product of the individual confidence values provides a global diagnosis on the abnormality of the whole lung. Zheng et al. [42] proposed a system that segments lung areas, identifies suspicious volumetric ILD lesions, computes five global features for each of them (size, contrast, average local pixel value fluctuation, mean of SFD, and GFD), and classifies the corresponding case into one of the three categories of severity (mild, moderate, and severe) by using a distance-weighted *k*-NN algorithm. Fukushima et al. [31] proposed the use of an ANN that combines 10 clinical parameters with 23 HRCT features in order to provide a final DDx. However, the used HRCT features

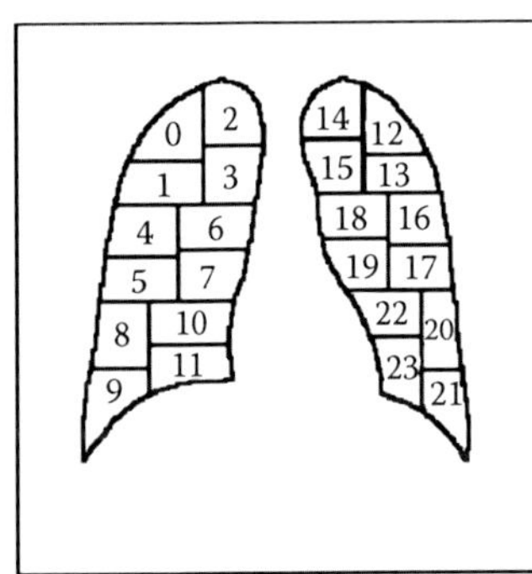

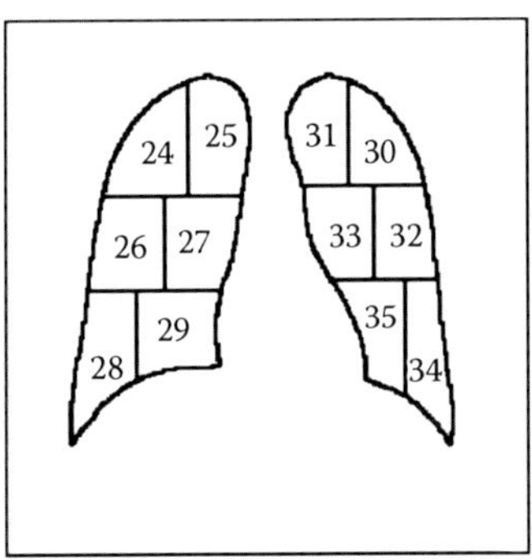

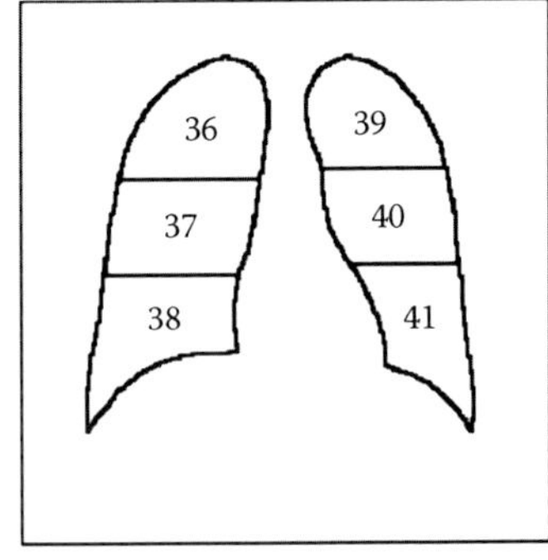

FIGURE 6.14
The proposed lung regions. The lung fields are subdivided into 24 regions. These regions are merged to form larger regions. In total, 42 regions—numbered 0 to 41—are used. (Data from Van Ginneken B. et al., *IEEE Trans. Med. Imaging*, 21, 139–149, 2002.)

were not computed automatically but rated manually by radiologists. Wang et al. [43], after classifying the VOIs of a case into normal/abnormal using RLM and GLCM features, used a simple rule to classify the case itself: If the number of VOIs reported as abnormal in a case is greater than a specified threshold, the case was considered as abnormal; otherwise, it was considered as normal.

6.3 Discussion

The development of a CAD system for the automatic analysis of CT scans with ILD pathologies constitutes a very challenging task with a high significance for the clinical practice. Despite the numerous challenges and after nearly two decades of relative investigation, the area of CAD on ILDs has definitely achieved some of the initial objectives, but there are still several remaining issues that need to be addressed.

Some of the proposed systems have already demonstrated their capabilities and potentials by proving the increased diagnostic abilities of the radiologists that use the specific CAD system as an advising additional observer. The provided assistance can be even more valuable in cases of remote hospitals where a chest expert might not be available, or in cases of new radiologists by supplementing their limited experience and helping toward their faster training. However, the progress of the research field also faced significant obstacles during the past few decades. Apart from the objective difficulties of the problem, the vagueness of the medical definitions and classifications of the involved pathologies caused inconsistencies among the various approaches and prevented from setting universally accepted targets. To this end, an international multidisciplinary consensus [2,3] attempted to clarify the area by defining a standardized terminology, classification, and diagnostic protocol for the ILDs. These efforts contributed toward a better communication within the scientific community facilitating the optimization of various CAD components and their adaptation to the specific problem of ILDs. However, the development of a complete CAD system for the assessment and DDx of ILDs is still far from feasible.

The different levels of lung segmentation comprise a very important preprocessing step for the success of a lung CAD system. Several methods have been proposed recently for the lung field segmentation in presence of ILD pathologies. Most of these methods are based on classic medical segmentation approaches adjusted to the special needs of cases with mild or severe ILDs. Since gray-level values proved insufficient for discriminating lung areas from the background, additional information was employed, originating from the specific texture of the abnormalities and the anatomical structure of the chest. Significant progress was also done with the localization and removal of the bronchovascular tree, which facilitates the quantification by reducing the false alarms. This progress is mainly due to the 3D imaging data provided by the new MDCT protocol that enabled the tracing of the airways and vessels in the 3D space. However, the segmentation of the lung lobes with ILD patterns was not addressed, although it may be proven valuable for the final diagnosis.

The ILD pattern quantification via the local texture classification is probably the most investigated subject in the relative literature. Many different feature sets and classification techniques were utilized producing promising results. The 3D images provided by MDCT enabled the use of 3D texture descriptors able to capture the actual 3D fibrotic structure. However, there is still a lot of space for exploring modern texture description and classification techniques in order to increase the local recognition accuracy. Moreover, the transition from the local classification to the global lung quantification needs to be investigated. The optimal shape and size of the used ROI/VOI as well as their overlap is still an open issue. Large regions are usually able to provide more information and increase the classification accuracy, whereas small regions guarantee a better resolution for the global lung analysis.

A vital issue for the task of ILD pattern classification is the data acquisition and annotation. Since the proposed methods rely on machine learning techniques, the quantity and quality of the used data are crucial for proving its actual capabilities. In many studies, the number of different considered cases is not sufficient, so inevitably the training set is not representative or it is correlated to the testing set. Another important problem is the correctness of the annotations because one single medical expert is not considered absolutely reliable due to the high interobserver variance. Hence, an assessment by two or three radiologists is often suggested. A very significant achievement for the entire field would be the establishment of common, publicly available databases, which could be used for the development, evaluation, and comparison of all available approaches. An attempt toward this direction has been done recently by Depeursinge et al. [44]. The provided database consists of a relatively large amount of HRCT scans of ILD cases along with the corresponding annotations of the patterns and various clinical parameters. Unfortunately, the database does not include the MDCT data.

Finally, a major challenge for the new generation of the CAD systems for ILDs is raised by the need for automatic DDx. Currently, most of the methods are limited to the quantification of the various ILD patterns in the lung volume. Very few works have attempted to provide an actual diagnosis and usually this involved the interaction of the radiologist. One important obstacle towards this direction use to be the vagueness of the medical classification for ILDs. However, currently the medical definition of the problem is definitely more explicit and the previous stages of segmentation and classification have reached a sufficiently mature point able to support an attempt for a final diagnosis considering at least the most common ILDs.

6.4 Conclusion

This survey provided an overview of the existing literature for the automatic identification and characterization of ILDs in CT images. The specific research area has made significant progress in the past two decades providing health-care professionals with valuable tools able to improve the diagnostic procedure and disease management. For the lung segmentation, new methods were proposed focusing on the processing of cases with ILD pathologies. To this end, the special textural characteristics of the interstitial patterns were considered along with any available anatomical information, often incorporated into a machine learning framework. Regarding the segmentation of the lung airways, the utilization of the relatively new MDCT imaging protocol enabled the use of advanced 3D region growing techniques that achieved height accuracies and facilitated the ILD pattern identification. Moreover, some methods were proposed for the segmentation of the vascular tree, although there are still several open issues to be addressed. Most of these methods used vessel enhancement filters based on the Hessian matrix analysis, whereas some machine learning approaches were also proposed based on 3D texture features. The localization and characterization of the ILD patterns is probably the most investigated task of the area.

Many different 2D or 3D textural feature sets were employed and combined with various classification schemes for the classification of sliding ROIs or VOIs, which then provide the global quantification of the ILD pathologies. However, novel textural descriptors and classifiers, emerging from the fields of computer vision and machine learning, could still be used in order to further enhance the performance. Finally, the major remaining challenge of the area is the development of a complete system that will provide a fully automatic DDx based on the results of the aforementioned image analysis steps along with any other available clinical and biochemical data. Although very few researchers have attempted to deal with the problem of DDx, the state of the existing literature seems now mature enough to provide reliable image analysis results that will support attempts toward the automatic ILD diagnosis.

References

1. British Thoracic Society and Standards of Care Committee. BTS guidelines on the diagnosis, assessment and treatment of diffuse parenchymal lung disease in adults. *Thorax* 1999; 54:S24–S30.
2. American Thoracic Society and European Respiratory Society. ATS/ERS international multidisciplinary consensus classification of the idiopathic interstitial pneumonias. *Am J Respir Crit Care Med*. 2002; 165(2):277–304.
3. Wells AU. Managing diagnostic procedures in idiopathic pulmonary fibrosis. *Eur Respir Rev*. 2013; 22(128):158–162.
4. Sluimer I. et al. Computer analysis of computed tomography scans of the lung: A survey. *IEEE Trans Med Imaging*. 2006; 25(4):385–405.
5. Uchiyama Y. et al. Quantitative computerized analysis of diffuse lung disease in high-resolution computed tomography. *Med Phys*. 2003; 30(9):2440–2454.
6. Kim KG. et al. Computer-aided diagnosis of localized ground-glass opacity in the lung at CT: Initial experience. *Radiology* 2005; 237(2):657–661.
7. Goo JM, Lee JW, Lee HJ, Kim S, Kim JH, Im JG. Automated lung nodule detection at low-dose CT: Preliminary experience. *Korean J Radiol*. 2003; 4:211–216.
8. Zavaletta VA. et al. High resolution multidetector CT-aided tissue analysis and quantification of lung fibrosis. *Acad Radiol*. 2007; 14(7):772–787.
9. Hu S. et al. Automatic lung segmentation for accurate quantitation of volumetric x-ray CT images. *IEEE Trans Med Imaging*. 2001; 20(6):490–498.
10. Sluimer IC. et al. Automated classification of hyperlucency, fibrosis, ground glass, solid, and focal lesions in high-resolution CT of the lung. *Med Phys*. 2006; 33(7):2610–2620.
11. Armato SG, Giger ML, Moran CJ, Blackburn JT, Doi K, MacMahon H. Computerized detection of pulmonary nodules on CT scans. *Radiographics* 1999; 19(5):1303–1311.
12. Korfiatis P. et al. Texture classification-based segmentation of lung affected by interstitial pneumonia in high-resolution CT. *Med Phys*. 2008; 35:5290–5302.
13. Wang J, Li F, Li Q. Automated segmentation of lungs with severe interstitial lung disease in CT. *Med Phys*. 2009; 36:4592–4599.
14. Sluimer I, Prokop M, van Ginneken B. Toward automated segmentation of the pathological lung in CT. *IEEE Trans Med Imaging*. 2005; 24(8):1025–1038.
15. Prasad MN. et al. Automatic segmentation of lung parenchyma in the presence of diseases based on curvature of ribs. *Acad Radiol*. 2008; 15(9):1173–1180.
16. Hua P. et al. Segmentation of pathological and diseased lung tissue in CT images using a graph-search algorithm. *IEEE ISBI*. 2011; 2:2072–2075.

17. Delorme S. et al. Usual interstitial pneumonia. Quantitative assessment of high resolution computed tomography findings by computer-assisted texture-based image analysis. *Invest Radiol.* 1997; 32(9):566–574.
18. Uppaluri R. et al. Computer recognition of regional lung disease patterns. *Am J Respir Crit Care Med.* 1999; 160(2):648–654.
19. Aykac D. et al. Segmentation and analysis of the human airway tree from 3D X-ray CT images. *IEEE Trans Med Imaging.* 2003; 22:940–950.
20. Xu Y. et al. MDCT-based 3-D texture classification of emphysema and early smoking related lung pathologies. *IEEE Trans Med Imaging.* 2006; 25(4):464–475.
21. Tschirren J. et al. Intrathoracic airway trees: Segmentation and airway morphology analysis from low-dose CT scans. *IEEE Trans Med Imaging.* 2005; 24(12):1529–1539.
22. Shikata H, Hoffman EA, Sonka, M. Automated segmentation of pulmonary vascular tree from 3D CT images. In *Medical Imaging 2004: Physiology, Function, and Structure from Medical Images*, edited by Amini AA and Manduca A, Proceedings of SPIE, Vol. 5369, SPIE, Bellinham, WA, 2004, pp. 107–116.
23. Korfiatis P, Karahaliou A, Kazantzi A, Kalogeropoulou C, Costaridou L. Towards quantification of interstitial pneumonia patterns in lung multidetector CT. *8th IEEE International Conference on BioInformatics and BioEngineering.* BIBE 2008, pp. 1–5, October 8–10, 2014. doi:10.1109/BIBE.2008.4696813.
24. Zhou C. et al. Automatic multiscale enhancement and segmentation of pulmonary vessels in CT pulmonary angiography images for CAD applications. *Med Phys.* 2007; 34(12):4567–4577.
25. Korfiatis P, Kalogeropoulou C, Karahaliou A, Kazantzi A, Costaridou L. Vessel tree segmentation in presence of interstitial lung disease in MDCT. *IEEE Trans Inf Technol Biomed.* 2011; 15(2):214–220.
26. Heitmann KR. et al. Automatic detection of ground glass opacities on lung HRCT using multiple neural networks. *Eur Radiol.* 1997; 7(9):1463–1472.
27. Chabat F. et al. Obstructive lung diseases: Texture classification for differentiation at CT. *Radiology* 2003; 228(3):871–877.
28. Sluimer I. et al. Computer-aided diagnosis in high resolution CT of the lungs. *Med Phys.* 2003; 30(12):3081–3090.
29. Song Y, Cai W, Zhou Y, Feng DD. Feature-based image patch approximation for lung tissue classification. *IEEE Trans Med Imaging.* 2013;32(4):797–808.
30. Park SO. et al. Feasibility of automated quantification of regional disease patterns depicted on high-resolution computed tomography in patients with various diffuse lung diseases. *Korean J Radiol.* 2009; 10(5):455–463.
31. Fukushima A. et al. Application of an artificial neural network to high-resolution CT: Usefulness in differential diagnosis of diffuse lung disease. *AJR Am J Roentgenol.* 2004; 183(2):297–305.
32. Vo KT. et al. Multiple kernel learning for classification of diffuse lung disease using HRCT lung images. *Conf Proc IEEE Eng Med Biol Soc.* 2010; 2010:3085–3088.
33. Rubner Y, Tomasi C, Guibas LJ. The earth mover's distance as a metric for image retrieval. *Int J Comp Vision.* 2000; 40:99–121.
34. Korfiatis P. et al. Texture-based identification and characterization of interstitial pneumonia patterns in lung multidetector CT. *IEEE Trans Inf Technol Biomed.* 2010; 14(3):675–680.
35. Haralick RM, Shanmugam KS, Dinstein I. Textural features for image classification. *IEEE Trans Syst Man Cybern.* 1973; SMC-3:610–621.
36. Galloway MM. Texture analysis using gray level run lengths. *Comput Graphics Image Process.* 1975; 4:172–179.
37. Mandelbrot BB. *Fractal Geometry of Nature.* Freeman, New York, 1982.
38. Do MN, Vetterli M. The contourlet transform: An efficient directional multiresolution image representation. *IEEE Trans Image Process.* 14(12):2091–2106.
39. Xu Y. et al. Computer-aided classification of interstitial lung diseases via MDCT: 3D adaptive multiple feature method. *Acad Radiol.* 2006; 13(8):969–978.

40. Mariolis I. Korfiatis P, Costaridou L, Kalogeropoulou C, Daoussis D, Petsas T. Investigation of 3D textural features' discriminating ability in diffuse lung disease quantification in MDCT. *IEEE International Conference on Imaging Systems and Techniques*, Thessaloniki, Greece, pp. 135–138, July 1–2, 2010. doi:10.1109/IST.2010.5548528.
41. Van Ginneken B. et al. Automatic detection of abnormalities in chest radiographs using local texture analysis. *IEEE Trans Med Imaging*. 2002; 21(2):139–149.
42. Zheng B. et al. Automated detection and classification of interstitial lung diseases from low-dose CT images. In *Medical Imaging 2004: Image Processing*, edited by Fitzpatrick JM and Sonka M, Proceedings of SPIE, Vol. 5370, SPIE, Bellinham, WA, 2004, p. 849.
43. Wang J. et al. Computerized detection of diffuse lung disease in MDCT: The usefulness of statistical texture features. *Phys Med Biol*. 2009; 54:6881–6899.
44. Depeursinge A, Vargas A, Platon A, Geissbuhler A, Poletti, PA, Müller H. Building a reference multimedia database for interstitial lung diseases. *Comput Med Imaging Graphics* 2012; 36(3):227–238.

7

Induced Optical Natural Fluorescence Spectroscopy for Giardia lamblia *Cysts*

Sarhan M. Musa and Kendall T. Harris

CONTENTS

7.1 Introduction

Today, fluorescence techniques is one of the most important research tools in cell biology and biochemistry. The spontaneous fluorescence of waterborne microorganisms [such as *Giardia lamblia, Paramecium,* and rotifer] without adding dyes is referred to in this chapter as optical natural fluorescence (ONF). The ONF spectroscopy is useful because it does not require any chemical manipulation for the sample, gives real-time results immediately from the sample reaction, and supplies information about the biological structure of the cell through its fluorophores. This technique is compatible with microscope optics and sensitive to weakly fluorescing microscopic samples. It can be used to distinguish the microorganisms from each other.

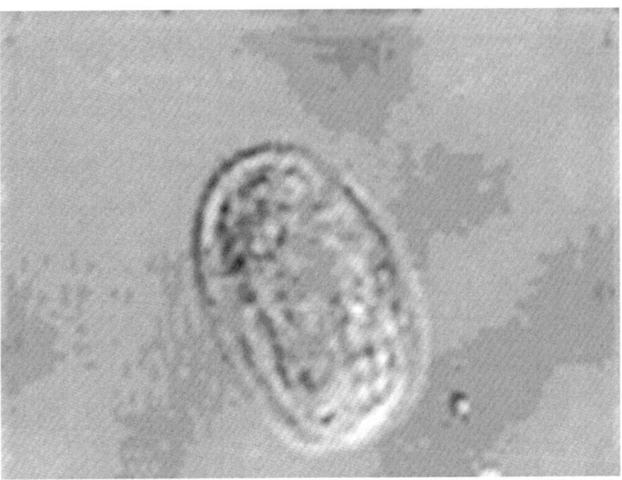

FIGURE 7.1
Giardia lamblia cyst.

Giardia lamblia is a common intestinal protozoan (which consists of a single cell for its body) that causes giardiasis (gastrointestinal disease, symptoms including diarrhea, nausea, and/or stomach cramps, can be fatal to immune-suppressed individuals), which has been distributed worldwide and implicated in recent outbreaks of illness in the United States [1–4]. *Giardia lamblia* is commonly found in surface waters (rivers, lakes, streams, etc.), especially where there is contamination from sewage or animal wastes [5]. It lives mainly in cyst forms that are resistant to disinfections. The cysts of *G. lamblia* are oval in shape and are approximately 8–14 μm in size [6–9] as in Figure 7.1. Because of its translucent appearance and low concentration in natural waters [10], *G. lamblia* is not easy to detect, but even a few cysts are enough to make a person severely sick.

There were a number of methods proposed for the detection of *G. lamblia* cysts, such as conventional microscopy, flow cytometry (FC) [11–13], UV–Vis spectroscopy method [14,15], enzyme-linked immunosorbent assay (ELISA) [16], and polymerase chain reaction (PCR) [17]. These methods are at various stages of development, some of them being incomplete. Many of them have problems for ascertaining *G. lamblia* cysts in water [2,5,18]. For example, conventional microscopy needs an experienced person dedicated to the observation and is inefficient in time and prone to error. The FC method relies on the ability of the instrument to analyze particles individually in a suspension, which means that the method needs to have a highly trained full-time operator. This method is limited in efficiency due to losses occurring during sample concentration, and the purified and concentrated samples are stained with antibodies specific to *G. lamblia* cysts. For UV–Vis spectroscopy method, the background particles such as algae and nonbiotic particulate cause spectral patterns that mask the deconvolution of the target organism fingerprint, and the antibody labeling is used to enhance the signal and allow differentiation. In the ELISA method, algae present in the sample can pick up the antibody reagents and give false-positive readings, which reduce the analytical sensitivity. In addition, the PCR method has limitation during the sample collection and elution. Organic matter and dissolved solids can be concentrated in the sample, and these compounds (such as humic substances) can interfere with the activity of the enzymes used in PCR.

The current state-of-the-art technique for detecting *G. lamblia* relies upon microscopic immunofluorescence assay method that uses antibodies directed against the cyst forms of the protozoa. This method uses specific antibodies labeled with fluorescent dye to tag *G. lamblia* cysts, which are then viewed through a UV fluorescence microscope [19–21]. To use this method, it is first necessary to determine the antigen of *G. lamblia* in order to make the antibody specific to the cyst membrane, although several commercial sources of antibodies for *G. lamblia* are already available. In practice, there is a slight possibility of cross-reactivity for the antibody because different organisms may share similar surface antigens [22]. Interference with the immunoassay signal may also arise from nonspecific binding of the antibody or autofluorescence from other microorganisms and the background [23–25]. While a number of commercially developed antibodies are highly specific to *G. lamblia,* it is impractical to deploy a large quantity of them for continuous field sensing. In practice, water samples need to be taken to a laboratory for the assay, which can delay the detection of cysts in time. At the same time, the viability of the cysts cannot be determined unambiguously without additional staining or other procedure [26,30,31]. Therefore, the limitations of the current methods prompt a need to develop a detection method that passively senses the natural characteristics of this microorganism.

Optical spectroscopy is one of the most scientific researches still going for a new development. There are several experimental setups that can be used to study the optical spectroscopy [27–29]. But, there is still a difficulty to investigate and use an experimental setup to study the ONF of microorganisms using laser.

The experimental setup we developed was successfully used to detect the ONF for *G. lamblia* cysts using laser. We will concentrate on the autofluorescence spectra of *G. lamblia* cysts that were measured using our new experimental setup. The excitation wavelength was focused on the blue side in the visible range due to the observation that blue light excites stronger fluorescence from *G. lamblia* than green light and visible light experiences less absorption by water and glass than true UV.

Unless otherwise mentioned, the ONF of *G. lamblia* refers to the data acquired from a cluster of cysts that were never exposed to laser before. The sample was always stored at refrigerator temperature, and then prepared and tested at room temperature. The background emission by the medium in the sample under the same excitation condition as *G. lamblia* was always subtracted from the raw data from *G. lamblia* to produce the net ONF. This chapter and the rest of the book will not investigate the viability of *G. lamblia.*

In this work, a fiber-optic fluorescence microscope is specifically designed to excite and collect the fluorescence from *G. lamblia* cysts. For the geometric convenience of the work, optical fibers are used for the transmission of the laser excitation to the sample and for collecting the fluorescence light emitted and bringing it to a spectrometer detector. This approach is also in concert with a potential ultimate application of these techniques, specifically the direct optical detection of microorganisms in reservoirs and other natural bodies of water.

Through this experimental setup, the laser-induced autofluorescence of *G. lamblia* cyst is extensively studied. The fluorescence microscope technique used in this work is not *in situ,* yet the result of this study on autofluorescence will facilitate the development of *in situ* fluorescence probes for the detection of *G. lamblia* cysts.

7.2 Experimental Setup

The schematic of the experimental setup used for studying ONF of microorganisms such as *G. lamblia* cysts or other microorganisms is shown in Figure 7.2. A continuous-wave (CW) argon-ion laser is used as an excitation source. This laser has a high-power output and is ideal for research such as fluorescence spectroscopy. Its output is selectable at 458, 476, 488, 496, and 514 nm. The wavelength 458 nm is used to excite the fluorescence of *G. lamblia* cysts because it induces the clearest ONF and least amount of noise among all the wavelengths.

In the setup, the laser output is low in height, so two plane mirrors are used to raise the laser beam. The reflected beam then goes through two equilateral dispersion prisms. The dispersion prism ensures that the spontaneous emission background from the laser (which may overlap with ONF) is minimized. Otherwise, it would be mixed with the monochromatic excitation and contribute to the scattering noise. The laser beam is then reflected by

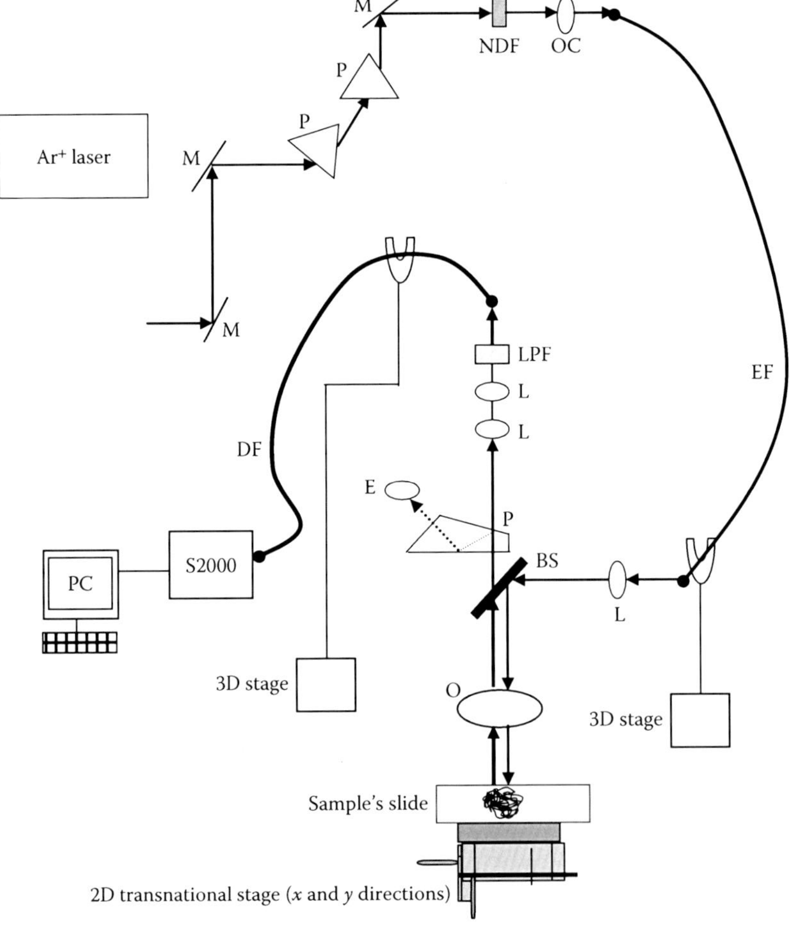

FIGURE 7.2
Schematic display of the experimental setup. BS, beam splitter; C, fiber coupler; DF; detection fiber; E, eyepiece (ocular); EF, excitation fiber; L, lens; LPF, long-pass filter; M, flat mirror; NDF, neutral density filters; OC, optical coupler; OL, objective lens; P, prism; PC, personal computer.

a third plane mirror and enters a miniature fiber optical coupler. Inside the coupler, there is a small lens [diameter 5 mm, focal number (f.l.) 10 mm] that focuses the laser into the core of one branch of a bifurcated multimode fiber [200 μm core, numerical aperture (NA) 0.22]. However, the laser output is linearly polarized, whereas the output from the fiber is unpolarized.

It is important to center the laser beam in the aperture of the fiber optical coupler and to tilt the beam at the right angle with respect to the coupler to maximize the coupling efficiency. Because of its large core, the fiber can easily provide good coupling efficiency. The output of the optical fiber becomes a clean Gaussian beam, which corrects any spatial nonuniformity in the original laser beam profile. Sometimes, neutral density (ND 0.3) filters are used in front of the fiber optical coupler in order to attenuate the laser intensity at the fiber output.

The light exiting the fiber is sent into a Nikon microscope through the epi-illumination port on the side of the microscope. The optics assembly of the illumination port is partially disassembled, leaving a single 100-mm focal length lens remaining inside the port. The excitation beam entering the microscope is first converged by this lens. An internal beam splitter window then reflects the collimated light downward along the optical axis of the microscope, with a reflection coefficient around 25%. The transmission spectrum of the beam splitter is shown in Figure 7.3.

The reflected light by the beam splitter is focused by the microscope objective lens on the microscope slide. The microscope objective lens used in most experiments has a power of 20× [Japanese standard (JIS), f.l. 8.55 mm, width diameter (w.d.) 3 mm] and an NA of 0.40, which corresponds to half of the angular aperture of the objective lens $\theta = 23.58°$.

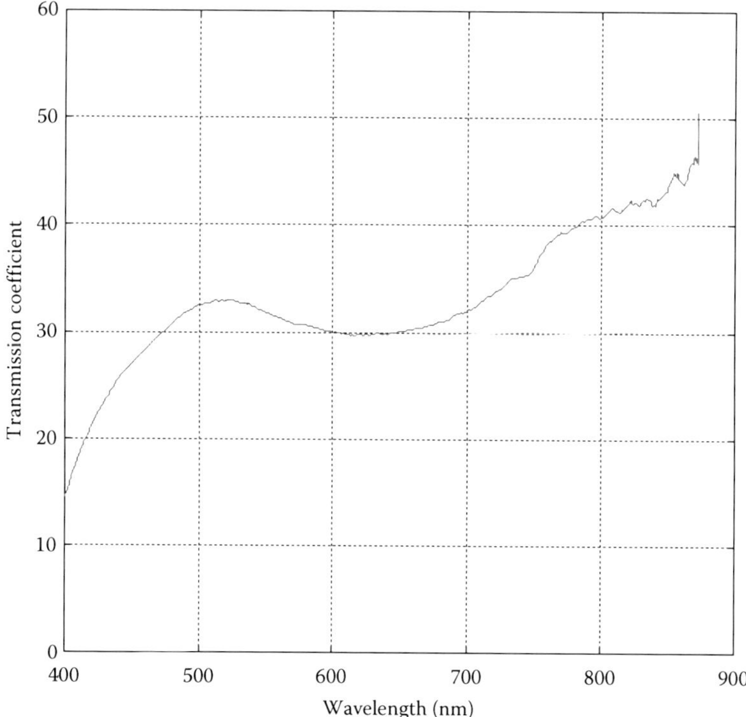

FIGURE 7.3
Measured transmission of the microscope beam splitter.

The objective lens of different powers, 20X (NA 0.4) and 5X (NA 0.1), often induces the same fluorescence spectra in shape; however, the intensity of the image increases with the power of the objective lens (lower NA means lower power). The standard objective lens used for the experiment is 20X. The targets that appear nonfluorescent with a power of 5X may appear weakly fluorescent with a power of 20X for the objective lens because 20X has a bigger NA of 0.4 and focuses the laser more tightly on the sample. Sometimes the objective lens of 5X is used in order to see the target in a small magnification.

The *G. lamblia* samples to be observed are prepared in a Fisher drop slide, which has a concavity of 0.5 mm depth and 18 mm diameter. The Fisher drop slide is covered with a slim cover glass (premium cover glass, size 24 × 50 mm). In addition, a flat glass microscope slide (plain 3 × 1 × 1 mm) with the slim cover glass can be used for other microorganism's samples to reduce the movement of the microorganisms.

The light emitted by the sample passes through the same objective lens of the microscope, transmits through the beam splitter, passes a prism in the microscope, and finally goes through a relay lens assembly to form a real image above the microscope. Before the image plane, the output light also goes through an optical long-pass filter (LPF; Schott GG 495, Schott OG 515, Corning 3-69, and Corning 3-72 filters for various laser wavelengths) that is placed in front of a detection fiber. The filter is necessary to ensure that the laser scattering signal does not swamp the detector for the relatively low-level fluorescence signal. The filter passes the fluorescence on the red side of the laser line and blocks unwanted laser scattering. The optical LPFs match the proper excitation wavelength: Corning 3-72 ~ 401 nm; GG 495 ~ 458 nm; Corning 3-69 or OG 515 (better than 3-69) ~496 nm. The transmission spectra of these LPFs are measured and the results are shown in Figure 7.4.

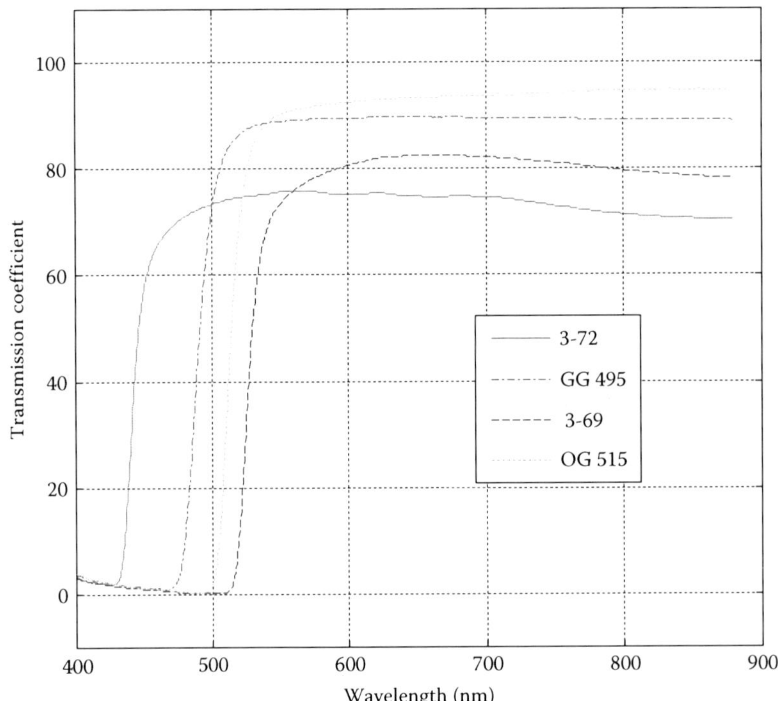

FIGURE 7.4
Measured transmission spectra of the LPFs: Corning 3-72, Schott GG 495, Corning 3-69, and Schott OG 515.

These are measured by using Ocean Optics white light source with ND 3.0 attenuation. The setting for a spectrometer channel is (6, 100) for [integration time (ms), average number]. Different LPF does change the shape of the spectra, but it does not change the decay (photobleaching) character of the microorganisms.

Another multimode fiber (400 μm core, NA 0.22) is aligned above the photographic port of the microscope and is coaxial with the optical axis of the microscope to detect the emission light of the object. The real image mentioned above is sent into the core of this fiber. The other end of this collecting fiber is connected to a fiber optical spectrometer S2000, which is interfaced with a computer that has LabVIEW software displaying the emission spectra integrated by the spectrometer's charge-coupled device (CCD) array detector. The integration time typically used with the spectrometer is 4 s. The S2000 spectrometer has high detection sensitivity (86 photons per count, 2.9×10^{-17} J per count, which is equivalent to 2.9×10^{17} W per count for 1 s integration). The spectrometer accepts the light transmitted through the multimode optical fiber and disperses it via a fixed grating across the CCD array, which is responsive from 187 to 879 nm (effective range). The approximate FWHM optical resolution (OR) is related to the dispersion (D) and the resolution (R) of the spectrometer by

$$\mathrm{OR} = D \times R \tag{7.1}$$

The dispersion is calculated from the spectral range of the grating (S) and the number of detector elements (N):

$$D = \frac{S}{N} \tag{7.2}$$

For the S2000 spectrometer, the spectral range of the grating is 650 nm, the number of the detector elements is 2048, and the resolution for a 25-μm slit is 4.2 pixels, so the calculated optical spectral resolution of the spectrometer is 1.3 nm [full width half maximum (FWHM)]. The acquired spectra are smoothed and the final spectral resolution is 16.6 nm.

A white light source is used to align both the excitation and collection fibers. A circular spot of size 25 μm representing the circle of illumination and another circular spot of size 40 μm representing the circle of collection are located at the center of the crosshairs in the microscope's field of view. The laser power coupled to the excitation fiber output is either 2.5 or 10 mW. The throughput of the microscope for the excitation is approximately 1%. Therefore, the direct excitation intensity on the slide sample is about 4–16 W/cm^2.

Before taking the measurements in the experimental setup, the microscope is first focused on an individual object (e.g., a *G. lamblia* cyst) by adjusting the height of the stage that carries the slide (*z*-direction) and observing through the eyepiece. The excitation and collection fibers are then translated along their longitudinal axes until they form sharp images on the same plane as the object. The sharp images are viewed through the eyepiece as sharp circular reflections of the fiber's illumination by the glass slide. The two fibers are then translated along their transverse axes until their images overlap concentrically on the crosshairs of the field.

The image of the excitation and detection fibers on the crosshairs of the eyepiece of the microscope is shown in Figure 7.5.

The target microorganisms are first translated away from the crosshair where the laser would be focused. The autofluorescence spectrum of the substrate (blank buffer solution or water) is then measured and saved as a background by choosing a store dark mode on the spectrometer software. The background spectrum is then automatically subtracted from

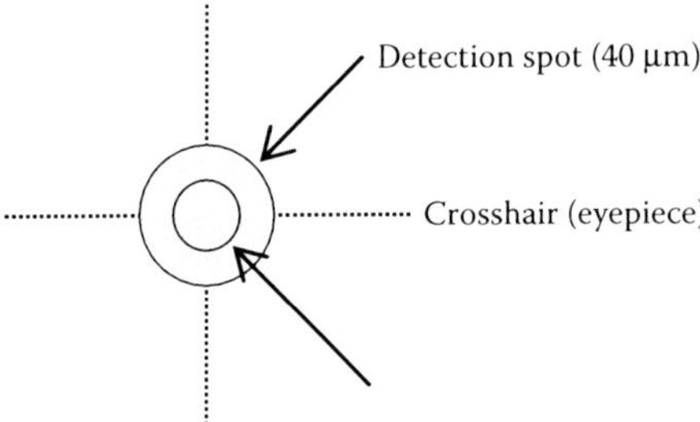

FIGURE 7.5
Image of the excitation and detection fibers on the crosshairs of the eyepiece of the microscope.

the signal (by using Scope-dark mode on the spectrometer software). The microorganisms are then moved to center on the crosshair while the laser is blocked. Subsequently, the signal spectra are saved as soon as the laser is unblocked. The measurements are always taken in a dark room at room temperature.

A CW semiconductor laser is used as an excitation source. The laser emits light of 401 nm from a 5-mW gallium nitride-based laser diode. The spectral line width is less than 1 nm. The experimental setup using the blue laser is similar to the setup in Figure 7.1 without using the two equilateral dispersion prisms.

7.3 Material and Discussion

The *G. lamblia* cysts introduced into the slide cavity for excitation and observations are normally prepared in a phosphate buffer saline solution (a solution of penicillin, streptomycin, and gentamicin). The sources of the cysts are experimentally infected gerbils. The cysts are purified from feces by sucrose and Percoll density gradient centrifugation. The storage condition for the sample is 4°C–6°C (i.e., refrigerator temperature). Suspensions of live cysts expire 21 days after the date of shipment.

It is important that the background signal of the sample to be small in order to obtain clear fluorescence spectra of large amplitude (after the subtraction of the background from the emission spectra). A large background will greatly reduce the signal-to-noise ratio of the fluorescence even with background subtraction. Note that when a shorter laser wavelength is applied, a stronger background from the autofluorescence of the sample substrate can occur. Therefore, for the 401 nm excitation wavelength, the background is higher than that for the 458 nm wavelength excitation.

When the white light and the fibers are used for the alignments, the fibers give several images that come from the front surface of the multimode fibers. These images can be explained by the *cloud theory*. The cloud theory for the image of the fiber on the slide can be expressed as follows: The reflections from planes above and below the focal plane create the cloudy (fuzzy) image, but the sharp image comes from the reflection from the plane of the focus as shown in Figure 7.6.

The cloud image (bad image) for the excitation fiber is about a factor of 5 larger than its sharp image (good image), and the cloud for the collection fiber is about a factor of 3 larger than its sharp image. The size of the cloud also depends on the type of the microscope slide used in the experiment. Slides with wells (cavities) give larger cloud image than flat slides.

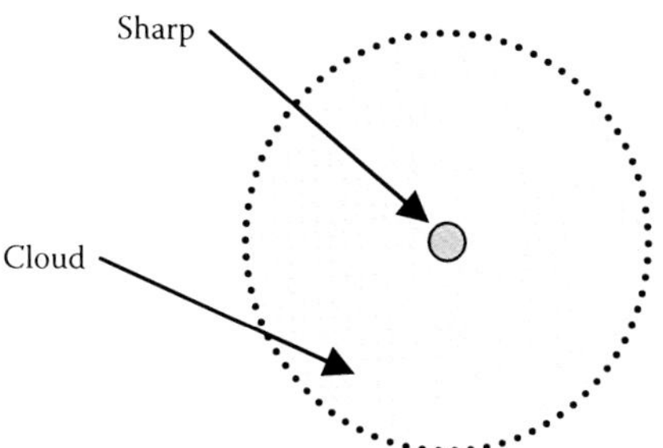

FIGURE 7.6
Cloud theory.

In order to minimize the background, the following steps must be followed: On the microscope, use the objective lens of a power of 5X to locate the protozoa (microorganisms) and a relatively clean background surrounding the protozoa. Then, use the objective lens of 20X and adjust the sample stage until the protozoa are on focus. Send white light into the two fibers. Look for their images on the focal plane of the protozoa. Move the fibers along their longitudinal axes. Multiple positions along these axes can be found, which form a sharp image of the same size, accompanied by cloudy images of different sizes. Choose the position that gives the largest relative size in cloudy image. This position spreads out the unwanted background fluorescence to a maximum degree and will give the minimum background in the collected signal.

The reason why there are bigger cloud images of the fiber on the well slide than on the flat slide is that the light goes deeper from the plane of focus (POF) on the well slide, which creates cloudier image, as shown in Figure 7.7.

A light beam (white light or laser) is brought to a point of focus by the objective lens at the level labeled POF. The fluorescence is emitted by the specimen from the point of focus (solid rays) and passes back through the objective lens, through the beam splitter to the fiber and the detector (spectrometer). However, autofluorescence will also be emitted by the substrate background from planes above and below the POF (dashed lines), but this light will be prevented from reaching the spectrometer by the detector fiber, which has a small core diameter. The more spreaded out the background image is, the less efficient

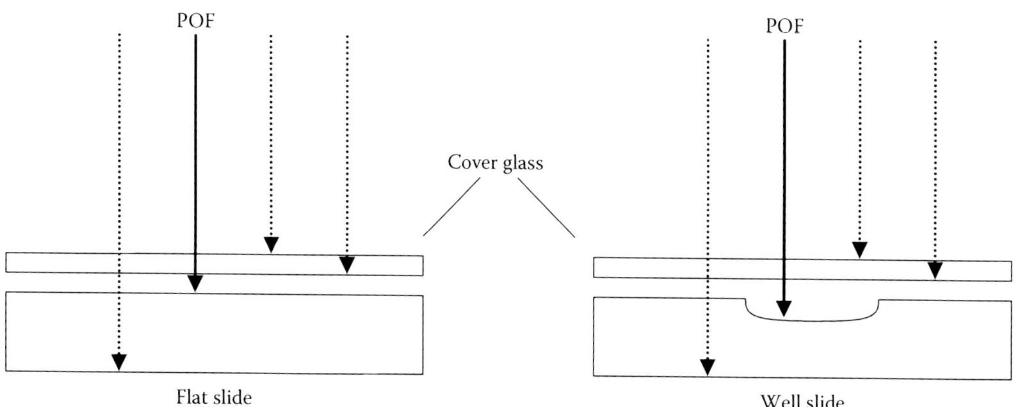

FIGURE 7.7
POF for flat and well slides.

the background can reach the spectrometer. Thus, only the light originating from the POF is most effectively collected by the detector. Note that *G. lamblia* cysts (and other microorganisms) can sit at different depths on the well slide, which can make the focus of the microscope and the focus of the fibers difficult. Therefore, a piece of rock or plant is sometimes focused on as a reference for the POF. The microscope for the setup is designed to illuminate the sample with the excitation light (laser) and collect the emitted light while excluding the laser scattering by using long pass filters.

7.4 Spectral Results and Discussion

7.4.1 ONF Spectra of a Cluster of *G. lamblia* Cysts Excited at 458 nm (Argon-Ion Laser Line)

The power of the laser coming out of the excitation fiber (Pf) was usually set at 10 mW. This power corresponded to an intensity of 16 W/cm^2 falling directly on the microscope slide that carried the *G. lamblia* sample. A Schott GG 495 LPF was always used in front of the detection fiber to block laser scattering.

The raw ONF spectra of *G. lamblia* of the same age are first plotted (after smoothing) to check for the data's reproducibility. Once they are reproducible in shape, the raw data are first averaged before any further processing. The result is then smoothed (across a window of 8.5 nm) and normalized according to the primary fluorescence peak height. The average spectra of *G. lamblia* of different ages (and from different samples) are then compared to see if the age of the cysts affects the emission spectrum.

7.4.1.1 Pf = 10 mW

Figure 7.8 shows the ONF data from young *G. lamblia* samples that were 1, 4, and 5 days old. The spectra overlap with each other in the major band, which peaks at 525 nm and has a FWHM of 75 nm. The negative peaks on the side in some of the spectra are caused by excessive subtraction of the laser scattering background.

Some additional peaks at 661 and 637 nm are believed to represent the fluorescence of pigments (such as chlorophyll) of algal impurities in the *G. lamblia* sample. These peaks are not an intrinsic part of *G. lamblia*'s ONF because they are not reproducible in magnitude relative to each other and to the major band.

Figure 7.9 shows the ONF spectra from slightly older *G. lamblia* that were 11, 12, 14, and 16 days old. Likewise, Figure 7.10 shows the ONF spectra from *G. lamblia* that were 25 and 27 days old; Figure 7.11 shows the ONF spectra from *G. lamblia* that were 40 and 44 days old; Figure 7.12 shows the ONF spectra from *G. lamblia* that were 50 and 54 days old. Finally, Figure 7.13 shows the ONF spectra from *G. lamblia* that were from 60 and 111 days old.

Each figure indicates that the ONF spectra of *G. lamblia* of a similar age are reproducible in shape under 458 nm excitation. Therefore, the spectra in each figure can be averaged to represent the typical spectrum of *G. lamblia* in each age group. As a test, the average spectrum from very old cysts (60–111 days old) is compared with that from very young ones (1–5 days old) in Figure 7.14. The result indicates that although the two groups of cysts are extremely different in age, their ONF spectra are very similar in shape. A similar conclusion may be drawn from *G. lamblia* of all other ages. Therefore,

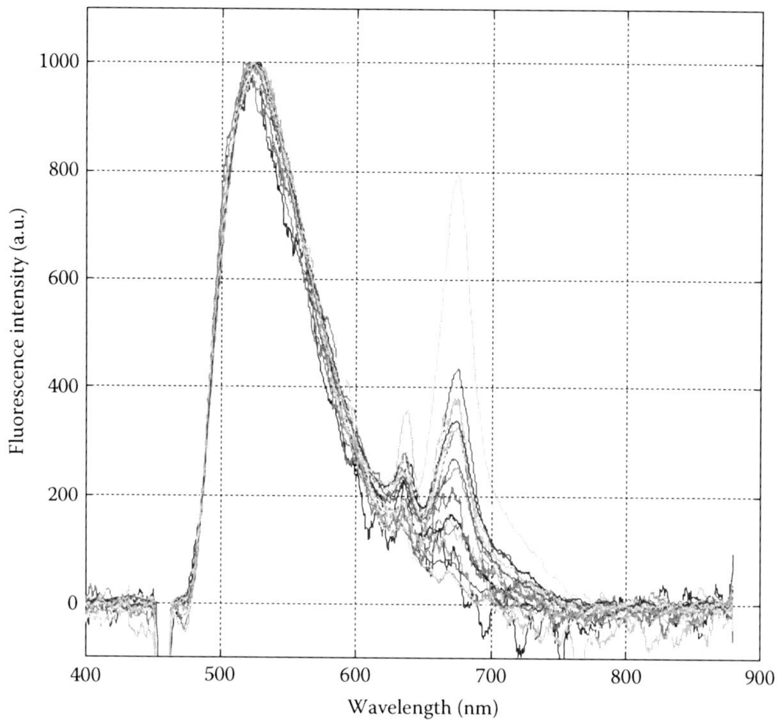

FIGURE 7.8
Comparison of ONF spectra of *G. lamblia* for samples 1, 4, and 5 days old. The excitation wavelength is 458 nm and the power out of the excitation fiber is 10 mW.

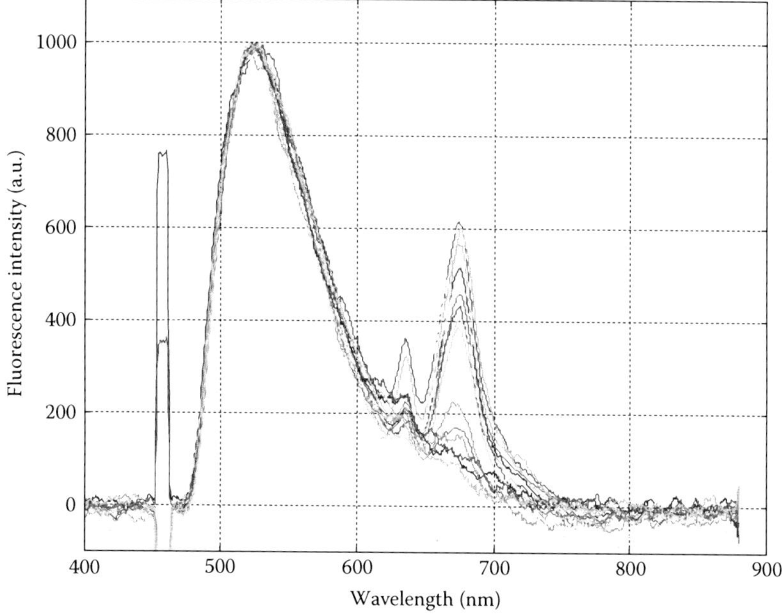

FIGURE 7.9
Comparison of ONF spectra of *G. lamblia* for samples 11, 12, 14, and 16 days old. The excitation wavelength is 458 nm and the power out of the excitation fiber is 10 mW.

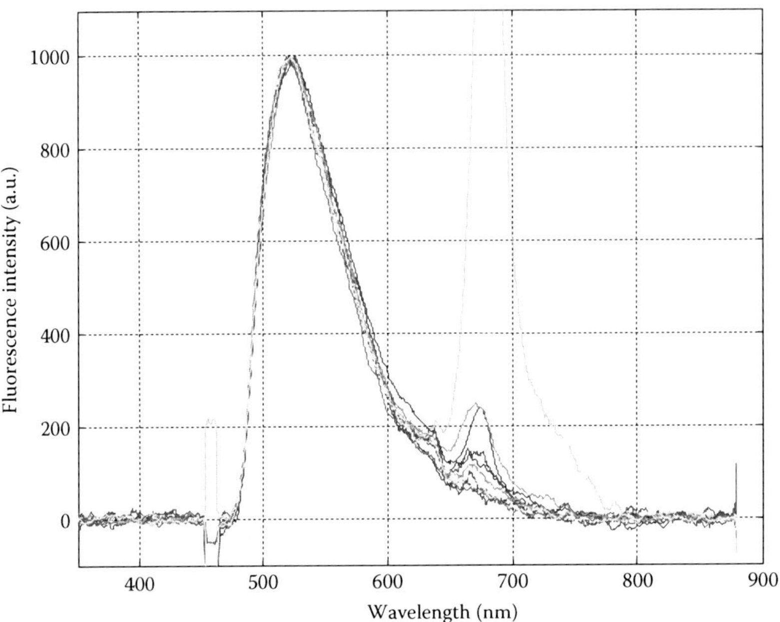

FIGURE 7.10
Comparison of ONF spectra of *G. lamblia* for samples 25 and 27 days old. The excitation wavelength is 458 nm and the power out of the excitation fiber is 10 mW.

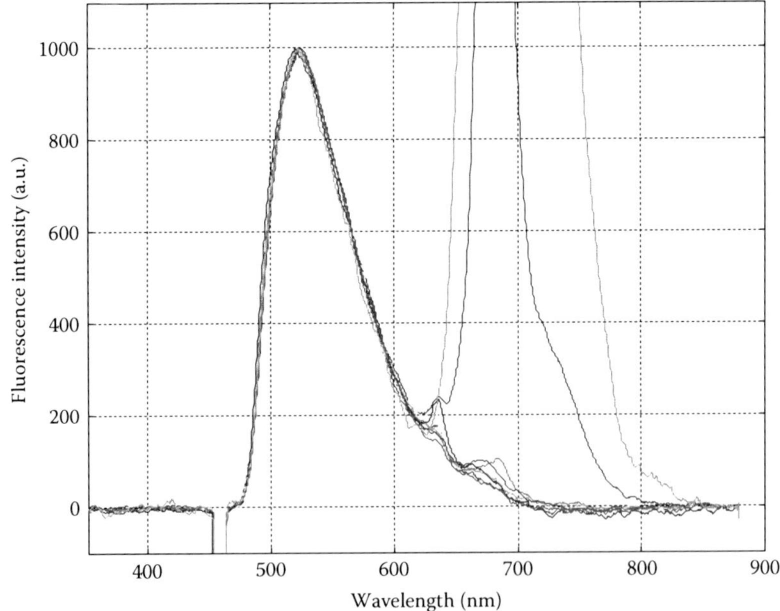

FIGURE 7.11
Comparison of ONF spectra of *G. lamblia* for samples 40 and 44 days old. The excitation wavelength is 458 nm and the power out of the excitation fiber is 10 mW.

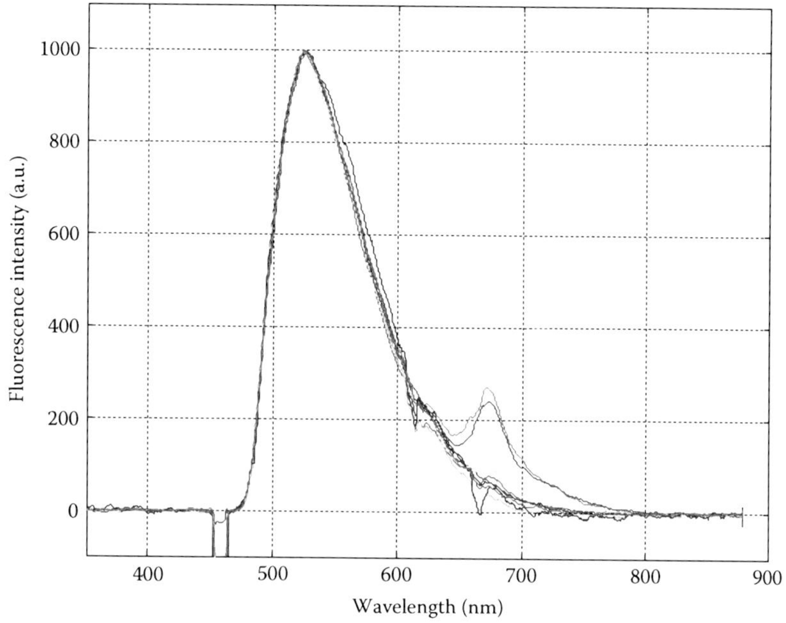

FIGURE 7.12
Comparison of ONF spectra of *G. lamblia* for samples 50 and 54 days old. The excitation wavelength is 458 nm and the power out of the excitation fiber is 10 mW.

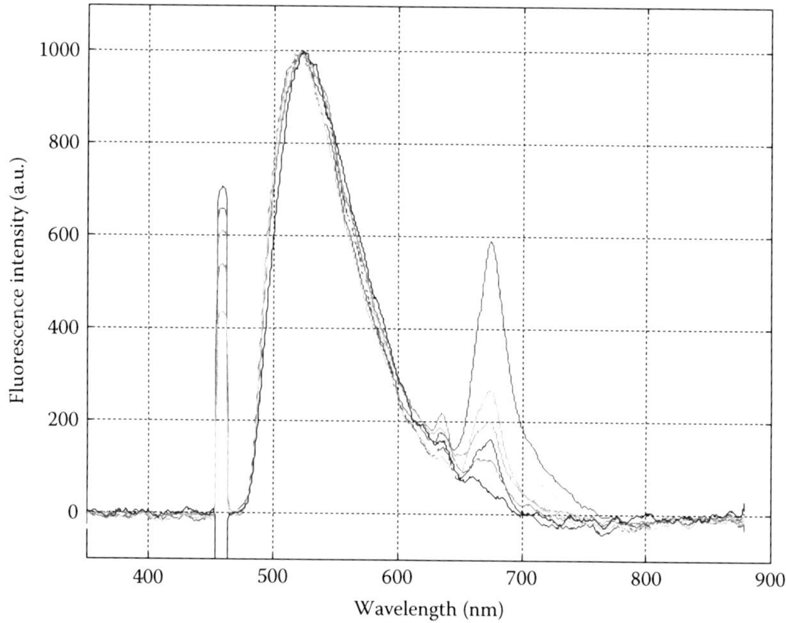

FIGURE 7.13
Comparison of ONF spectra of *G. lamblia* for samples 60 and 111 days old. The excitation wavelength is 458 nm and the power out of the excitation fiber is 10 mW.

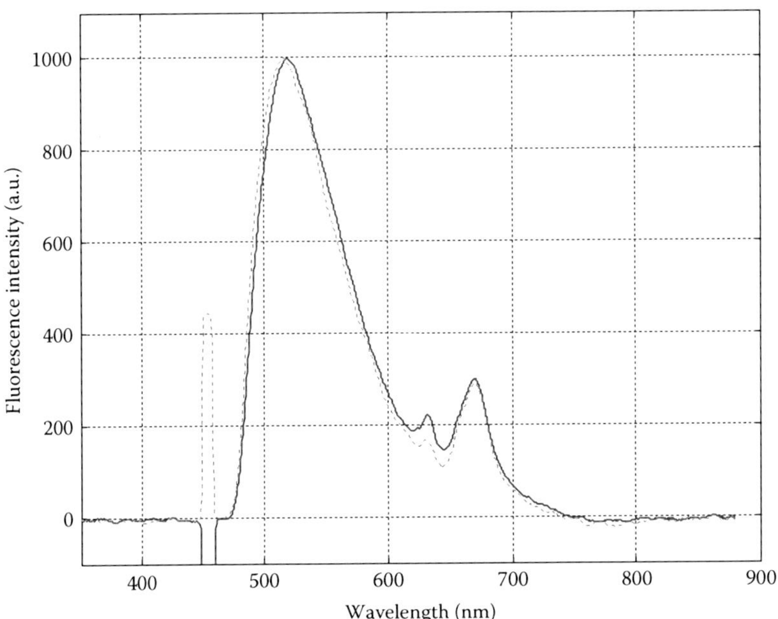

FIGURE 7.14
Comparison of the averaged ONF spectra of *G. lamblia* for very young ones (1–5 days old) and very old ones (60–111 days old). The excitation wavelength is 458 nm and the power out of the excitation fiber is 10 mW.

the age of the *G. lamblia* sample does not seem to have any effect on the ONF spectral shape. Consequently, the entire set of ONF spectra from *G. lamblia* of all age groups can be averaged, as shown in Figure 7.15. The resulting spectrum has an FWHM of 83 nm and a primary peak at 519 nm.

Besides the age, the preparation condition for the sample was also varied to see its effect on the ONF of *G. lamblia*. Figure 7.16 shows a group of ONF spectra from *G. lamblia* cysts that were initially stored at refrigerator temperature until it was 45 days old, and then aged over night at room temperature. It is believed that the room-temperature condition can greatly quicken the aging of *G. lamblia*. During this aging process, the sample was kept wet. This figure shows that the ONF spectra from aged *G. lamblia* cysts also overlap with each other. An average of these spectra is then compared with that from a fresh sample, as shown in Figure 7.17. The two spectra still overlap. This result confirms that the ONF spectral shape does not change even if the sample has aged enormously.

In conclusion, the age of *G. lamblia* does not change its ONF spectrum.

7.4.1.2 Pf Higher or Lower than 10 mW

To find out whether the excitation intensity affects the spectral result of *G. lamblia*, the power of the laser out of the excitation fiber (Pf) was also varied. For example, Figures 7.18 through 7.20 show a group of ONF spectra from *G. lamblia* (1 day old), where the power out of the fiber was 20, 5, and 2.5 mW, respectively. Similar to the result obtained at 10 mW, each figure shows that the spectral shape is reproducible at a fixed excitation intensity. The average of the spectra at each excitation power is then taken and plotted together in Figure 7.21. The result proves that the excitation intensity used in this experiment has no effect on the spectral shape of ONF of *G. lamblia*.

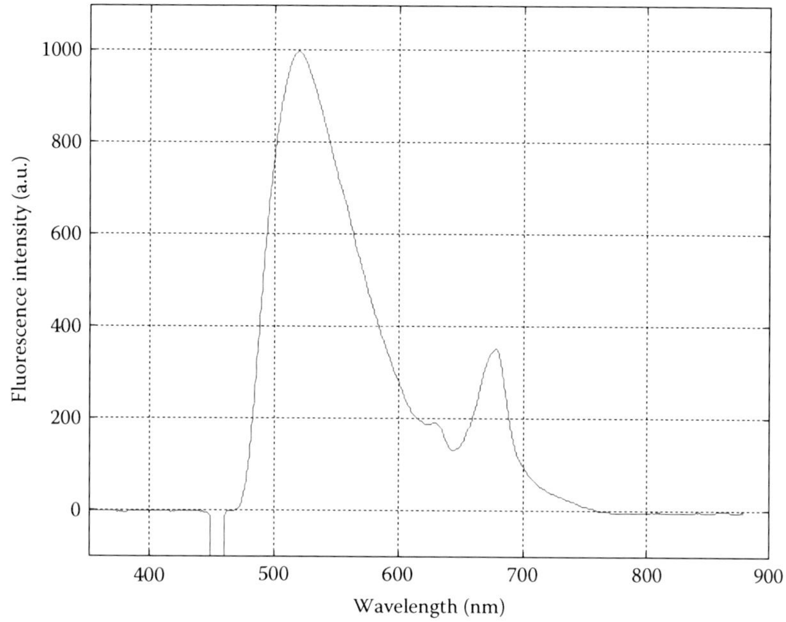

FIGURE 7.15
An average of ONF spectra of *G. lamblia* for samples 1–111 days old. The excitation wavelength is 458 nm and the power out of the excitation fiber is 10 mW.

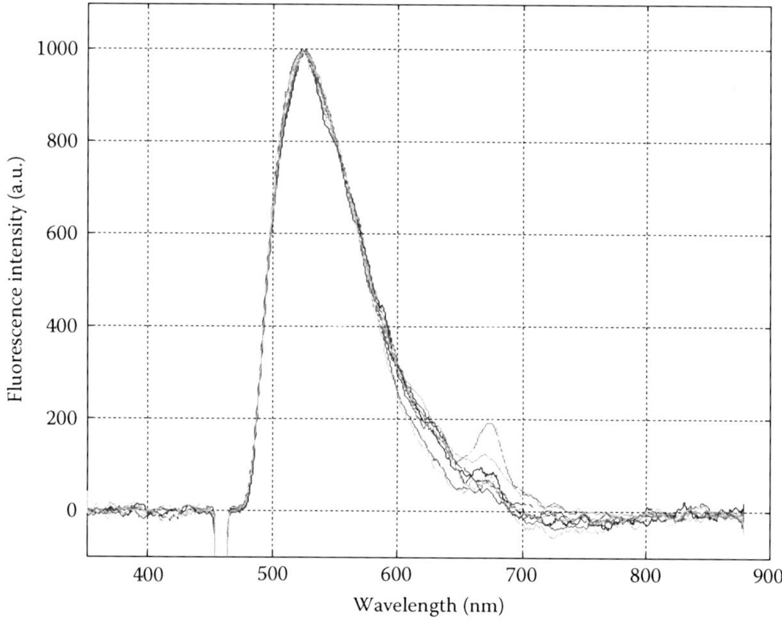

FIGURE 7.16
Comparison of the ONF spectra of *G. lamblia* for samples stored at refrigerator temperature until it was 45 days old and then aged over night at room temperature. The excitation wavelength is 458 nm and the power out of the excitation fiber is 10 mW.

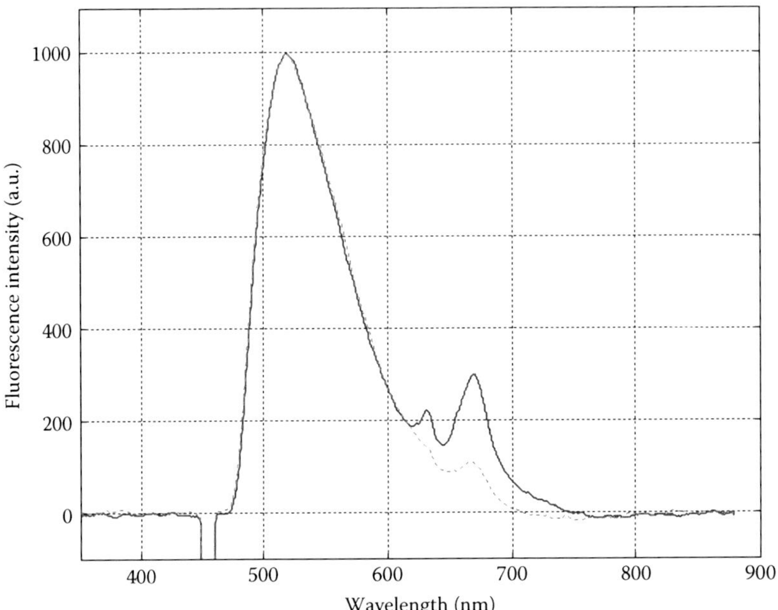

FIGURE 7.17
Comparison of the averaged ONF spectra of *G. lamblia* for a fresh sample and a sample aged over night at room temperature. The excitation wavelength is 458 nm and the power out of the excitation fiber is 10 mW.

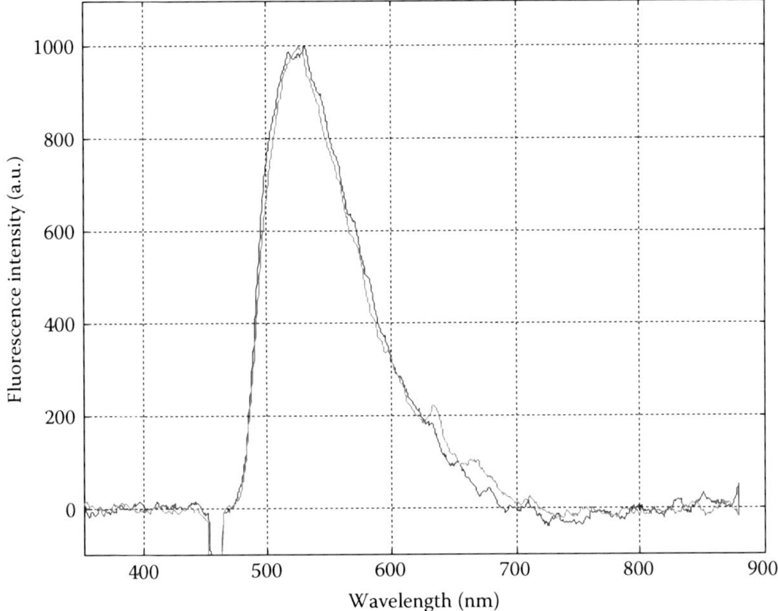

FIGURE 7.18
Comparison of the ONF spectra of *G. lamblia* for a sample 1 day old. The excitation wavelength is 458 nm and the power out of the excitation fiber is 20 mW.

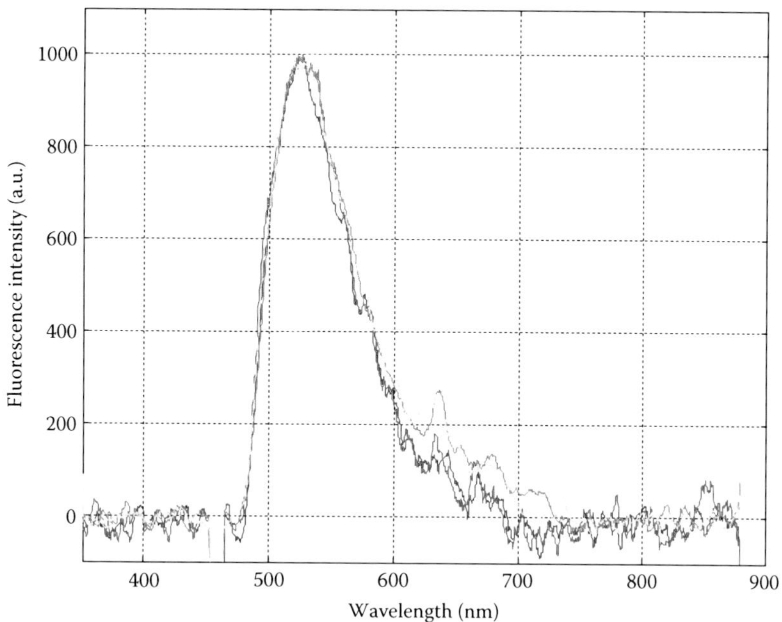

FIGURE 7.19
Comparison of the ONF spectra of *G. lamblia* for a sample 1 day old. The excitation wavelength is 458 nm and the power out of the excitation fiber is 5 mW.

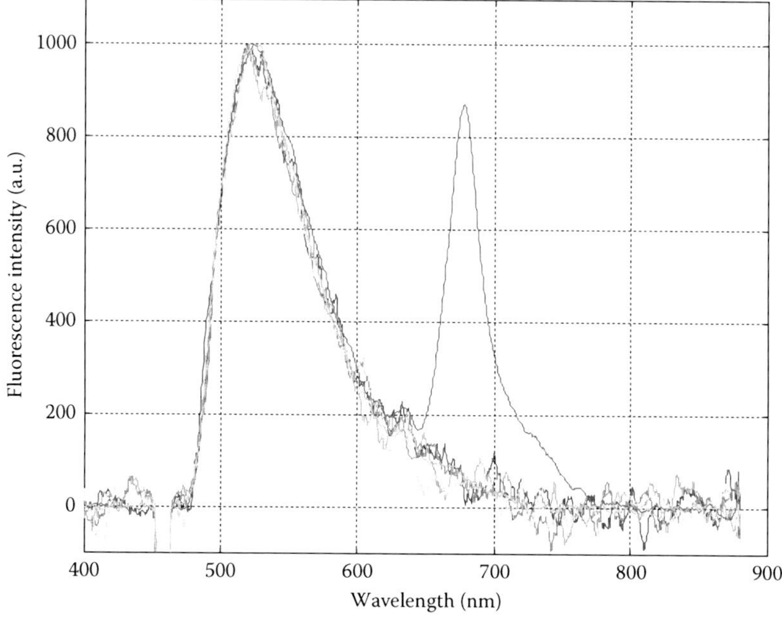

FIGURE 7.20
Comparison of the ONF spectra of *G. lamblia* for a sample 1 day old. The excitation wavelength is 458 nm and the power out of the excitation fiber is 2.5 mW.

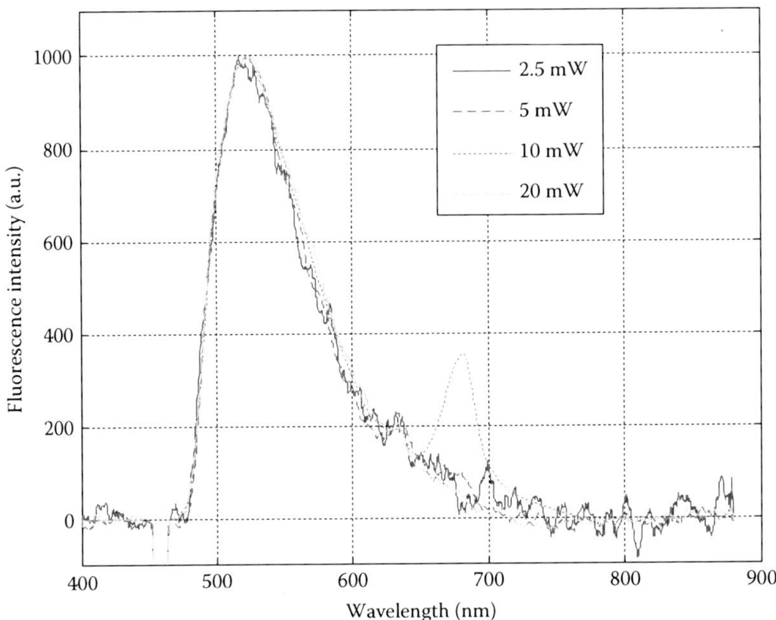

FIGURE 7.21
Comparison of the averaged ONF spectra of *G. lamblia*. The power out of the excitation fiber is 2.5 (solid), 5 (dash dot), 10 (dotted), and 20 mW (dashed) excited at 458 nm.

7.4.2 ONF Spectra of Single *G. lamblia* Cyst Excited at 458 nm (Argon-Ion Laser Line)

The ONF spectra from a single cyst are measured in the same manner as that from a cluster of cysts. Although it is easier to get higher emission intensity from a cluster of *G. lamblia* cysts, it is equally important to study the ONF from a single *G. lamblia* cyst in order to estimate the detection limit of the experimental setup. The signal-to-noise ratio of a single cyst spectrum will serve this purpose. In addition, it is necessary to compare the spectral shape of the single cyst with that of a cluster to ensure that the ONF spectrum does not depend on the number of cysts excited.

7.4.2.1 Pf = 10 mW

Figure 7.22 shows a group of ONF spectra from a single young *G. lamblia* cyst that was 1 and 4 days old. Likewise, Figure 7.23 shows for cyst that was 12 and 14 days old; Figure 7.24 shows for cyst that was 26 and 44 days old; Figure 7.25 shows for cyst that was 60 and 111 days old. Each of these figures indicates that, similar to a cluster of cysts, single *G. lamblia* cyst at each age gives reproducible ONF in shape under 458 nm excitation.

The average spectrum from a single cyst that was 1–4 days old is compared with that for 12–14-day-old cyst, as shown in Figure 7.26. Likewise, the average spectrum for 1–4-day-old cyst is compared with that for 60–111-day-old cyst, as shown in Figure 7.27. These tests indicate that, like a cluster of cysts, single cyst emission is independent of age.

Consequently, the data from single *G. lamblia* cyst of all ages may be averaged together, as shown in Figure 7.28. The resulting spectrum has an FWHM of 84 nm and a peak at

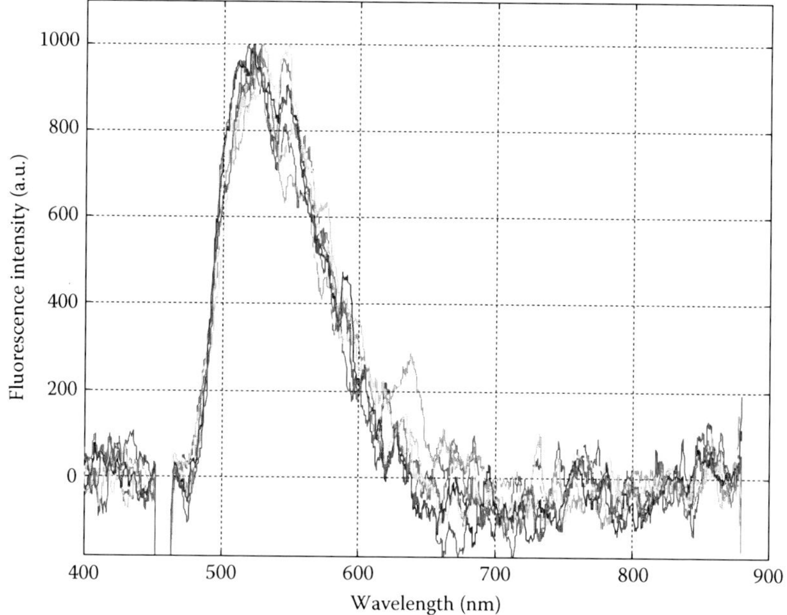

FIGURE 7.22
Comparison of ONF spectra of single *G. lamblia* for samples 1 and 4 days old. The excitation wavelength is 458 nm and the power out of the excitation fiber is 10 mW.

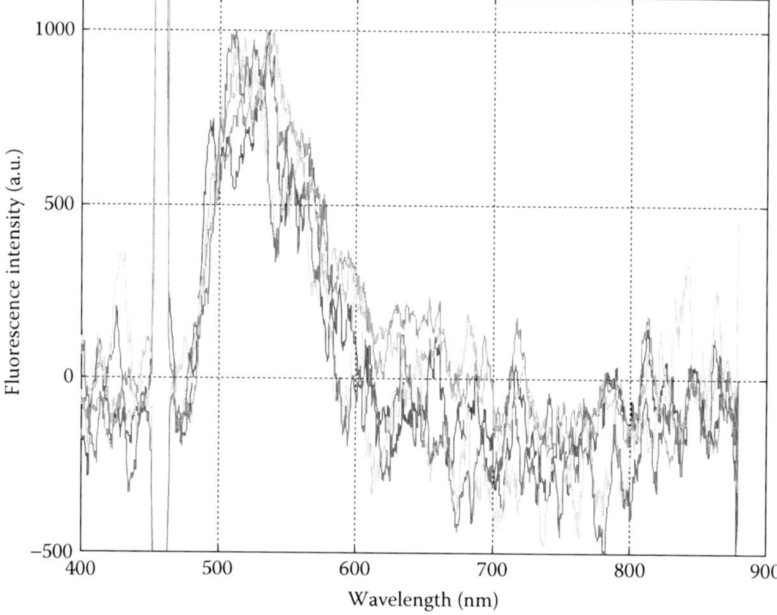

FIGURE 7.23
Comparison of ONF spectra of single *G. lamblia* for samples 12 and 14 days old. The excitation wavelength is 458 nm and the power out of the excitation fiber is 10 mW.

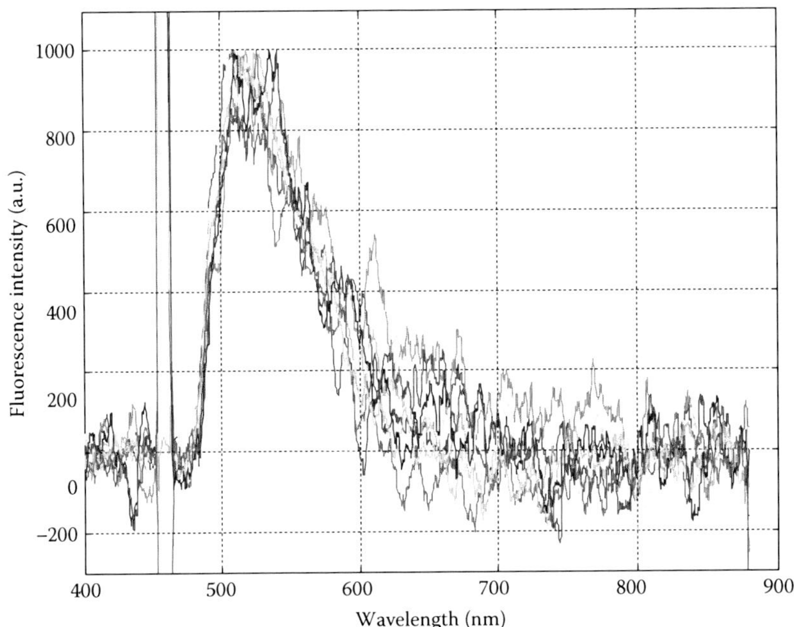

FIGURE 7.24
Comparison of ONF spectra of single *G. lamblia* for samples 26 and 44 days old. The excitation wavelength is 458 nm and the power out of the excitation fiber is 10 mW.

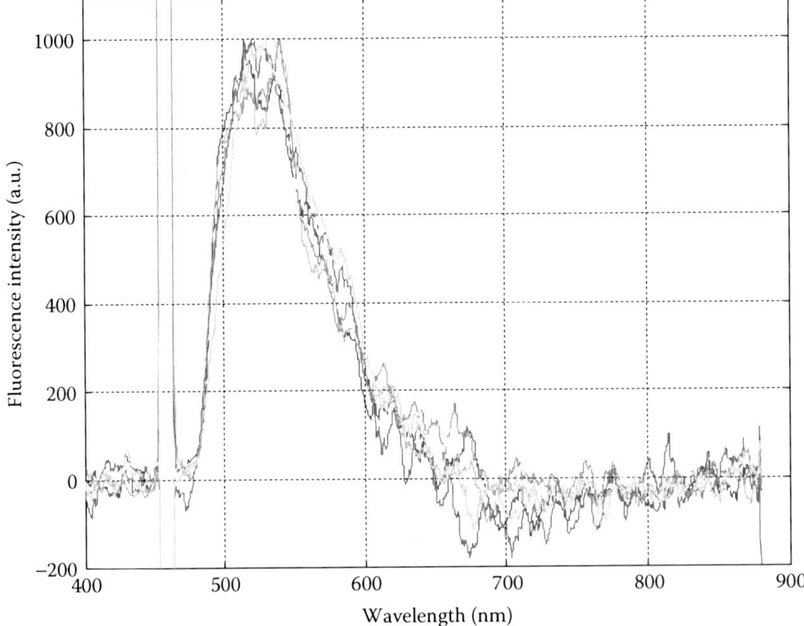

FIGURE 7.25
Comparison of ONF spectra of single *G. lamblia* for samples 60 and 111 days old. The excitation wavelength is 458 nm and the power out of the excitation fiber is 10 mW.

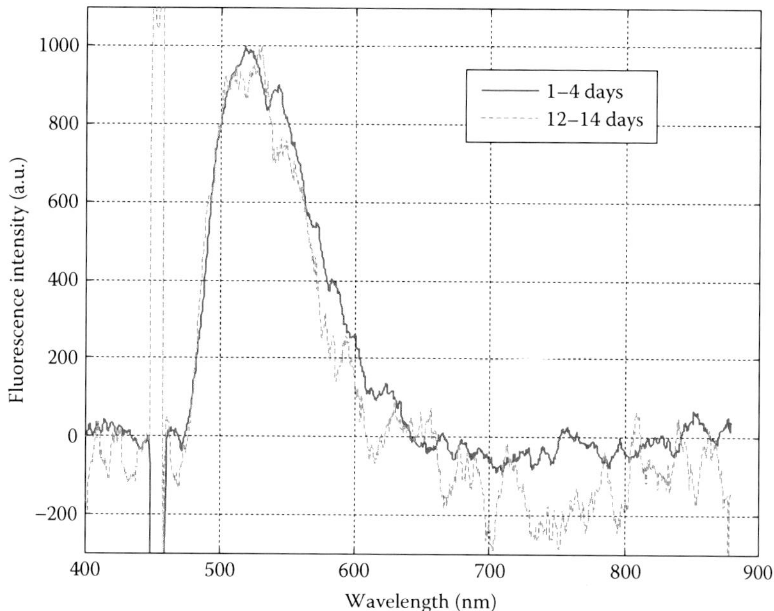

FIGURE 7.26
Comparison of the averaged ONF spectra of single *G. lamblia* for a sample 1–4 days old (solid) and the other one 12–14 days old (dash dot). The excitation wavelength is 458 nm and the power out of the excitation fiber is 10 mW.

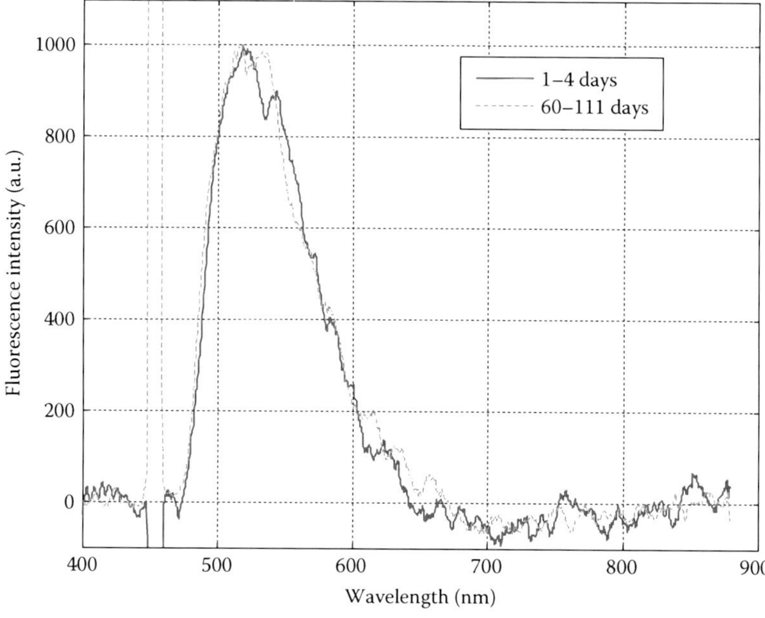

FIGURE 7.27
Comparison of the averaged ONF spectra of single *G. lamblia* for a sample 1–4 days old (solid) and the other one 60–111 days old (dash dot). The excitation wavelength is 458 nm and the power out of the excitation fiber is 10 mW.

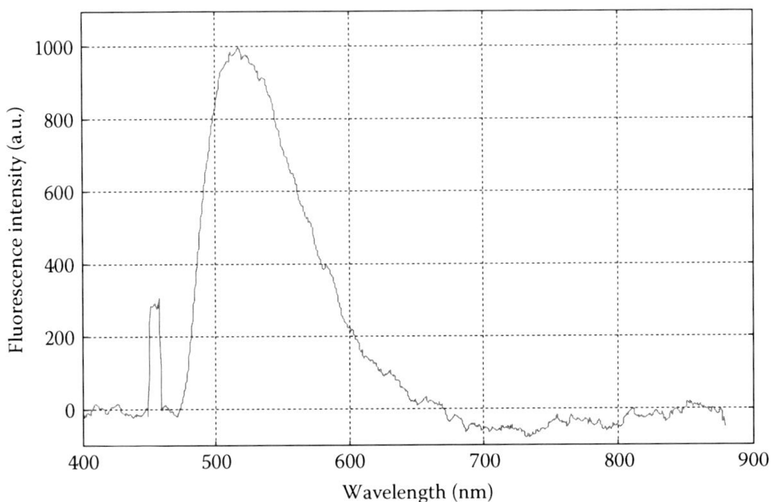

FIGURE 7.28
An average of ONF spectra of single *G. lamblia* for a sample 1–111 days old. The excitation wavelength is 458 nm and the power out of the excitation fiber is 10 mW.

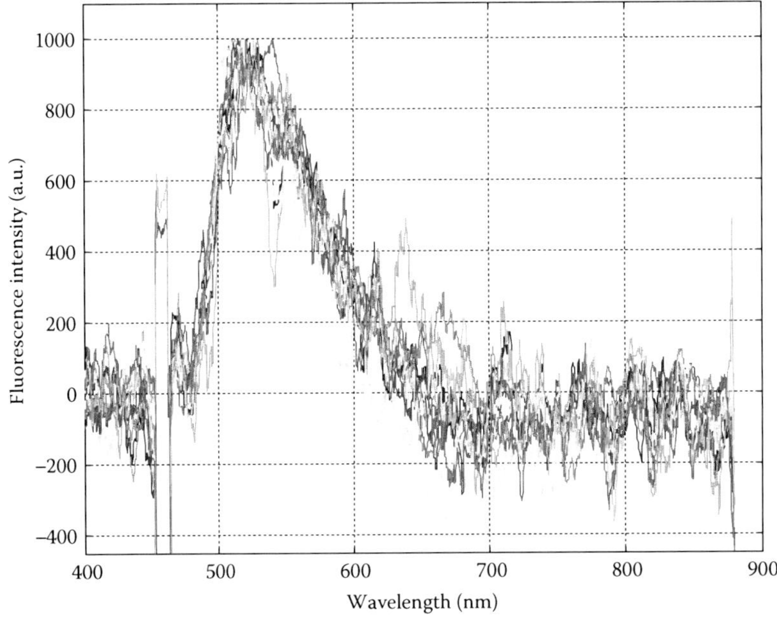

FIGURE 7.29
Comparison of the ONF spectra of single *G. lamblia* for a sample aged over night at room temperature. The excitation wavelength is 458 nm and the power out of the excitation fiber is 10 mW.

517 nm. Compared to the average of ONF from a cluster of cysts, the spectrum has an FWHM of 83 nm and a peak at 519 nm.

However, Figure 7.29 shows a group of ONF spectra from single *G. lamblia* cyst that was stored at room temperature over night. This figure shows that the ONF of aged single cyst is also reproducible in shape. The average spectrum from an aged cyst is then compared with that from a fresh cyst stored at refrigerator temperature, as shown in Figure 7.30.

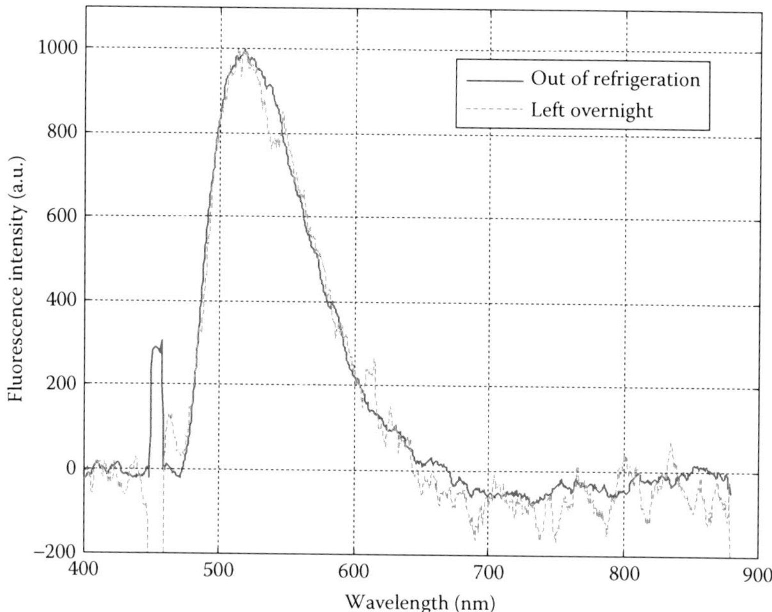

FIGURE 7.30
Comparison of the averaged ONF spectra of single *G. lamblia* for a fresh sample (solid) and a sample aged over night at room temperature (dash dot). The excitation wavelength is 458 nm and the power out of the excitation fiber is 10 mW.

Like that from a cluster of cysts, the spectral shape of ONF from a single *G. lamblia* cyst is not dependent on its age.

7.4.2.2 Pf = 2.5 mW

Figure 7.31 shows a group of ONF spectra from a single *G. lamblia* cyst that was 58 days old. Because the spectra are reproducible in shape, they can be averaged together. This average spectrum is then compared to that from a cluster of *G. lamblia* cysts, as shown in Figure 7.32.

7.4.3 ONF Spectra of a Cluster of *G. lamblia* Cysts Excited at 401 nm (Blue Diode Laser)

The 401 nm emission of a blue diode laser was used to excite the ONF spectra from *G. lamblia* of different ages. The power out of the excitation fiber was 2.2–2.5 mW and the LPF used was Corning 3-72. The spectral data are processed in the same way as those obtained through 458 nm excitation.

Figure 7.33 shows a group of ONF spectra from *G. lamblia* that were 32 and 36 days old. The spectrum has an FWHM of 79 nm approximately and a primary peak at 497 nm. The secondary peaks at 678 and 636 nm are again assigned to the pigments of algal impurities in the *G. lamblia* sample. These peaks basically agree with their counterparts at 661 and 637 nm in the data from 458 nm excitation. This figure shows that the ONF spectra of *G. lamblia* of the same age are reproducible in shape under 401 nm excitation. Thus, these spectra can be averaged together.

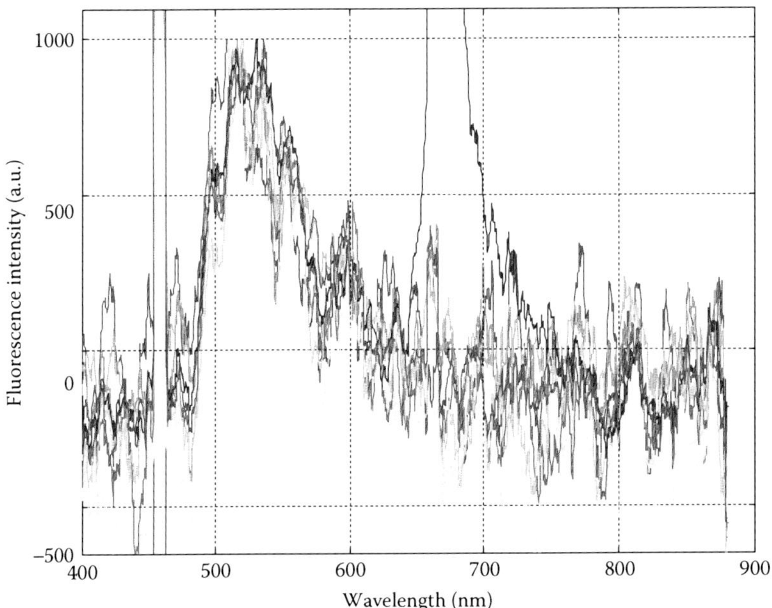

FIGURE 7.31
Comparison of the ONF spectra of single *G. lamblia*. The excitation wavelength is 458 nm and the power out of the excitation fiber is 2.5 mW.

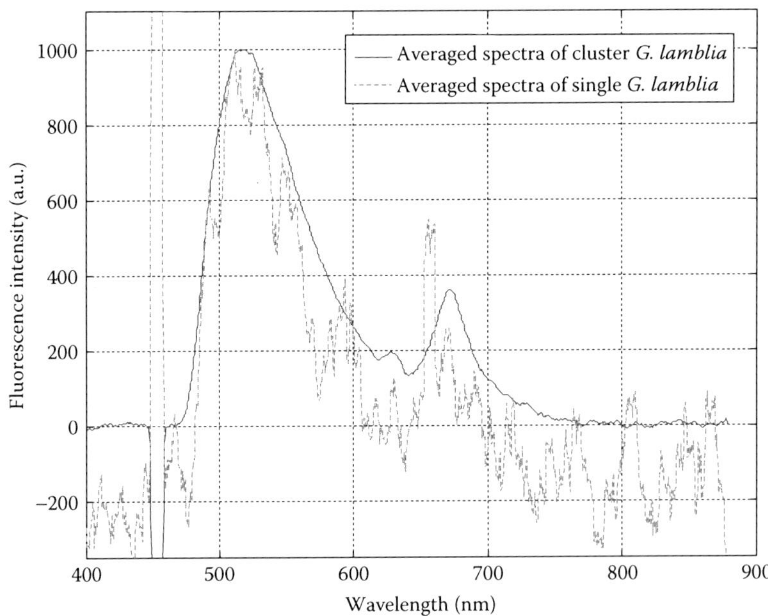

FIGURE 7.32
Comparison of averaged ONF spectra of cluster *G. lamblia* (solid) and single *G. lamblia* (dash dot). The excitation wavelength is 458 nm and the power out of the excitation fiber is 2.5 mW.

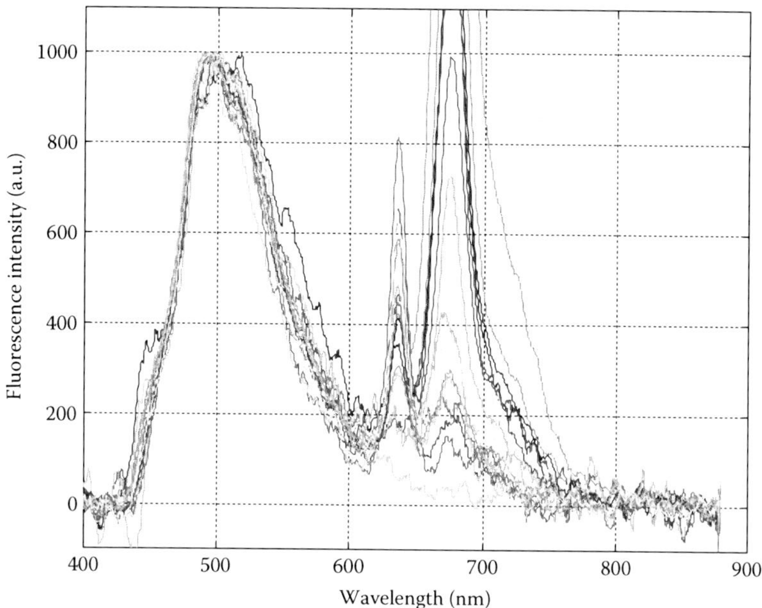

FIGURE 7.33
Comparison of ONF spectra of *G. lamblia* for samples 32–36 days old. The excitation wavelength is 401 nm and the power out of the excitation fiber is 2.2–2.5 mW.

Likewise, Figure 7.34 shows a group of ONF spectra from *G. lamblia* that were 27 days old. The spectra are again reproducible in shape. They can also be averaged together. Figure 7.35 then plots the average of the two set of data together. Because the two average spectra agree in shape, all the raw data may be averaged, as shown in Figure 7.36.

7.4.4 ONF Spectra of Single *G. lamblia* Cyst Excited at 401 nm (Blue Diode Laser)

Figure 7.37 shows a group of ONF spectra of single *G. lamblia* cyst that was 32 and 36 days old, respectively. This spectrum shows that the ONF spectra of single *G. lamblia* of the same age are also reproducible in shape under 401 nm excitation. Thus, these data can be averaged together.

Figure 7.38 shows a group of ONF spectra of single *G. lamblia* cyst that was 91 days old. These data are also reproducible in shape and can be averaged.

The average spectra of these two sets of data are then compared in Figure 7.39. The result shows that the aging of single *G. lamblia* cyst does not change its ONF spectral shape under 401 nm excitation. This result is similar to the 458 nm excitation result. Because all of the single cyst data (from 401 nm excitation) overlap in shape, they can be averaged together, as shown in Figure 7.40.

7.4.5 ONF Spectra of *G. lamblia* Cysts Excited at 496 nm (Argon-Ion Laser Line)

When the excitation wavelength was 496 nm, the power out of the excitation fiber was varied between 10 and 2.5 mW. The LPF used was Schott OG 515. The spectral data are processed in the same way as those obtained through 458 nm excitation.

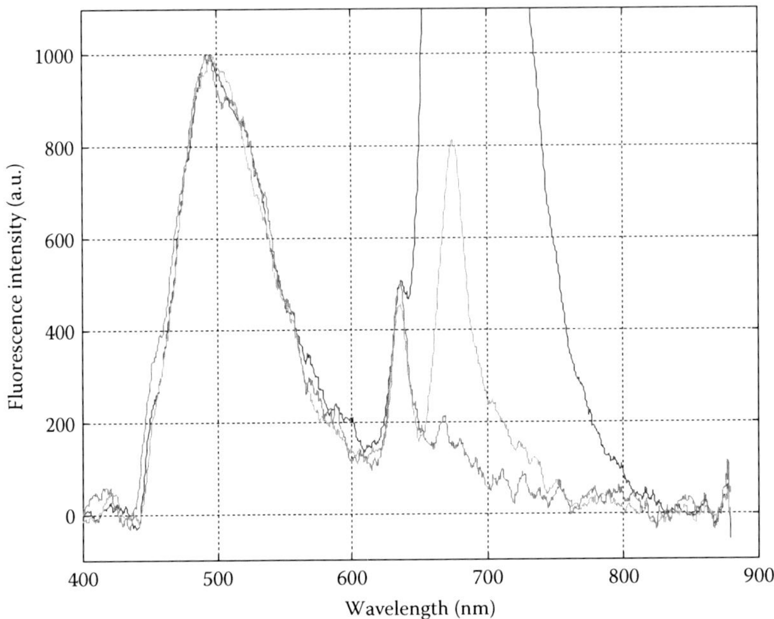

FIGURE 7.34
Comparison of ONF spectra of *G. lamblia* for samples 27 days old. The excitation wavelength is 401 nm and the power out of the excitation fiber is 2.2–2.5 mW.

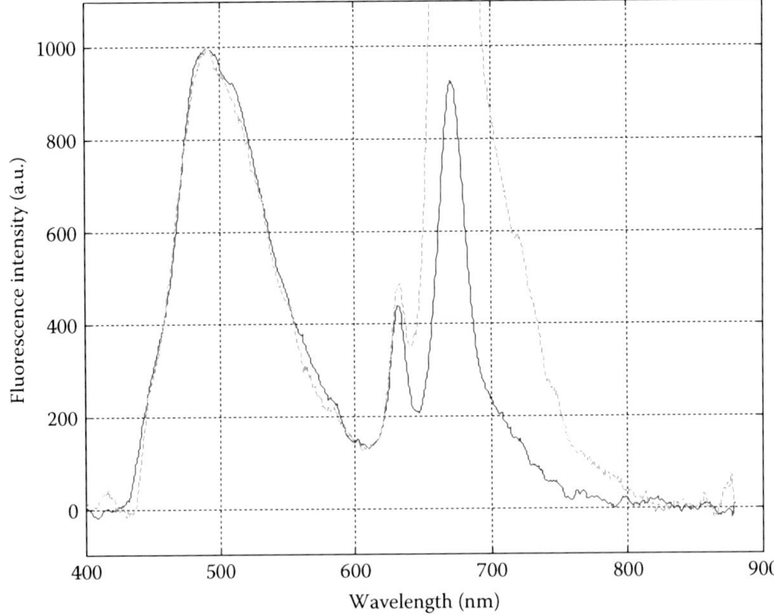

FIGURE 7.35
Comparison of the averaged ONF spectra of *G. lamblia* for samples 32 and 36 days (solid) old and 27 days old (dash dot). The excitation wavelength is 401 nm and the power out of the excitation fiber is 2.2–2.5 mW.

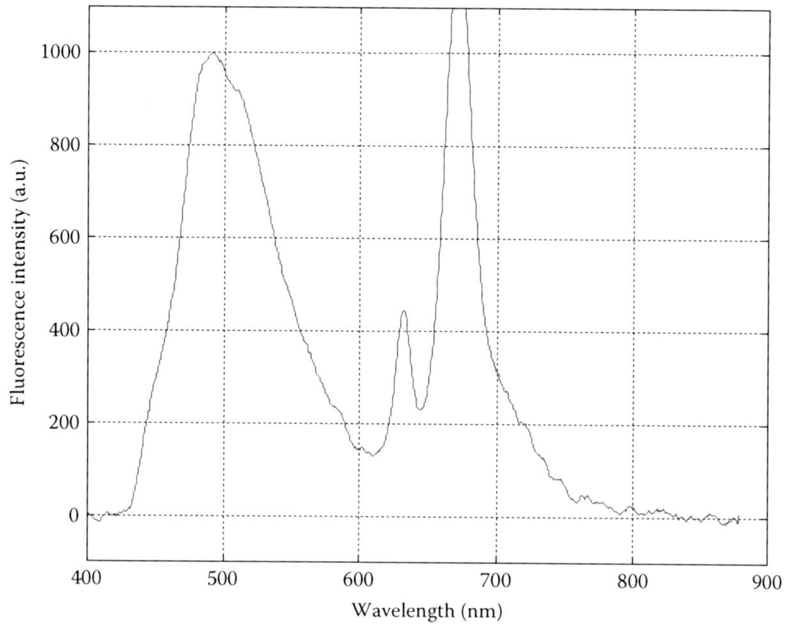

FIGURE 7.36
Averaged ONF spectra of *G. lamblia* for samples 32–36 days old and 27 days old. The excitation wavelength is 401 nm and the power out of the excitation fiber is 2.2–2.5 mW.

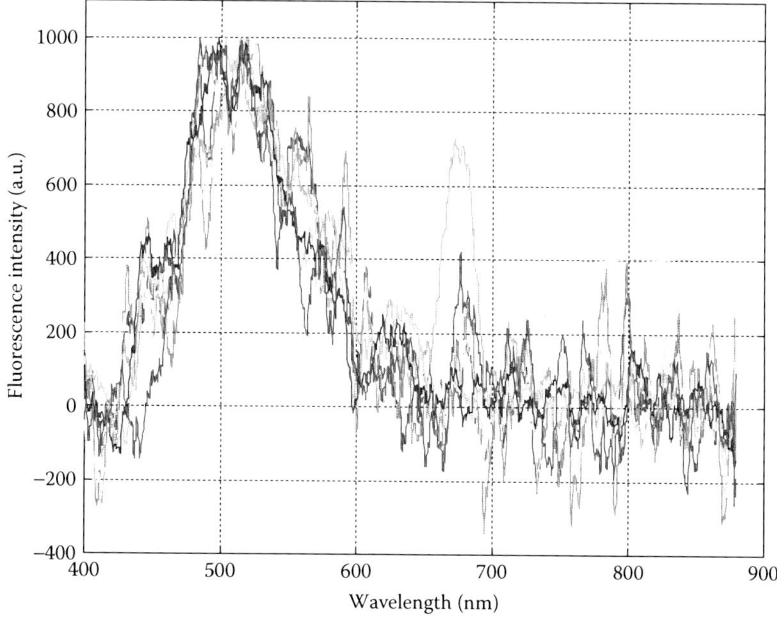

FIGURE 7.37
Comparison of ONF spectra of single *G. lamblia* for samples 32–36 days old. The excitation wavelength is 401 nm and the power out of the excitation fiber is 2.2–2.5 mW.

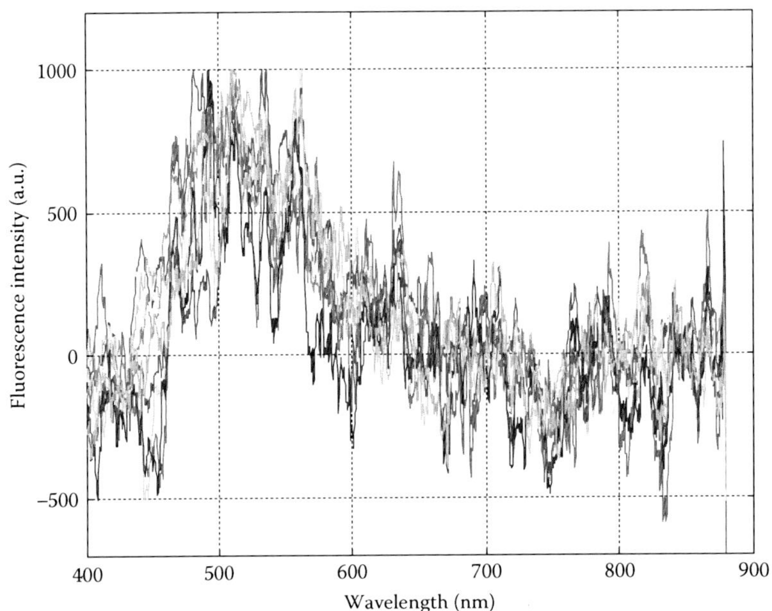

FIGURE 7.38
Comparison of ONF spectra of single *G. lamblia* for samples 90 days old. The excitation wavelength is 401 nm and the power out of the excitation fiber is 2.2–2.5 mW.

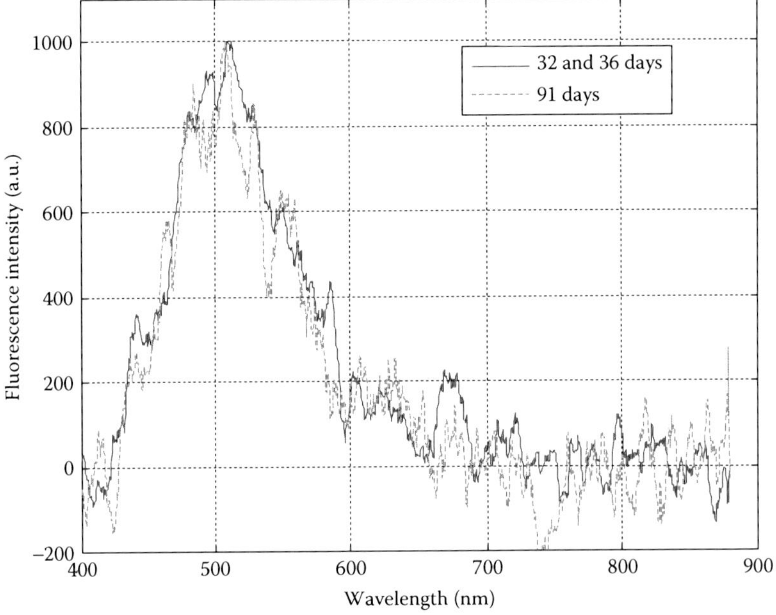

FIGURE 7.39
Comparison of the averaged ONF spectra of single *G. lamblia* for samples 32 and 36 days old (solid) and the other ones 91 days old (dash dot). The excitation wavelength is 401 nm and the power out of the excitation fiber is 2.2–2.5 mW.

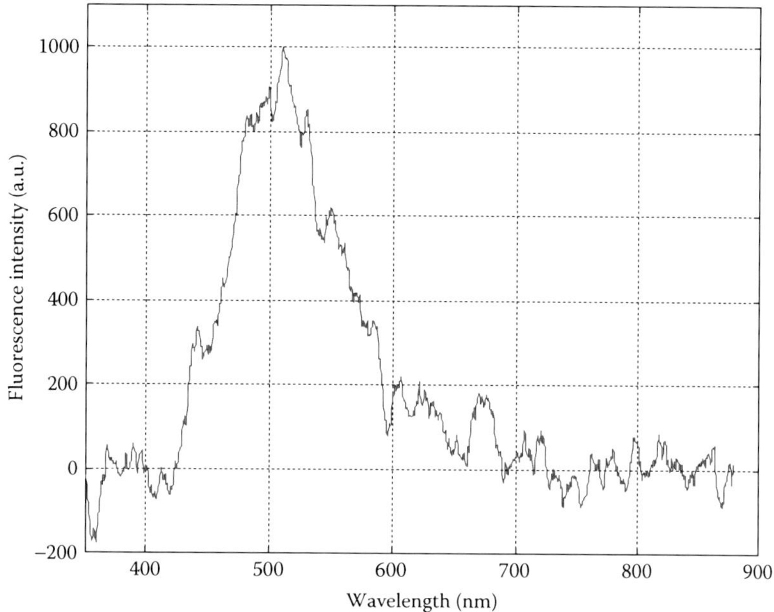

FIGURE 7.40
An averaged of the ONF spectra of single *G. lamblia* for samples 32 and 36 days old and the other ones 91 days old. The excitation wavelength is 401 nm and the power out of the excitation fiber is 2.2–2.5 mW.

In Figure 7.41, a group of ONF spectra of *G. lamblia* under 496 nm excitation are plotted. The approximate age of the sample was 2 days. The power out of the fiber was also set at 10 mW. This spectrum shows that the ONF of *G. lamblia* is reproducible in shape under 496 nm excitation. Thus, all the data in Figure 3.34 can be averaged. The result is shown in Figure 7.42. The average spectrum has an FWHM of 88 nm approximately and a primary peak at 646 nm. The secondary peaks that appear at approximately 666 and 631 nm are also attributed to pigments of algal impurities in the *G. lamblia* sample.

Under a lower excitation power of 2.5 mW, a group of ONF spectra of *G. lamblia* were measured and the data plotted in Figure 7.43. The approximate age of the sample was 5 days. These data are again reproducible in shape. Thus, they can be averaged and compared with the data obtained at higher excitation power. Figure 7.44 plots these averages together. The result shows that the excitation power does not affect the shape of ONF spectra under 496 nm excitation.

7.4.6 Miscellaneous Results of the Observed ONF Spectra of *G. lamblia* Cysts

Here we present very useful miscellaneous results of ONF spectra for *G. lamblia* cysts based on the changes of different excitation wavelengths. Also, we show the ONF data for a cluster of *G. lamblia* cysts.

1. Figure 7.45 compares the ONF spectra of *G. lamblia* when the excitation wavelength is changed from 458 to 401 nm. The individual data are divided by the transmission spectrum of the LPFs. Likewise, Figure 7.46 compares the ONF spectra of *G. lamblia* when the excitation wavelength is changed from 458 to 496 nm. Figure 7.47 shows all of three excitation wavelengths (401, 458, and 496 nm) together.
2. Some typical ONF data from a cluster of *G. lamblia* cysts, as well as their associated background spectra, are shown in Figure 7.48 for two excitation wavelengths

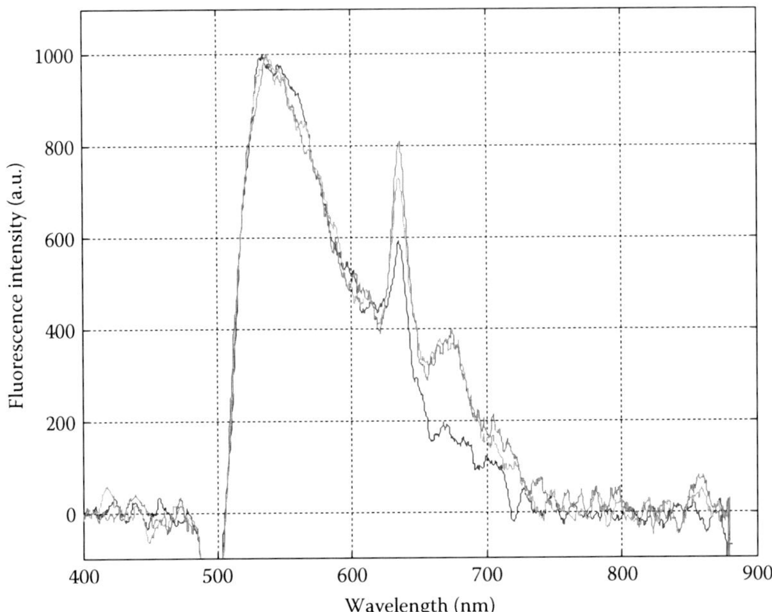

FIGURE 7.41
Comparison of ONF spectra of *G. lamblia* for samples 2 days old. The excitation wavelength is 496 nm and the power out of the excitation fiber is 10 mW.

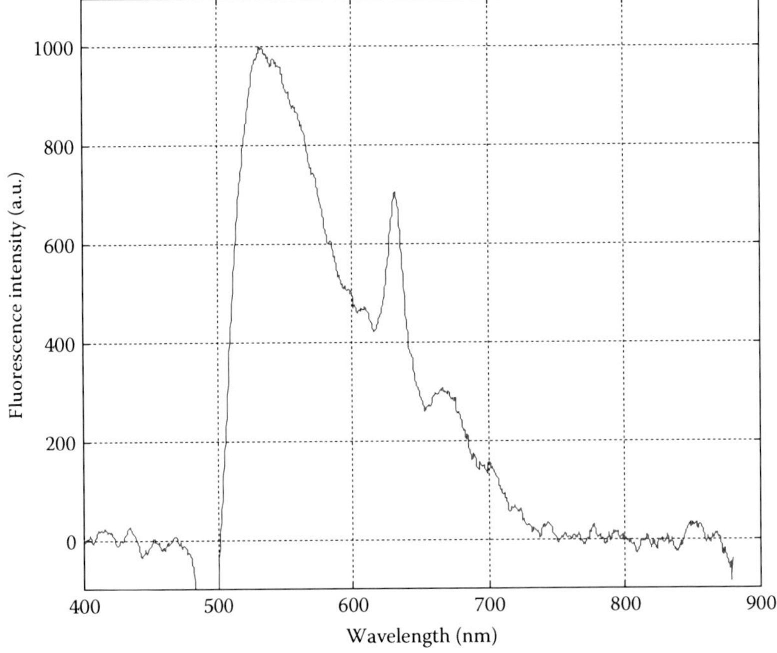

FIGURE 7.42
An averaged of ONF spectra of *G. lamblia* for samples 2 days old. The excitation wavelength is 496 nm and the power out of the excitation fiber is 10 mW.

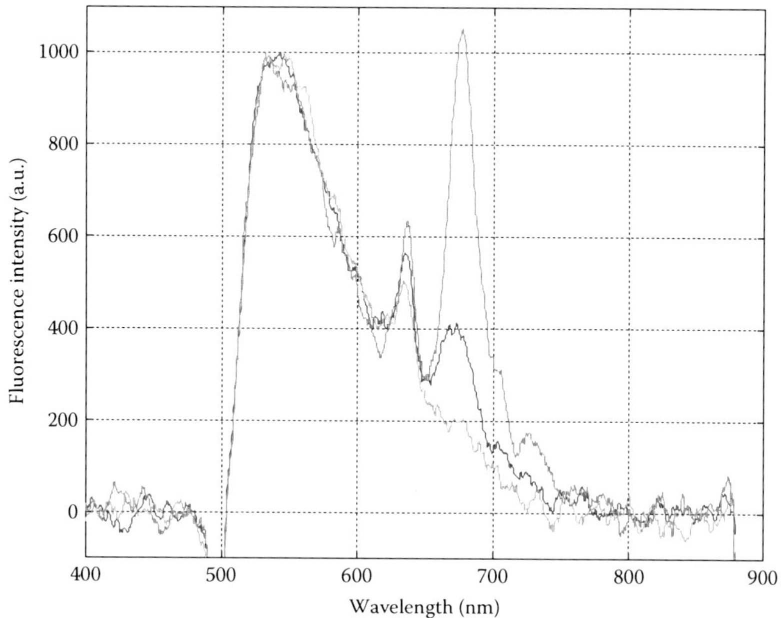

FIGURE 7.43
Comparison of ONF spectra of *G. lamblia* for samples 5 days old. The excitation wavelength is 496 nm and the power out of the excitation fiber is 2.5 mW.

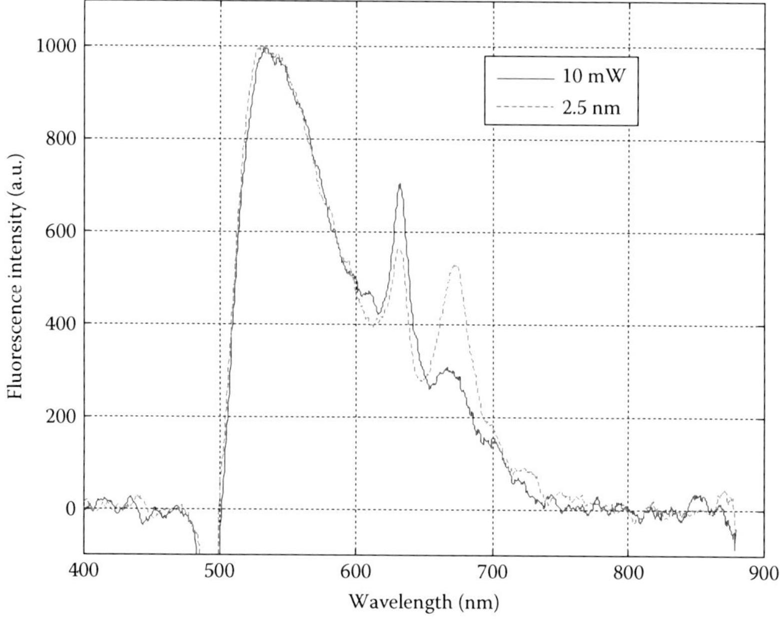

FIGURE 7.44
Comparison of the averaged ONF spectra of *G. lamblia*. When the power out of the excitation fiber is 10 (solid) and 2.5 (dashed) mW excited at 496 nm.

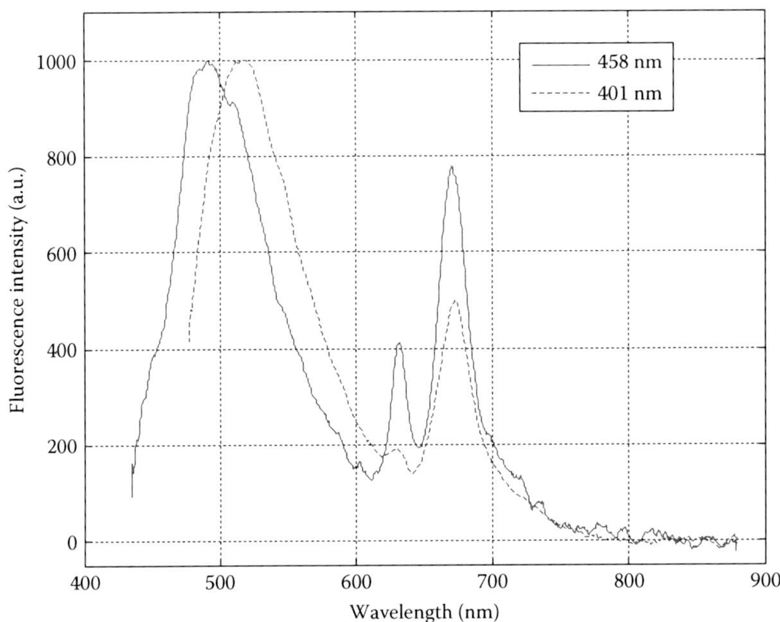

FIGURE 7.45
Comparison of the ONF spectra of *G. lamblia* when the excitation wavelength changes from 458 (dashed) to 401 (solid) nm. The individual data have been divided by the transmission spectrum of the LPFs, respectively.

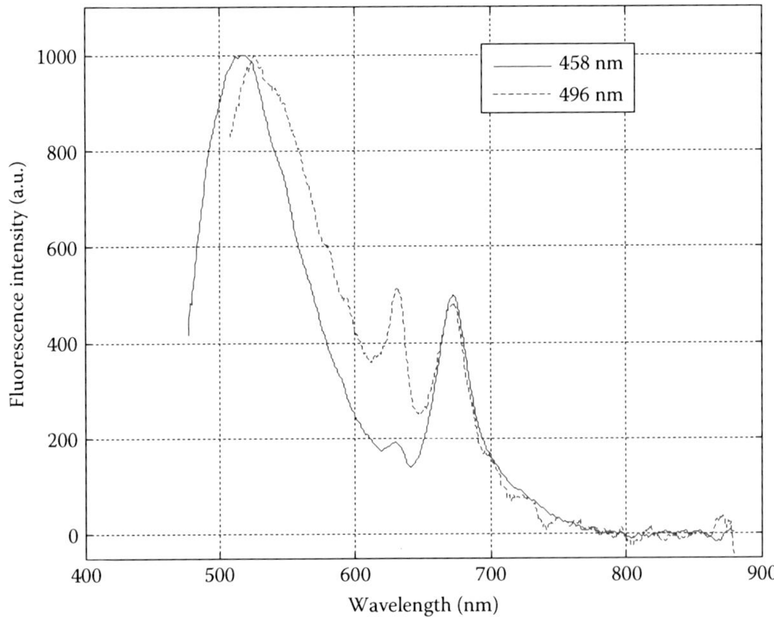

FIGURE 7.46
Comparison of the ONF spectra of *G. lamblia* when the excitation wavelength changes from 458 (solid) to 496 (dashed) nm. The individual data have been divided by the transmission spectrum of the LPCs, respectively.

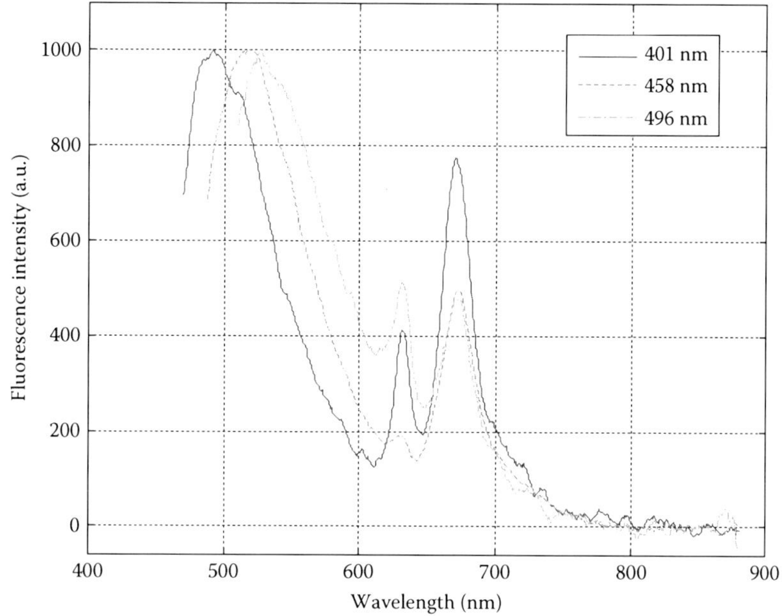

FIGURE 7.47
Comparison of the ONF spectra of *G. lamblia* when the excitation wavelength 401 nm (solid), 458 nm (dashed dot) nm, and 496 nm (dotted). The individual data have been divided by the transmission spectrum of the LPFs respectively.

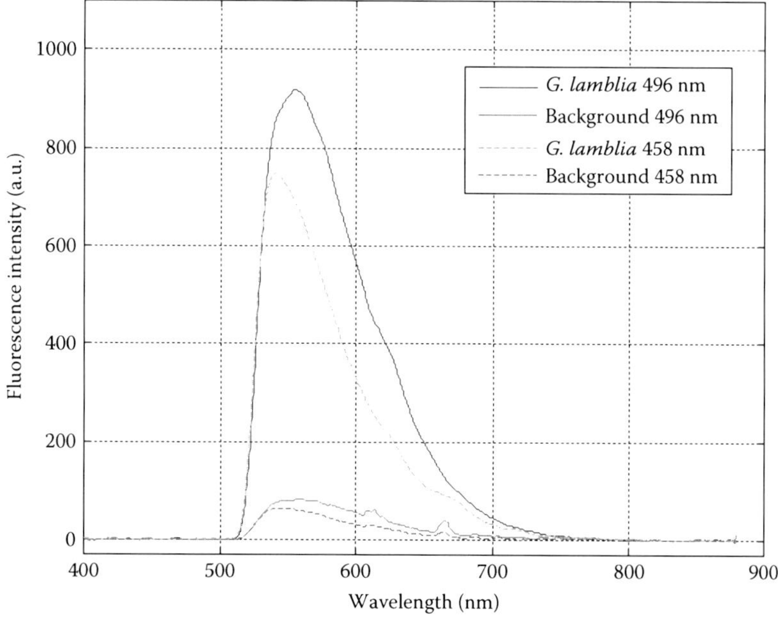

FIGURE 7.48
Comparison of the ONF spectra of *G. lamblia* and their associated background spectra for two excitation wavelengths 458 and 496 nm.

at 458 and 496 nm. This figure gives an example of the magnitude of *G. lamblia* fluorescence intensity compared to the sample background.

3. Figure 7.49 shows a group of background spectra of the *G. lamblia* sample at 458 nm excitation, when the power out of the excitation fiber was set at 10 mW. Figure 7.50 shows the ONF spectra of the background of *G. lamblia* under 401 nm excitation with the power out of the excitation fiber at 2.2–2.5 mW.
4. The excitation power was changed to check the linearity of the fluorescence data. As shown in Figure 7.51, the integrated fluorescence intensity of *G. lamblia* is linear with respect to the laser power, which confirms that the excitation intensity used is low enough.
5. A pair of cross-polarizers were used as an alternative method to reject laser scattering, which was assumed to be more polarized than the spontaneous emission of fluorescence. When one polarizer polarizes the laser output from the excitation fiber, another one was aligned at 90° in front of the collection fiber so that the laser scattering reaching the detector was minimized. The transmission spectrum of these linear polarizers was measured and shown in Figure 7.52.

The ONF spectrum of *G. lamblia* obtained through the above method is then compared with that obtained from the standard method (LPF), as shown in Figure 7.53. This figure shows that under 458 nm excitation, there is a reduction of the ONF intensity on the blue side of the spectrum by the LPF GG 495. The distortion is noticeable but insignificant for the shape of the spectrum.

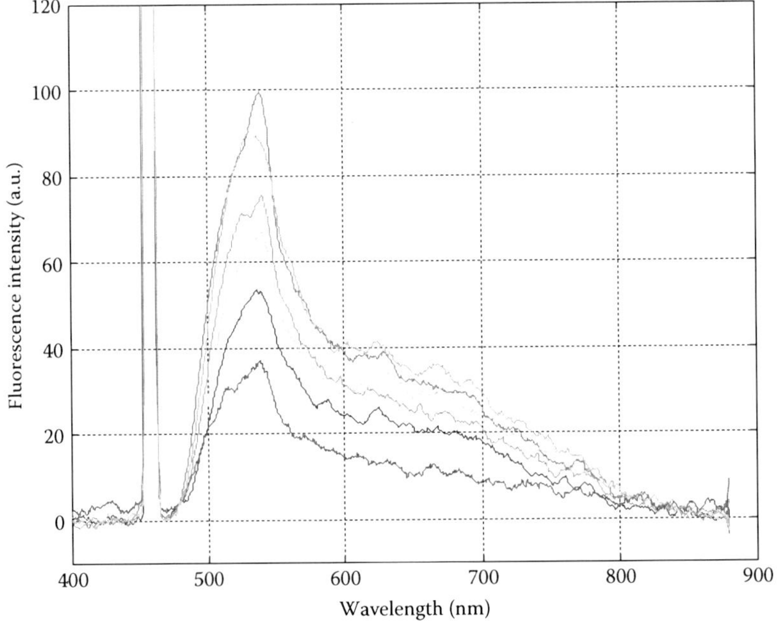

FIGURE 7.49
A group of background spectra of *G. lamblia* sample at 458 nm excitation with power out of the excitation fiber of 10 mW.

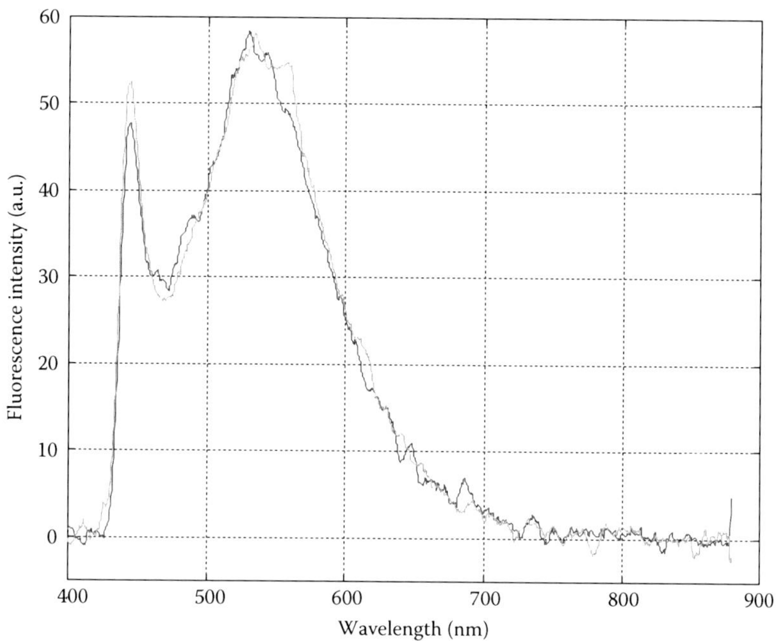

FIGURE 7.50
A group of background spectra of *G. lamblia* sample at 458 nm excitation with power out of the excitation fiber is 10 mW.

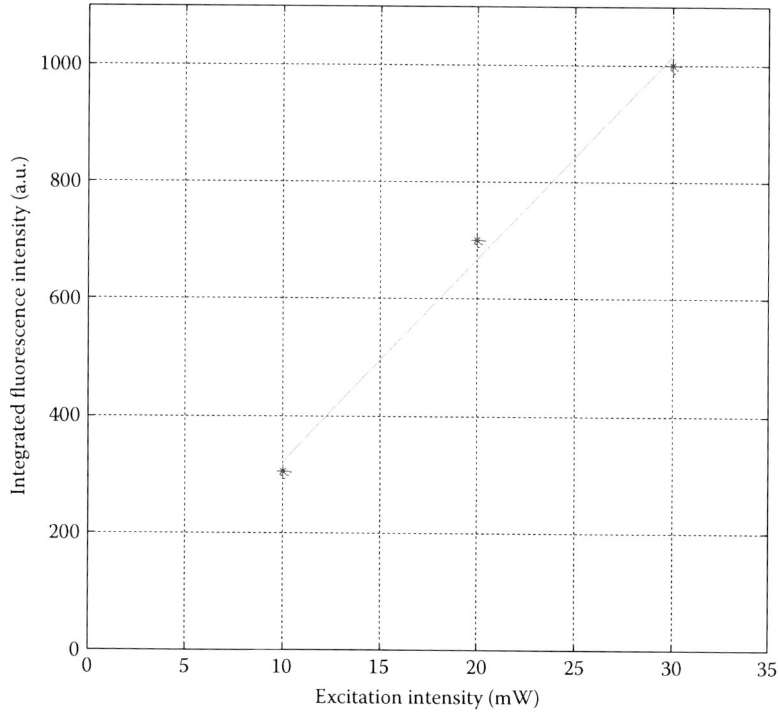

FIGURE 7.51
Integrated ONF intensity of *G. lamblia* (linear) with respect to the laser power.

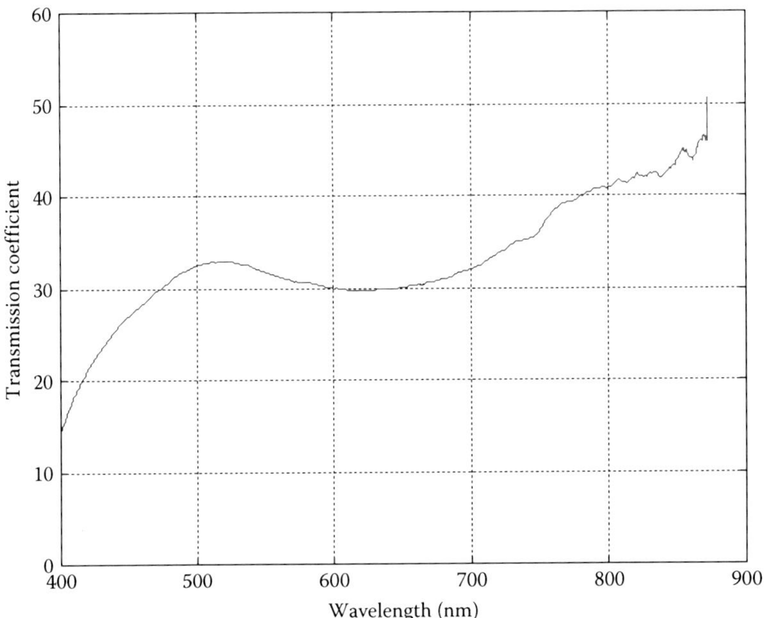

FIGURE 7.52
Measured transmission spectrum of the linear polarizer.

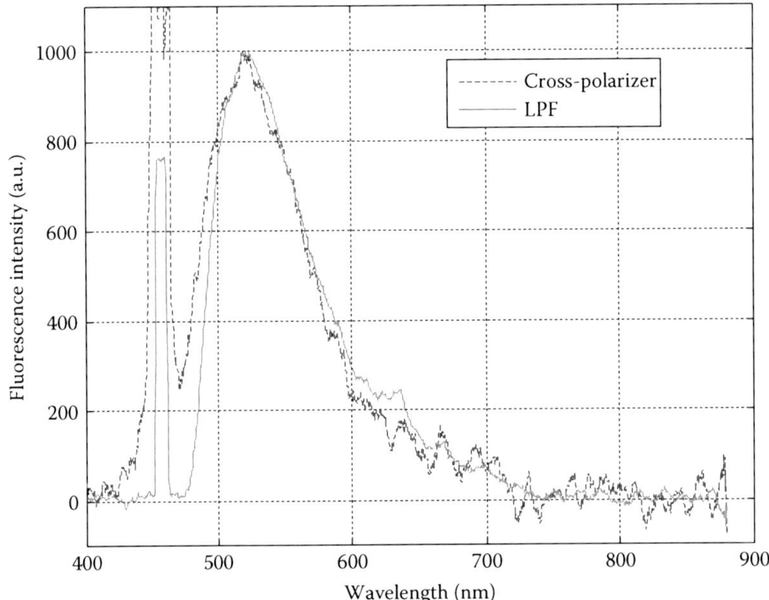

FIGURE 7.53
Comparison of the ONF spectra of *G. lamblia* sample using cross polarizers method and the LPF method.

7.5 Conclusion

This chapter reports the ONF spectra of *G. lamblia* cysts excited and detected using new experimental setup. The experimental setup above was successfully used to study the ONF of *G. lamblia* cysts. The main components consist of optical multimode fibers used as waveguides for illumination and detection, a highly sensitive spectrometer, LPFs, and suitable CW blue excitation sources.

The laser excitations used are of blue wavelength (458 and 401 nm). At 458 nm excitation, the peak of the ONF for *G. lamblia* is located at 525 nm; at 401 nm excitation, the peak of the ONF is located at 496 nm; and finally, at 496 nm excitation, the peak of the ONF is located at 532 nm approximately.

References

1. G. F. Craun, "Waterborne outbreaks of giardiasis: Current status," In: *Giardia and Giardiasis*, S. L. Erlandsen and E. A. Meyer Eds., Plenum Press, New York, 1984, pp. 243–261.
2. P. Willis and B. Hammond, *Advances in Giardia Research*, University of Calgary Press, Calgary, Canada, 1987.
3. A. Ashendorff, M. Principe, A. Seeley, J. Laduca, L. Beckhardt, W. Faber, and J. Mantus, "Watershed protection for New York City's supply," *J. Am. Water Works Assoc.* 89, 75–88, 1997.
4. G. F. Craun, "Surface water supplies and health," *J. Am. Water Works Assoc.* 80, 40–52, 1988.
5. S. L. Erlandsen and E. A. Meyer, Giardia *and Giardiasis*, Plenum press, New York, 1984.
6. G. J. Tortora, B. R. Funke, and C. L. Case, *Microbiology: An Introduction*, 4th ed., Benjamin Cumming, Redwood City, CA, 1992, pp. 313–319.
7. M. W. LeChevallier, W. D. Norton, and T. B. Atherholt, "Protozoa in open reservoirs," *J. Am. Water Works Assoc.* 89, 84–96, 1997.
8. S. T. Bagley, M. T. Auer, D. A. Stern, and M. J. Babiera, "Sources and fate of *Giardia* cysts and cryptosporidium oocysts in surface waters," *J. Lake Reserv. Manag.* 14, 379–392, 1998.
9. M. W. Lechevallie et al., Giardia *and* Cryptosporidium *in Water Supplies*, AWWARF, Denver, CO, 1991.
10. D. A. Stern, "Monitoring for *Cryptosporidium* spp. and *Giardia* spp. and human enteric viruses in the watersheds of the New York City water supply system," *Proceedings, Watershed'96 Moving Ahead Together Technical Conference and Exposition*, Water Environment Federation, Baltimore, MD, June 8–12, 1996.
11. G. Vesey et al., "Application of flow cytometry methods for the routine detection of *Cryptosporidium* and *Giardia* in water," *Cytometry* 16, 1–6, 1994.
12. G. Vesey et al., "Detection of specific microorganisms in environmental samples using flow cytometry," *Methods Cell Biol.* 42, 489, 1994.
13. R. Hoffman, J. Standridge, A. Prieve, J. Cucunato, and M. Bernhardt, "Using flow cytometry to detect protozoa," *J. Am. Water Works Assoc.* 89, 104–111, 1997.
14. I. C. Bacon et al., "Quantitative classification of *Cryptosporidium* oocysts and *Giardia* cysts in water using UV/VIS spectroscopy," *BIOS: Biomedical Optics Conference*, The Moscone Center, San Francisco, CA, 1995.
15. K. Patten et al., "Rapid methods for on-line detection of *Cryptosporidium* oocysts and *Giardia* cysts," *Proceedings of AWWA WQTC*, San Francisco, CA, 1994.
16. D. L. Cruz, A. Armah, and M. Sivaganesan, "Detection of *Giardia* and *Cryptosporidium* spp. in source water samples by commercial enzyme immunoassay kits," *Proceedings of AWWA WQTC*, San Francisco, CA, 1994.

17. C. L. Mayer and C. J. Palmer, "Evaluation of PCR, nested PCR, and fluorescent antibodies for detection of *Giardia* and *Cryptosporidium* species in wastewater," *Appl. Environ. Microbiol.* 62, 2081, 1996.
18. W. Jakubowski, S. Boutros, W. Faber, R. Fayer, W. Ghiorse, M. LeChevallier, J. Rose, S. Schaub, A. Singh, and M. Stewart, "Environmental methods for *Cryptosporidium*," *J. Am. Water Works Assoc.* 87, 107–121, 1996.
19. C. Water and P. Hibler, "An overview of the techniques used for detection of *Giardia* cysts in surface water," In: *Advances in Giardia Research*, University of Calgary Press, Calgary, Canada, 1988, pp.197–204.
20. S. K. Stephen, J. L. Riggs, P. D. Dileanis, and T. J. Suk, "Isolation and detection of *Giardia* cysts from water using direct immunofluorescence," *J. Am. Water Resour. Assoc.* 22, 843–845, 1986.
21. J. L. Riggs, K. W. Dupuis, K. Nakamura, and D. P. Spath, "Detection of *Giardia lamblia* by immunofluorescence," *Appl. Environ. Microbiol.* 45, 698–700, 1983.
22. W. Jakubowski, Giardia *Methods Workshop*, Denver, CO, 1984, pp. 1–12.
23. M. W. LeChevallier et al. "Evaluation of the immunofluorescence procedure for detection of *Giardia* cysts and *Cryptosporidium* oocysts in water," *Appl. Environ. Microbiol.* 61, 690, 1995.
24. E. C. Nieminski, F. W. Schaefer III, and J. E. Ongerth, "Comparison of two methods for detection of *Giardia* cysts and *Cryptosporidium* oocysts in water," *Appl. Environ. Microbiol.* 61, 1714–1719, 1995.
25. Anon, "Development of performance evaluation sample preparation protocols for *Giardia* cysts and *Cryptosporidium* oocysts," US EPA Contract No. 68-C3-0365, USEPA Water Docket, 1996.
26. L. Thiriat, F. Sidaner, and J. Schwartzbrod, "Determination of *Giardia* cyst viability in environmental and faecal samples by immunofluorescence, fluorogenic dye staining and differential interference contrast microscopy," *Lett. Appl. Microbiol.* 26, 237–242, 1998.
27. R. Kingslake, *Applied Optics and Optical Engineering*, Vol. IV, Optical Instruments Part I, Academic Press, New York, 1967.
28. R. Kingslake, *Applied Optics and Optical Engineering*, Vol. III, Optical Components, Academic Press, New York, 1965.
29. D. Malacara, *Geometrical and Instrumental Optics*, Vol. 25, Methods of Experimental Physics, Academic Press, Boston, MA, 1988.
30. B. R. Dixon, M. Parenteau, C. Martineau, and J. Fournier, "A comparison of conventional microscopy, immunofluorescence microscopy and flow cytometry in the detection of *Giardia lamblia* cysts in beaver fecal samples," *J. Immunol. Methods* 202, 27–33, 1997.
31. H. A. Lindquist, A. Dufour, L. Wymer, and F. Schaefer III, "Criteria for evaluation of proposed protozoan detection methods," *J. Microbiol. Methods* 37, 33–43, 1999.

8

Strong Interaction between Nanophotonic Structures for Their Applications on Optical Biomedical Spectroscopy and Imaging

Y. Yi, P. Pignalosa, K. Broderick, B. Liu, and H. Chen

CONTENTS

8.1 Introduction

With the rapid development of integrated photonics technology, nanoscale computational photonics are having a major impact on next-generation biomedical engineering and sciences. This chapter describes three representative directions:

1. Computational nanophotonics for multiple metallic nanoparticle sensing and detection. The unique optical properties of the integrated microring resonator mode have been utilized to enhance the light signal from potentially biomolecule-tagged metallic nanoparticles. Extremely low-concentration biomolecule sensing and detection can be achieved.
2. Photonic resonator with nanoscale notch for high-sensitivity label-free biomolecule quantification, single-bacterium detection, and single-virus sensing. The basic

working mechanism of photonic resonator and the optical properties of whispering gallery mode (WGM) are explained; the future lab-on-a-chip single-biomolecule sensing using the integrated photonic resonator are also described.

3. Bio-inspired nanophotonic structure from nature and its applications to next-generation biomedical imaging materials and devices. How we applied computational optical biomedical spectroscopy to enhance biomedical imaging materials efficiency has been demonstrated, which is potential to develop novel imaging techniques for early detection, screening, diagnosis, and image-guided treatment of life-threatening diseases and cancer with higher resolution and lower doses.

8.1.1 Metallic Nanoparticles on Microring Resonator for Bio-Optical Detection and Sensing

Optical resonator has generated worldwide interests in the detection and sensing field. For the relatively high-Q microring resonators, a small change in the refractive index can be detected from the shift of resonance wavelength. Recently, the splitting of the resonance modes has been observed, which is caused as a consequence of coupling of clockwise and anticlockwise propagating modes. This phenomenon has been proposed for various applications, such as photonic molecules [1–7]. Nanoparticles have been heavily used in the optical detection and sensor area, as fast, noninvasive, and potentially label-free techniques are becoming more important for bio-sensing, gas sensing, chemical sensing, and nanomedicine fields. For example, metal nanoparticles are used as contrast agents in biomolecule sensing. Semiconductor nanoparticles are used as single-photon emitters in quantum information processing and as fluorescent markers for biological processes. Nanoshells with special engineering methods are used for cancer therapies and photothermal tumor ablation. Polymer nanoparticles are employed as calibration standards and probes in biological imaging in functionalized form [8–11]. The synergy between microring resonator and nanoparticle is becoming more important with the rapid progress of nanophotonics field.

The influence on microring resonators by *dielectric* nanoparticles has been intensively studied recently. Fiber tip is used to study the resonance mode profile, especially from the splitting of resonance mode. The mean resonance mode wavelength shift, splitting bandwidth as well as their dependence on dielectric nanoparticle size and position were also studied by many groups [12,13].

Optical sensor (including biosensor, chemical sensor, and gas sensor) based on WGM microring resonator has generated worldwide interest in this emerging filed involving dielectric nanoparticles or dielectric bio-layer [14–23]. However, the microsphere is made manually, as a result, large-scale manufacture of optical sensor devices is very challenging. The integrated microring resonator does not have this limit as we can use semiconductor microelectronics process, and millions of devices can be fabricated and integrated on a single chip with nanoscale precision.

Metallic nanoparticles have also been used intensively on detection and sensing, especially in the form of surface plasmon polariton (SPP), as the surface mode generated at the interface between the dielectric surface and the metallic surface is strongly confined at the interface, which can be utilized for many potential applications [24]. Recently, in the conventional four-port microring resonator configuration, it has been found that the metallic nanoparticles can be used as strong scatters on the microring resonators, which induce large reflection signals at an output port, which is normally dark [25,26].

The interaction between metallic nanoparticles and microring resonators has revealed many interesting phenomena and has not been emphasized much in previous studies as *dielectric* nanoparticles.

8.2 On-Chip Microring Resonator Device Structure and Simulation Method

In this section, we have numerically demonstrated a unique result by Au nanoparticles, when it is adsorbed at the edge of microring resonator. Compared to the resonance position without any Au nanoparticle, it was found that there is a *blue* shift for the resonance peak, which is opposite to the resonance wavelength shift direction when the dielectric nanoparticles are adsorbed onto the microring resonator. Due to the unique refractive index properties of Au, the number and position effects are also appealing and investigated in detail.

We used the conventional four-port microring resonator configuration on a chip, as illustrated in Figure 8.1. The microring resonator is 4 μm in diameter, the ring waveguide width is 200 nm as a single-mode waveguide, and the thickness of the microring resonator and bus waveguide is 250 nm. The two bus waveguides are evanescently coupled to the microring resonator, with the coupling gap of 100 nm. In this work, we simulated the Si microring resonator and the coupled waveguide system, with SiO_2 as the bottom cladding and air as the top cladding. The Au nanoparticle was placed at the outside edge of the microring resonator. The refractive indices of Si and SiO_2 are 3.48 and 1.45, respectively, and the dispersion relation of Au around wavelength of 1.55 μm is used [27].

We used finite-difference time-domain (FDTD) method in three dimensions to simulate the four-port microring resonator with or without Au nanoparticles [28]. Due to the small size of nanoparticles, a fine grid size as small as 2 nm and sufficient long evolution time steps are used to check the reliability of the simulation until the optimized grid size and time steps are found to reduce the required memory and simulation time. Perfectly matched layer absorbing boundary condition is used for the entire simulation window (10 μm × 10 μm). The bus waveguide was excited with a Gaussian pulse, which covers the wavelength window around 1.55 μm; the detected signal at the Drop port was Fourier transformed to obtain the Drop port versus wavelength information.

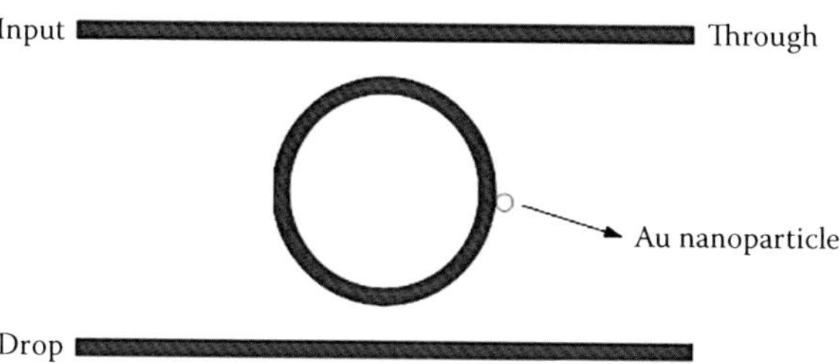

FIGURE 8.1
The on-chip four-port microring resonator configuration. The core of the microring resonator and the two bus waveguides is Si (n = 3.48), with SiO_2 bottom cladding (n = 1.45) and air top cladding. The waveguide width is 200 nm and satisfies the single-mode condition. The Au nanoparticle is adsorbed on the microring resonator.

8.3 Multiple Au Nanoparticle Effects on Microring Resonator and Simulation Results

Similar to the dielectric nanoparticles, the metallic nanoparticles will make the resonance wavelength shift and broadening of the splitting bandwidth. As illustrated in Figure 8.2, the main difference of the Au nanoparticle is its smaller real refractive index than air and very large imaginary part (absorption part), which leads to the *blue* shift of the resonance mode wavelength position. As the nanoparticle numbers reach a certain number, the splitting of the resonance begins to appear within our simulation resolution. For the 80 nm Au nanoparticle on the microring resonator with the nanoparticle number increasing from one to two, it is observed that both splitting modes are *blue* shifted. This phenomenon is unique as it provides us a very convenient approach to distinguish the dielectric nanoparticles and the Au nanoparticles, both of which are used extensively for sensing and nanomedicine field. Furthermore, the intensity of the Drop port is reduced rapidly with the increasing number of Au nanoparticles (30 in this case), which represents its large absorption characteristics at this wavelength. For sensing applications using Au nanoparticles or other metallic nanoparticles, it is inferred from this work that there is a limit for the number

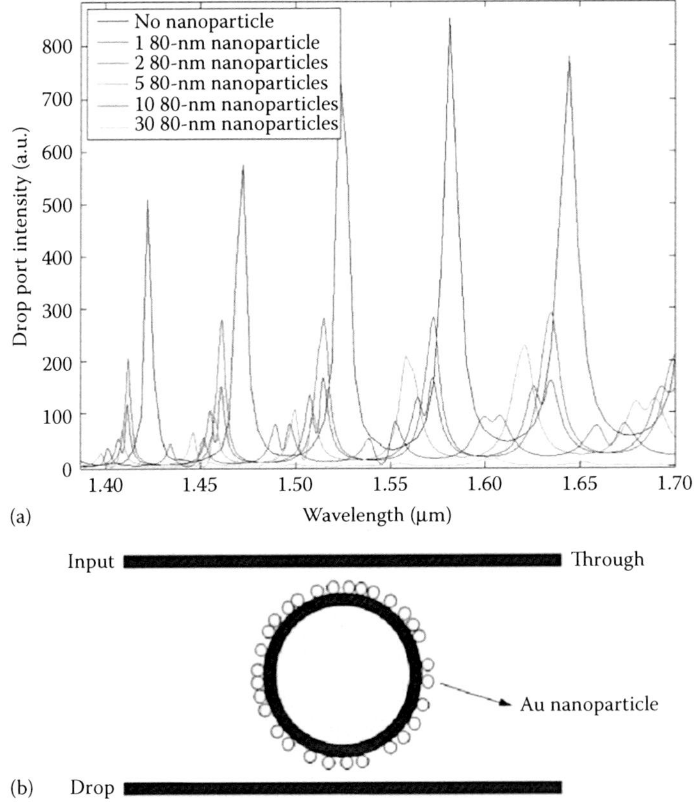

FIGURE 8.2
(See color insert.) (a) The Drop port intensity with different numbers of 80 nm-sized Au nanoparticles adsorbed on the microring resonator. (b) The example of multiple Au nanoparticles randomly distributed on the microring resonator.

of metallic nanoparticles adsorbed on the microring resonator, as there is normally large absorption for metallic nanoparticles. When the number reaches a certain point—critical number (30 Au nanoparticles in this case), the interaction between metallic nanoparticles and microring resonator is becoming so strong that they completely degrade the resonance—the Q is strongly degraded and the intensity at the Drop port is approaching zero.

For using Au nanoparticle and Drop port as a detection mechanism, the dependence on position is also important as the Au nanoparticles are possible to be adsorbed randomly on the microring resonator, the relatively position independence is necessary. For this purpose, for 30 Au nanoparticles with 80 nm in size, we have randomly distributed the 30 Au nanoparticles on the microring resonator and compared the Drop intensity in Figure 8.3. It is shown that the intensity at the Drop port for three random positions is on the same order and this result demonstrates the relatively independence of the Au nanoparticle position on the microring resonator.

Metallic nanoparticle size uniformity is very important for practical sensing and detection, as the nanoparticle size normally has a distribution around the target nanoparticle size which we would like to use. To study the effect of uniformity of the size of the Au nanoparticles on the performance of the integrated microring resonator, for five Au nanoparticles with 80 nm in size, we have randomly chosen the nanoparticle size that has certain distribution around 80 nm. Figure 8.4 shows the comparison between the uniform size of nanoparticles and nanoparticles with certain distribution, the overall signal from the Drop port is almost the same, although there is some small difference. The result demonstrates the robustness of our sensing mechanism using Au nanoparticles that can tolerate certain nonuniformity of Au nanoparticles.

We also studied the dependence on Au nanoparticle size, which is illustrated in Figure 8.5. Au nanoparticles with 10, 40, and 80 nm in size are compared at the Drop port with different numbers of Au nanoparticles. It is observed that the intensity at the Drop port for 10 nm Au nanoparticles is reduced in a much slower pace than that for the 40 and 80 nm Au nanoparticles. The reduction of the intensity is mainly caused by the strong absorption

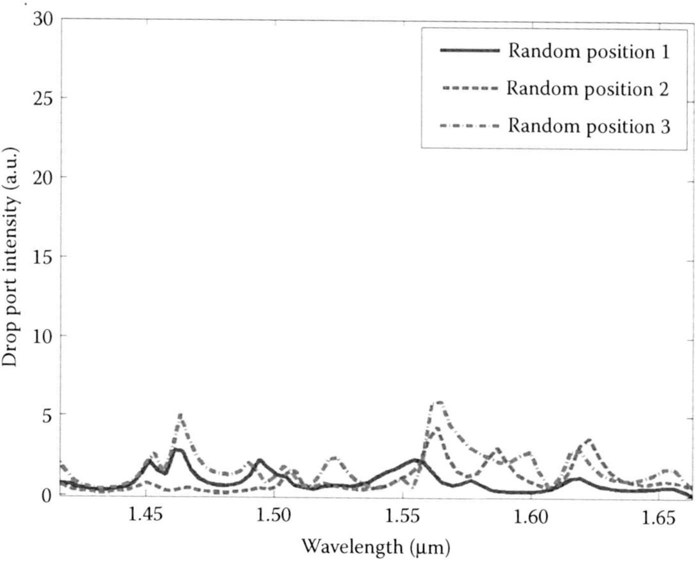

FIGURE 8.3
(See color insert.) The Drop port intensity with three random positions of 30 Au nanoparticles of 80 nm in size.

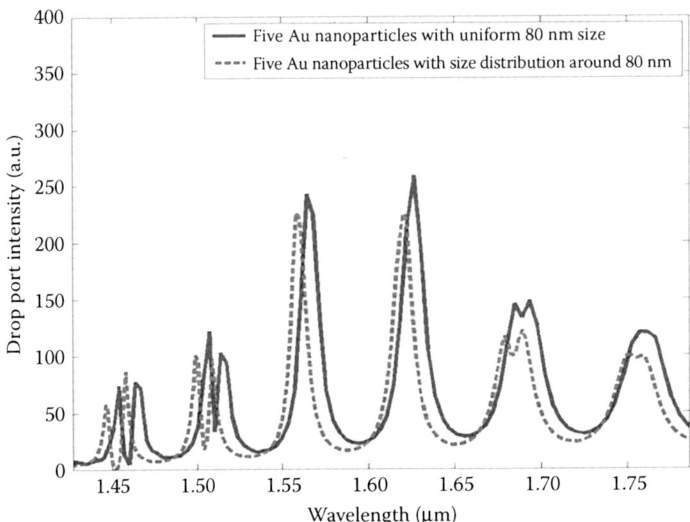

FIGURE 8.4
The Drop port intensity with Au nanoparticle size distribution, five Au nanoparticles with uniform 80 nm in diameter (solid line) and five Au nanoparticles with random size distribution around 80 nm in size.

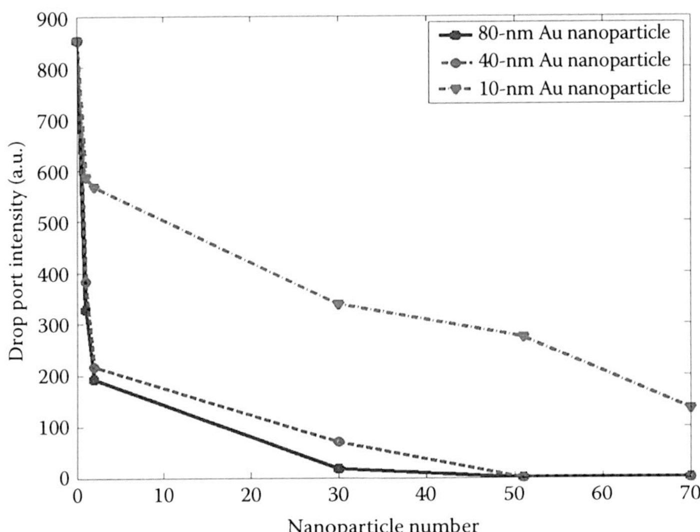

FIGURE 8.5
The Drop intensity vs. nanoparticle number for different size of Au nanoparticles.

of Au nanoparticles around 1.55 μm; the penetration depth of Au is about 45 nm. The dependence on nanoparticle size for the Drop port reveals the correlation between the penetration depths, nanoparticle size, and resonance mode evanescent tail length, which might be utilized to measure the Au nanoparticle size. For optical sensing and detection purposes, the optimized Au nanoparticle size should meet two requirements: One is relatively large shift when the nanoparticle is adsorbed on the microring resonator, and the other is the slow Q degradation ratio when more Au nanoparticles are adsorbed on the microring resonator. Based on the results in Figures 8.2 and 8.5, the optimized size of Au nanoparticles is estimated as 40 nm, which is around the penetration depth.

In sum, we have numerically demonstrated the unique optical response behavior for Au nanoparticles in the four-port microring resonator configuration, which can be utilized for single-nanoparticle detection and related applications. The blue shift of the resonance mode position at the Drop port due to the Au nanoparticles is drastically different from that due to dielectric nanoparticles, which are widely used for various applications in detection, sensing, and biomedical field. For sensing, the unwanted nanoparticles adsorbed on the microring resonator are mostly dielectric nanoparticles, which may be mixed with true signal; the unique *blue* shift by Au nanoparticles could be utilized to differentiate from that by dielectric nanoparticles. Therefore, it can be utilized to increase the signal-to-noise ratio. Due to large absorption of Au, the number of nanoparticles will reach a critical number before the microring resonator can still maintain an effective optical resonator device. The results on position and size dependence suggest the robustness to use Au nanoparticles for future applications in detection, sensing, and nanomedicine field.

8.3.1 Strong Coupling between On-Chip Notched Microring Resonator and Nanoparticle

With the rapid progress of nanotechnology, many nanoscale photonic devices as small as 30 nm have been realized, which are very promising to achieve manipulation of photons at a chip scale and have broad applications in renewable energy (photovoltaic cells, solid-state lighting), telecommunications, and biomedical field [29]. Recently, it has been found that the electromagnetic modes of certain photonic devices are very similar to the electronic wave function of molecules [2,30]. One of the most interesting examples is microring resonators; when we arrange two or more microring resonators together within optical coupling regime, the electromagnetic modes of the whole structure are very similar to the bonding (symmetric) or antibonding (antisymmetric) electronic wave function modes formed in molecules [1,6,31,32]. It is interesting to study the photonic molecule of various structures using optical techniques and it may further improve our understanding of the real molecular structures. Nanoparticles have played a key role in nanophotonics and have found many applications in medicine, drug delivery, solar cells, and sensors. It is also an important tool for the study of many nanoscale structures and is used to interact with nanoscale devices, as critical information could be obtained to understand their characteristics. Recently, nanoparticles have been heavily used in the optical sensor area, as fast, noninvasive, and potentially label-free techniques are becoming more important for bio-sensing, gas sensing, and chemical sensing. Single nanoparticle detection is one of the ultimate goals for a sensing device and represents sensing at the extreme. In the recent years, many novel methods have been utilized to realize nanoparticle detection. For example, metal nanoparticles are used as contrast agents in biomolecule sensing; semiconductor nanoparticles are used as single-photon emitters in quantum information processing and as fluorescent markers for biological processes; nanoshells with special engineering methods are used for cancer therapies and photothermal tumor ablation; and polymer nanoparticles are employed as calibration standards and probes in biological imaging in functionalized form [8–11]. The synergy between the photonic molecules and nanoparticles provides us with a unique opportunity, as we can utilize the special photonic molecular modes to interact with the nanoparticles to achieve single nanoparticle detection; inversely, we can also utilize nanoparticles to study photonic molecules and their properties.

In this work, we used on-chip notched microring resonator as a photonic molecular example, as a variety of types of optical microring resonators were investigated and is a natural photonic molecule to use (Figure 8.6a); for the nanoparticle, we used atomic force microscope (AFM) tip to simulate a single nanoparticle, where the small tip can be either dielectric materials (Si, GaAs, Si_3N_4, etc.) or metallic materials (Au, Ag, Al, etc.). We achieved, for the first time, strong coupling between an on-chip notched microring resonator and a single nanoparticle. Specifically, we used the nanoscale notch (~100 nm) in the microring resonator with diameter around 4 μm to strongly interact with the AFM tip. The intentionally created notch in the microring resonator causes the splitting of the original ring resonance mode and formation of bonding (symmetric) photonic states and antibonding (antisymmetric) states. The AFM tip can be positioned inside the notch. The strong coupling between core electromagnetic modes in the notch and the tip causes the bonding photonic modes to shift in nanometer scale, whereas there is almost no shift for the antibonding photonic modes. It confirms the photonic molecular mode characteristics generated by the notched microring resonator. The result suggests the potential to deeply study the photonic molecular mode characteristics. Furthermore, we found the unique and very different shift behavior of the splitting modes from the dielectric Si tip and the metallic Au tip, which can be a critical detection and sorting mechanism for the different type of nanoparticle systems.

In the conventional approach to study the splitting modes using microspheres, because the interaction between the fiber tip (or nanoparticle) and the evanescent mode of the microsphere is very small, it leads to a very small change in the effective refractive index of the

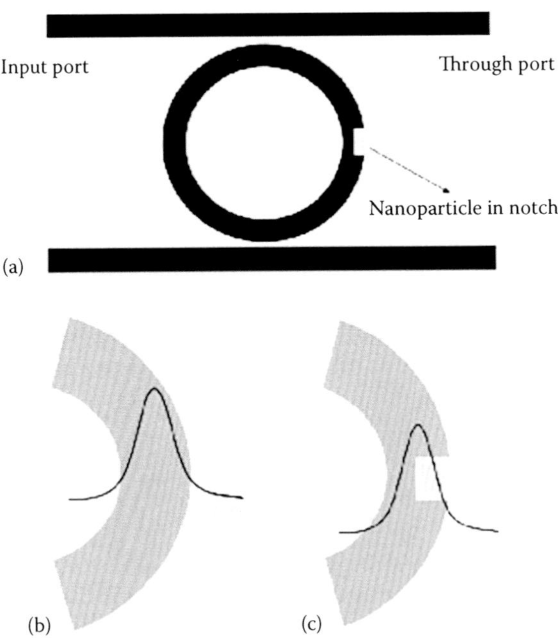

FIGURE 8.6
(a) The notched microring resonator and the coupling to the bus waveguide with input and output ports. The bus and ring waveguide width is 200 nm. The notch size is around 100 nm × 100 nm. (b) The evanescent coupling between the resonance mode and the nanoparticle. (c) The notch in the microring resonator. The nanoparticle is in the notch and coupling to the core of the resonance mode. The black curve represents the single-mode behavior for the ring waveguide. The ring waveguide is a single-mode waveguide at 1.53 μm.

sphere, which shifts the wavelength position of the peaks and causes small splitting in the resonator transmission (or reflection) spectrum; these changes are typically on the order of picometers in size [12]. In order to detect such small shifts, one must normally use an expensive tunable narrow-linewidth laser source to scan the relevant spectral region of the resonator output spectrum. Furthermore, the resonator itself must be designed to yield a very narrow linewidth, so that the small peak shifts and splitting can be detected. This requires a high finesse (free spectral range divided by linewidth), or equivalently high-quality factor (operating wavelength divided by linewidth ~10^8), which translates to low-loss waveguides in the resonator and weak coupling between the resonator and the fiber tip. Here, we demonstrated strong coupling between on-chip notched microring resonator and nanoparticle, where the nanoparticle can be placed inside the notch. Compared to the microring resonator without notch, as illustrated in Figure 8.6b, where the interaction strength between the nanoparticle and the resonator mode field is relatively weak and only a small portion of the field (evanescent tail) is interacting with the nanoparticle, the notch provides access to the peak of the electromagnetic field localized in the core, so that when a nanoparticle is placed there, the strong core field, rather than the weak evanescent cladding field, overlaps the nanoparticle and therefore produces an enhanced response, as shown in Figure 8.6c.

The on-chip photonic device configuration of a microring resonator with a notch on the ring, and with two bus waveguides coupled to the ring, has been fabricated [25,33–36]. The 100 nm notch in the ring was fabricated by e-beam lithography. We analyzed the case of a 100 nm-long notch with a 20 nm-diameter dielectric nanoparticle Si tip and 20 nm-diameter metallic Au particle inside the notch. The thickness of Si waveguide is 220 nm. Figure 8.7a is the image of the 100 nm-long notch in the microring resonator using portable scanning electron microscope (SEM) (Hitachi TM-1000). The microring resonator is 4.0 μm in diameter with the waveguide width of 200 nm. The core material of the ring is Si with refractive index 3.48 at around 1.53 μm wavelength, and the bottom cladding material is SiO_2 with refractive index 1.46. We used Silicon-On-Insulator (SOI) wafer for the small microring resonator fabrication, and an e-beam is utilized to fabricate the 100 nm × 100 nm notch at the edge of the Si microring resonator. The refractive index of the Au nanoparticle tip is $0.54 + 9.58i$ at 1.53 μm. Figure 8.7b is the SEM image within the coupling gap area, which shows the clear 100 nm gap between the bus waveguide and the notched microring resonator. A tunable laser from 1480 to 1580 nm is used to couple the light from the tapered optical fiber to the Si waveguide; a Ge detector is put at other end to collect the

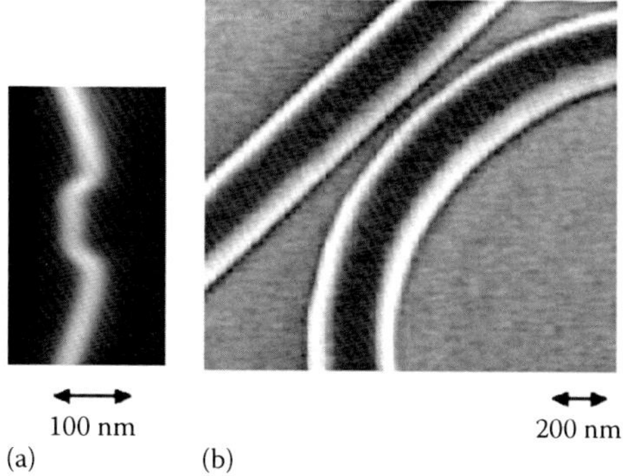

FIGURE 8.7
(a) The image of the 100 nm-sized notch by portable SEM. (b) The SEM image with clear 100 nm coupling gap between the notched microring resonator and the single-mode bus waveguide.

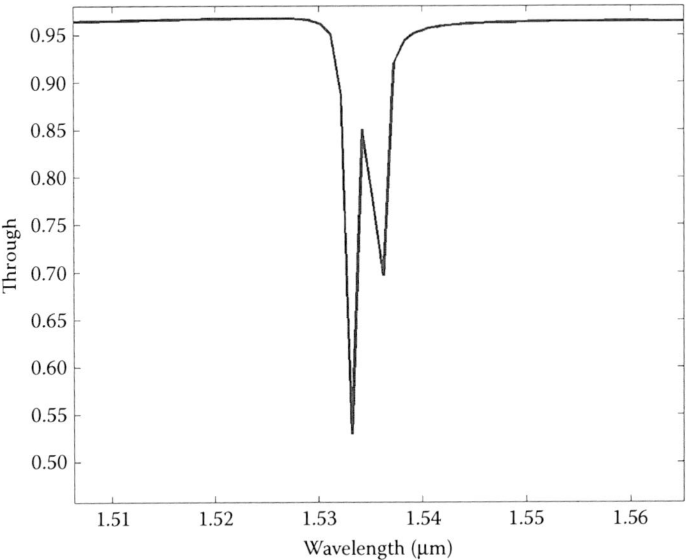

FIGURE 8.8
The transmission at the Through port is around 1.53 μm, the splitting is around 3 nm, the bonding photonic mode is at shorter wavelength 1.534 μm, and the antibonding photonic mode is at longer wavelength 1.537 μm.

through port signal. Figure 8.8 shows the measurement result for the "Through" port of bus waveguide for the microring resonator with notch at the edge. We can clearly see the splitting of the original resonance modes at around 1.53 μm, which represents the bonding photonic mode at shorter wavelength (1.534 μm) and the antibonding photonic mode at longer wavelength (1.537 μm). The splitting width is almost 3 nm; this nanometer scale splitting bandwidth is much larger than the picometer splitting bandwidth we normally see for the fiber tip close to the microsphere resonator.

We used a Si nanoparticle tip of 20 nm in diameter and put it inside the center of 100 nm-sized notch of the microring resonator using portable AFM; transmission at the "Through" port of the coupled waveguide is shown at Figure 8.9a. It is observed that the bonding mode is *red* shifted by 1.1 nm, whereas the position of the antibonding mode is almost unchanged, leading to a smaller splitting bandwidth due to the red shift of the bonding mode. Next the Si nanoparticle tip is replaced with a 20 nm Au nanoparticle tip and also put inside the notch center of the microring resonator. We can see the dramatically different shift with the Au nanoparticle for the bonding state, a 1 nm *blue* shift is observed instead, and the antibonding state remains unchanged, causing larger splitting bandwidth than the original splitting due to the notch with just air, as shown in Figure 8.9b. The results reveal the drastically different photonic mode properties between the symmetric and antisymmetric states and suggest that we can utilize the nanoparticle to study the photonic molecule characteristics. We can understand the red shift induced by the Si nanoparticle by considering an effective index increase in the notch due to the substitution of air with high-index Si; the same reason is for Au nanoparticle, as the real part is less than that of air. We can also understand the fact that nanoparticles have little effect on the antibonding mode by considering that they are placed at the center of the notch and close to the zero-field node of the antibonding mode. It can be clearly seen that the strong coupling effect of a nanoparticle placed in a

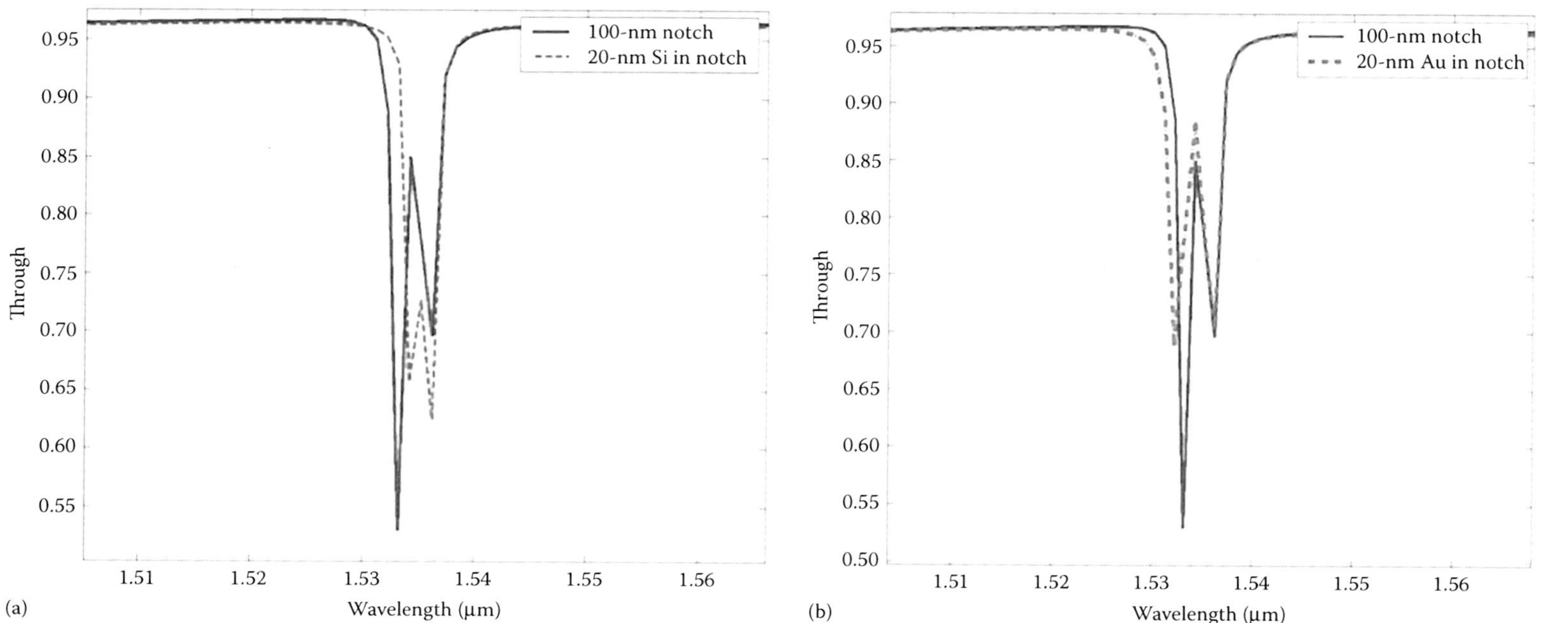

FIGURE 8.9
(See color insert.) (a) Si nanoparticle tip of 20 nm in diameter is put inside the 100 nm-sized notch of the microring resonator. The Through port from the coupled waveguide is shown. It is observed that the bonding mode wavelength (resonance at shorter wavelength) is *red* shifted with 1.1 nm, whereas the wavelength of the antibonding mode (resonance at longer wavelength) is almost unchanged; the splitting bandwidth is smaller due to the red shift of the bonding mode. (b) Au nanoparticle tip of 20 nm in diameter is put inside the notch size of the microring resonator. We see the dramatic different shift from the Au nanoparticle for the bonding state (resonance at shorter wavelength). The 1.0 nm *blue* shift is observed instead and the antibonding state (resonance at longer wavelength) is still unchanged; the splitting bandwidth is larger than the original splitting due to the notch with air.

notch is fundamentally different from the evanescent coupling effect of a nanoparticle placed in the evanescent field of the resonator, the latter coupling approach is commonly used in High-Q microring resonator.

The different wavelength shift between a dielectric Si nanoparticle and a metallic Au nanoparticle, as well as the splitting bandwidth narrowing with the dielectric Si compared to the widening with the metallic Au nanoparticle, provides us with a unique, *self-referencing* mechanism to distinguish these different types of nanoparticles. This is very important for the biosensor area utilizing nanoparticles, as Au nanoparticles are often used for tagging, but other dielectric nanoparticles other than the analyte may be present and cause misleading sensing signals; Au nanoparticle's unique effects on the shift direction and the splitting bandwidth widening will enable us to distinguish the real signal from the background dielectric nanoparticle noise and greatly enhance the signal-to-noise ratio. The intentionally fabricated notch in the microring resonator provides us with a localized position to trap the nanoparticle. Based on the strong splitting and wavelength shift when the nanoparticle is localized in the notch, we can identify whether the nanoparticle stays within the fabricated notch and differentiate the nanoparticle position. The future study on the electromagnetic force applied on the nanoparticle close to the notch will be interesting to help us to understand the mechanic response from the notch and how to deliver the nanoparticle to the notch.

In sum, we have experimentally demonstrated the strong coupling between on-chip nanoscale notched microring resonator and nanoparticle, in which a notch is introduced in the resonator to provide access to the core field, which is drastically different from previous studies using evanescent coupling to microsphere. Placing a nanoparticle in the notch produces a much stronger response than simply placing the nanoparticle in contact with the exterior of the core. In the exemplary case of a dielectric silicon and metallic gold nanoparticle placed in a notch, we demonstrated that nanoparticle induces a large wavelength splitting (~nm) and a very different shift in the resonant modes of the resonator. This is a significant improvement over the smaller wavelength shifts and splitting (~pm) observed in earlier experiments where the particle was placed outside the core of a conventional microsphere resonator and lowers the requirement for very high-Q resonator devices. Note that the utility of this approach is not limited to microring resonators used as examples here, but can be extended to other types of resonator geometries, such as racetracks and polygons. This work provides us a unique way to achieve single nanoparticle detection and sorting with thousands of times signal enhancement. The nature of the on-chip microring resonators will also make large-scale integration on a single chip possible.

8.3.2 Giant Efficiency Enhancement of Biomedical Imaging Scintillation Materials Using Bio-Inspired Integrated Nanostructures

Biomedical imaging has become one of the most relied-upon tools in health care for diagnosis and treatment of human diseases. The evolution of medical imaging from plain radiography (radioisotope imaging), to X-ray imaging, to computer-assisted tomography (scans), to ultrasound imaging, and to magnetic resonance imaging has led to revolutionary improvements in the quality of health care available today to our society. In order to develop novel imaging techniques for early detection, screening, diagnosis, and image-guided treatment of life-threatening diseases and cancer, there is a clear need for extending imaging to much higher resolution level; information at fine resolution levels can lead to the detection of the early stages of the formation of a disease or cancer or early molecular changes during intervention or therapy. Radiation detection has an important position in medical imaging. Compared to semiconductor detectors, radiation detector and gas proportional scintillation detectors have obvious advantages in the high-energy particle detection, mainly because scintillation detector has high stopping power for the high-energy radiation [37]. Scintillator material in the scintillation detector is the core part of the detector for detection purposes. High-energy particles or rays are absorbed by the scintillator materials in the detector, which emits pulsed light; this pulse of light is generally in the visible region or near UV region. These flashes of light are converted to electrical signals by optoelectronic converter (such as photomultiplier tubes and avalanche photodiodes) in order to achieve high-energy particles or radiation detection. Scintillation detector for radiation detection is usually required to achieve energy resolution, time resolution, and position resolution, which means that the scintillator materials must meet certain requirements. An ideal scintillator should have the following characteristics: high light yield, fast decay time, high density, and emission spectra consistent with the photodetector response, stable physical and chemical properties, high radiation hardness, and low cost [38].

Inorganic scintillators are widely used in modern medical imaging modalities as converter for the X-rays and γ-radiation that are used to obtain information about the interior of the body, such as X-ray and CT and positron emission tomography scans. One key problem in the development of the next-generation medical imaging systems is the enhancement of the energy resolution of the detectors. This parameter is influenced by the statistical fluctuations of the light output of the scintillators, that is, by the number of photons that are detected when a particle deposits its energy in the scintillator. The light output of the scintillator depends not only on the absolute number of generated photons but also on the geometrical shape of the material, its transmission properties at the wavelength of scintillation, and its refractive index. Especially in tiny detector crystals with small aspect ratio, a significant fraction of photons is lost before conversion into an electronic signal in the photodetector. This effect increases the statistical fluctuations of the light output and therefore deteriorates the resolution of medical imaging. Most of the high-density scintillators have a high refractive index, so most of the light is trapped in the crystal, only 10%–30% of the light from the scintillator can enter into the photodetector, and the majority of light could not be effectively extracted, which seriously affected the detection system's efficiency and detection sensitivity.

The past half-century has witnessed the discovery of many new scintillator materials and numerous advances in our understanding of the basic physical processes governing the operation of inorganic scintillators. Properties of both intrinsic and activated scintillator materials, crystalline and amphorous, have been considered. Several fundamental limits of scintillator performance have been examined together with the prospects for discovering better scintillators guided by first-principles theoretical calculations of the active processes in scintillation. Unfortunately, no single material can meet all the above requirements. For different detection purposes, we need to consider the most important requirements, while lowering secondary requirements. For example, scintillation light yield is the most important for medical imaging, in order to maximize detector sensitivity so that we can reduce patient radiation dose [39]. In high-energy physics experiments or high count-rate time-resolved radiation environment, it is required to have a faster luminescence decay time [40].

Although some traditional methods, such as the use of rough or polished surface of the crystal surface and insertion of the moderate refractive index materials into the middle of the oil to achieve better light coupling efficiency, the increase in light extraction efficiency is still very limited.

For the past 20 years, micro–nanophotonics research has made rapid developments. Combining the micro–nanostructure and spontaneous emission of radiation detection materials, we can design much better radiation detection materials and devices that have more superior performance and can meet the requirements much better. The use of micro–nanophotonics can help us to explore new structural flashing materials, such as artificial micro–nanostructures, photonic crystals [41], optical microcavity [42], and surface plasmon materials [43]. Artificial micro–nanostructure can improve the extraction efficiency of light-emitting materials and is becoming more mature. However, the idea to apply this novel photonic structure for scintillator has just started; the reported work is very limited.

8.4 Results

In this work, we have investigated a novel class of nanoscale devices based on photonic structures that function as efficient light extraction devices for scintillator materials and devices [44–51]. We have studied the light output enhancement characteristics of Lu_2SiO_5:Ce thin film scintillator materials; similar mechanism can be extended to various types of scintillator materials, such as γ-CuI single crystal and doped Tb^{3+} glass. We have designed light extraction structures based on their optical properties by using numerical simulation and optimized the sample preparation process, so that the overall efficiency of these scintillator materials is increased significantly. Specifically, we have achieved more than 70% external quantum efficiency by utilizing photonic light extraction structures in order to realize higher detection efficiency and sensitivity.

Figure 8.10 is the illustration of the moth's eye (Figure 8.10a) and its micro/nanostructures (Figure 8.10b and c). With the increasing interest in reducing the front surface reflection, intensive effort focuses on the development of surface structures and process methodologies to achieve broadband antireflection [52–54]. Nanostructure comprising an array of circular protuberances (corneal nipples) on the facet lenses in moth's eye (Figure 8.10b and c) is widely utilized to achieve this purpose. The optical action of the corneal nipple array provides a significant reduction of the reflectance of the facet lens surface. Accordingly, it increases the transmittance, and therefore the function of the nipple array was suggested to enhance the light sensitivity of moth's eyes. The corneal nipple array functions as an impedance matching device. Microscope images and AFM studies have revealed the moth's eye structures with many nipples on the surface of moth's eye; the nipple is like a pyramid structure, with a graded optical refractive index varying from higher substrate index to air, so the broadband reflectance can be achieved using the graded index structure. This bio-inspired nanostructure is only one of many fascinating photonic structures in nature. Instead of broadband antireflection properties, this structure can also be utilized to enhance the light extraction of the scintillator materials.

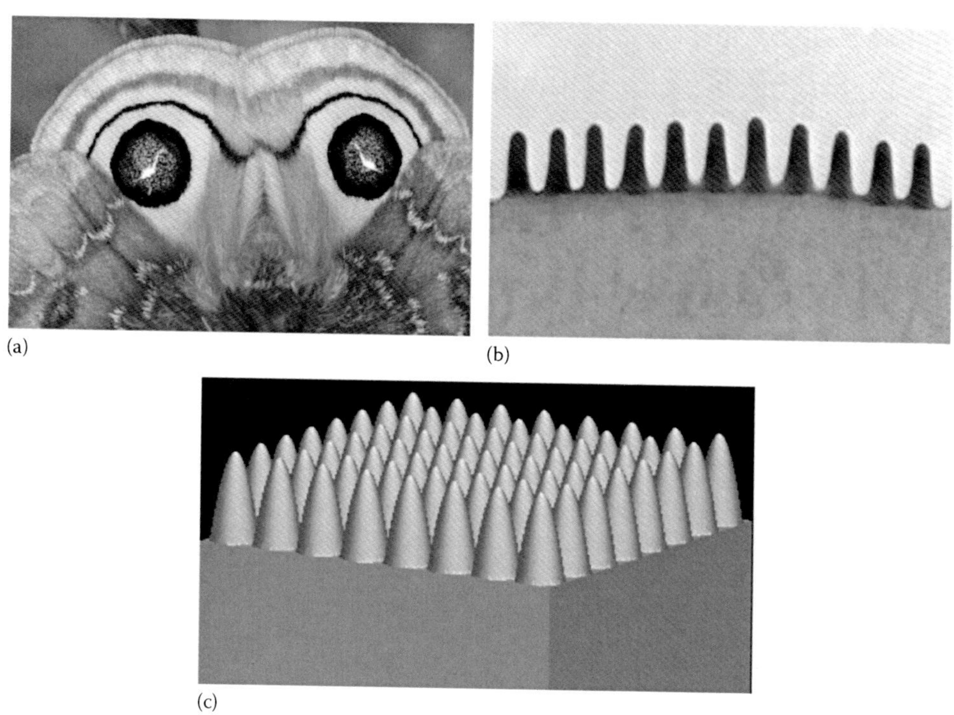

FIGURE 8.10
(See color insert.) Illustration of moth's eye structures. (a) The image of the moth's eye. (b) The nanoscale features under SEM. The typical size is as follows: height 250 nm and pitch 300 nm. (c) The 3D view of the moth's eye structure.

We have further improved the bio-inspired moth's eye structure for the light output enhancement; specifically, we have intentionally added some roughness on the sidewall of the original pyramid shape, which is due to the results of our fabrication technique, and the comparison is illustrated in Figure 8.11a and b. Figure 8.11a shows the conventional moth's eye structure with flat sidewall profile; Figure 8.11b shows the improved bio-inspired moth's eye structure, with a certain degree roughness on the sidewall of the pyramid. The light enhancement results are shown in Figure 8.11c: The blue curve is the reference structure with only Lu_2SiO_5:Ce thin film; the green curve is the light enhancement with conventional smooth moth's eye structure illustrated in Figure 8.11a; and the red curve is the light enhancement with improved bio-inspired moth's eye structure with sidewall roughness illustrated in Figure 8.11b. It can be clearly seen that the light enhancement factor from the improved bio-inspired moth's eye structure is significantly larger than that from pure moth's eye structure, with the enhancement as large as 2.7 when the periodicity is 400 nm. The external quantum efficiency is almost 86%, which demonstrates that most of the light from the Lu_2SiO_5:Ce thin film has come out of the high-index emission layer.

We have combined self-assembly and reactive-ion etching (RIE) methods to fabricate the improved bio-inspired moth's eye-like nanophotonic structures. As illustrated in Figure 8.12a, the fabrication processes start from the coating of nominal 40-nm SiO_2 nanoparticles, which was obtained from Nissan Chemical, Tokyo, Japan. Figure 8.12b is the SEM image of the high-index Si_3N_4 light extraction structure. It is very interesting to

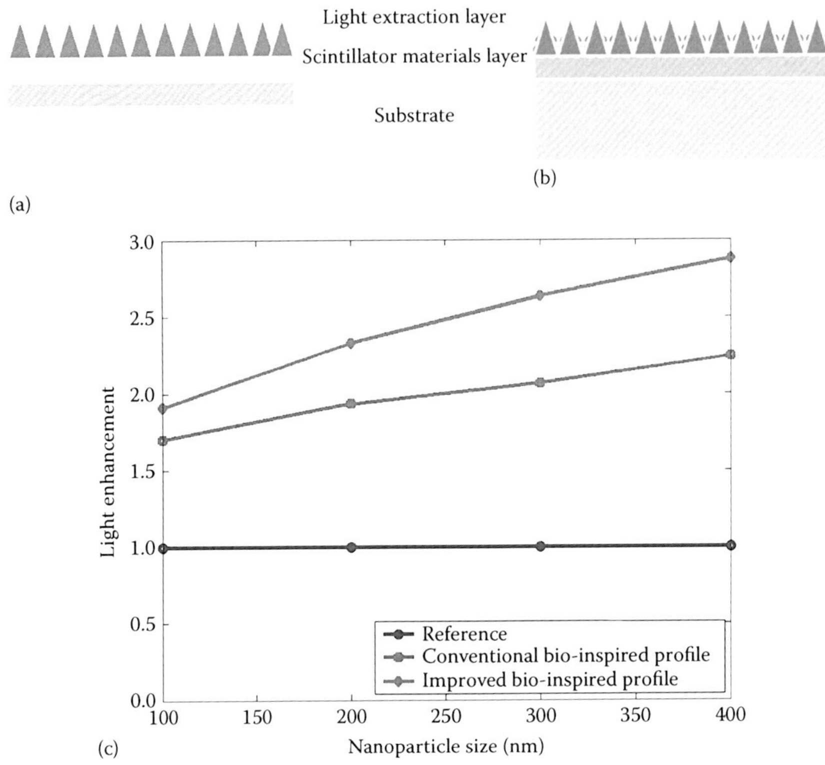

FIGURE 8.11
(See color insert.) The simulation device structure for two different bio-inspired photonic structures. (a) The conventional moth's eye structure with flat sidewall. (b) The improved bio-inspired moth's eye structure with intentionally designed roughness on the sidewall. (c) The FDTD simulation results, which show an enhancement for both bio-inspired light extraction structure, whereas the improved bio-inspired moth's eye structures show more light output enhancement.

note the similarity to the moth's eye structure (compared to Figure 8.10). In addition to the conventional smooth sidewall moth's eye structure, we notice the obvious *roughness* on the sidewall of each pyramid structure, and the roughness is at the same scale as the pyramid structure itself. Although our biomimetic structure is very similar to the moth's eye structures, the bumps (roughness at the similar scale) on the sidewall of the pyramid show new features, which will further enhance the light extraction properties. Furthermore, using self-assembly of nanoparticles as mask is superior to conventional e-beam nanofabrication techniques. For e-beam, it is very difficult to obtain large-area nanoscale patterns; with self-assembly of nanoparticles, we can potentially achieve reasonable large-area nanoscale patterns with real sample size as demonstrated in our experiment.

The X-ray mammographic unit was employed to demonstrate the light enhancement of the Lu_2SiO_5:Ce thin film with bio-inspired moth's eye-like nanophotonic structures. As illustrated in Figure 8.13a, the experiment was performed on a General Electric (New York) Senographe DMR and X-ray mammographic unit with molybdenum anode target and molybdenum filter. Tube voltage was checked using an radiation measurements incorporated (RMI) model 240 multifunction meter. Incident exposure rate measurements were performed using a Radcal 2026C ionization chamber dosimeter

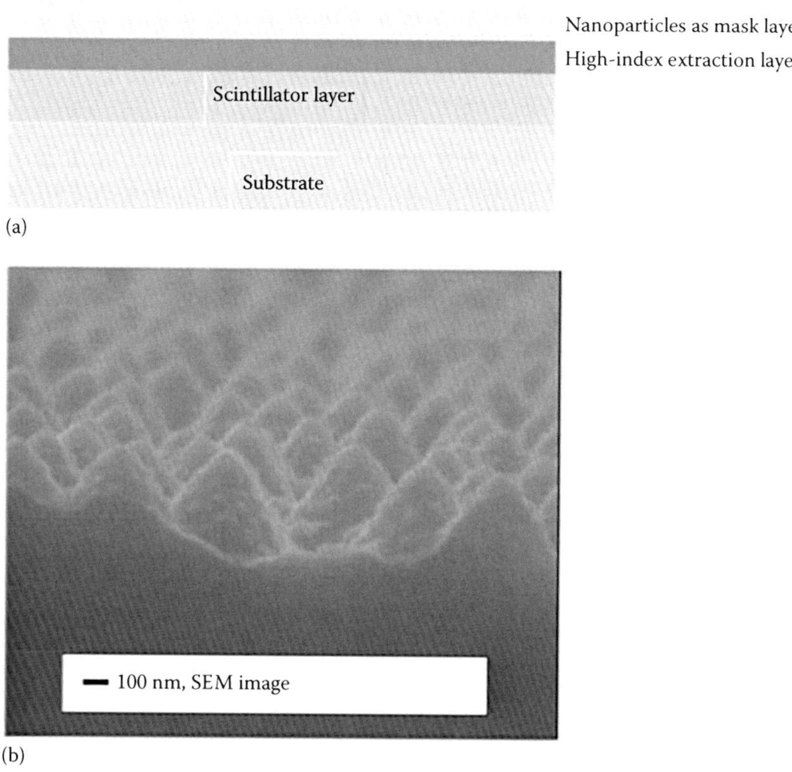

FIGURE 8.12
Using self-assembly of SiO_2 nanoparticles as a mask and RIE method to fabricate the improved bio-inspired moth's eye nanostructures on the Lu_2SiO_5:Ce thin-film scintillator materials. (a) The self-assembly of SiO_2 nanoparticles on the top of high-index Si_3N_4 light extraction layer, which is deposited on Lu_2SiO_5:Ce thin film. (b) The SEM image of the improved bio-inspired moth's eye nanostructures with certain degree *roughness* on the sidewall, which shows interesting *nano-on-nano* features.

(Radcal Co., Monrovia, CA). The output signal was measured by performing X-ray exposure and emitted light energy flux measurements. Emitted light energy flux measurements were performed using an experimental setup comprising a light integration sphere (Oriel model 70451, Newport Corporation, Irvine, CA) coupled to a photomultiplier (EMI 9798B, Thorn EMI, London, UK) connected to a Cary 401 (Cary Instruments Company, Palo Alto, CA) vibrating reed electrometer. The photomultiplier was coupled to the output port of the integrating sphere to reduce experimental errors due to illumination nonuniformities. The screen was positioned at the input port of the integrating sphere, whereas the photomultiplier was adapted at the output port. The photocathode of the photomultiplier (extended S-20) was directly connected to a Cary 401 vibrating reed electrometer by bypassing all dynodes. In this manner, photocurrent instability and electronic noise amplification due to photomultiplier's dynode high voltage were avoided. Figure 8.13b shows our results on the comparison between the two Lu_2SiO_5:Ce thin films: The blue curve is the light output from the referenced Lu_2SiO_5:Ce thin film without any light extraction structures, and the green curve is the light output from the Lu_2SiO_5:Ce thin film with improved bio-inspired moth's eye-like photonic structures, as illustrated in Figure 8.12b. The peak light intensity at round

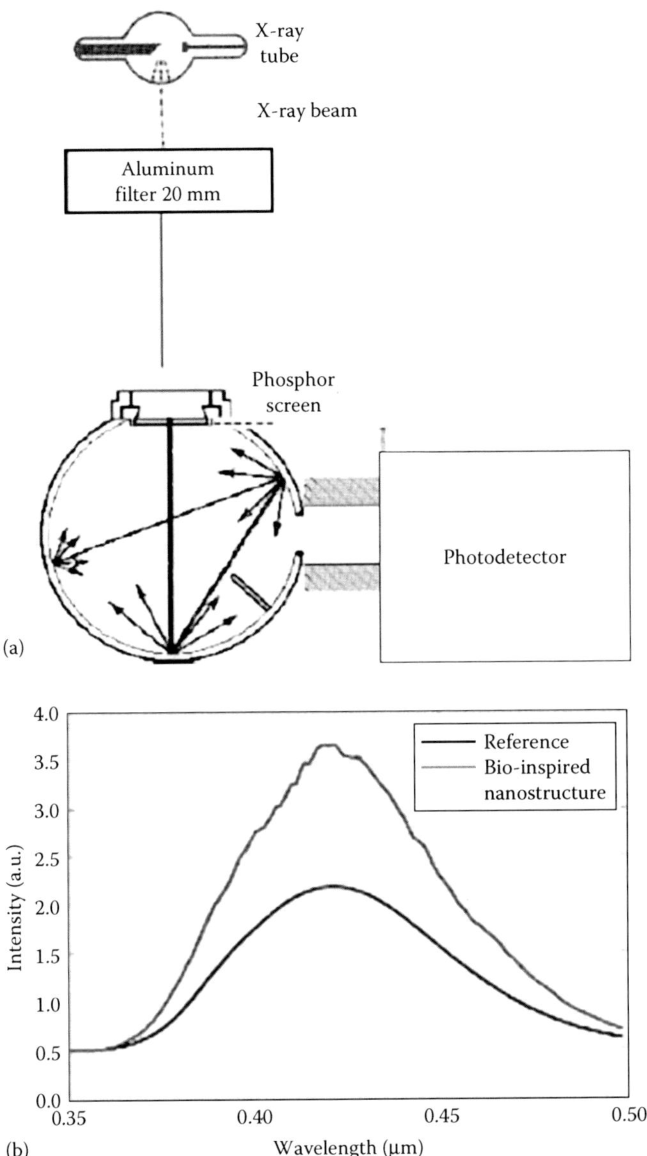

FIGURE 8.13
(See color insert.) The X-ray mammographic unit was employed to demonstrate the light enhancement of the Lu_2SiO_5:Ce thin film with bio-inspired moth's eye-like nano photonic structures. (a) A General Electric Senographe DMR and X-ray mammographic unit with molybdenum anode target and molybdenum filter. (b) The comparison between the two Lu_2SiO_5:Ce thin films. The blue curve is the light output from the referenced Lu_2SiO_5:Ce thin film without any light extraction structures and the green curve is the light output from the Lu_2SiO_5:Ce thin film with improved bio-inspired moth's eye-like photonic structures.

420 nm was increased almost 1.75 times under the same dosage. Our results have demonstrated the significant light output increase using the improved bio-inspired photonic structures, which may be utilized to reduce the radiation dosage of the medical imaging or increase the image quality under the same or smaller radiation dosage for better and early-stage diagnosis.

8.5 Discussion

In sum, we have investigated a novel class of nanoscale photonic structures that function as efficient light extraction devices for scintillator materials. Our work can be extended to study the characteristics of various types of scintillator materials, such as γ-CuI single crystal, doped Tb^{3+} glass, and Lu_2SiO_5:Ce thin film, which was demonstrated in this chapter. It is promising to design light extraction structures based on their optical properties, so that the overall efficiency of these scintillator materials will be increased significantly. Furthermore, we have utilized the unique properties of the improved bio-inspired nanoscale photonic structures to enhance the light output significantly; the future study on the control of beam shape is also very promising for us to control the angular behavior of the light, which will make smaller pixel, higher imaging resolution possible. Our work has opened a door to study and understand a new class of nanophotonic structures based on scintillator materials and devices. In future works, we will exploit the basic imaging characteristics of the nanophotonic structures to determine the feasibility of ultimate goal for next-generation radiation detection scintillator materials for lower dose, higher resolution medical imaging applications. Ultimately, we will explore the possibility to achieve the biomedical imaging at the cellular and molecular biology level.

8.6 Methods

8.6.1 Simulations

To demonstrate the advantage of the improved bio-inspired moth's eye structure for the scintillator materials, over the conventional structure, we have used the FDTD method to study the light enhancement properties of the two structures. The device size is 100 μm × 100 μm, the grid size is 10 nm, and the perfect boundary condition is used. Dipole sources with all polarizations (x-dipole, y-dipole, and z-dipole) are used and averaged for the final simulation results. Figure 8.2a and b shows the device structure for Lu_2SiO_5:Ce thin film on glass substrate: The film thickness is 500 nm, the high-index light extraction material is Si_3N_4 with refractive index 2.0, and the height of the film is 300 nm.

8.6.2 Fabrications

The Lu_2SiO_5:Ce thin film, with nominal thickness of 500 nm, was deposited using sol–gel technique. As illustrated in Figure 8.3a, the fabrication processes start from the coating of nominal 400 nm SiO_2 nanoparticles, which were obtained from Nissan Chemical. The nanoparticle solution was diluted in 1-methoxy-2-propanol to produce suspensions having different percentage by weight solids content. The 400 nm SiO_2 nanoparticles were coated on the high-index Si_3N_4 light extraction layer (whose refractive index is 2.0), which was previously deposited on the Lu_2SiO_5:Ce thin film. The nanoparticle layer on top of the Si_3N_4 film was used as a mask layer and the RIE method was then applied to etch the exposed Si_3N_4 film. After etching, the remaining SiO_2 nanoparticles were removed subsequently.

8.7 Summary

These works on computational photonics with experimental demonstration at the nanometer scale for biomedical engineering and sciences represent the future scientific trend in these forefront areas. Utilizing current nanophotonics technology (design, simulation, fabrication, and measurement techniques), it is possible to study these novel biomedical materials and devices in ways that were unimaginable a decade or two ago.

Acknowledgments

We thank the University of Michigan (Ann Arbor, Michigan), Microsystems Technology Laboratories and Center for Materials Science and Engineering at MIT (Cambridge, Massachusetts), and 3M's Central Research Laboratories (St. Paul, Minnesota) for their support to provide access to the facility and measurement equipments.

References

1. A. Francois and M. Himmelhausa, Optical biosensor based on whispering gallery mode excitations in clusters of microparticles, *Appl. Phys. Lett.* 92, 141107, 2008.
2. B. E. Little, S. T. Chu, and H. A. Haus, Second-order filtering and sensing with partially coupled traveling waves in a single resonator, *Opt. Lett.* 23, 1570, 1998.
3. M. Bayer, T. Gutbrod, J. P. Reithmaier, A. Forchel, T. L. Reinecke, P. A. Knipp, A. A. Dremin, and V. D. Kulakovskii, Optical modes in photonic molecules, *Phys. Rev. Lett.* 81, 2582, 1998.
4. A. Forchel, M. Bayer, J. P. Reithmaier, T. L. Reinecke, and V. D. Kulakovskii, Semiconductor photonic molecules, *Phys. E* 7, 616, 2000.
5. K. J. Vahala, Optical microcavities. *Nature* 424, 839–846, 2003.
6. Q. Song, H. Cao, S. T. Ho, and G. S. Solomon, Near-IR subwavelength microdisk lasers, *Appl. Phys. Lett.* 94, 061109, 2009.
7. M. L. Gorodetsky, A. D. Pryamikov, and V. S. Ilchenko, Rayleigh scattering in high-Q microspheres, *Opt. Lett.* 17, 1051, 2000.
8. Z. Yuan, B. E. Kardynal, R. M. Stevenson, A. J. Shields, C. J. Lobo, K. Cooper, N. S. Beattie, D. A. Ritchie, and M. Pepper, Electrically driven single-photon source, *Science* 295, 102, 2002.
9. M. Bruchez, M. Moronne, P. Gin, S. Weiss, and A. Paul Alivisatos, Semiconductor nanocrystals as fluorescent biological labels, *Science* 281, 2013, 1998.
10. C. Loo, A. Lin, L. Hirsch, M. Lee, J. Barton, N. Halas, J. West, and R. Drezek, Nanoshell-enabled photonics-based imaging and therapy of cancer, *Technol. Cancer Res. Treat.* 3, 33, 2004.
11. R. Wiese, Analysis of several fluorescent detector molecules for protein microarray use, *Luminescence* 18, 25, 2003.
12. A. Mazzei, S. Gotzinger, L. de S. Menezes, G. Zumofen, O. Benson I, and V. Sandoghdar, Controlled coupling of counterpropagating whispering-gallery modes by a single Rayleigh scatterer: A classical problem in a quantum optical light, *Phys. Rev. Lett.* 99, 173603, 2007.
13. M. Borselli, T. J. Johnson, and O. Painter, Beyond the Rayleigh scattering limit in high-Q silicon microdisks: Theory and experiment, *Opt. Exp.* 13, 1516, 2005.

14. F. Vollmer and S. Arnold, Whispering-gallery-mode biosensing: Label-free detection down to single molecules, *Nat. Methods* 5(7), 591–596, 2008.
15. X. Fan, I. M. White, S. I. Shopova, H. Zhu, J. D. Suter, and Y. Sun, Sensitive optical biosensors for unlabeled targets: A review, *Anal. Chim. Acta* 620(1/2), 8–26, 2008.
16. S. Arnold, M. Khoshsima, I. Teraoka, S. Holler, and F. Vollmer, Shift of whispering-gallery modes in microspheres by protein adsorption, *Opt. Lett.* 28(4), 272–274, 2003.
17. A. M. Armani, R. P. Kulkarni, S. E. Fraser, R. C. Flagan, and K. J. Vahala, Label-free, single-molecule detection with optical microcavities, *Science* 317(5839), 783–787, 2007.
18. M. Loncar, Molecular sensors: Cavities lead the way, *Nat. Photonics* 1(10), 565–567, 2007.
19. D. Evanko, Incredible shrinking optics, *Nat. Methods* 4(9), 683, 2007.
20. F. Vollmer, S. Arnold, and D. Keng, Single virus detection from the reactive shift of a whispering-gallery mode, *Proc. Natl. Acad. Sci. U.S.A.* 105(52), 20701–20704, 2008.
21. S. A. Wise and R. A. Watters, Bovine serum albumin (7% solution) (SRM 927d), NIST Gaithersburg, MD, 2006.
22. W. E. Moerner and D. P. Fromm, Methods of single-molecule fluorescence spectroscopy and microscopy, *Rev. Sci. Instrum.* 74(8), 3597–3619, 2003.
23. J. B. Jensen, L. H. Pedersen, P. E. Hoiby, L. B. Nielsen, T. P. Hansen, J. R. Folkenberg, J. Riishede et al. Photonic crystal fiber-based evanescent-wave sensor for detection of biomolecules in aqueous solutions, *Opt. Lett.* 29(17), 1974–1976, 2004.
24. A. Polman and H. A. Atwater, Plasmonics: Optics at the nanoscale, *Mater. Today* 8(1), 56, 2005.
25. B. Koch, Y. Yi, J. Zhang, S. Znameroski, and T. Smith, Reflection-mode sensing using optical microring resonators, *Appl. Phys. Lett.* 95, 201111, 2009.
26. B. Koch, L. Carson, C. Guo, C. Lee, Y. Yi, J. Zhang, M. Zin, S. Znameroski, and T. Smith, Hurricane: A simplified optical resonator for optical-power-based sensing with nanoparticle taggants, *Sens. Actuators* B 147, 573–580, 2010.
27. E. D. Palik, in *Handbook of Optical Constants of Solids,* edited by E. D. Palik, Academic, Orlando, FL, 2, 519, 1985.
28. A. Taflove and S. C. Hagness, *Computational Electrodynamics: The Finite Difference Time Domain Method,* Artech House, Inc., Salt Lake City, UT, 2005.
29. W. D. Li and S. Y. Chou, Solar-blind deep-UV band-pass filter (250–350 nm) consisting of a metal nano-grid fabricated by nanoimprint lithography, *Opt. Exp.* 18(2), 931, 2010.
30. S. V. Boriskina, Spectrally engineered photonic molecules as optical sensors with enhanced sensitivity: A proposal and numerical analysis, *J. Opt. Soc. Am.* B 23, 1565, 2006.
31. A. Yariv, Y. Xu, R. K. Lee, and A. Scherer, Coupled-resonator optical waveguide: A proposal and analysis, *Opt. Lett.* 24, 711–713, 1999.
32. M. L. Gorodetsky, A. D. Pryamikov, and V. S. Ilchenko, Rayleigh scattering in high-Q microspheres, *J. Opt. Soc.* Am. B 17, 1051, 2000.
33. M. Borselli, T. J. Johnson, and O. Painter, Beyond the Rayleigh scattering limit in high-Q silicon microdisks: theory and experiment, *Opt. Exp.* 13, 1515, 2005.
34. A. Gondarenko, J. S. Levy, and M. Lipson, High confinement micron-scale silicon nitride high Q ring resonator, *Opt. Exp.* 17, 11366, 2009.
35. E. S. Hosseini, S. Yegnanarayanan, A. H. Atabaki, M. Soltani, and A. Adibi, High quality planar silicon nitride microdisk resonators for integrated photonics in the visible wavelength range, *Opt. Exp.* 17, 14543, 2009.
36. T. Barwicz, M. A. Popovic, P. T. Rakich, M. R. Watts, H. A. Haus, E. P. Ippen, and H. I. Smith, Microring-resonator-based add-drop filters in SiN: fabrication and analysis, *Opt. Exp.* 12, 1437, 2004.
37. D. L. Bailey, J. S. Karp, and S. Surti, *Positron Emission Tomography,* Springer-Verlag, London, 2005.
38. S. E. Derenzo, M. J. Weber, E. Bourret-Courchesne, and M. K. Klintenberg, The quest for the ideal inorganic scintillator, *Nucl. Instrum. Methods Phys. Res. A* 505, 111, 2003.
39. B. Liu and C. Shi, Progress of medical scintillators, *Chin. Sci. Bull.* 47, 1057, 2002.
40. M. J. Weber, Inorganic scintillators: Today and tomorrow, *J. Lumin.* 100, 35, 2002.

41. E. Yablonvitch, Inhibited spontaneous emission in solid-state physics and electronics, *Phys. Rev. Lett.* 58, 2059, 1987.
42. K. J. Vahala, Optical microcavity, *Nature* 424, 839, 2003.
43. J. Pendry, Playing tricks with light, *Science* 285, 1687, 1999.
44. M. Kronberger, E. Auffray, and P. Lecoq, Probing the concepts of photonic crystals on scintillating materials, *IEEE Trans. Nuclear Sci.* 55, 1102, 2008.
45. M. Kronberger, E. Auffray, and P. Lecoq, The concepts of photonic crystals on scintillating materials, *IEEE Nucl. Sci. Symp. Conf. Rec.* M06-179, 3914, 2008.
46. A. Knapitsch, E. Auffray, C. W. Fabjan, J.-L. Leclercq, P. Lecoq, X. Letartre, and C. Seassal, Photonic crystals: A novel approach to enhance the light output of scintillation based detectors, *Nucl. Instrum. Methods Phys. Res. A* 628, 385, 2011.
47. N. Ganesh, W. Zhang, P. C. Mathias, E. Chow, J. A. N. T. Soares, V. Malyarchuk, A. D. Smith, and B. T. Cunningham, Enhanced fluorescence emission from quantum dots on a photonic crystal surface, *Nat. Nanotechnol.* 2, 515, 2007.
48. M. Boroditsky, T. F. Krauss, R. Coccioli, R. Vrijen, R. Bhat, and E. Yablonovitch, Light extraction from optically pumped light-emitting diode by thin-slab photonic crystals, *Appl. Phys. Lett.* 75, 1036, 1999.
49. H. Ichikawa and T. Baba, Efficiency enhancement in a light-emitting diode with a two dimensional surface grating photonic crystal, *Appl. Phys. Lett.* 84, 457, 2004.
50. S. Fan, P. R. Villeneuve, J. D. Joannopoulos, and E. F. Schubert, High extraction of spontaneous emission from slabs of photonic crystals, *Phys. Rev. Lett.* 78, 3294, 1997.
51. P. Gao, M. Gu, X. L. Liu, B. Liu, S. Huang, and C. Ni., X-ray excited luminescence of cuprous iodide single crystals: On the nature of red luminescence, *Appl. Phys. Lett.* 95, 221904, 2009.
52. C. H. Sun, P. Jiang, and B. Jiang, Broadband moth-eye antireflection coatings on silicon, *Appl. Phys. Lett.* 92, 061112, 2008.
53. S. A. Boden and D. M. Bagnall, Tunable reflection minima of nanostructured antireflective surfaces, *Appl. Phys. Lett.* 93, 133108, 2008.
54. H. Sai, Y. Kanamori, K. Arafune, Y. Ohshita, and M. Yamaguchi, Light trapping effect of submicron surface textures in crystalline Si solar cells, *Prog. Photovolt. Res. Appl.* 15, 415, 2007.

9

Surface Plasmon Resonance-Based Devices: Simulations, Design, and Applications

Mina Ray, Mahua Bera, Kaushik Brahmachari, Sharmila Ghosh, and Sukla Rajak

CONTENTS

9.1 Introduction

The collective oscillations of free charges in a material due to an applied electromagnetic field are responsible for plasmonic phenomena occurring at optical and telecommunication frequencies. The essential property of any plasmonic material is that it should have negative real permittivity, which would be provided by the free electrons in the material. In presence of an electric field satisfying certain desirable properties, the free carriers in the metal tend to move in such a manner so as to screen this field. The Drude free-carrier model reveals that this screening effect is responsible for negative permittivity. Surface plasma waves (SPWs) represent a particular mode of guided waves governed by Maxwell's equations and propagate along the metal–dielectric interface, whereas surface plasmon polaritons (SPPs) are the quantization of these waves. They can also be looked upon as the transverse electromagnetic (TEM) wave propagating along the interface of metal and dielectric having real dielectric permittivity of opposite signs. Using the attenuated total reflection (ATR) coupler method and satisfying the phase-matching condition, surface plasmon resonance (SPR) occurs. Otto and Kretschmann are the two well-known prism coupling-based configurations for the SPR measurement [1,2]. These techniques are widely used for chemical and biological sensing [3,4].

A theoretical investigation on sensitivity comparison of prism and grating coupler-based plasmonic sensors in angular and wavelength interrogation modes was reported by Homola et al. [5]. Plasmonic sensors that are made from silica glass restrict their operation within the visible range. However, plasmon excitation in the infrared (IR) region is found to be very advantageous for sensing purposes and it involves the use of high-index coupling prism materials such as chalcogenide [6] and silicon [7]. Due to favorable IR transmission properties and high refractive index (RI), they give narrower resonance curves, which further help in precise determination of SPR dip position in the IR region of wavelength. Modeling of silicon and chalcogenide material-based plasmonic sensor using IR light was reported earlier [8–10]. Studies on plasmonic sensor comprising high-index coupling prism materials were also reported [11,12].

In addition to the conventional three-layer configuration, modified four-layer structures are also used with the aim of achieving certain additional advantages. Due to the introduction of an intermediate dielectric layer, two decoupled modes excite at the two opposite boundaries of the metal films [13]. Further, these two modes are designated as the symmetric surface plasmon (S-SP) or long-range surface plasmon (LRSP) mode and the antisymmetric surface plasmon (A-SP) or short-range surface plasmon (SRSP) mode according to their magnetic intensity distribution and propagation length, respectively. Theoretical and experimental investigations of LRSP and SRSP modes [14–16] claim to provide an improved measurement accuracy compared to conventional three-layer Kretschmann configuration. Optimum dielectric and metal thicknesses provide higher resolution for designing a nanoplasmonic sensor [17,18]. Simultaneous excitations of symmetric and antisymmetric modes differentiate surface RI change from interfering bulk RI change. In this context, the issues of cross-sensitivity and limit of detection for an efficient sensor have been analyzed previously [19–21]. Dual-mode SPR sensor using bimetallic nanofilms gives certain advantages over the single metal layer and also provides flexibility in the optimization procedure [22–24]. Tailoring of the structure of individual metal films of bimetallic loaded dielectric plasmonic structure realizing equi- and nonequireflectance modes has also been proposed, which may further be utilized for realizing plasmonic trinary logic using binary di-bit images [25]. Moreover, optical light modulation, optical interconnects, and second-harmonic generation are the important aspects of the applications in view of the present scenario [26–28].

Admittance loci method has been used in thin-film modeling [29] and design of plasmonic structures [30–32]. Prism material dependency as well as sensing application of plasmonics based on admittance loci analysis has been reported earlier [33,34]. Gupta and Kondoh [35] worked on tuning and sensitivity enhancement of plasmonic sensor. Work on the design of plasmonic biosensor has been reported earlier [36,37]. A plasmonic structure consisting of Ag–Au bimetallic alloy film has also been studied and designed using the admittance loci method [38]. The performance of different metals in fiber optic SPR sensor is discussed in detail in Reference 39. Design of a plasmonic structure using a ceramic prism material [40] and also a nanocomposite film [41] has been recently reported.

Coupling of waveguide resonance and plasmonic resonance provides an enhanced measurement precision in biological or chemical nanoplasmonic sensing [42]. Coupled plasmon waveguide resonance (CPWR) structure has a structural modification in comparison with the LRSP–SRSP structure, a positional interchange in the metallo-dielectric block. Progressive multiple resonances in the CPWR structure characterize the waveguide material and can be used as a thickness monitoring device in a multilayer composite nanoplasmonic structure [43]. Also different types of coupled nanoplasmonic structures are compared for the performance of biosensors [44]. Waveguide-coupled SPR (WCSPR) structure produces multiple resonance dips, which can further detect more than one optical property of the biomolecular layer [45]. Model analysis of the WCSPR structure using homo- and hetero-bimetallic nanofilms provides an overview of angular figure of merit (AFOM) for the performance of an efficient nanoplasmonic sensor [46]. Also the parametric influence of the multilayer metallo-dielectric (MMD) nanoplasmonic structure was investigated in order to improve the performance of biological and chemical sensors in our previous work [47].

Surface plasmon waves find a large number of application areas such as near-field microscopy, nanoscale imaging, medicine, nanobiotechnology, and environmental monitoring due to their enhancement and resonance characteristics in the transverse plane with respect to the metal–dielectric interface. Another field related to near-guided-wave SPR (NGW-SPR) excitation is important for near-field microscopy, medical and biological imaging, optical biosensors, ultrasensitive detectors, and biomedical spectroscopy [48]. Sensitivity and field enhancement offered by SPR are very much effective for these purposes [49]. Phase-sensitive SPR is also an important new area that provides higher inherent accuracy as well as increased sensitivity [50].

For coupling between the applied electromagnetic field and SPPs, which results in resonance, the frequency of the light must be lower than the screened plasma absorption frequency of the conductor. According to Drude free electron model, the screened plasma frequency $\omega_{ps} = \sqrt{ne^2/\mu\varepsilon_0\varepsilon_\infty}$, where n is the charge carrier density, μ is their effective mass, ε_0 is the permittivity of free space, and ε_∞ is a high-frequency dielectric constant [49,51]. For noble metals such as gold and silver, ω_{ps} lies in the ultraviolet (UV) to visible regions of optical frequencies. This is the main reason behind the use of noble metals for most SPW studies in the visible wavelength region. However, due to interband electronic transitions in the metals and large losses in the visible and UV spectral ranges, their uses are limited. The metals even with the highest conductivities suffer from large losses due to their very large negative real permittivity in the near-IR and visible optical frequencies [52,53], which remained a major obstacle in the design and fabrication of efficient plasmonic devices. New plasmonic materials have the promise of overcoming this major bottleneck and enabling high-performance devices. Also, new plasmonic materials allow greater flexibility in the design of a device owing to the moderate magnitude of the real part of permittivity in such materials. Alternative plasmonic materials have two other major advantages: They can exhibit tunable optical properties [54] and they can be compatible with standard fabrication and integration procedures [55]. Clearly, alternative plasmonic materials have significant advantages over conventional metals for plasmonic and metamaterial (MM) designs.

As alternatives to conventional metals for the production of transparent electrodes for optoelectronic device applications and solar cells [56–58], transparent conducting oxides (TCOs) (such as metal-doped In_2O_3 and ZnO) and transition metal nitrides (TMNs) (TiN, ZrN, and HfN) are greatly used. Most optically transparent and electrically conducting oxides and TMNs are binary or ternary compounds, containing one or two metallic elements. The electrical resistivity of these alternative materials should be ~10^{-5} Ωcm and their extinction coefficient k in the optical visible range could be lower than 0.0001. These TCO and TMN materials have a wide optical bandgap of ~3 eV.

Alternative plasmonic materials in the near-IR and visible ranges can be classified into several categories such as semiconductor-based materials [59], intermetallics [60], ceramics [61], and organic materials [62]. Semiconductor-based oxides and TMNs can be used as alternative plasmonic materials in the near-IR and visible ranges, respectively. These materials have advantages over the other types, since oxides enable low-loss all-semiconductor-based plasmonic and MM devices in the near-IR frequencies, whereas metal nitrides are complementary metal–oxide–semiconductor compatible and provide alternatives to gold and silver in the visible frequencies. Here, we study their optical properties in the context of plasmonic applications.

Furthermore, a novel analysis technique together with simulations was proposed, which detects the phase shift between the p-polarized and s-polarized light associated with the SPR [63]. From a differential phase measurement, the RI of the sensing medium can be obtained. An interferometric phase-based sensor has been shown as a novel and potentially very sensitive phase detection technique. The presence of SPR enhances the phase change at the sensing surface. We also theoretically compared this technique with the conventional one by varying the concentration of sugar solution resulting in a subsequent change of RI of the sample. Again better performance of the phase-based sensor and better system accuracy can be expected using more sophisticated instrumentation.

A technique has also been proposed based on image analysis of the surface plasmon excitation at the metal–dielectric interface of inside silver-coated fused silica capillary glass tube. Chemical deposition technique has been used for the deposition of silver. Angular interrogation in Kretschmann-like configuration is realized by nonradial transverse illumination of this cylindrical dielectric–metal–dielectric structure with a He–Ne laser source. Here the uniform film deposition of the inside surface of the capillary is not that crucial except within the transversely illuminated working area concerned. Moreover, the proposed technique has been validated experimentally for sensing different aqueous dielectric samples inserted inside the tube [64]. Significant improvement in simplicity of measurement, lower cost, and no index-matching fluid are the major advantages of this interferogram approach over many existing and published sensing methods already developed.

An experimental approach for investigation on the observation of SPR at the metal–dielectric interface of silver-coated tapered light guiding glass rods of different dimensions has also been reported [65]. Angular interrogation in Kretschmann-like configuration of this nonplanar (cylindrical) structure is used to locate the resonance dip in the reflectance measurement.

9.2 Theoretical Background for Simulation

Numerical approach is based on the ATR coupler method. The detailed mathematical analysis for a multilayer structure has been reported in our earlier research article [18]. Here only the generalized formulation will be discussed.

9.2.1 Phase Matching

To excite the SPs, the wavevector of the incident light in the coupling glass prism (pr) must phase match with the wavevector of the SPs at the metal–dielectric interface. Therefore, the resonance condition of the light in the prism (pr) with the surface plasmon (sp) at the metal–dielectric interface (Kretschmann–Raether configuration) is

$$K_x^{\mathrm{pr}} = K_x^{\mathrm{sp}} \tag{9.1}$$

$$\sqrt{\varepsilon_1}\frac{\omega}{c}\sin\theta_1 = \frac{\omega}{c}\sqrt{\frac{\varepsilon_2\varepsilon_3}{\varepsilon_2+\varepsilon_3}} \tag{9.2}$$

where:

ε_1, ε_2, and ε_3 are the dielectric permittivities of prism, metal, and dielectric layers, respectively

θ_1 is the incident angle in the prism

ω is the angular frequency of the incident light

9.2.2 Dispersion Relation and Drude Model

For our general analysis, we have taken high RI SF11 glass prism as the coupling device and fused silica as the dielectric or waveguide material. The RI of these two materials varies with the wavelength according to the following Sellmeier dispersion formula:

$$n^2(\lambda) = 1 + \frac{B_1\lambda^2}{\lambda^2 - C_1} + \frac{B_2\lambda^2}{\lambda^2 - C_2} + \frac{B_3\lambda^2}{\lambda^2 - C_3} \tag{9.3}$$

where:

the coefficients B_1, B_2, B_3, C_1, C_2, and C_3 have certain numeric values

λ depicts the wavelength in micrometers

The refractive indices for any wavelength within the range between 0.2483 and 2.3254 μm can be calculated using specified constants.

According to the Drude model, the dielectric function (ε_m) of any metal can be written as

$$\varepsilon_{\mathrm{m}}(\lambda) = \varepsilon_{\mathrm{mr}} + i\varepsilon_{\mathrm{mi}} = 1 - \frac{\lambda^2\lambda_c}{\lambda_p^2(\lambda_c + i\lambda)} \tag{9.4}$$

where:
λ_p denotes the plasma wavelength
λ_c denotes the collision wavelength

9.2.3 Theoretical Background for Admittance Loci-Based Analysis

A bioplasmonic structure can be designed using the admittance loci method. In this method, the admittance of a plasmonic structure is considered, which starts from the sample and ends at the front surface of the structure. We consider a three-layer bioplasmonic structure consisting of prism, gold metal film, and biosample with refractive indices represented by n_{pr}, n_m, and n_{sample}, respectively.

For metallic thin film, the phase is given by

$$\delta_m = \left(\frac{2\pi}{\lambda}\right) d_m \left(n_m^2 - k_m^2 - n_{pr}^2 \sin^2\theta_i - 2in_m k_m\right)^{1/2} \tag{9.5}$$

where:
n_m and k_m are the real and imaginary parts of the complex RI of the metallic thin film, respectively
λ is the wavelength of incident light

The chalcogenide glass (2S2G) RI dispersion relation is given by [6]

$$n_{pr}(\lambda) = 2.24047 + 2.693 \times 10^{-2}\lambda^{-2} + 8.08 \times 10^{-3}\lambda^{-4} \tag{9.6}$$

For silicon, the Sellmeier RI dispersion relation is given by [7]

$$n_{pr}(\lambda) = \sqrt{11.6858 + \frac{0.939816}{\lambda^2} + \frac{0.000993358}{\lambda^2 - 1.22567}} \tag{9.7}$$

The values of plasma wavelength and collision wavelength of gold (Au) metal are taken from the literature [39]. The RI dispersion for three blood groups in the form of Cauchy expression is given by [36]

$$n_{sample} = 1.357 + \frac{A}{\lambda^2} + \frac{B}{\lambda^4} \tag{9.8}$$

where:
λ is the wavelength of incident light
A and B are Cauchy coefficients having different values for different blood groups

Therefore, we can write the admittance of a plasmonic structure as

$$Y = \frac{\eta_{sample}\cos\delta_m + i\eta_m \sin\delta_m}{\cos\delta_m + i(\eta_{sample}/\eta_m)\sin\delta_m} \tag{9.9}$$

where:
η_m and η_{sample} are the admittances of metallic thin film and sample, respectively.

The reflectance of a plasmonic structure is given by

$$R = \left(\frac{\eta_{pr} - Y}{\eta_{pr} + Y}\right)\left(\frac{\eta_{pr} - Y}{\eta_{pr} + Y}\right)^* \tag{9.10}$$

where:

η_{pr} is the admittance of incident medium (prism material)

The performance of a plasmonic structure can be studied by plotting isoreflectance contours along with the admittance loci plot of the structure in the admittance diagram. These isoreflectance contours are the circles with centers on real axis, centers and radii being given by $(\eta_{pr}(1+R)/(1-R),0)$ and $2\eta_{pr}(R)^{1/2}/(1-R)$, where η_{pr} is the admittance of the incident medium (prism) and R is the reflectance.

At an oblique incidence for p-polarized light, the modified optical admittances are given by

For dielectric sample

$$\eta_{sample} = \frac{y_{sample}\cos\theta_i}{\cos\theta_{sample}} \tag{9.11}$$

where:

$y_{sample} = n_{sample} y_f$ is the optical admittance of the sample (y_f is the admittance of free space and its value is unity in Gaussian units and n_{sample} is the RI of the sample medium concerned)

θ_i is the angle of incidence

θ_{sample} is the angle corresponding to sample

For metallic thin film

$$\eta_m = \frac{(n_m - ik_m)^2\cos\theta_i}{\left(n_m^2 - k_m^2 - n_{pr}^2\sin^2\theta_i - 2in_m k_m\right)^{1/2}} \tag{9.12}$$

Theoretically, as from Equation 9.2, the propagation constant of a surface plasmon wave (SPW) propagating at the interface between a metal and a dielectric sample can be rewritten as

$$k_{SPW} = k_0\sqrt{\frac{\varepsilon_m n_{sample}^2}{\varepsilon_m + n_{sample}^2}} \tag{9.13}$$

The real part of propagation constant of an SPW wave is given by

$$\text{Re}(k_{SPW}) \cong k_0\sqrt{\frac{\varepsilon_{mr} n_{sample}^2}{\varepsilon_{mr} + n_{sample}^2}} \tag{9.14}$$

The coupling condition is given by

$$k_0 n_{pr}\sin\theta_i = k_0\sqrt{\frac{\varepsilon_{mr} n_{sample}^2}{\varepsilon_{mr} + n_{sample}^2}} \tag{9.15}$$

where:

ε_m is the dielectric constant of the metallic thin film

ε_{mr} is the real part of dielectric constant of the mentioned metallic thin film

k_0 is the free space wavenumber

The sensitivity of a plasmonic sensor under consideration is given by [5]

$$S = \frac{d\theta_{SPR}}{dn_{sample}} = \frac{[\varepsilon_{mr}/(\varepsilon_{mr} + n_{sample}^2)]^{3/2}}{\sqrt{n_{pr}^2 - [\varepsilon_{mr} n_{sample}^2/(\varepsilon_{mr} + n_{sample}^2)]}} \tag{9.16}$$

where:
$d\theta_{SPR}$ is the small change in SPR angle corresponding to small change in sample RI, dn_{sample}

9.2.4 General Solution of *N*-Layer Model Using Characteristic Transfer Matrix Method

The reflectance and transmittance of the generalized *N*-layer model can be computed by solving Maxwell's equation and Fresnel reflection and transmission coefficient [66]. Our approach will be based on the characteristic transfer matrix method (CTM). The generalized formula for the characteristic matrix for the *k*th layer of a multilayer nanoplasmonic structure is given by [67]

$$M_k = \begin{pmatrix} \cos\beta_k & -i\sin\beta_k/q_k \\ -iq_k \sin\beta_k & \cos\beta_k \end{pmatrix} \tag{9.17}$$

where:

$$q_k^{TM} = \sqrt{\left(\frac{1}{\varepsilon_k}\right)} \cos\theta_k \text{ and } q_k^{TE} = \sqrt{\varepsilon_k} \cos\theta_k \tag{9.18}$$

β_k is the phase factor

$$\beta_k = \left(\frac{2\pi}{\lambda}\right) n_k \cos\theta_k \left(z_k - z_{k-1}\right) \tag{9.19}$$

The reflection coefficient for the multilayer nanoplasmonic structure is given by

$$r = \frac{\left[\left(M_{11} + M_{12}q_N\right)q_1 - \left(M_{21} + M_{22}q_N\right)\right]}{\left[\left(M_{11} + M_{12}q_N\right)q_1 + \left(M_{21} + M_{22}q_N\right)\right]} \tag{9.20}$$

where:

$$M_{ij} = \left(\prod_{k=2}^{N-1} M_k\right)_{ij}, \quad i,j = 1,2 \tag{9.21}$$

where:
i, j denote the row and column indices, respectively
k denotes the layer index
q_1 and q_N are calculated for the incident medium (prism) and the final medium (analyte)
N denotes total number of layers under consideration

Now the reflectance of a multilayer system is given by

$$R = |r|^2 = rr^* \tag{9.22}$$

where:

* denotes the complex conjugate

r is the complex amplitude reflection coefficient

$$r = R^{1/2} e^{i\phi_r} \tag{9.23}$$

and r can also be written in the functional form as

$$r = |r|(\lambda, \theta, n, d) e^{i\phi_r(\lambda, \theta, n, d)} \tag{9.24}$$

where:

λ, θ, n, d are the working wavelength, incident angle, RI, and thickness of the layers of the plasmonic structure, respectively

ϕ_r is the phase shift of the reflected wave

$$\phi_r = \arg(r) = \tan^{-1}\left[\frac{\text{Im}(r)}{\text{Re}(r)}\right] \tag{9.25}$$

At the phase-matching condition, when the energy of the incident light beam gets totally transferred to the surface plasmons (SPs), the reflectance of the multilayer plasmonic structure goes to zero and produces a reflection dip. Mathematically, from the differential calculus, the condition of minimum of any function can be written as from Equation 9.22

$$\left(rr^*\right)' = 0 \tag{9.26}$$

and

$$\left(rr^*\right)'' \rangle 0 \tag{9.27}$$

where:

the symbols ′ and ″ denote the first and second derivatives, respectively, with respect to any dependent variables (λ, θ, n, d) as the case may be

Hence, from Equation 9.26

$$\frac{r'}{r} = -\frac{r^{*\prime}}{r^*} \tag{9.28}$$

Moreover,

$$\phi_r{}' = -i\left(\frac{r'}{r} - \frac{|r|'}{|r|}\right) \tag{9.29}$$

and

$$\frac{r'}{r} = (\pm i)\frac{|r|'}{|r|} \quad \text{at } \phi_r' = 0 \tag{9.30}$$

According to the differential calculus, $\phi_r(\theta_{SPR}/\lambda_{SPR})$ is positive or negative, when $\phi_r(\theta/\lambda)$ is increasing or decreasing in a suitably restricted neighborhood of $\theta_{SPR}/\lambda_{SPR}$.

When r goes through the point $r = 0$ in the complex plane, the parameter d passes through its optimum value and the reflectance of the multilayer structure posses its minimum value (zero) with a phase jump at that point.

To solve the sign ambiguity and to remove the discontinuity in the phase, it is needed to add or subtract integer multiples of 2π as needed. This method is called phase unwrapping as the phase is integrated or unwrapped along a line or path counting the 2π discontinuities and scanning all the values lying between 0 and 2π.

When the reflectance of SPR acquires its minimum value, the electromagnetic field posses a maximum value at the metal–dielectric interface. Thus, the electric (E) or magnetic (H) field enhancement factor is the ratio of the field intensity at the metal–dielectric interface and the incident field intensity in the prism. Therefore, the transmission coefficient for magnetic field is

$$t_H = \frac{2q_1}{\left(M_{11} + M_{12}q_N\right)q_1 + \left(M_{21} + M_{22}q_N\right)} \tag{9.31}$$

and the corresponding transmittance or magnetic field enhancement factor is

$$T_H = F_H = \left|t_H\right|^2 \tag{9.32}$$

Similarly, the transmission coefficient for electric field is

$$t_E = \frac{\mu_N n_1}{\mu_1 n_N} t_H \tag{9.33}$$

Assuming the permeabilities $\mu_N = \mu_1 \cong 1$,

$$t_E = \frac{n_1}{n_N} t_H \tag{9.34}$$

and the corresponding transmittance or electric field enhancement factor is

$$T_E = F_E = \left|t_E\right|^2 \tag{9.35}$$

Detailed reflectivity calculations of four-layer and five-layer nanoplasmonic structures based on the Fresnel equations have been reported in our earlier publications [18,43]. Here also a four-layer structure has been investigated. From the generalized N-layer model, the characteristic matrix for such a model will be

$$M = M_1 * M_2 \tag{9.36}$$

where:

M_1 and M_2 are the characteristic matrices for the metal and dielectric, respectively

The same formulation can be applied for five- and six-layer structures also. Using this characteristic matrix and calculating q and β for the corresponding layers, the reflectance, transmittance, and phase can be computed for multilayer structures.

9.3 Computational Approach and Simulation Model

Three-layer Kretschmann configuration consists of 50 nm gold (Au) metal layer deposited on a high RI glass prism and the sensing medium is in contact with the metal layer as shown in the schematic diagram in Figure 9.1a. Reflectance contour plot in simultaneous angular and wavelength interrogation of this configuration is also demonstrated in Figure 9.1b along with the two-dimensional (2D) reflectance, phase, and E-field enhancement in angular and wavelength regimes in the inset.

Simulated admittance loci plots of a plasmonic structure (SF11–Au–water) are shown in Figure 9.1c, which demonstrates that the starting admittance (the position of which is dependent

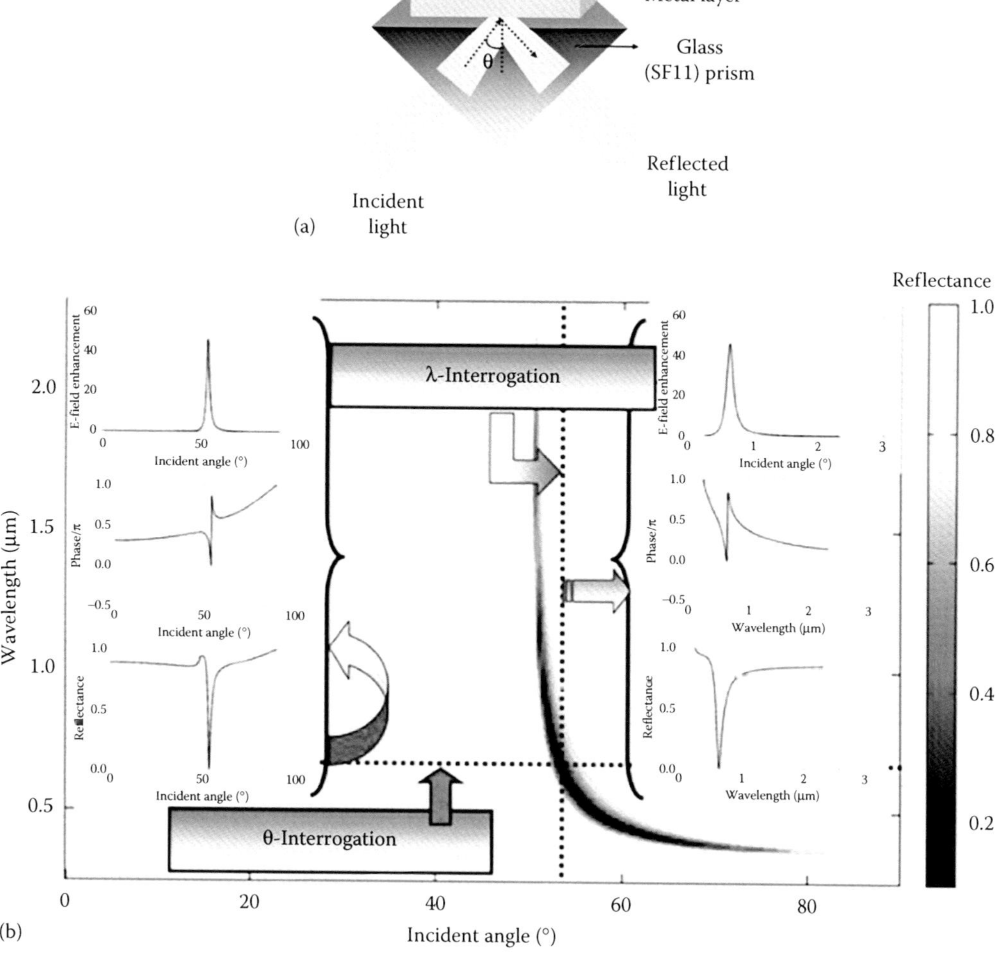

FIGURE 9.1
(a) Schematic diagram of three-layer Kretschmann configuration. (b) Reflectance contour plot in simultaneous angular and wavelength interrogation along with the 2D reflectance, phase, and E-field enhancement in both the angular and wavelength regimes in the inset. (c) Admittance loci plots of a plasmonic structure comprising SF11 prism–Au–water at four different wavelengths. (Data from Bera, M. and Ray, M., *IEEE Photon. Technol. Lett.*, 25, 1965–1968, 2013.)

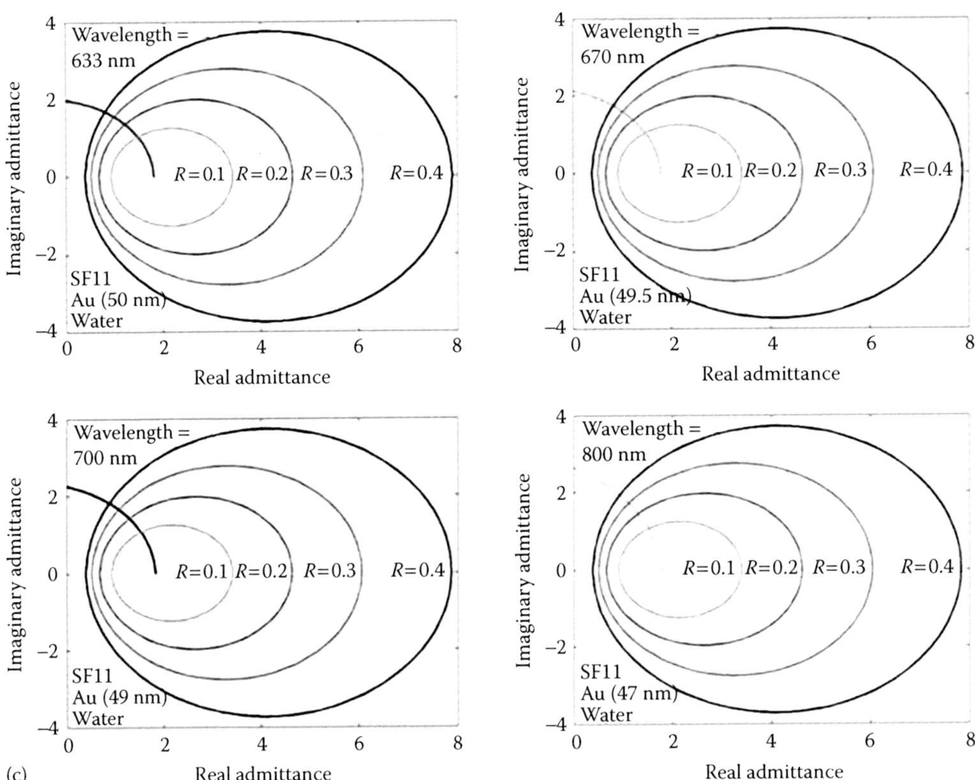

FIGURE 9.1
Continued

on the RI of the sample) for the gold film is on the imaginary axis, and the locus point moves from a point on the imaginary axis corresponding to the initial value of gold film thickness and ends at a point (the value of which is close to RI of prism) on the real axis corresponding to the final value of gold film thickness at a particular angle of incidence under a fixed wavelength. It also corresponds to near-zero reflectance for that particular angle of incidence and gold metal film thickness. If we had made this locus to intercept the real axis of the admittance diagram exactly at the RI of the incident medium (prism) depending upon the prism material we are using, the excitation of SPs would have been achieved with maximum efficiency. In all these cases, we have optimized the gold metal film thickness values for each wavelength so as to ensure that the respective metal loci end with the real admittance value closer to the value of RI of the prism material, which in turn ensures the most efficient excitation of SPs.

Our computational approach is mainly based on the Fresnel equation and CTM method. Numerical simulations cover two main areas such as admittance loci and resonance parameter-based analysis in both the angular and wavelength regimes and use three- and four-layer structures as schematically indicated in Figure 9.2. Different prisms are used as coupling devices for investigation in the visible and IR regions. We have also incorporated mono- and bimetallic layers with different metal films in the MMD structure. Moreover, metal oxides are also good replacements for noble metals in nanoplasmonic sensors. The SPR measurement technique can detect the optical properties of biological and chemical materials. Here, mainly optical properties of some biological materials such as protein, human blood samples, and hemoglobin are investigated.

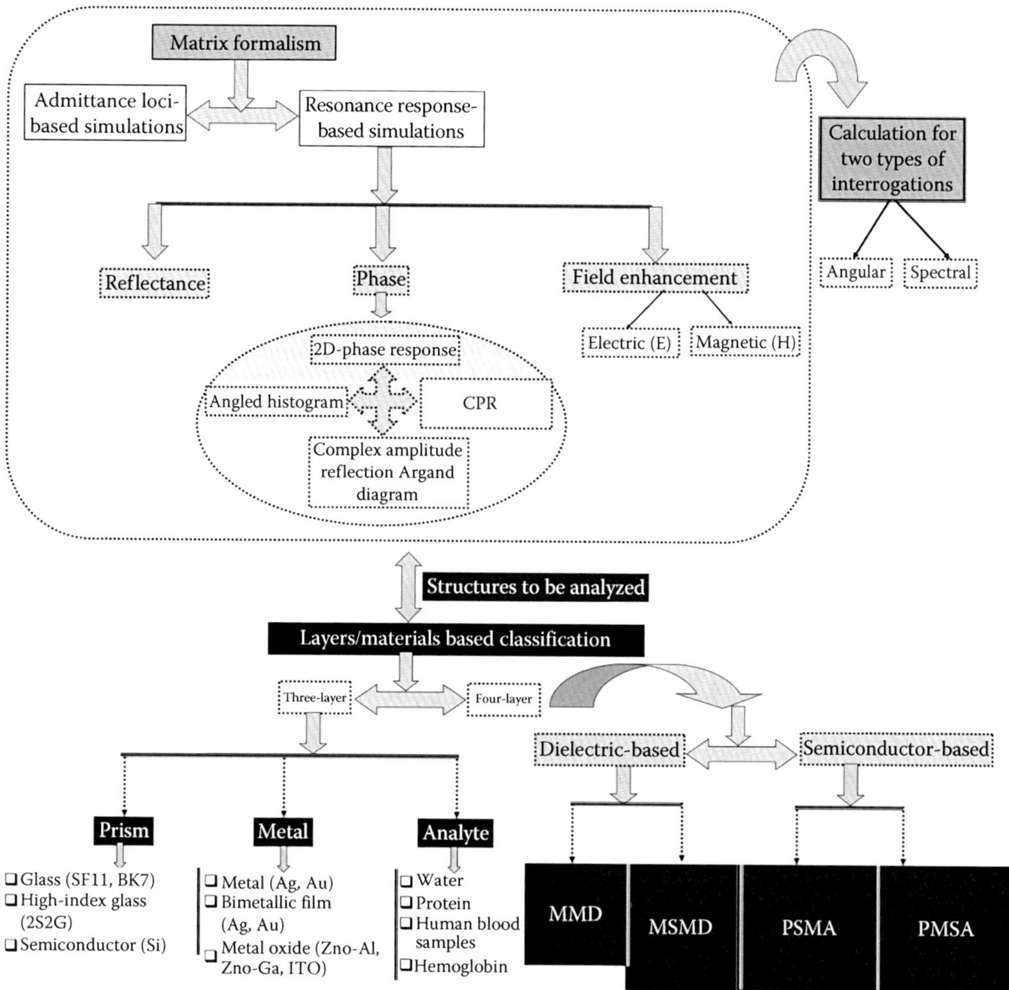

FIGURE 9.2
A schematic overview of simulation model and structure-based classification.

9.4 Admittance Loci-Based Simulations (Using High-Index Prism)

As high-index material offers certain advantages when sensing application is concerned, this section is focused on this issue. The admittance loci method is used to design and analyze the bioplasmonic structure concerned with SPR-based sensing with the main emphasis given on the role of high-index prism materials. To investigate the effect of high-index coupling prism materials in such structures, different prism materials are used in a Kretschmann-type bioplasmonic structure consisting of prism–gold metal film–sample. This section will first describe the admittance loci method and then its use to compare the influence of silicon and chalcogenide (2S2G) prism materials on the design of bioplasmonic structure.

In this work, the admittance loci method is used to design the high-index chalcogenide and silicon prism material based bioplasmonic structure at 700 nm wavelength. The admittance

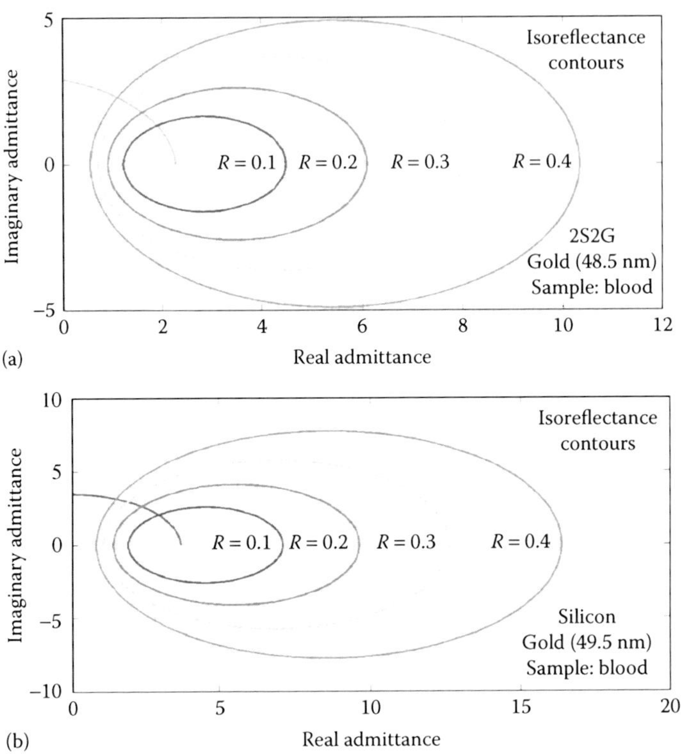

FIGURE 9.3
Admittance loci plots of the bioplasmonic structure with isoreflectance contours for (a) 2S2G and (b) silicon prism materials, respectively.

loci plots along with isoreflectance contours of the bioplasmonic structure consisting of prism (incident medium), gold metal film and biosample using incident p-polarized light for two different prism materials: 2S2G and silicon with human blood as a biosample are shown in Figure 9.3a and b.

In Figure 9.3a, we have simulated the admittance loci plot of a bioplasmonic structure using human blood as a sample and chalcogenide (2S2G) as a prism material (incident medium), where the locus point moves from 2.889i on the imaginary axis corresponding to gold film thickness of 0 nm and ends at 2.258 (which is close to RI of 2S2G at 700 nm wavelength = 2.3291 at an angle of incidence = 38.98°) on the real axis corresponding to gold film thickness of 48.5 nm. If we had made this locus to intercept the real axis of the admittance loci plot exactly at the RI of the incident medium (prism material) depending upon the prism type we are using (e.g., 2S2G, silicon), the plasmonic excitation would have been much more efficient.

Similarly, in Figure 9.3b, the admittance loci plots for silicon prism material have been shown. The calculated values of admittances, angle of incidence, wavelength and optimized gold metal film thickness are tabulated in Table 9.1 for two prism materials. In all these cases, we have optimized the gold metal film thickness for each prism material so as to ensure that the respective metal loci end with the real admittance value close to the value of RI of the prism material, which in turn ensures the most efficient excitation of SPs.

TABLE 9.1

Admittance-Related Parameters for Different Prism Material Configurations

Prism Materials	Thickness of Gold Metal Film (nm)	Starting Imaginary Admittance	End Admittance	Angle of Incidence (°)	Wavelength (nm)
2S2G (RI = 2.3291)	48.5	2.889i	2.258, –0.01106i	38.98	700
Silicon (RI = 3.6881)	49.5	3.467i	3.679, –0.001312i	23.36	700

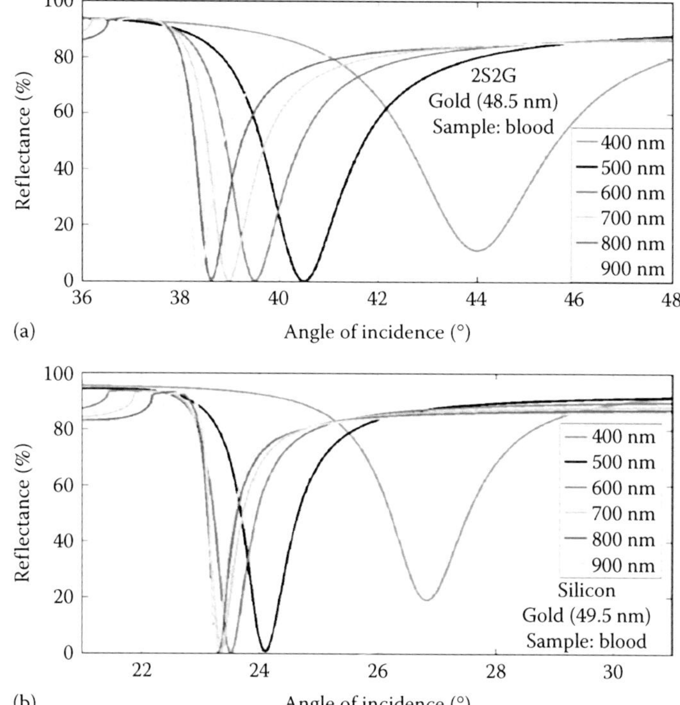

FIGURE 9.4
SPR curves of the bioplasmonic structure at different wavelengths for (a) 2S2G and (b) silicon prism materials.

Figure 9.4a and b shows the angular interrogation-based SP sensing curves for 2S2G and silicon prism materials at different wavelengths as indicated in the figure legends. It can be seen from these plots that SPR occurs at lower incidence angle with an increase in prism material RI. The width of the SPR curve is higher for 2S2G prism material and lower for silicon prism material, and also these SPR curves become much sharper at higher wavelength than at lower wavelength.

Figure 9.5 depicts the comparison of wavelength interrogation-based SPR curves for prism materials, such as chalcogenide (2S2G) and silicon, with human blood sample. Here, gold metal film thickness is optimized for each prism material in order to achieve minimum reflectance.

Figure 9.6 shows the SPR momentum-matching condition with different prism materials, such as 2S2G and silicon, at 700 nm wavelength. It is obtained by simulating the left-hand side (propagation constant of incident light) and the right-hand side (propagation

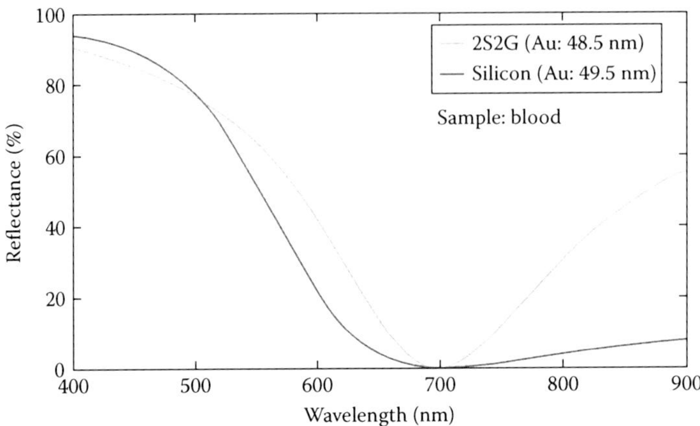

FIGURE 9.5
SPR curves of the bioplasmonic structure with optimized gold metal film thickness for 2S2G and silicon prism materials.

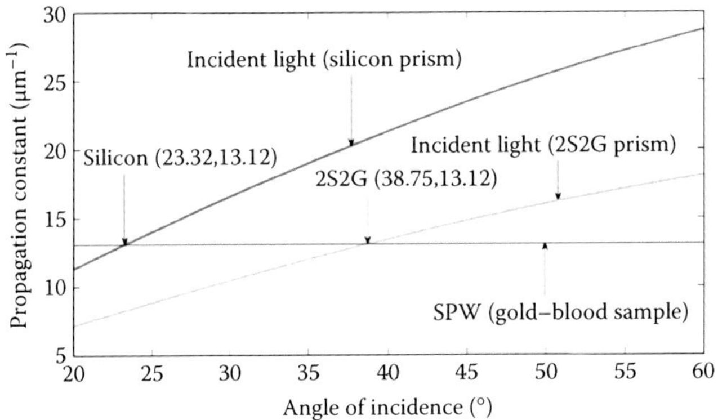

FIGURE 9.6
Propagation constant plots of the bioplasmonic structure for 2S2G and silicon prism materials.

constant of SPW) of Equation 9.15. This simulation is done at 700 nm wavelength. The intersections of curves correspond to the propagation constant of incident light and the propagation constant of SPW that satisfy Equation 9.15. From this, it can also be concluded that for a fixed plasmon-active metal film and biosample, the prism RI varies inversely with the SPR angle; therefore, the SPR angle is lower for higher index prism material and vice versa.

The dynamic range is very crucial parameter for designing a bioplasmonic structure. It is defined as the range of dielectric samples that can be sensed as governed by the SPR condition. The maximum value of the RI of the sample that can be sensed by a particular prism material-based bioplasmonic configuration at a particular wavelength can be obtained using Equation 9.15, for which $\sin(\theta_{SPR})$ is just less than 1. Figure 9.7 shows the plot of dynamic range with variation of wavelength for different prism materials under consideration. This plot reveals that the dynamic range is higher for high-index prism materials such as silicon and comparatively lower for prism materials such as 2S2G.

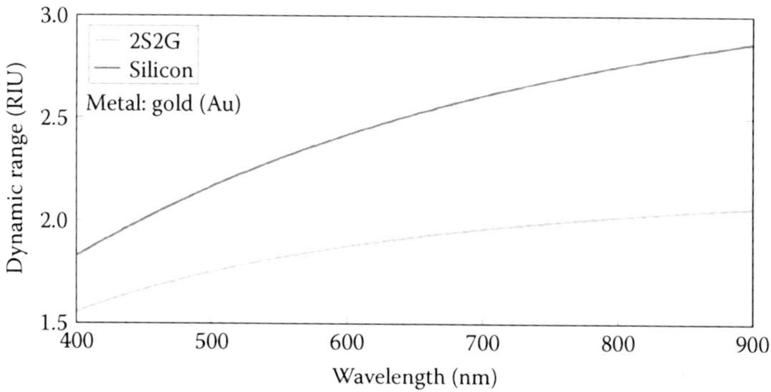

FIGURE 9.7
Dynamic range plot of the bioplasmonic structure with variation of wavelength for 2S2G and silicon prism materials.

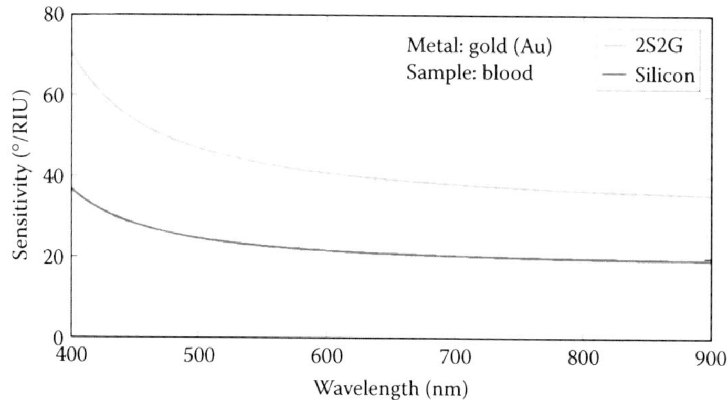

FIGURE 9.8
Sensitivity plot of the bioplasmonic structure with variation of wavelength for 2S2G and silicon prism materials.

Sensitivity plot with variation of wavelength for two prism materials is shown in Figure 9.8. It is seen that the 2S2G prism material-based bioplasmonic structure shows higher sensitivity than the silicon prism material-based bioplasmonic structure for all wavelengths. It is evident that sensitivity is higher while operating the bioplasmonic structure at lower wavelength (400 nm) compared to higher wavelength (900 nm), and also it does not vary much at higher wavelengths. It can be concluded that at lower wavelengths the sensitivity is governed by an occurrence of singularity when $|\varepsilon_{mr}| = n_p^2 n_{sample}^2/(n_p^2 - n_{sample}^2)$ and the sensitivity at higher wavelengths mainly depends on the RI contrast between the prism and the biosample and decreases with increasing contrast [5]. Actually, the optical constants of the metal-supporting SPW are more pronounced at lower wavelengths near the singularity in sensitivity and for bioplasmonic structures that use metals with low value of $|\varepsilon_{mr}|$ the singularity occurs at higher wavelengths [5]. The values of RI of different materials are tabulated in Table 9.2, which shows that both real and imaginary parts increase with increasing wavelength, whereas the RI values of human blood sample decrease with increasing wavelength.

Therefore, the prism material should be chosen depending on the application concerned with a compromise between the sensitivity and the dynamic range. For sensing broad RI range of samples, it is not preferable to reduce the RI of the prism. However, if the ranges of RI of samples are fixed and known, low-index prism material can be used. The lowest

TABLE 9.2
RI Data at Different Wavelengths

Wavelength (nm)	RI of Gold Metal	RI of Blood Sample
400	(0.0586 + 2.1549i)	1.4298
500	(0.0881 + 2.7947i)	1.3968
600	(0.1244 + 3.4167i)	1.3820
700	(0.1673 + 4.0286i)	1.3742
800	(0.2167 + 4.6339i)	1.3696
900	(0.2725 + 5.2342i)	1.3667

value of RI of prism material needs to be chosen for which the denominator of Equation 9.16 is real for all sample RI values [35].

The SPR angle is the angle at which reflectance becomes minimum and the full width half maximum (FWHM) is the width of the SPR curve that corresponds to the angular width of the curve for half the value of reflectance relative to the dip of reflectance minima. Detection accuracy (DA) is defined as the reciprocal of FWHM of the SPR curve. Parameters, such as SPR angle, FWHM, and DA, have also been investigated with variation of wavelength, which are shown in Figure 9.9a–c.

It can be concluded that the SPR angle is highest for prism materials such as 2S2G and lowest for prism materials such as silicon. Lower value of the SPR angle allows us to operate the silicon prism-based bioplasmonic structure at the low incident angle region compared to the 2S2G glass-based bioplasmonic structure and also the value of FWHM is lowest for prism materials such as silicon and highest for prism materials such as 2S2G and vice versa for DA, which enables us determine SPR dip position more accurately.

9.5 Resonance Response-Based Simulations

This section covers the resonance parameter-based simulations such as reflectance, phase, and field enhancement in spectral interrogation for three- and four-layer semiconductor-based nanobioplasmonic structures.

9.5.1 Semiconductor Nanofilm-Based Four-Layer Plasmonic Structures

This section describes the four-layer structures incorporating an additional semiconductor nanofilm in the conventional three-layer structure as depicted schematically in Figure 9.10a and b. The structure consists of SF11L glass prism as a coupling device, 50 nm gold layer as an excitation layer, and 10 nm Ge layer as a semiconductor nanolayer that also acts as a protective layer for the gold surface. It is observed that for the positional interchange of the Ge nanolayer and the gold layer, the characteristics of the SPR sensor changes. These three nanoplasmonic structures are referred to as prism–metal–analyte (PMA), prism–metal–semiconductor–analyte (PMSA), and prism–semiconductor–metal–analyte (PSMA) structures. For the spectroscopic analysis by the SPR sensor, resonance characteristics have been studied in the spectral interrogation mode. For biosensing analysis, different biological samples such as carbohydrates, nucleic acid (DNA), and different protein samples have been considered.

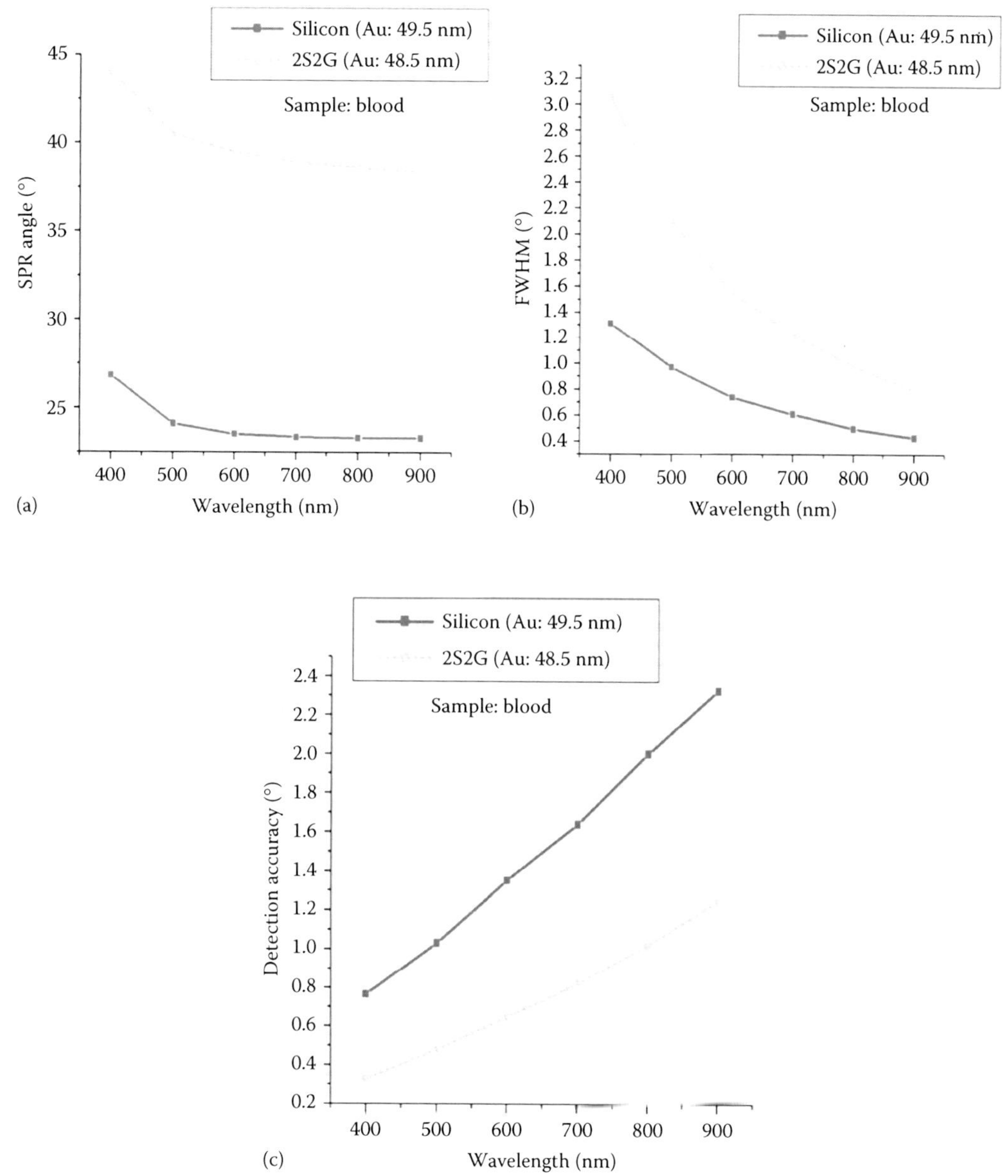

FIGURE 9.9
Plot of (a) SPR angle, (b) FWHM and (c) DA with variation of wavelength for different prism materials.

9.5.1.1 Evanescent Field Enhancement in Spectral Interrogation

Figure 9.11a shows the enhancement of evanescent field of the PMSA structure over the PMA structure due to presence of high RI semiconductor nanolayer over gold surface. The resonance position is shifted toward higher wavelength range. Here the high RI semiconductor layer also acts as a protective surface for the gold layer, which increases the stability of the SPR sensor and makes more light confinement toward the sensing region. However, Figure 9.11b shows that the enhancement of evanescent field of the

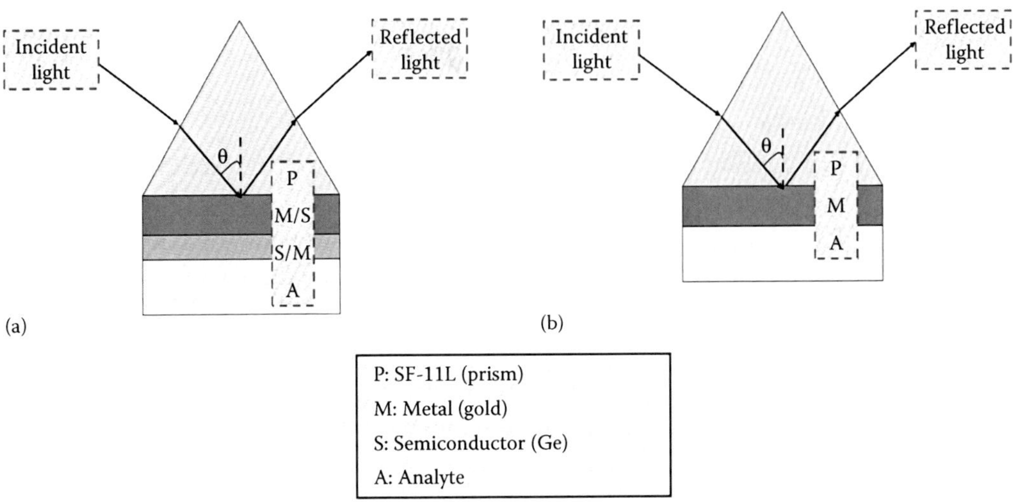

FIGURE 9.10
(a) Four-layer and (b) three-layer nanoplasmonic structures used for SPR excitations.

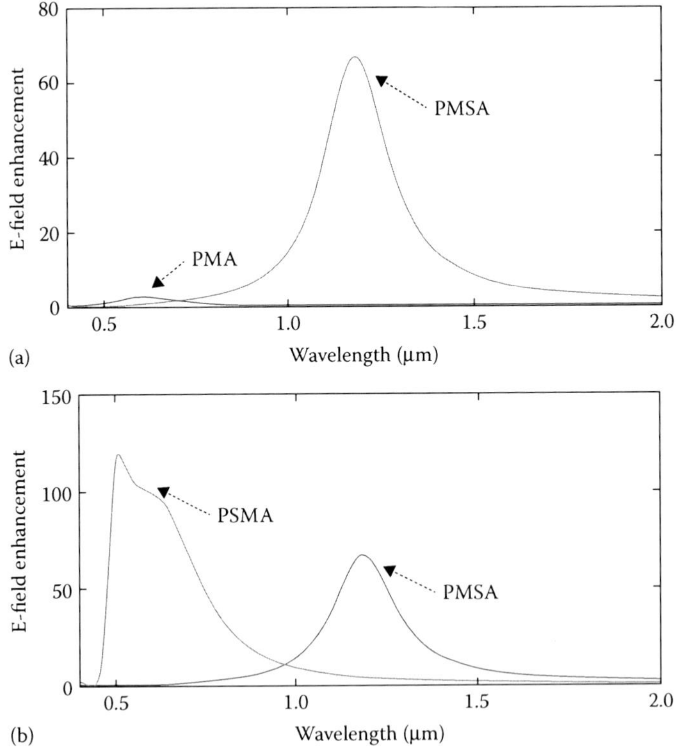

FIGURE 9.11
Evanescent field enhancement in spectral interrogation for (a) PMA and PMSA and (b) PSMA and PMSA structures, respectively.

PSMA structure is much larger than that of the PMSA structure. Due to presence of high RI semiconductor material below the glassy prism, there is more light confinement toward the sensing region of the SPR sensor. Hence, the evanescent field is enhanced at the metal–dielectric interface, which further increases the sensitivity and overall performance of the SPR sensor due to higher interaction volume of the evanescent field with the analyte.

9.5.1.2 Reflectance and Phase Response

Figure 9.12 represents the reflectance curves and phase plot of the PMA structure over the PMSA and PSMA structures, which shows that broader resonance curve and larger FWHM are achieved for the PMSA and PSMA structures than for the PMA structure. It can be noticed that dual resonance occurs for the PSMA structure. Moreover, DA and sharpness of resonance curve for the PMA structure are much higher than the PSMA and PMSA structures. Similarly, the PMA structure provides sharper phase jump across resonance than the PMSA and PSMA structures.

In such a nanoplasmonic structure, the thickness of the additional semiconductor nanolayer (10 nm) is less than that required to support the TM-guided mode. Hence, it is referred as NGW-SPR configuration, which is important for the applications such as near-field microscopy, optical and biomedical spectroscopy, and imaging [68].

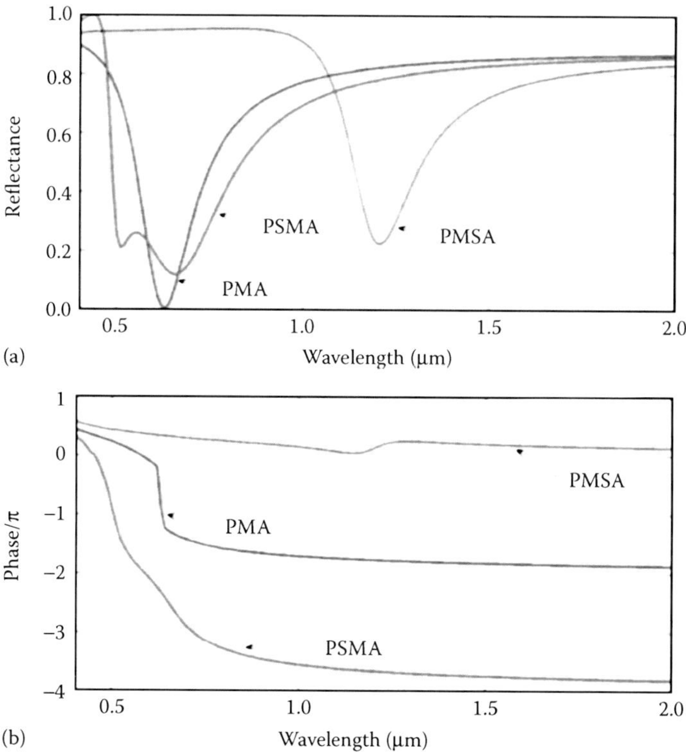

FIGURE 9.12
(a) Reflectance curves and (b) phase plots in spectral interrogation for PMA, PMSA, and PSMA structures.

9.5.1.3 Protein-Based Biosensing

For biomedical and spectroscopic analyses using SPR-based sensors, we have considered different concentrations of protein (in water) as sensing media. For sensing of biological molecules, living cells can be considered to be composed of protein. For simulation purpose of biosensing, 5% carbohydrates, nucleic acids, and different concentrations of proteins in water are considered. The RI of water (1.333) is considered as a reference medium and Table 9.3 shows the RI [69] and relative spectral shift of resonance for sensitivity calculation using different biosamples.

Figure 9.13a and b shows the plots of reflectance and shift in resonance wavelength for the PMA structure. Due to increase of RI of biosamples, resonance position is also shifted

TABLE 9.3
Spectral Sensitivity for Different Biosamples for PMA Structure

Biosamples	RI of Biosamples in RIU (n)	RI Change with respect to Water, δn (RIU)	Shift of Spectral Resonance with respect to Water, $\delta\lambda$ (μm)	Spectral Sensitivity, S_λ (μm/RIU)
Carbohydrate	1.3395	0.0065	0.0305	4.69
Nucleic acid (DNA)	1.341	0.0080	0.0389	4.86
Serum globulin (horse)	1.3423	0.0093	0.046	4.95
Amino acid (tryptophan)	1.3455	0.0125	0.0642	5.136

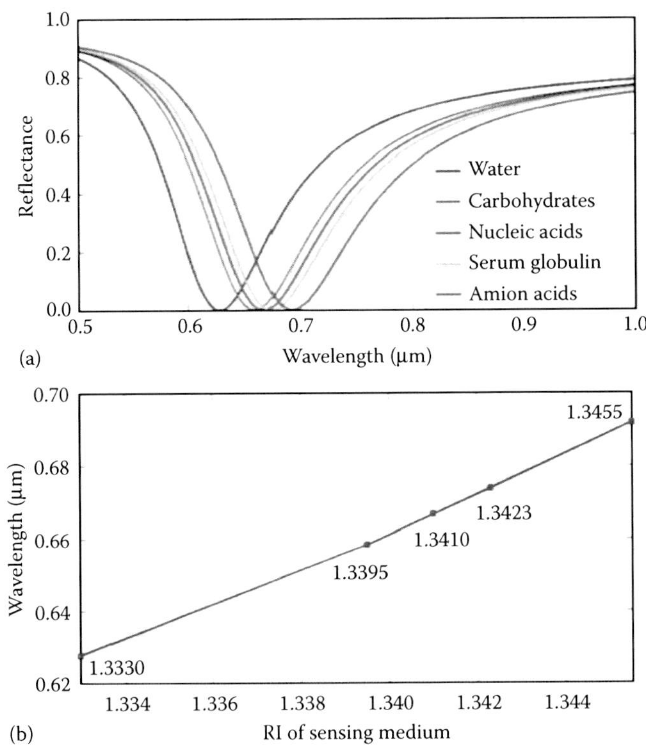

FIGURE 9.13
Plot of (a) reflectance curves and (b) shift of resonance wavelength in spectral interrogation for different biosamples as sensing media for the PMA structure.

toward higher wavelength with an increase in FWHM of the SPR curve and a decrease in steepness of phase jump. The same plots for the PSMA structure are depicted in Figure 9.14a and b along with the sensitivity calculation tabulated in Table 9.4. Figure 9.15a–c shows the phase plots for the PMA, PMSA and PSMA structures, respectively. It is seen that for the PMSA structure, resonance position is shifted toward higher wavelength than that for the PSMA compared to the PMA structure, but for the PSMA structure, the sharpness of phase jump decreases.

Figure 9.16a–c shows the differential phase plots for the PMA, PSMA and PMSA structures in the wavelength interrogation mode for different biosamples. Here, differential

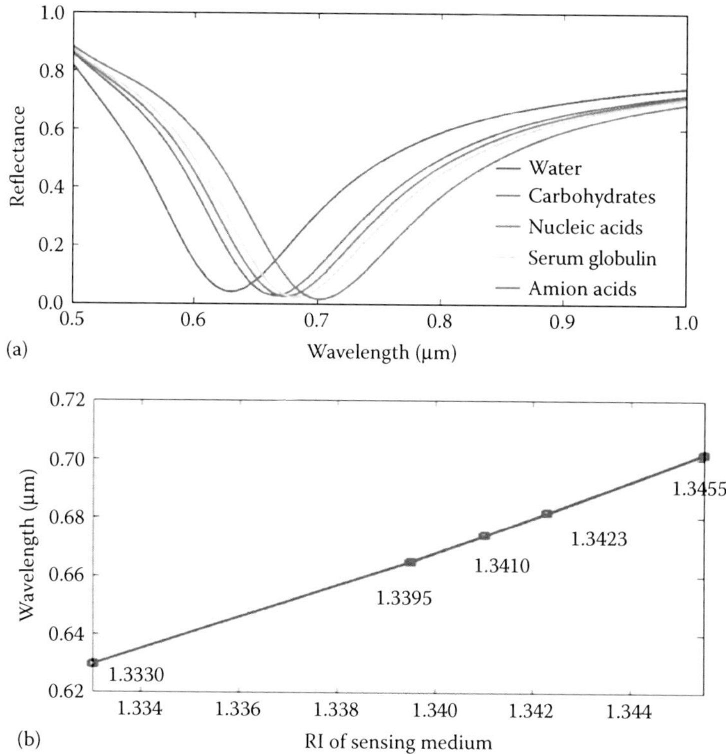

FIGURE 9.14
Plot of (a) reflectance curves and (b) shift of resonance wavelength in spectral interrogation for different biosamples as sensing media for the PSMA structure.

TABLE 9.4
Spectral Sensitivity for Different Biosamples for PSMA Structure

Biosamples	RI of Biosamples, n (RIU)	RI Change with respect to Water, δn (RIU)	Spectral Shift of Resonance with respect to Water, $\delta\lambda$ (μm)	Spectral Sensitivity, S_λ (μm/RIU)
Carbohydrate	1.3395	0.0065	0.0352	5.41
Nucleic acid (DNA)	1.341	0.0080	0.0442	5.526
Serum globulin (horse)	1.3423	0.0093	0.052	5.591
Amino acid (tryptophan)	1.3455	0.0125	0.0721	5.768

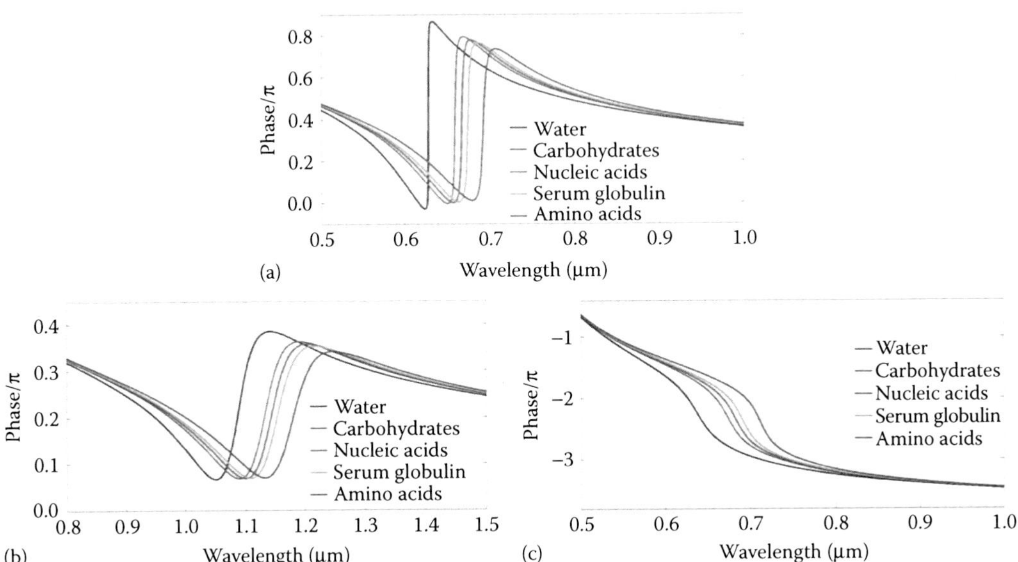

FIGURE 9.15
(See color insert.) Phase plots for (a) PMA, (b) PMSA, and (c) PSMA structures in spectral interrogation with different biosamples.

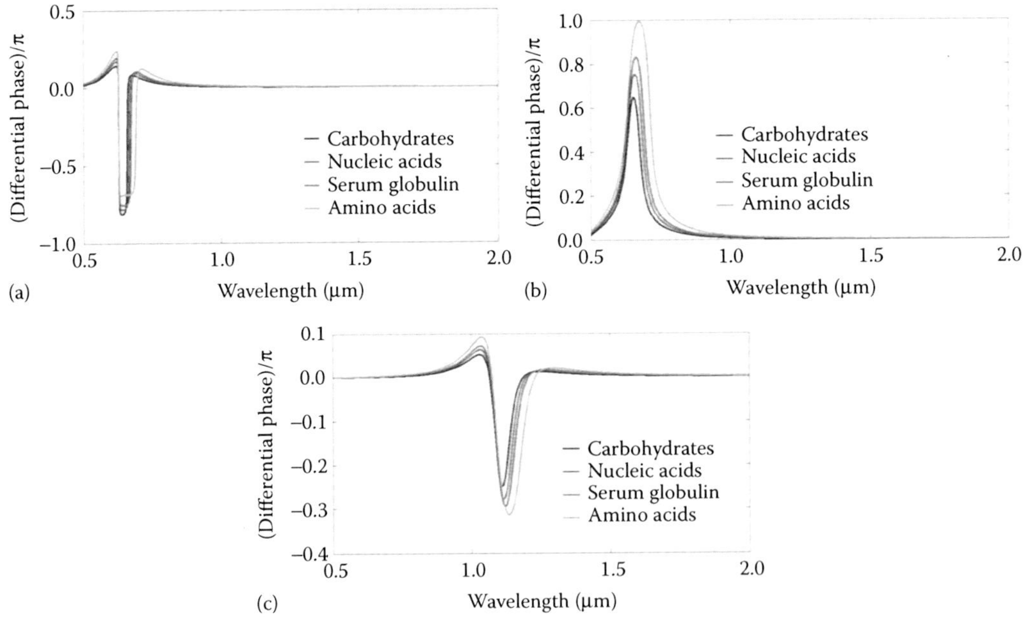

FIGURE 9.16
(See color insert.) Differential phase plots for (a) PMA, (b) PSMA, and (c) PMSA structures in spectral interrogation for different biosamples.

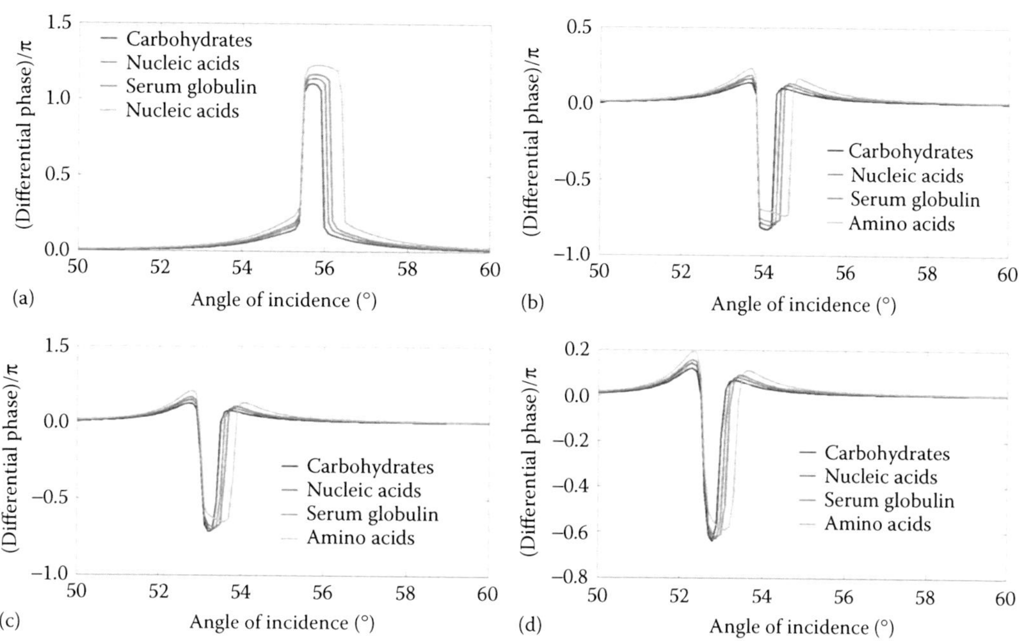

FIGURE 9.17
(See color insert.) Differential phase plots for the PMA structure for specific wavelengths of (a) 550 nm, (b) 633 nm, (c) 700 nm, and (d) 750 nm in angular interrogation mode for different biosamples.

phase is defined as the difference of phase for two different sensing media. For differential phase measurements, the RI of water is considered as reference and the phase of different biosamples is subtracted from the phase of water. For each case, differential phase is normalized by π to make it unitless. If the PMA and PSMA structures are compared, then it can be concluded that due to introduction of additional nanolayer of semiconductor material over the glassy prism, the position for differential phase curves remain at the same location, but the sign of differential phase plots changes from negative to positive. For the PMSA structure, due to presence of semiconductor nanolayer over the metal surface sharpness of differential phase curve decreases compared to that for PMA and PSMA structures. Again differential phase position is also shifted toward higher wavelength ranges.

Differential phase curves are also demonstrated in angular interrogation for four specific wavelengths as depicted in Figure 9.17. It is clearly seen that for the PMA structure as wavelength increases from 550 nm toward 750 nm the differential phase position is shifted toward lower angular ranges along with a decrease in sharpness of the differential phase curve and the sign of differential phase changes from positive to negative.

Figure 9.18a–d depicts the same plots for the PSMA structure. Here, the sign of phase jump remains positive. Whereas in case of the PMSA structure, Figure 9.19a–d shows that as wavelength increases from 750 nm toward 1050 nm range the differential phase position is shifted toward lower angular position with an increase in sharpness of the dip and the sign changes to negative. It also shows that the differential phase curve for the PMSA structure gives satisfactory performance in the IR wavelength region only.

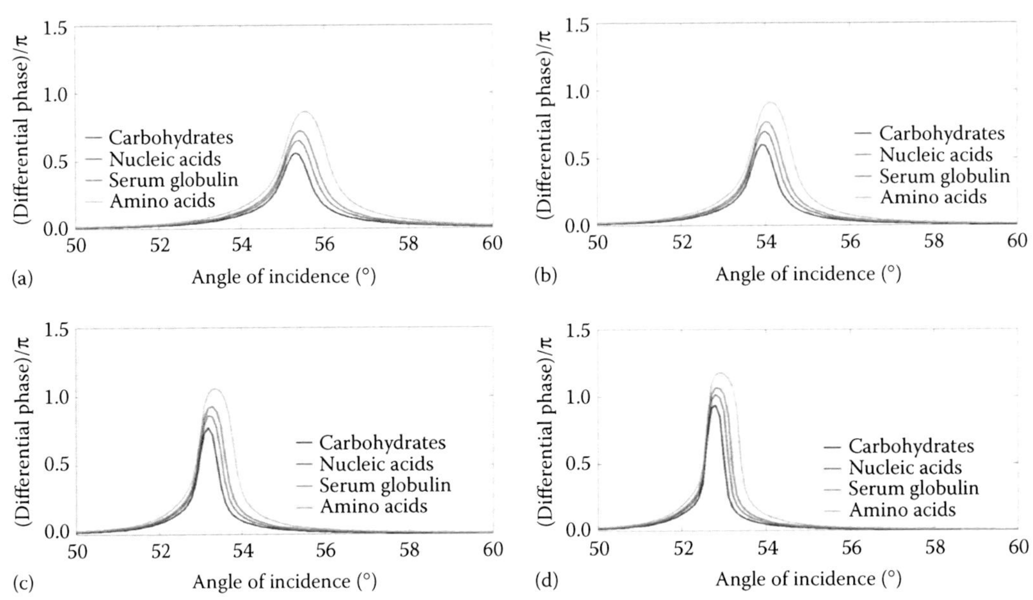

FIGURE 9.18
(See color insert.) Differential phase plots for the PSMA structure for specific wavelengths of (a) 550 nm, (b) 633 nm, (c) 700 nm, and (d) 750 nm in angular interrogation mode for different biosamples.

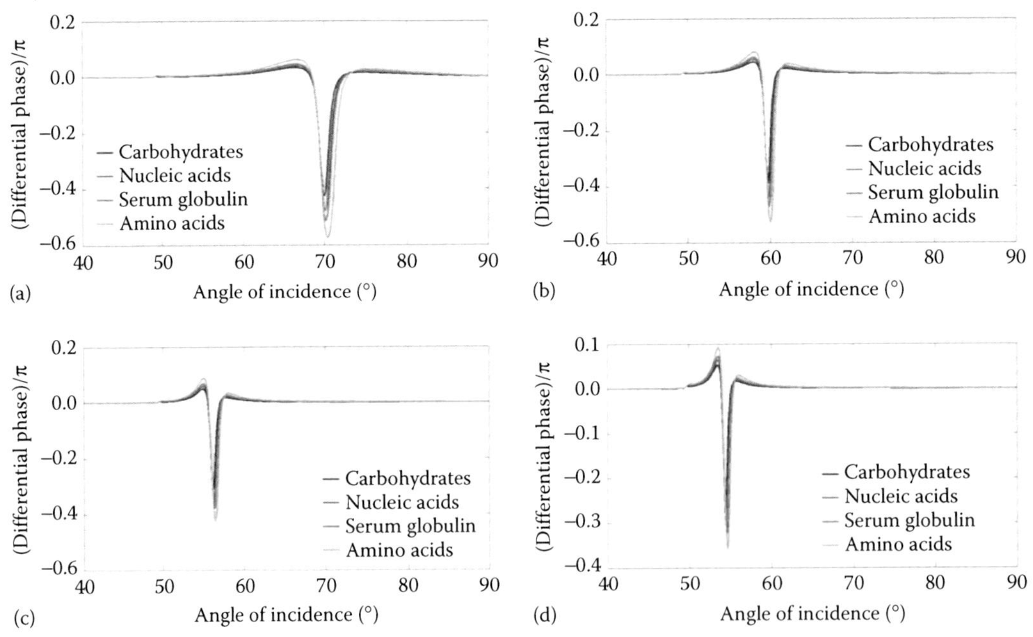

FIGURE 9.19
(See color insert.) Differential phase plots for the PMSA structure for specific wavelengths of (a) 750 nm, (b) 850 nm, (c) 950 nm, and (d) 1050 nm in angular interrogation mode for different biosamples.

9.6 Phase Response of MMD-Coupled Nanoplasmonic Structure

Phase-related SPR measurement technique is of great interest due to its increased sensitivity in comparison with the conventional SPR measurement technique in angular and wavelength regimes. Comparison of the sensitivity of angular, wavelength, and phase interrogation gives some insight into the use of phase measurement technique [70]. An interferometric method for the detection of phase shifts of reflected light under the SPR condition due to the change of the RI of biological or chemical samples with an optimized structural parameter has been proposed and analyzed theoretically and experimentally by Nikitin and coworkers [71,72]. Furthermore, they also explored the interferometric SPR phase imaging technique [73]. Simulation study on the five-layer SPR biosensor based on phase detection has also been reported for measuring the real part of RI of absorptive samples [74]. A theoretical scheme based on phase interrogation using Bragg grating consists of a five-layer structure having a substrate, a waveguide layer, and a buffer that separates the thin SPP supporting metal layer and the waveguide layer, which has been proposed for the measurement of phase change in transmitted light through a grating due to RI change of the analyte in contact with the metal layer in Mach–Zehnder RI interferometer [75]. Increased sensitivity has also been reported based on phase analysis of an SPR biosensor using multipass interferometry [76].

SPR phase interrogation technique can be used to detect the temperature change where the resonance behavior has been analyzed by the closed loop of symmetric Lorentzian amplitude reflection coefficient in complex plane for different wavelengths [77] and also for measurement of the real part of RI of any absorbing material [78]. Visualization of SPR phase imaging [79] and wavelength shift detection using the heterodyne SPR interferometry [80] have also been reported. Moreover, the theoretical [81] and experimental [82] investigations of microscopic phase-sensitive SPR biosensor are based on the measurement of differential phase between two cylindrical vector beams having an ultrawide dynamic range. Recently, Huang et al. [83] have reviewed quite a number of phase measurement techniques used in SPR sensor.

Our main focus in this section is to numerically analyze the phase-dependent resonant behavior of two types of the nanoplasmonic structure, MMD and multilayer-swapped metallo-dielectric (MSMD), in angular and wavelength interrogation along with the reflectivity and field enhancement plots. Also the parametric influence of the metallo-dielectric block and the significance of positional swapping have also been demonstrated in terms of differential reflectance and phase [84]. Circular phase response (CPR) and angled histogram give a detailed information about the resonant phase-dependent behavior of both the structures. Contour plots of phase, reflectivity, and E-field enhancement for simultaneous angular and wavelength interrogation have been analyzed. Moreover, differential reflectance and phase have demonstrated for biomolecular interaction. Direct imaging of differential phase-based interferometric technique provides certain advantages over the conventional SPR-based interferometric technique.

Silver (Ag) and gold (Au) are the two ideal metal films for the excitation of SPR in the visible wavelength range. Au is chemically stable and also provides higher angular and wavelength shift than Ag. But silver provides narrower resonance curve in angular and spectral regimes, and hence provides higher DA though with poor chemical stability. Thus, to accommodate the advantages of the two metal layers, the concept of bimetallic nanofilm has come. Optimization of metal thickness in coupled nanoplasmonic structure with mono- and bimetallic nanofilms plays an important role in the evaluation of the performance of the sensor. Bimetallic loaded dielectric plasmonic structure provides better measurement accuracy. Also the modulation of symmetric and antisymmetric modes provides equi- and nonequireflectance dual resonance dips for the variation of thickness ratio of individual metal film in a bimetallic composite. The imaging of the modulation of dual dips can be used as plasmonic trinary logic. Also metal–dielectric–metal (MDM) structure is very efficient for biological and chemical sensors due to the coupling of waveguide and plasmonic resonance.

9.6.1 MSMD Nanoplasmonic Structure

The schematic diagrams of MMD and MSMD structures are demonstrated in Figure 9.20a. In our investigations, SF11 glass prism is used as the coupling device. In case of the MMD structure, 50 nm Au layer is used along with an additional 500 nm dielectric silica layer and the whole structure is placed in the aqueous (n_w = 1.332) environment. In case of the MSMD structure, the metallo-dielectric block only has a positional interchange and all other parameters remain the same. The wavelength dependence of the dielectric permittivity of the materials used in our analysis is also shown in Figure 9.20b and c. The dielectric function of SF11 glass prism and silica can be calculated from the Sellmeier dispersion equation [85]. Also the dielectric permittivity of metal can be calculated from the Drude model [86]. The constants for the dispersion equation and the Drude formula are given in the table as shown in Figure 9.20g. The resonant behavior of the MMD and MSMD structures is analyzed to show the distinct and significant behavior due to the swapping of metallo-dielectric block (Figure 9.20d–f). Phase is also a very important parameter along with the reflectance and field enhancement. Reflectance dip or field enhancement peak is always associated with a phase jump at a particular angle and wavelength in spectral and angular interrogation, respectively. In case of the MMD structure, coupling of waveguide and plasmonic resonance provides two reflectance dips, phase jumps, and E-field enhancement peaks. The dip at the lower angle is due to the waveguide resonance (WGR) mode and the dip at the higher angle is due to the conventional SPR mode. Also with the increase of the thickness of waveguide silica layer, the number of the WGR mode will increase progressively along with the almost unchanged plasmonic resonance due to the unchanged parameter of the plasmon generating thin metal layer. These progressive multiple WGRs can be further utilized for the characterization of dielectric material in a multilayer nanocomposite. In case of the MSMD structure due to the introduction of the intermediate dielectric layer between the glass prism and the metal layer, two decoupled symmetric and antisymmetric modes are produced. Symmetric SPs have lower attenuation and larger propagation length, hence termed as long-range surface plasmons. However, A-SPs have larger attenuation and smaller propagation length, hence termed as short-range surface plasmons. The dip at the lower angle is due to the LRSPR, which occurs at the upper metal–dielectric interface, and the dip at the higher angle is due to the SRSPR, which occurs at the lower metal–dielectric interface.

9.6.1.1 Phase-Dependent Resonant Behavior: Angular and Spectral Dependence of 2D Reflectance and Phase, CPR, and Angled Histogram

The graphical analysis of phase-dependent resonant behavior has been carried out in detail to show the significance of the positional swap of the metallo-dielectric block. The phase is plotted in degrees in order to show the actual phase jump without using any normalization procedure or unwrapping algorithm. The reflectance curves as well as the corresponding phase plots have been demonstrated in order to indicate the resonance in angular and wavelength regimes as shown in Figure 9.21 for the MMD structure depicting the WGR and SPR modes. Figure 9.21 also demonstrates the CPR plot along with the envelope (red circle) of the amplitude of complex reflection coefficient of the MMD structure. All the arrows (blue arrow) are emanating from the origin and represent the phase value (ϕ_r) in degrees. The length of the arrow represents the modulus of the complex reflection coefficient ($|r|$)

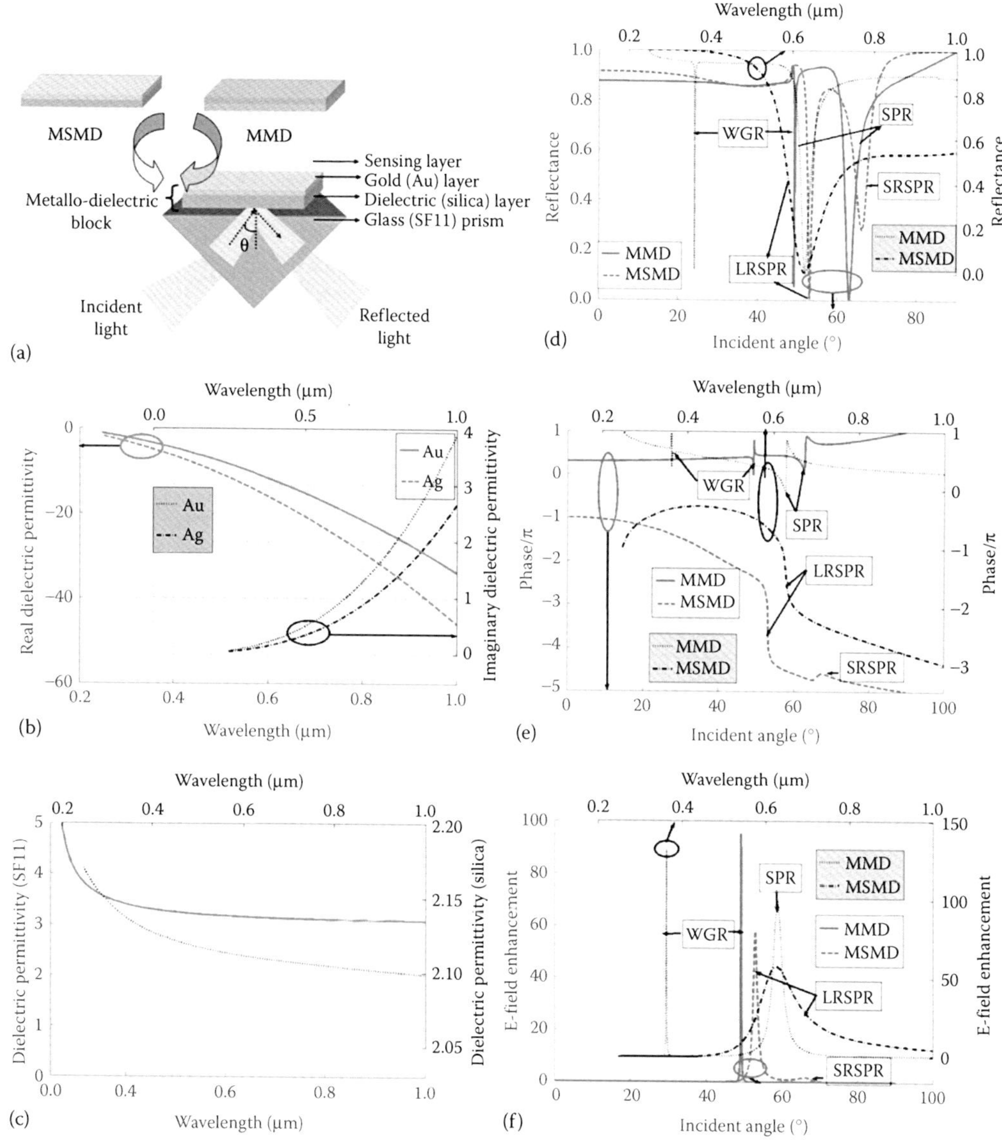

FIGURE 9.20
(See color insert.) (a) Schematic diagram of MMD and MSMD nanoplasmonic structures. Spectral dependence of dielectric permittivity for (b) Ag and Au layer (Drude formula) and (c) SF11 glass prism and silica dielectric layer (Sellmeier dispersion relation) along with the 2D angular and spectral interrogation of (d) reflectance, (e) phase and (f) E-field enhancement, and (g) constants of the Sellmeier dispersion and Drude model. (With kind permission from Springer Science + Business Media: *Plasmonics*, Circular phase response-based analysis for swapped multilayer metallo-dielectric plasmonic structures, 9, 2014, 237–249, Bera, M. and Ray, M.)

Material used	Constants of dispersion equation					
	B_1	B_2	B_3	C_1	C_2	C_3
BK7 glass	1.03961212	0.231792344	1.01046945	0.00600069867	0.0200179144	103.560653

Metal used	Parameters of Drude formula	
	λ_p (μm)	λ_c (μm)
Gold	0.16826	8.9342
Silver	0.14571	17.614

(g)

FIGURE 9.20
Continued

involving both real and imaginary parts. The reflectivity dip is associated with a phase jump for both the modes. The envelope of the complex amplitude reflection coefficient crosses the origin when the minimum reflectance value is zero and is also associated with a Heaviside phase jump. One can notice that both the WGR and SPR modes cross the origin because both the modes acquire zero reflectance value. But, as the FWHM of the WGR mode is much narrower than the conventional SPR mode, the WGR mode envelop in the CPR plot enclosing the origin traces a more closed circular path than that of the conventional SPR mode. Moreover, angled histogram of the phase plot is also demonstrated here, which is a distribution of values according to their numeric range. Figure 9.21a–e indicates different zones of the plots that correlate the points in 2D phase and reflectance, CPR, and angled histogram plots. The angular interrogation has been carried out at the wavelength of 633 nm and the spectral interrogation has been done by fixing the angle at 49.25° corresponding to WGR resonance in the angular interrogation plot.

Similar analysis for the MSMD structure is depicted in Figure 9.22. In this case, angular interrogation (Figure 9.22a) is done at the same working wavelength, 633 nm. However, the wavelength interrogation is done for two separate resonance angles corresponding to LRSPR and SRSPR as depicted in Figure 9.22b and c, respectively. In case, if one chooses the single angle in the neighborhood of these two resonances, the structural parameters must be optimized in order to obtain the zero reflectivity. For the simplicity of understanding, the parameters are kept the same, considering only the layer-wise swapping of the metallo-dielectric block. In this case, also dual resonance dips along with phase jumps are obtained: a zero minimum reflectivity dip with a Heaviside phase jump for LRSPR and a nonzero dip with a less steep phase jump for SRSPR. Thus, the envelope of the complex amplitude reflection coefficient of LRSPR crosses the origin in the CPR plot, whereas SRSPR does not. Moreover, the FWHM of LRSPR is much narrower than that of SRSPR. In the CPR plot of angular interrogation as shown in Figure 9.22d, the magnitudes indicating arrows also depict the fact that SRSPR has a higher reflectance value than LRSPR, which has a closed loop enclosing the origin. The same effect is also observed in angled histograms. The CPR plots of wavelength interrogation of LRSPR and SRSPR as shown in Figure 9.22e and f, respectively, indicate the similar phenomena as in the angular regime.

9.6.1.2 Simultaneous Angular and Spectral Interrogation: Reflectance, Phase, and E-Field Enhancement

Simultaneous interrogation of angle and wavelength provides a better overview of the coupling of two modes such as WGR and SPR for the MMD structure and LRSPR and SRSPR for the MSMD structure. Separate contour plots for the WGR and SPR modes are demonstrated for both reflectance and phase in simultaneous angular and spectral regimes as shown in Figure 9.23a and c. Similarly, in case of the MSMD structure, separate contour plots for LRSPR and SRSPR are also denoted in reflectance and phase plots.

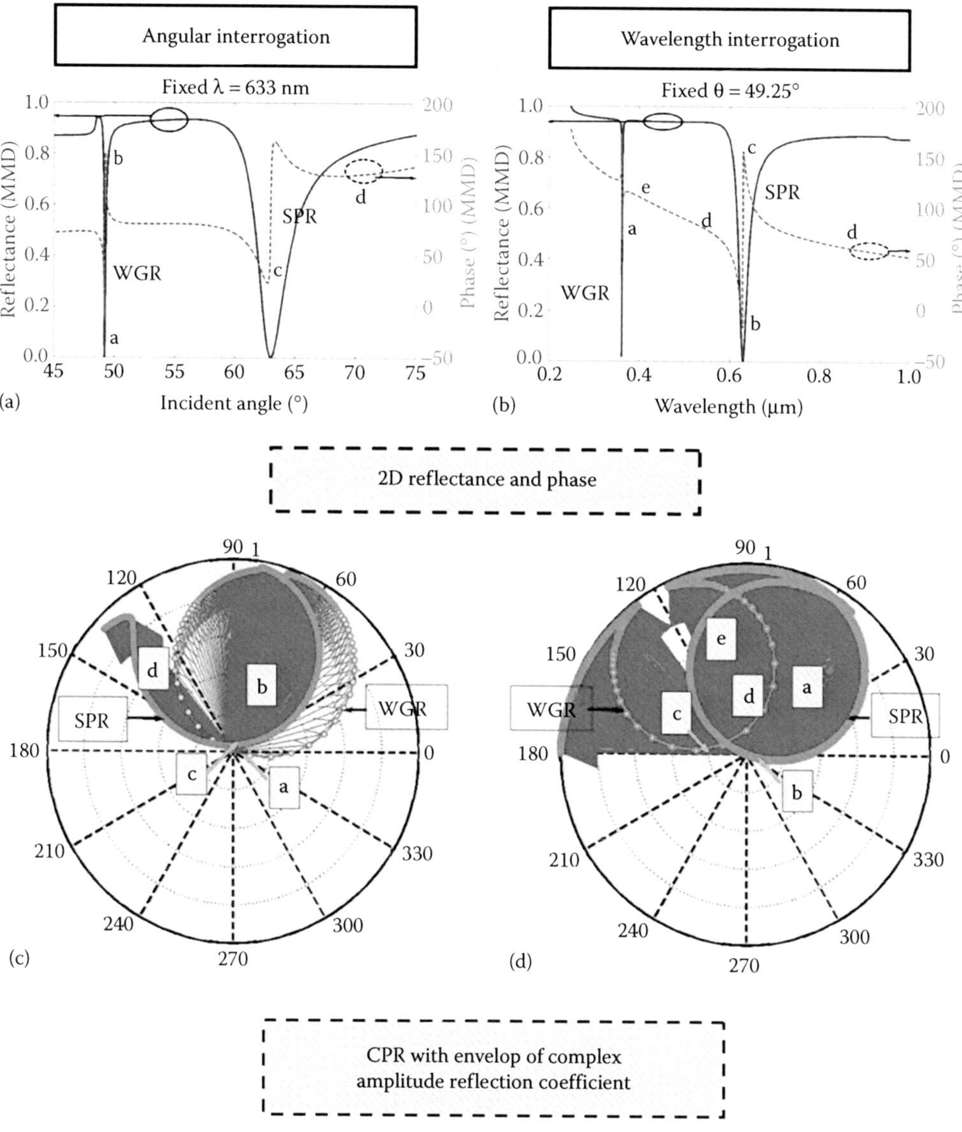

FIGURE 9.21
(See color insert.) Simulated graphical representation for angular (left) and spectral (right) interrogation showing 2D phase and reflectance (a,b), CPR (c,d), and angled histogram (e,f) plots for MMD nanoplasmonic structure. (With kind permission from Springer Science + Business Media: *Plasmonics*, Circular phase response-based analysis for swapped multilayer metallo-dielectric plasmonic structures, 9, 2014, 237–249, Bera, M. and Ray, M.)

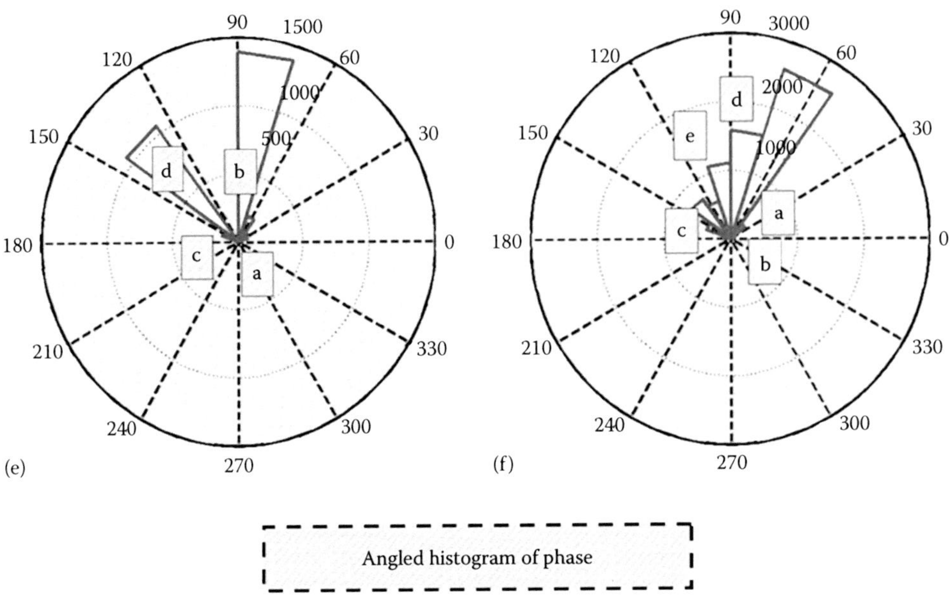

FIGURE 9.21
Continued

Moreover, contour plots of E-field enhancement in simultaneous interrogation for both the MMD and MSMD structures are shown in Figure 9.24a and b, respectively. In case of the MMD structure, the WGR mode is much enhanced than the plasmonic mode due to the coupling of energy from the Au–silica interface to the silica waveguide layer. Due to the total internal reflection in the silica layer, the WGR mode occurs at a particular resonance angle and wavelength. In case of the MSMD structure, the LRSPR mode is much enhanced than the SRSPR mode.

9.6.1.3 Optimization of Metal and Dielectric Layer Thickness in MMD and MSMD Structures

Optimization of metal and dielectric layer thicknesses is an important issue in order to design a highly sensitive nanoplasmonic sensor. 2D reflectance and phase are demonstrated for the variation of metal layer thickness for a fixed angle and wavelength chosen from angular and spectral interrogation along with the contour plot with simultaneous variation of wavelength and metal layer thickness for the MMD and MSMD structures as shown in Figure 9.25a and b, respectively. From Figure 9.25a and b, it is clear that reflectance acquires its minimum value at 50 nm metal layer thickness for both the structures and phase shows its highest slope at that particular optimized metal thickness. From the phase contour plots, the phase jump indicates the optimized metal layer thickness and the operating wavelength region. The thickness of the dielectric layer plays an important role in this optimization procedure. But due to the swapping of the position of the dielectric layer, a different significant phenomenon occurs. In case of the MMD structure, the number of reflectance dips or phase jumps increases with the an increase of the thickness of the dielectric silica layer as shown in Figure 9.25c. One of the thicknesses of the dielectric layer can be chosen where the reflectance dip acquires its minimum value according to the application. In case of the MSMD structure, the reflectance dip modulates when the

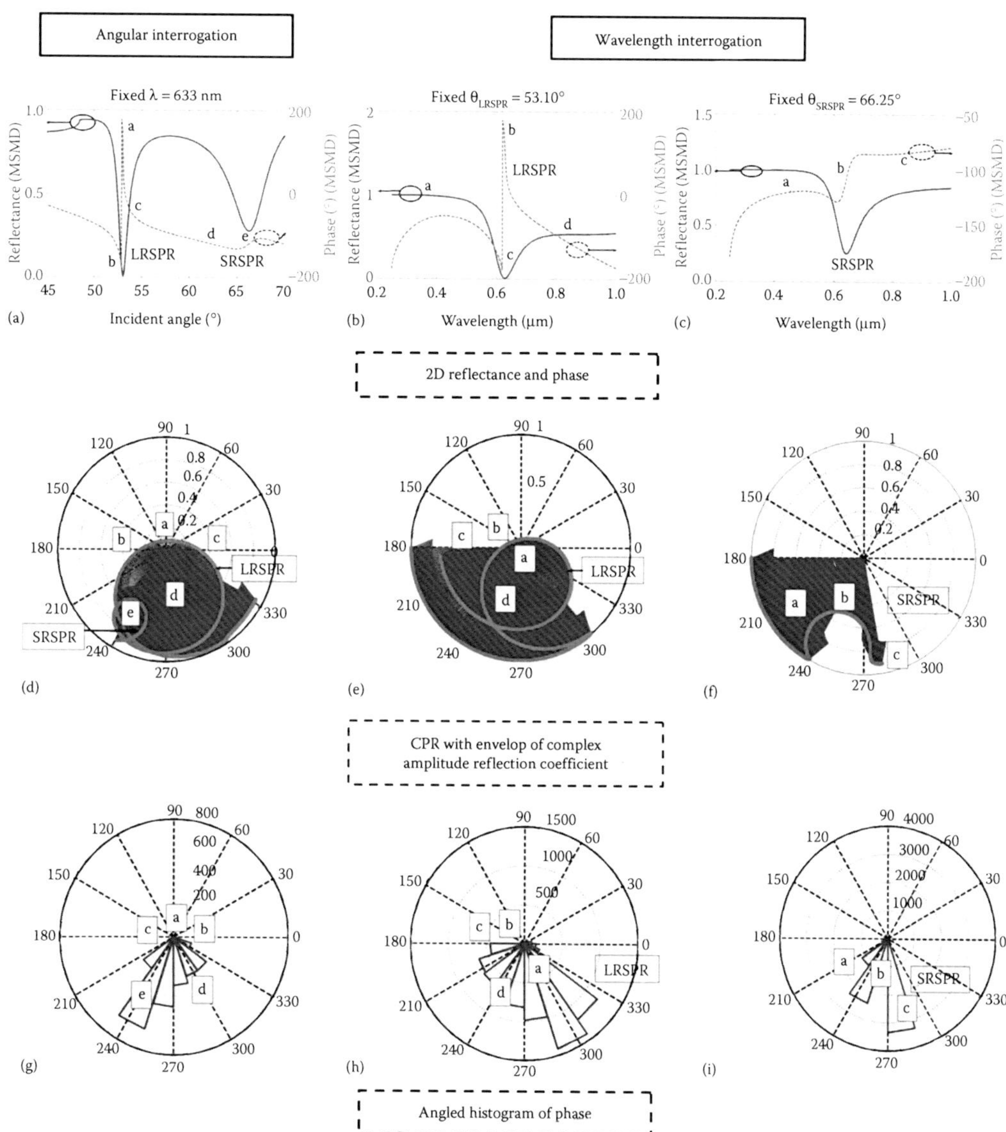

FIGURE 9.22
(See color insert.) Simulated graphical representation for angular (left) and spectral (middle: LRSPR and right: SRSPR) interrogation showing 2D phase and reflectance (a–c), CPR (d–f), and angled histogram (g–i) plots for MSMD nanoplasmonic structure. (With kind permission from Springer Science + Business Media: *Plasmonics*, Circular phase response-based analysis for swapped multilayer metallo-dielectric plasmonic structures, 9, 2014, 237–249, Bera, M. and Ray, M.)

thickness of the dielectric layer changes as shown in Figure 9.25d. 2D phase and contour also demonstrate the phenomenon and selects the wavelength region.

The performance of chemical and biological sensing can be improved using the optimized structural parameters. Reflectance or phase jump is very sensitive to the change of the RI of the sensing medium in angular and wavelength regimes. Direct difference of reflectance as a result of change of the RI of the sensing medium is evaluated. This differential reflectance is plotted in angular and spectral interrogation for the RI change of

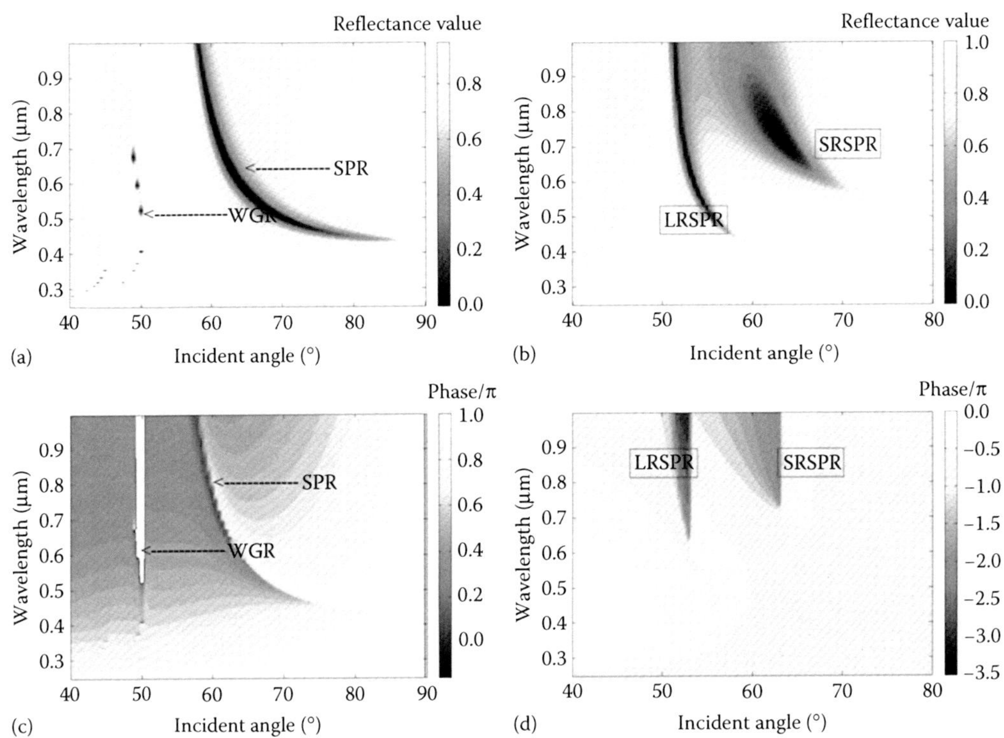

FIGURE 9.23
(See color insert.) Simultaneous angular and spectral interrogation: (a and b) reflectance contour plot and (c and d) phase contour plot for MMD (a and c) and MSMD (b and d) structures, respectively. (With kind permission from Springer Science + Business Media: *Plasmonics*, Circular phase response-based analysis for swapped multilayer metallo-dielectric plasmonic structures, 9, 2014, 237–249, Bera, M. and Ray, M.)

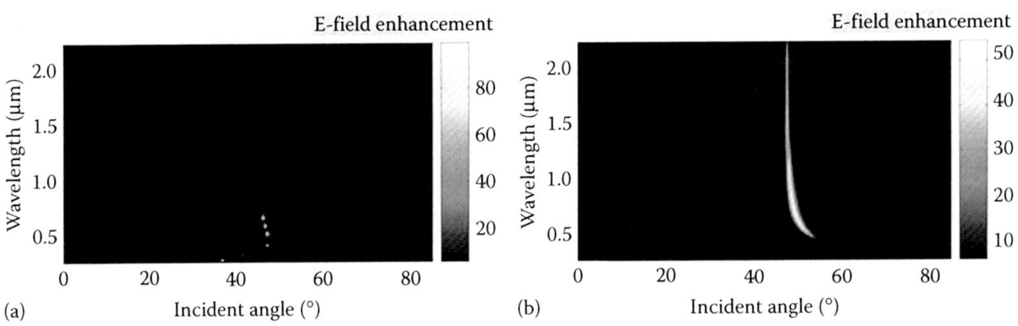

FIGURE 9.24
(See color insert.) Simultaneous angular and spectral interrogation for E-field enhancement of (a) MMD and (b) MSMD structures. (With kind permission from Springer Science + Business Media: *Plasmonics*, Circular phase response-based analysis for swapped multilayer metallo-dielectric plasmonic structures, 9, 2014, 237–249, Bera, M. and Ray, M.)

50% hemoglobin ($n_h = 1.342$ at 632.8 nm) with respect to water ($n_w = 1.332$ at 632.8 nm). The differential reflectance is demonstrated for different metal layer thicknesses in both the interrogation modes for both the MMD and MSMD structures as shown in Figures 9.26 and 9.27, respectively. From these plots, it is also evident that the metal thickness must be chosen in the neighborhood of 50 nm for optimized performance.

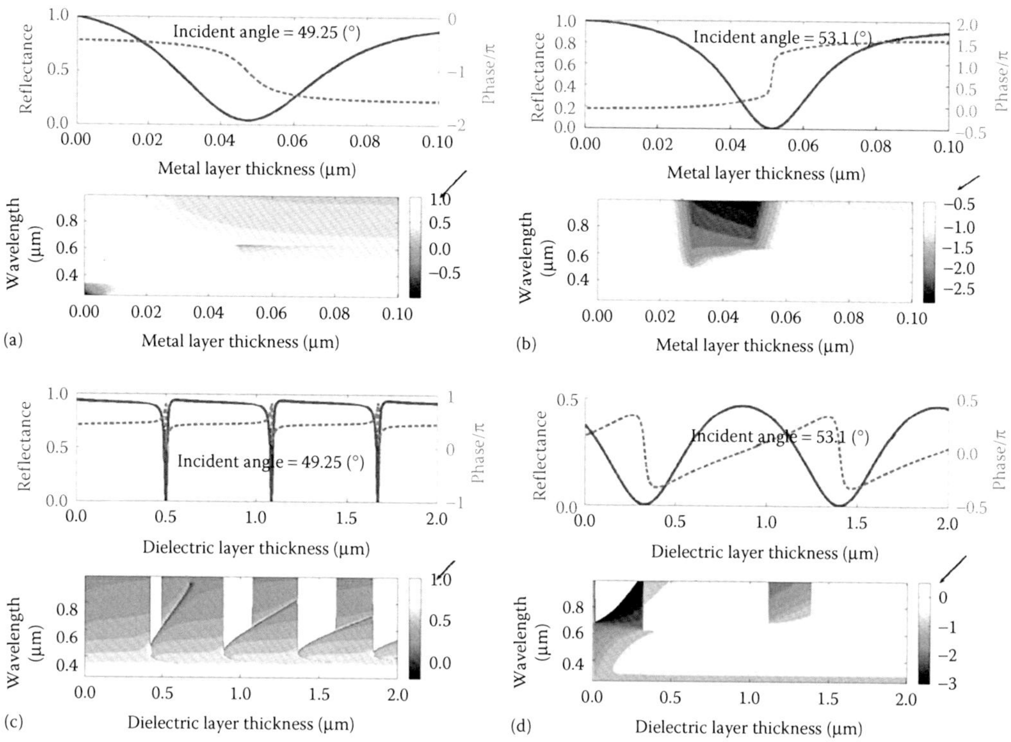

FIGURE 9.25
Contour plots of phase with the variation of wavelength and metal layer thickness for (a) MMD and (b) MSMD structures and also with the variation of wavelength and dielectric layer thickness for (c) MMD and (d) MSMD structures along with the 2D plots of reflectance and phase with the variation of metal and dielectric layer thickness. (With kind permission from Springer Science + Business Media: *Plasmonics*, Circular phase response-based analysis for swapped multilayer metallo-dielectric plasmonic structures, 9, 2014, 237–249, Bera, M. and Ray, M.)

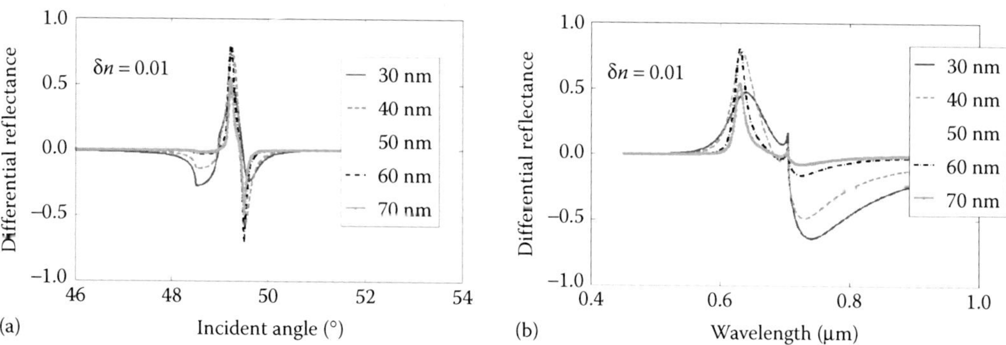

FIGURE 9.26
Differential reflectance of the MMD structure for different metal thicknesses in (a) angular and (b) spectral regimes for the change of the RI of the sensing medium $\delta n = 0.01$.

9.6.1.4 *Sensitivity Issues: Differential Reflectance and Phase Measurement*

Resonance parameters of SPR phenomenon are very sensitive to the change of the optical properties of the sensing medium. Resonance angle or wavelength is shifted significantly due to the change of the RI of the sensing medium. Thus, SPR measurement technique is widely used in biological and chemical sensing as well as real-time biomolecular

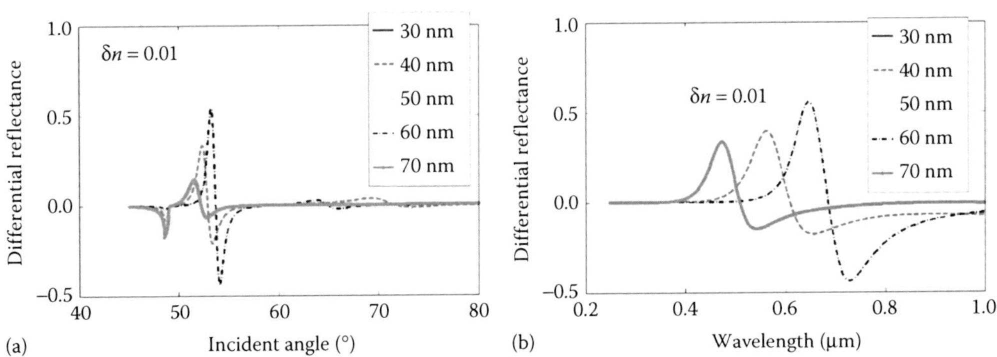

FIGURE 9.27
Differential reflectance of the MSMD structure for different metal thicknesses in (a) angular and (b) spectral regimes for the change of the RI of the sensing medium $\delta n = 0.01$.

interaction, which is also studied in this context successfully. In the conventional SPR measurement technique, the shift of the angle or wavelength is measured for the change of the RI of the sensing medium. Moreover, phase jump associated with SPR phenomenon is also very sensitive due to the RI change of dielectric medium adjacent to the plasmon generating thin metal film. Comparison of the phase detection SPR measurement technique with the conventional SPR technique suggests that the former method provides better measurement precision [70]. However, the differential phase measurement technique is very much robust rather than the measurement of the shift of the phase jump in angular and wavelength regimes. Thus, differential phase sensitivity can be defined as the ratio of the differential phase ($\delta\phi$) and the RI change (δn) of the sensing medium [87]:

$$(S_n)_\phi = \frac{\delta\phi}{\delta n} \tag{9.37}$$

where differential phase can be defined as

$$\delta\phi = \phi_s - \phi_r \tag{9.38}$$

ϕ_s and ϕ_r are the phases of the nanoplasmonic structure for sample under test and the reference medium. The change of RI can be written as

$$\delta n = n_s - n_r \tag{9.39}$$

Similarly, n_s and n_r are the RI of the sample and reference medium.

Furthermore, the phase sensitivity (S_n) can be defined in terms of the measurement of shift of the phase jump in angular and wavelength interrogation as the ratio of the angle or wavelength shift ($\delta\theta_{res}$ or $\delta\lambda_{res}$) due to the small change of the RI (δn) of the sensing medium:

$$S_n = \frac{\delta\theta_{res}\left(\delta\lambda_{res}\right)}{\delta n} \tag{9.40}$$

Phase jump shifts have been demonstrated in angular and wavelength interrogation for the RI change δn of the order of 10^{-2}, 10^{-4}, and 10^{-6} as shown in Figure 9.28.

As the photodetectors cannot directly measure the phase information of light, an interferometric technique is to be used to retrieve the phase information from the interference

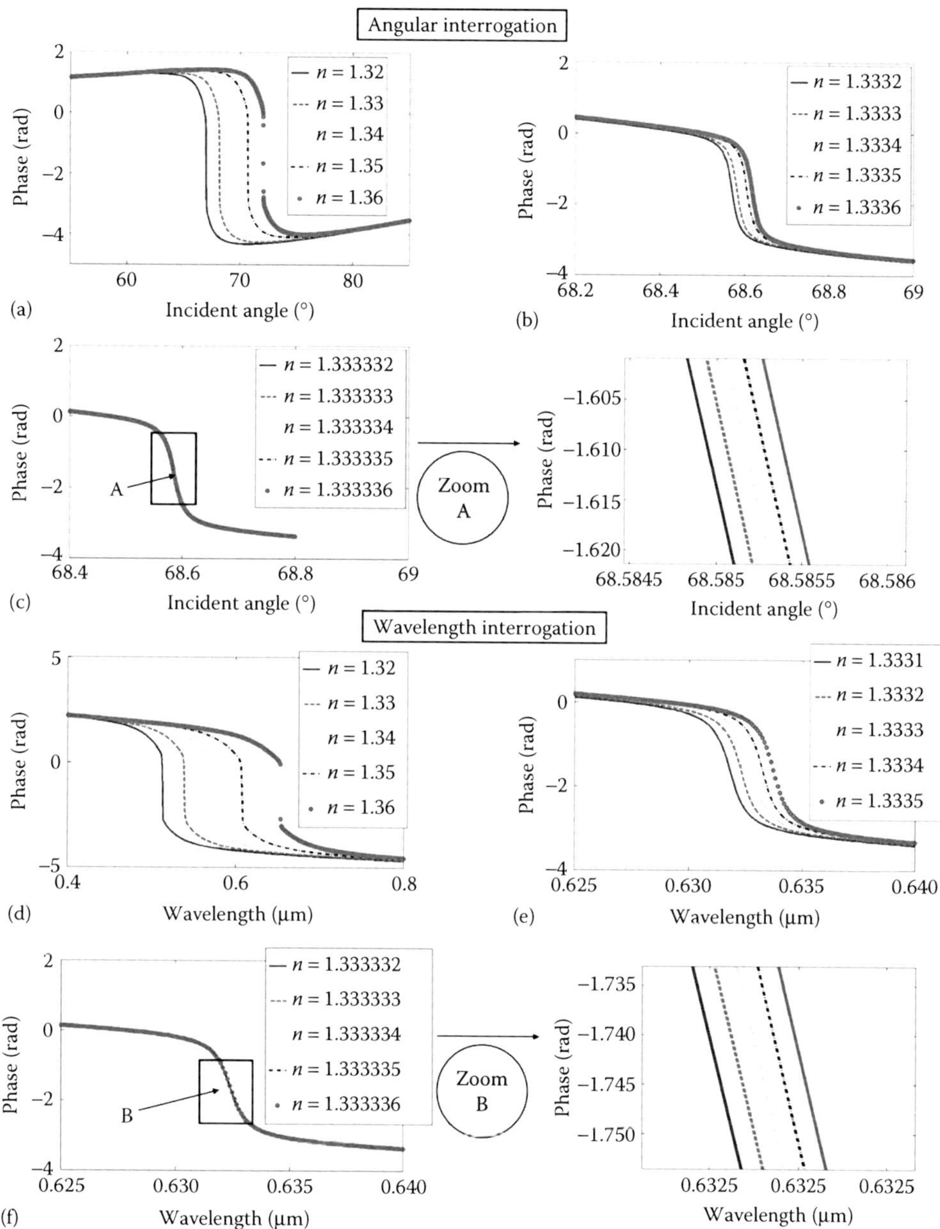

FIGURE 9.28
(See color insert.) Phase jump shifts due to the change of the RI of the sensing medium of the order of (a and d) 10^{-2}, (b and e) 10^{-4}, and (c and f) 10^{-6} in angular and wavelength interrogation. (With kind permission from Taylor & Francis: *J. Mod. Opt.*, Resonance parameters based analysis for metallic thickness optimization of a bimetallic plasmonic structure, 61, 2014, 182–196, Bera, M., Banerjee, J., and Ray, M.)

between the reference and the signal beam. A novel interferometric setup has been proposed in our previous work as shown in Figure 9.29a, which can simultaneously measure the differential phase between a reference medium (water) and a sample medium. Here, beam splitter 1 divides the incident collimated p-polarized light into the reference beam and the signal beam. Two identical right-angle prism-based SPR structures with bimetallic

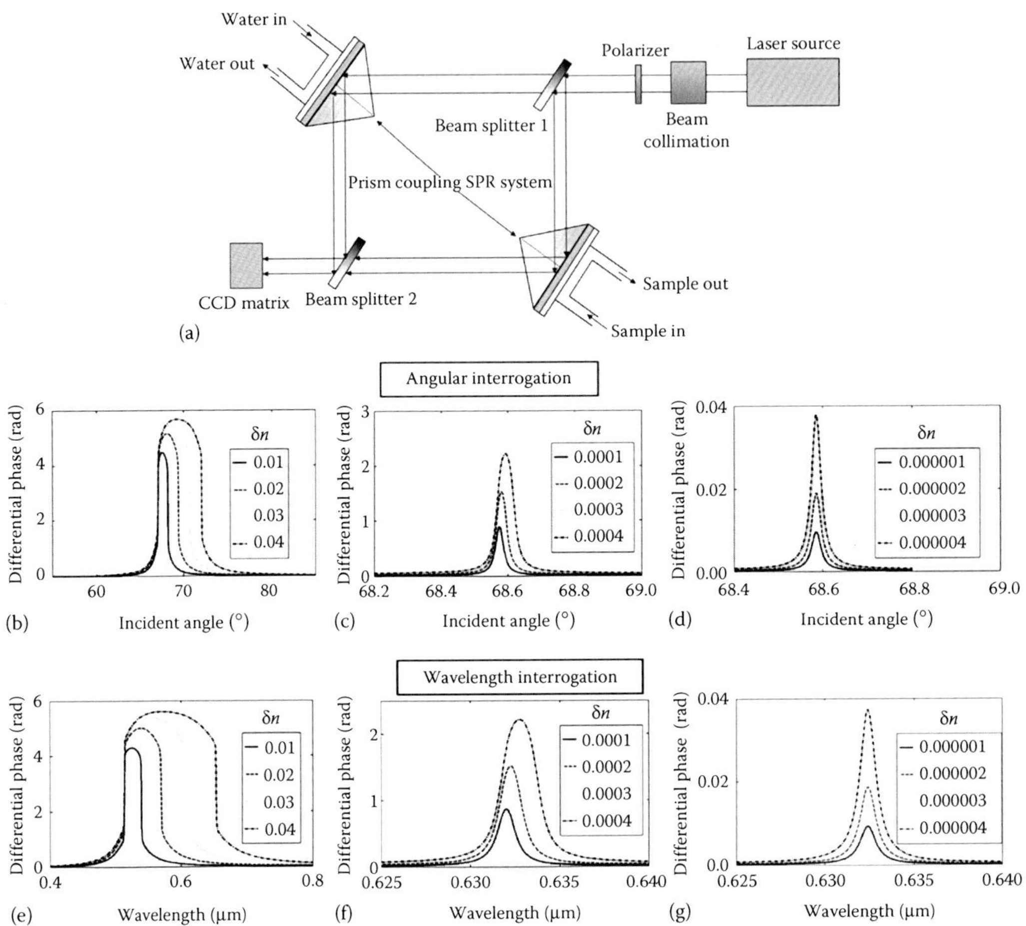

FIGURE 9.29
(a) Proposed interferometric setup for the direct measurement of differential phase with δ*n* of the order of (b and e) 10^{-2}, (c and f) 10^{-4}, and (d and g) 10^{-6} in angular and wavelength interrogation. (With kind permission from Taylor & Francis: *J. Mod. Opt.*, Resonance parameters based analysis for metallic thickness optimization of a bimetallic plasmonic structure, 61, 2014, 182–196, Bera, M., Banerjee, J., and Ray, M.)

nanofilm are situated in the two arms of the setup with water as a reference sensing medium in one arm and different test sample in the other. Two SPR-generated reflected beams from the reference and the sample media having almost comparable intensity are then made to interfere using the beam splitter 2. Thus, the good contrast interference pattern observed at the charge-coupled device (CCD) matrix can directly provide the information of the phase difference (δϕ) of the two SPR-generated beam due to the change of RI of the sensing medium (δ*n*). Theoretically simulated differential phase for different samples with respect to water is shown in Figure 9.29b–g. From the three sets of sensing media with δ*n* of the order of 10^{-2}, 10^{-4}, and 10^{-6}, it is also evident that both the maximum value and the FWHM of the differential phase are increased due to the increased RI change between the sensing media.

Figure 9.30 shows the three-dimensional (3D) bar diagram for the comparison of phase sensitivity and differential phase sensitivity in angular and wavelength regimes for three different order of RI change. It is clearly evident that the differential phase sensitivity gives better result than the conventional phase sensitivity calculated from the shift of the phase jump.

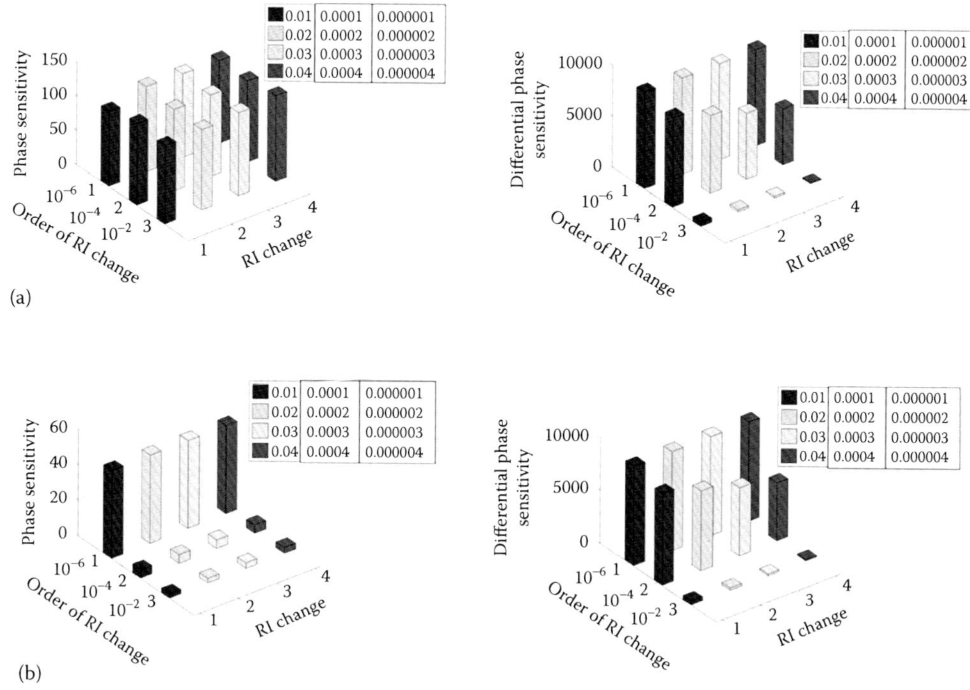

FIGURE 9.30
Comparison of phase sensitivity and differential phase sensitivity for the same three order of RI change in angular (a) and wavelength (b) interrogation. (With kind permission from Taylor & Francis: *J. Mod. Opt.*, Resonance parameters based analysis for metallic thickness optimization of a bimetallic plasmonic structure, 61, 2014, 182–196, Bera, M., Banerjee, J., and Ray, M.)

However, in our investigation, direct difference of reflectance as well as phase has been demonstrated in angular interrogation due to the change of the RI of the sensing medium. SPR-based phase detection technique can be implemented by interferometric method. Direct imaging of differential phase is more convenient than the detection of shift of the angle or wavelength. Differential reflectance and phase are plotted for different concentrations (g/l) of deoxygenated hemoglobin mentioned in Table 9.5 with respect to water (n_w = 1.332) at 632.8 nm wavelength for the MMD and MSMD structures as shown in

TABLE 9.5
RI of Deoxygenated Hemoglobin for Different Concentrations at 632.8 nm

Deoxygenated Hemoglobin	
Concentration (g/l)	**RI at 632.8 nm**
0	1.334
20	1.337
40	1.341
60	1.343
80	1.346
100	1.349

Source: Zhernovaya, O. et al., *Phys. Med. Biol.*, 56, 4013–4021, 2011.

Figures 9.31 and 9.32, respectively. Moreover, the maximum value of differential reflectance and phase is shown with the variation of RI change in Figure 9.33.

In order to study the spectroscopic dependence of differential reflectance and phase for both the MMD and MSMD structures, the RI of 50 g/l concentration of hemoglobin with respect to water is considered for different wavelengths in the visible region

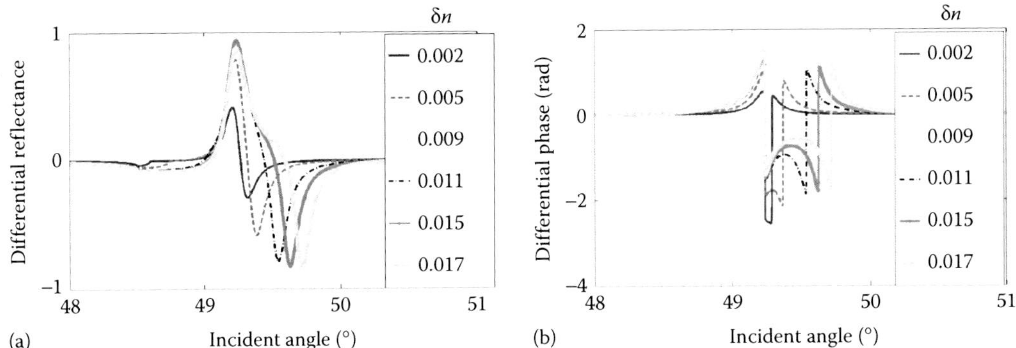

FIGURE 9.31
(a) Differential reflectance and (b) differential phase in angular interrogation for the MMD structure (considering WGR only) due to the change in RI for different concentrations of hemoglobin with respect to water.

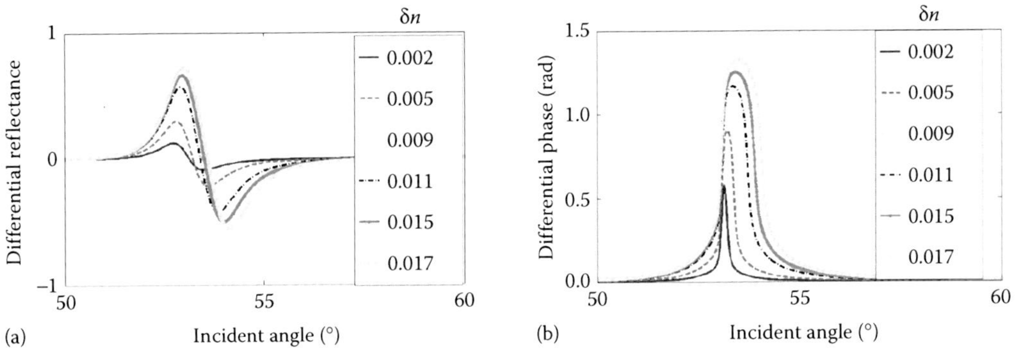

FIGURE 9.32
(a) Differential reflectance and (b) differential phase in angular interrogation for the MSMD structure (considering LRSPR only) due to the change in RI for different concentrations of hemoglobin with respect to water.

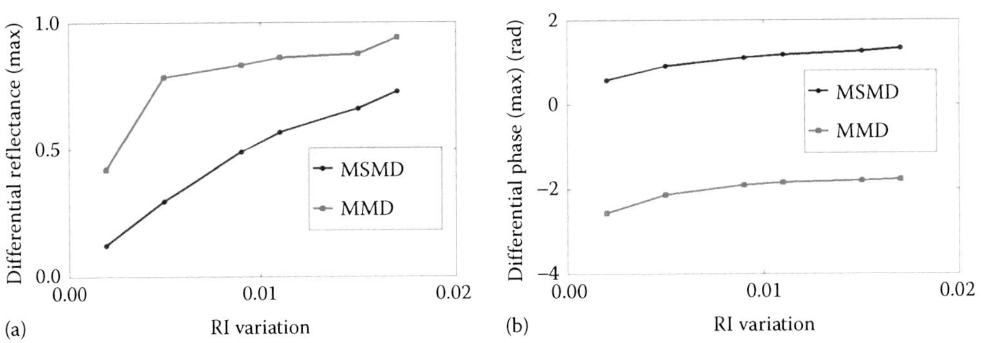

FIGURE 9.33
Maximum of (a) differential reflectance and (b) differential phase with the RI variation for both the MMD and MSMD nanoplasmonic structures.

TABLE 9.6
RI of Deoxygenated Hemoglobin (Concentration 50 g/l) and Water at Different Wavelengths in the Visible Wavelength Region

Wavelength (nm)	RI of Water	RI of Deoxygenated Hemoglobin (50 g/l)
401.5	1.343	1.353
435.8	1.340	1.352
468.1	1.337	1.348
546.1	1.334	1.345
587.6	1.333	1.343
589.3	1.333	1.343
632.8	1.332	1.342
656.3	1.332	1.341
706.3	1.330	1.340

Source: Zhernovaya, O. et al., *Phys. Med. Biol.*, 56, 4013–4021, 2011.

as demonstrated in Table 9.6. As the structures are optimized at 632.8 nm wavelength, the differential reflectance and phase give better result than the other wavelengths (Figures 9.34 and 9.35). In our previous work, AFOM is investigated in complex plane with consideration of sensitivity and DA. Moreover, it is compared in the neighborhood of the wavelength of He–Ne laser source for both the structures, and the MSMD structure provides better result than the MMD structure as the sensing medium is in contact with the plasmon generating thin metal film in MSMD configuration [84].

9.7 MSMD Nanoplasmonic Structure with Bimetallic Nanofilm: Improvement of Measurement Accuracy of LRSPR and SRSPR

As the MSMD structure is more useful in sensing application, bimetallic layer is used instead of single metal layer as shown in the schematic diagram (Figure 9.36a). Combination of silver (Ag) and gold (Au) layer, where Ag is in contact with 300 nm silica layer and Au is in contact with the sensing medium, is used with a total thickness of 50 nm. This bimetallic loaded dielectric-coupled nanoplasmonic structure also produces LRSPR and SRSPR but with narrower FWHM due to the Ag layer as Ag produces higher ratio of the imaginary part of dielectric permittivity to the real part, and hence produces higher measurement accuracy. The resonance curve of this structure in angular interrogation and the corresponding pseudocolor image representation are demonstrated in Figure 9.36b at 632.8 nm wavelength.

9.7.1 Comparison of Mono- and Bimetallic Loaded Dielectric MSMD Plasmonic Structures

The reflectivity curves in angular interrogation along with the image representation are shown in Figure 9.37 depicting LRSPR and SRSPR for mono- and bimetallic configurations. Performance of a bimetallic nanofilm having 25 nm Ag and 25 nm Au layers is

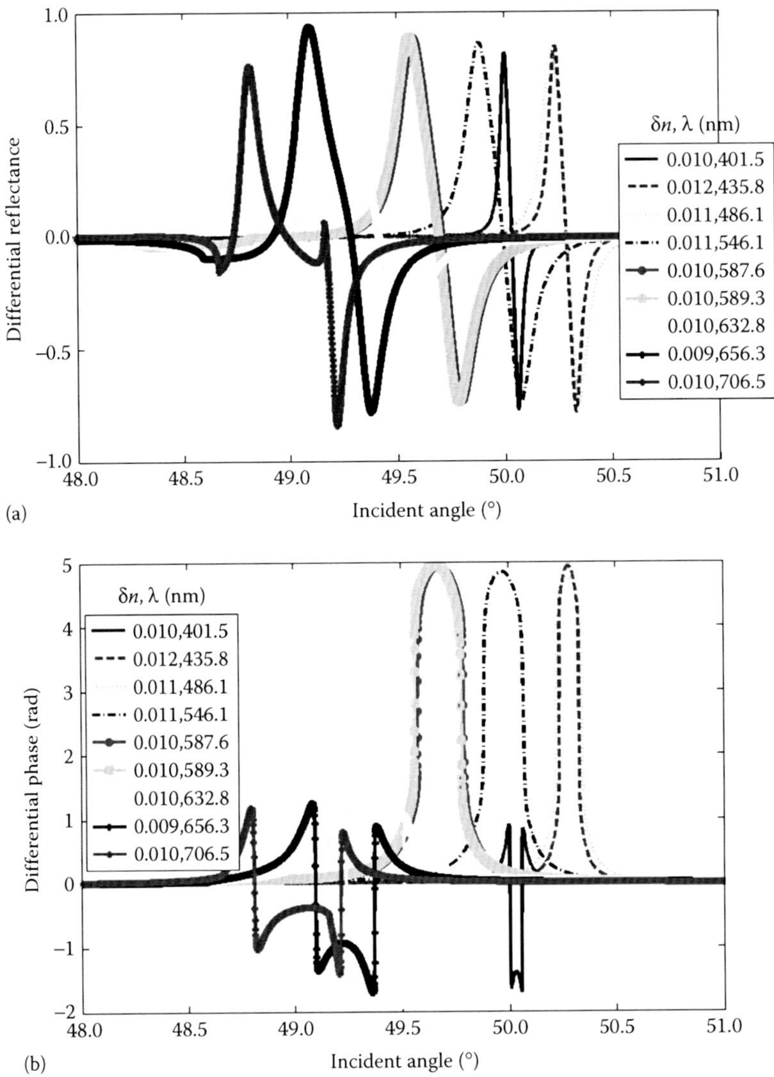

FIGURE 9.34
(a) Differential reflectance and (b) differential phase with angle for the change of RI of 50 g/l concentration of hemoglobin with respect to water for different wavelengths in the visible region for the MMD structure.

compared with a single 50 nm Au and 50 nm Ag film with a fixed silica layer thickness of 300 nm as in Figure 9.37a and c. This particular choice of the bimetallic film with a thickness ratio of 1 and a total thickness of 50 nm is quite justified as it simultaneously produces zero minimum reflectivity for both LRSPR and SRSPR.

In order to obtain equireflectance dips for the monometallic configurations, both the metal and dielectric layer thicknesses are optimized as indicated in Figure 9.37b and d. From this comparison, one can conclude that although Ag monometallic configuration gives narrower resonance compared to that of Au as well as the bimetallic nanofilm, the latter is a better choice compared to the monometallic configurations due to the chemical instability of Ag and broader response of Au.

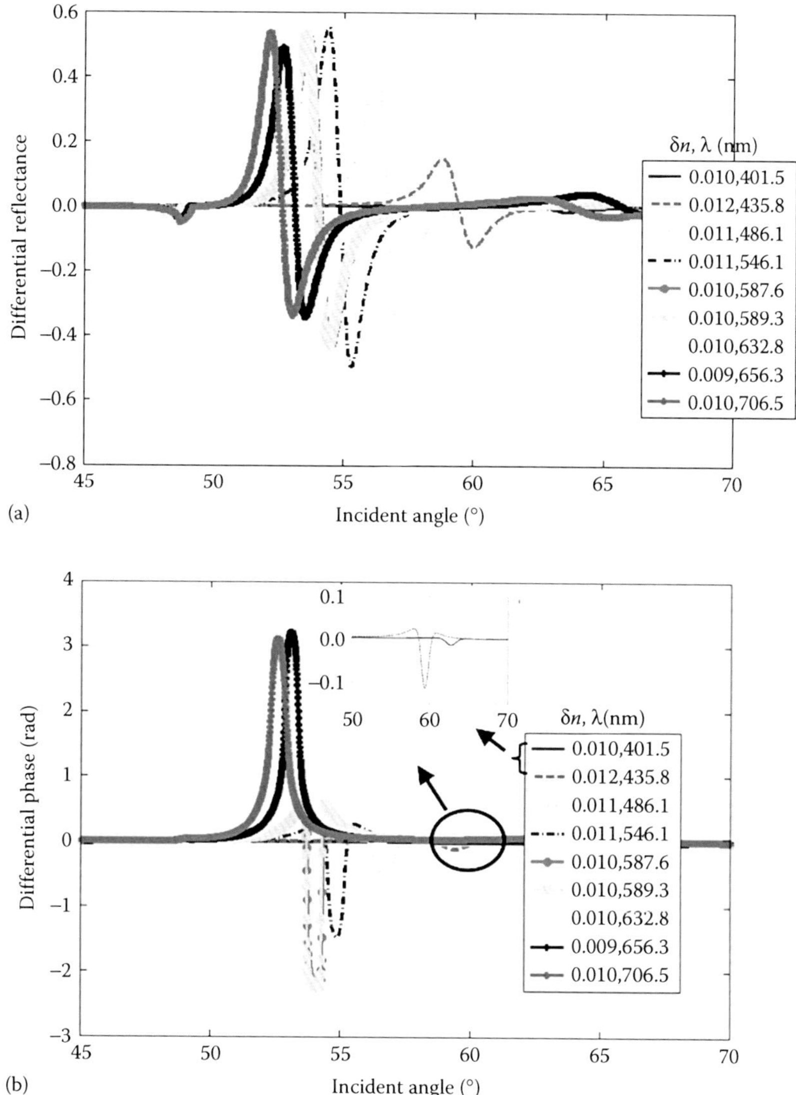

FIGURE 9.35
(a) Differential reflectance and (b) differential phase with angle for the change of RI of 50 g/l concentration of hemoglobin with respect to water for different wavelengths in the visible region for the MSMD structure.

9.7.2 Simultaneous Angular and Spectral Sensitivity

Reflectance contour with simultaneous angular and spectral interrogation can be used for the demonstration of simultaneous sensitivity for two different analytes, ethanol ($n_e = 1.36042$) and water, as shown in Figure 9.38 with a conventional three-layer SPR model using 50 nm gold metal layer. This contour plot provides the information about the simultaneous angular $(S_n)_\lambda$ sensitivity in degrees per refractive index unit (RIU) and spectral $(S_n)_\theta$ sensitivity in nanometers per RIU for fixed wavelength and fixed angle, respectively.

Dual resonance dip for LRSPR and SRSPR of the above-mentioned bimetallic loaded dielectric plasmonic structure can differentiate the interfering bulk RI change from the surface

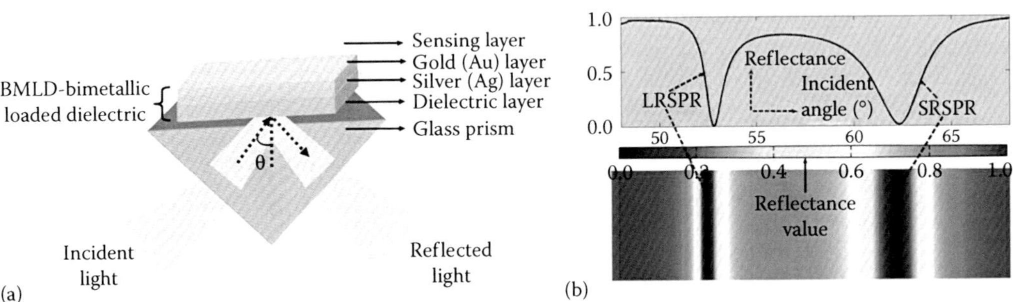

FIGURE 9.36
(a) Schematic diagram of a bimetallic loaded dielectric plasmonic structure and (b) reflectance curve with the pseudocolor image representation of LRSPR and SRSPR in angular interrogation. (With kind permission from IEEE/Photonics Society: *IEEE Photon. Technol. Lett.*, Equi-reflectance dual mode resonance using bimetallic loaded dielectric plasmonic structure, 25(20), 2013, 1965–1968, Bera, M. and Ray, M.)

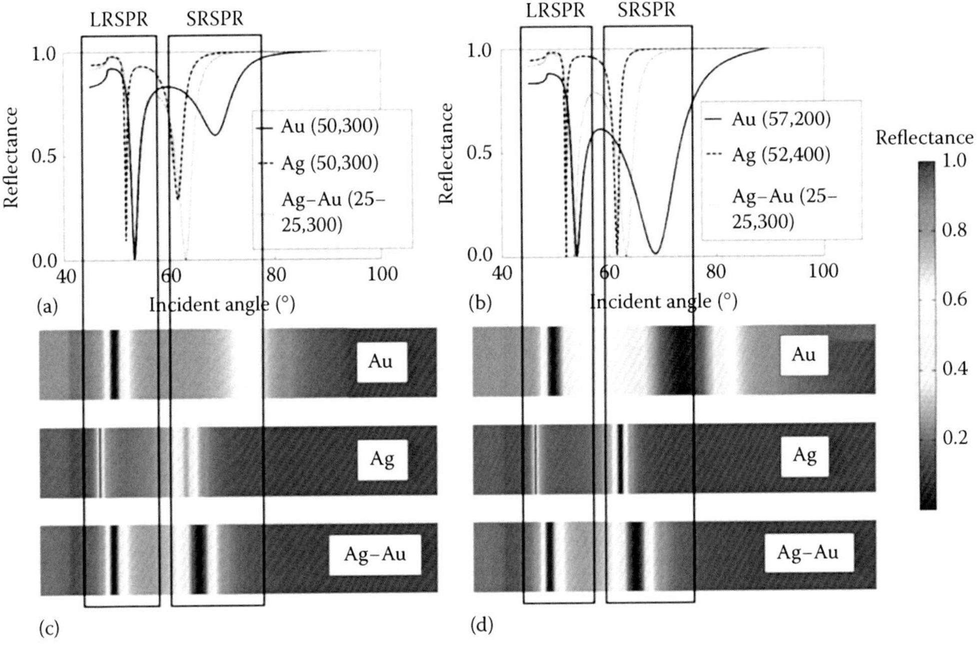

FIGURE 9.37
Comparison of mono- and bimetallic structures: (a) reflectance plots in angular interrogation mode and (c) corresponding image representations for fixed total metal and dielectric thickness. The same plots in (b) and (d) are for optimized metal and dielectric thickness. The values in the parentheses indicate the thickness of metal and dielectric in nanometers. (With kind permission from IEEE/Photonics Society: *IEEE Photon. Technol. Lett.*, Equi-reflectance dual mode resonance using bimetallic loaded dielectric plasmonic structure, 25(20), 2013, 1965–1968, Bera, M. and Ray, M.)

biomolecular interaction. Reflectance contour with simultaneous angle and wavelength modulation for two different analytes, water (n_w = 1.332) and acetone (n_a = 1.3578), is shown in Figure 9.39. Curved trajectories of the reflectivity contour $R(\theta, \lambda)$ of LRSPR is shifted significantly in comparison with SRSPR due to the change of the RI of the analyte (δ_n in RIU). Lower angle and higher wavelength always give higher spectral and angular sensitivity, respectively, as shown in the table (inset of Figure 9.39a) and vice versa (inset of Figure 9.39b).

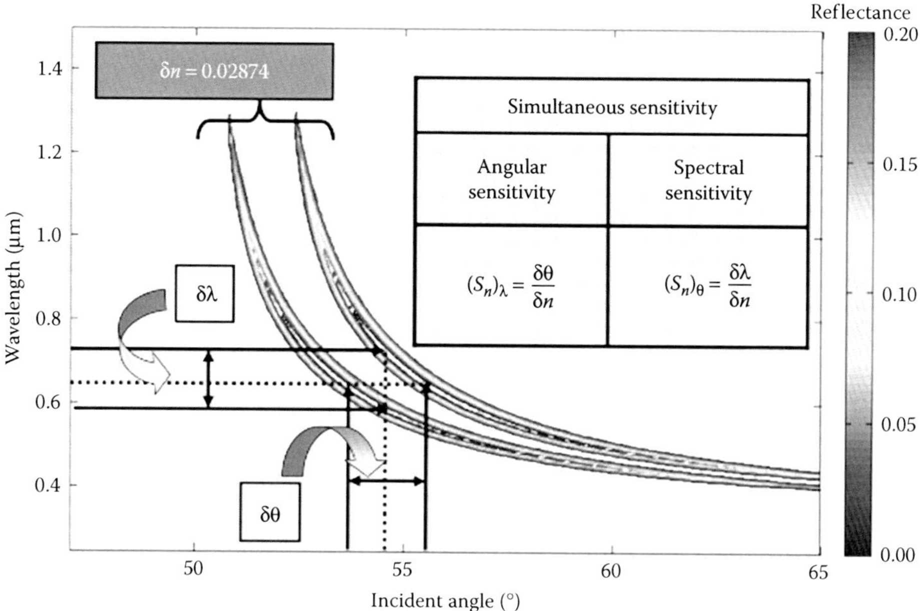

FIGURE 9.38
Graphical representation of simultaneous angular and spectral sensitivity in the conventional three-layer SPR model.

9.7.3 Modulation of Coupling Strength of Symmetric and Antisymmetric Modes: Realization of Logic Operation

The minimum reflectance value of dual resonance modes is plotted with the variation of thickness ratio (d_{Ag}/d_{Au}) of silver and gold metal layers with different total thicknesses of the bimetallic block such as 40, 50, and 60 nm as shown in Figure 9.40. The intersection points denote the equireflectance modes showing the minimum reflectance value along with the corresponding thickness ratio and the other points denote the nonequireflectance modes as shown in the insets of Figure 9.40. From this investigation, it is evident that a total thickness of 50 nm and a thickness ratio of 1 are always desirable for nanoplasmonic sensing as the dual dips provide the equireflectance modes and simultaneous zero minimum reflectance.

These kinds of structures also have other nanophotonic applications for which the equi- and nonequireflectance dual modes are equally important. Special layered geometry and tailoring of the thickness of individual metal film in a bimetallic block as shown in Figure 9.41 demonstrates a proposed plasmonic trinary logic using binary di-bit images. Structured layers, realizing four different Ag/Au thickness ratios within the same bimetallic loaded dielectric (BMLD) structure, can be used as four-input/four-output channels. The corresponding four outputs when analyzed via the imaging module giving the images of LRSPR and SRSPR, as shown in Figure 9.41, can be utilized for realizing di-bit logic operations such as 01, 11, 00, and 10. Here the thickness ratios of the individual metal films are 2, 2.55, 3, and 5 for these four combinations of logic operations with a total thickness of the bimetallic nanofilm 60 nm. It is also possible to modulate the value of reflectivity of LRSPR and SRSPR by varying the thickness ratio of the individual metal film for achieving the different logic outputs. These binary-coded di-bit images can be further utilized for the realization of a more advantageous multivalued logic such as trinary logic, with each di-bit representing a trinary number as shown in Figure 9.41.

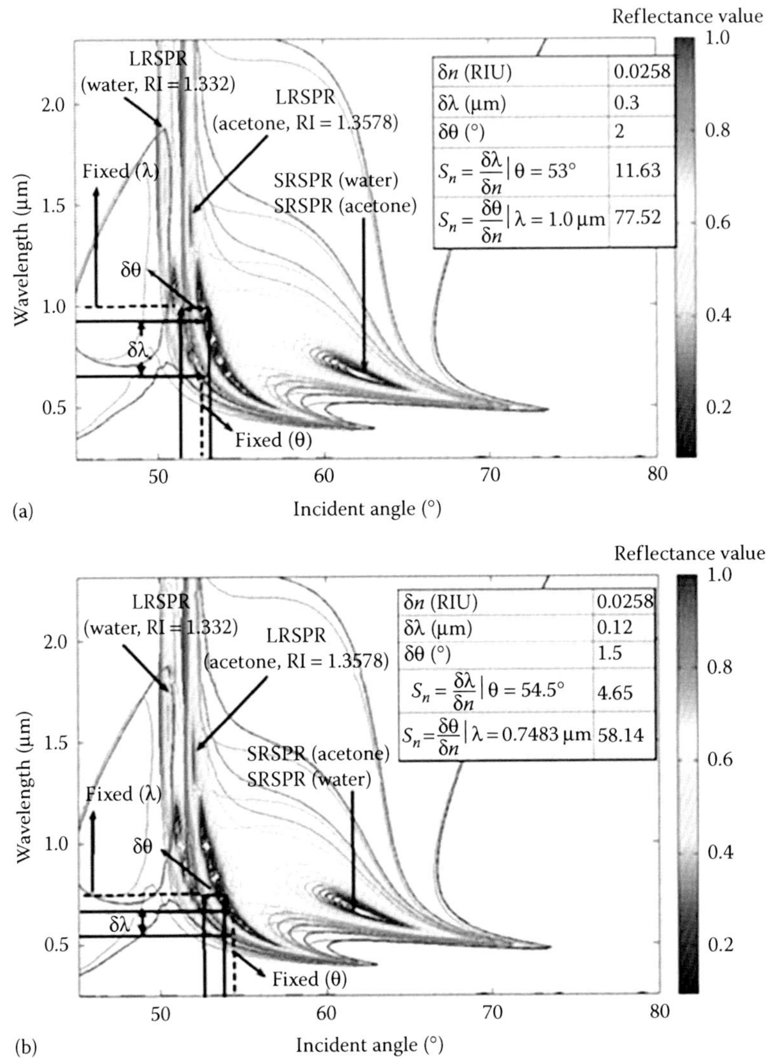

FIGURE 9.39
Simultaneous sensitivity representation using reflectance contour plot with angular and wavelength interrogation calculated for a change in RI of the analyte = 0.0258 RIU for two typical cases: (a) higher wavelength (1.0883 μm), lower angle (52.5°); (b) lower wavelength (0.6383 μm), higher angle (54°). Insets show respective parameters and calculated sensitivities for the two cases under consideration. (With kind permission from IEEE/Photonics Society: *IEEE Photon. Technol. Lett.*, Equi-reflectance dual mode resonance using bimetallic loaded dielectric plasmonic structure, 25(20), 2013, 1965–1968, Bera, M. and Ray, M.)

9.8 Coupled Nanoplasmonic MDM Structure

Another coupled plasmonic structure, termed as waveguide-coupled surface plasmon resonance structure, consists of a silica waveguide layer sandwiched between the two metal layers and the whole MDM block is situated on the high RI SF11 glass prism and the sensing medium is in contact with the upper metal layer as shown in Figure 9.42. Coupling of

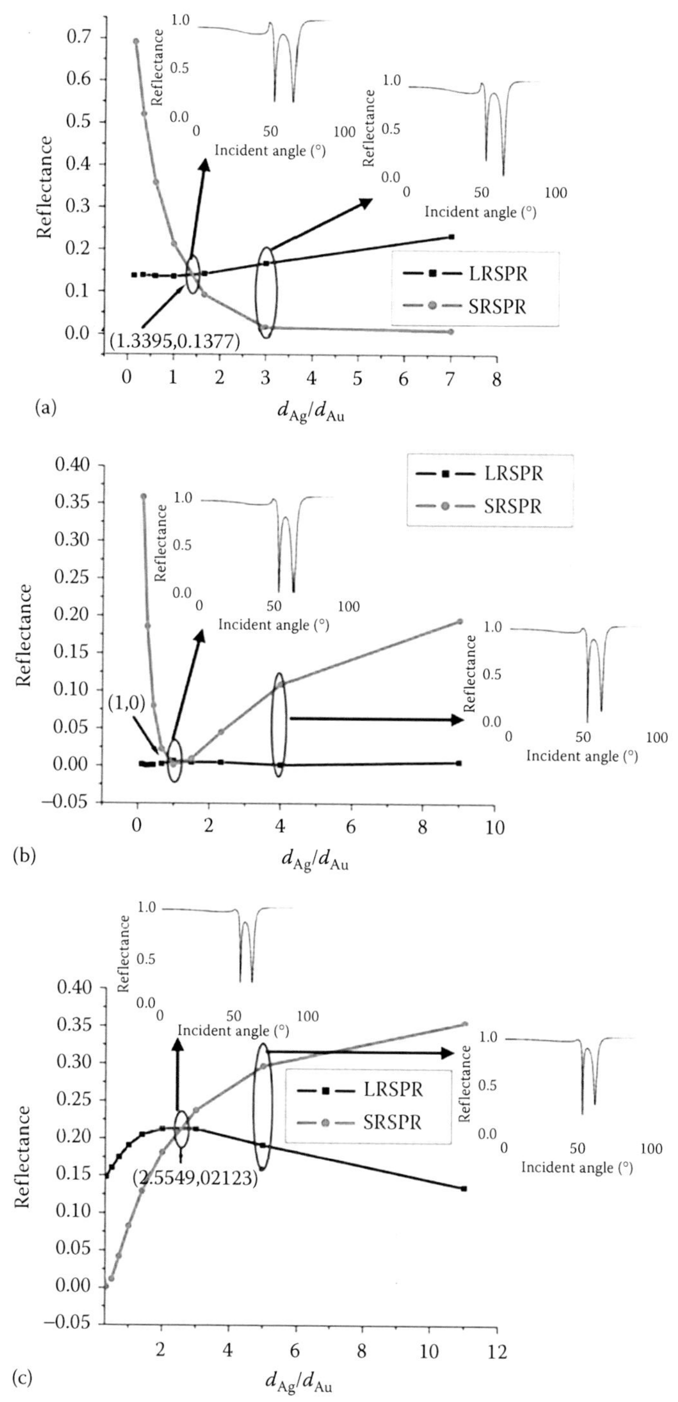

FIGURE 9.40
Minimum reflectance with the variation of d_{Ag}/d_{Au} of LRSPR and SRSPR for different total thicknesses of the bimetallic nanofilm: (a) 40 nm; (b) 50 nm; (c) 60 nm. The insets also depict the equireflectance and nonequireflectance plots for two different thickness ratios. (With kind permission from IEEE/Photonics Society: *IEEE Photon. Technol. Lett.*, Equi-reflectance dual mode resonance using bimetallic loaded dielectric plasmonic structure, 25(20), 2013, 1965–1968, Bera, M. and Ray, M.)

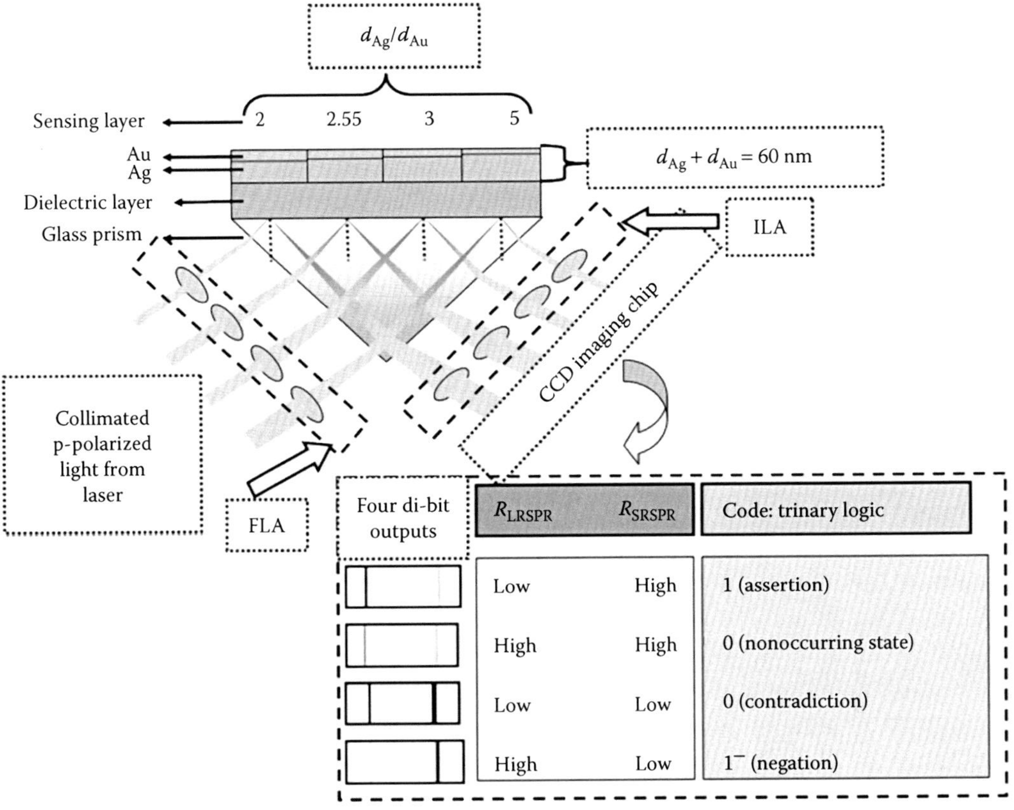

FIGURE 9.41
Proposed realization of trinary logic using di-bit image outputs with nonequireflectance (01, 10) and equireflectance (00, 11) dual modes, where 0 and 1 correspond to low and high reflectance values relative to some threshold, respectively. FLA, focusing lens array; ILA, imaging lens array. (With kind permission from IEEE/Photonics Society: *IEEE Photon. Technol. Lett.*, Equi-reflectance dual mode resonance using bimetallic loaded dielectric plasmonic structure, 25(20), 2013, 1965–1968, Bera, M. and Ray, M.)

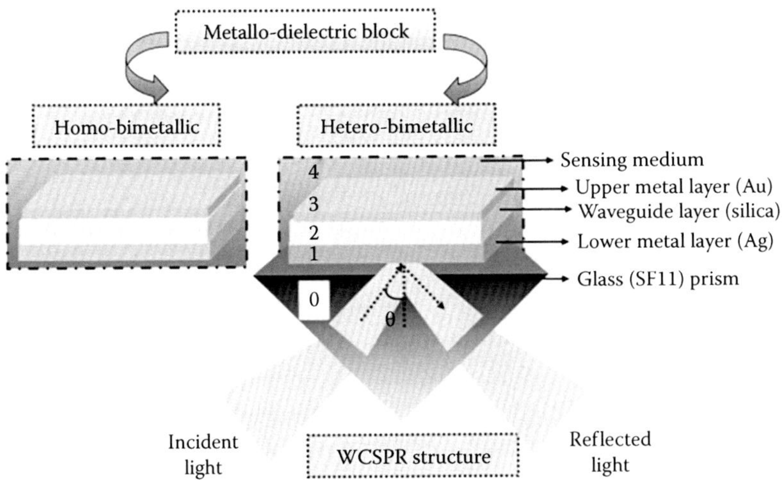

FIGURE 9.42
Schematic diagram of the MDM-coupled nanoplasmonic structure.

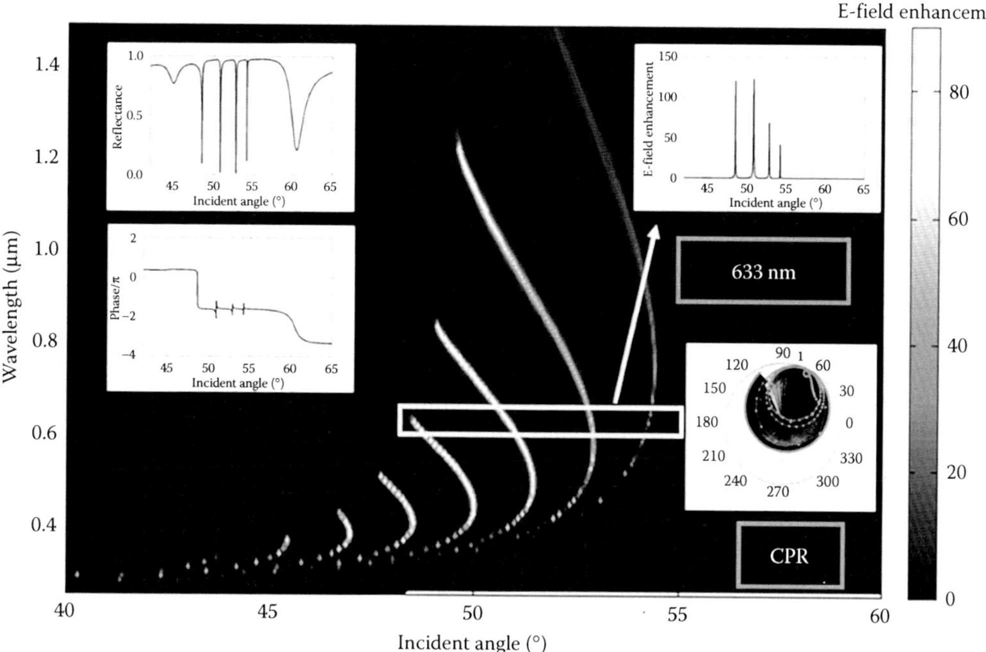

FIGURE 9.43
E-field enhancement contour plot of the MDM structure with 2000 nm silica layer sandwiched between 40 nm Ag and 10 nm Au layers in simultaneous angular and spectral interrogation along with the 2D plot of reflectance, phase, E-field enhancement, and CPR in angular interrogation with a fixed wavelength of 633 nm.

waveguide and plasmonic resonance provides higher DA, and hence improve the performance of the nanoplasmonic sensor. Moreover, the number of waveguide resonance will increase as the thickness of silica waveguide layer is increased as shown in the E-field contour in Figure 9.43. In order to have a better overview of the coupling phenomenon, 2D plots of resonance parameters are shown in the inset.

9.9 TCO-Based Plasmonic Resonance

So far, we have discussed about the structures based on the use of noble metals as plasmon generating film. In this section, we explore the use of TCO films as a replacement of these pure metals. Among all the TCO materials, indium–tin–oxides (ITOs) provide wide opportunities for realizing SPP excitations in the IR wavelength window. A work on SPP propagation in TCO surfaces was reported [89]. Frazen et al. [90–95] extensively explored the observation of broad SPP-like resonances in ITO-coated glass slides in the near IR. Their results showed that SPWs can be easily excited on an ITO surface [96] and the properties of the SPWs can be accurately described by the Drude free electron model. However, the scarcity and high price of pure indium necessitate the replacement of ITO by alternative oxides. Thus, zinc oxide (ZnO) thin films having advantages of low cost, resource availability, and non toxicity draws special attention of recent investigations on the plasmonic study.

Intrinsic ZnO is an n-type wide bandgap semiconductor with a hexagonal wurzite structure. Due to the low carrier density, pure stoichiometric ZnO has very high resistivity.

Depending upon the doping concentrations of Gr III (Al and Ga), conductivity of doped ZnO can be improved significantly. Therefore, having low losses, ZnO:Al could become promising low-loss alternative to ITO in the near IR, as aluminium-doped ZnO (AZO) has low resistivity of ~10^{-4} Ω cm.

Since the carrier density in TCO (10^{20}–10^{21} cm^{-3}) is much less than that in metals (10^{22}–10^{23} cm^{-3}), the plasmon frequency in TCO is in the IR region, as can be obtained from the Drude free electron model with screened plasma frequency $\omega_{ps} = \sqrt{ne^2/\mu\varepsilon_0\varepsilon_\infty}$, where n is the charge carrier density, μ is their effective mass, ε_0 is the permittivity of free space, and ε_∞ is a high-frequency dielectric constant [51].

In AZO ($ZnAl_2O_4$), a large doping level increases the mobility of the carriers through increased scattering but does not affect the carrier concentration. This causes the zero crossover of real permittivity to shift toward longer wavelengths. This happens in pulsed laser-deposited (PLD) AZO films with 2%wt of Al_2O_3. Evidently, lower Al_2O_3 concentration of about 0.8%wt produces AZO films with real part of permittivity ε′ crossover at wavelength smaller than 1.5 μm [97]. Hence, AZO is a promising low-loss alternative material for plasmonics at this wavelength.

At the interface of ITO film and BK7 glass, the SPPs are excited in the near-IR wavelength (1.45–1.59 μm) region [96]: important optical wavelengths for plasmonic applications such as molecular vibrational spectroscopes and light telecommunications [98,99], whereas aluminum-doped ZnO is not found to support SPP propagation in this wavelength window [100].

In this study, theoretical investigation on SPP propagation in TCO thin films, transition metal nitrides and also the constraints under which they give better performance in the IR and visible optical frequency range are discussed. For the study, different thicknesses of TCO films such as ITO, ZnO:Al, ZnO:Ga, and TiN coated on BK7 glass slides are considered and comparative analysis supported by MATLAB simulations is presented in Section 9.9.1.

9.9.1 Simulation Results and Analysis

9.9.1.1 ITO Film

The SPP resonance curves are shown in Figure 9.44 for ITO as a function of frequency and incident angle from 45° to 69° with 3° increments in the 1.55–2.5 μm wavelength window. Figure 9.44 consists of nine subplots with variation of thickness of TCO film. Here, from the theoretical data, we have found that the angle of incidence at which resonance occurs is 42° for both ITO and ZnO:Al. In the whole frequency range of interest, 30 nm thick ITO film shows two minimum reflectivity positions (dips) observed at 2122 cm^{-1} and 7692 cm^{-1}. The second dip at 7692 cm^{-1}(0.954 eV) corresponds to screened bulk plasmon polariton (SBPP) resonance, although this energy is slightly lower than that theoretically calculated $\omega_{ps} = \omega_p/\sqrt{\varepsilon_\infty} = 7771$ cm^{-1}(0.964 eV). With every increment of θ, the SBPP resonant frequency does not alter, although the reflectivity increases to some extent, which means absorbance of the incident light decreases as evident from the resulting change of profile. With an increased thickness, for 100 nm ITO, at resonant frequency, the reflectivity dip (in between 6000 and 8000 cm^{-1}) almost approaches zero and the incident light is efficiently coupled to plasmons.

With further increase of the film thickness, the SBPP resonance appears to behave differently. For 150 nm-thick film layer of ITO, the resonant frequency shifts to higher energies for lower incident angle (7639 cm^{-1} for 45°) and to lower energies for higher incident angle (7056 cm^{-1} for 54°), and beyond that the resonant energy is found to be almost independent

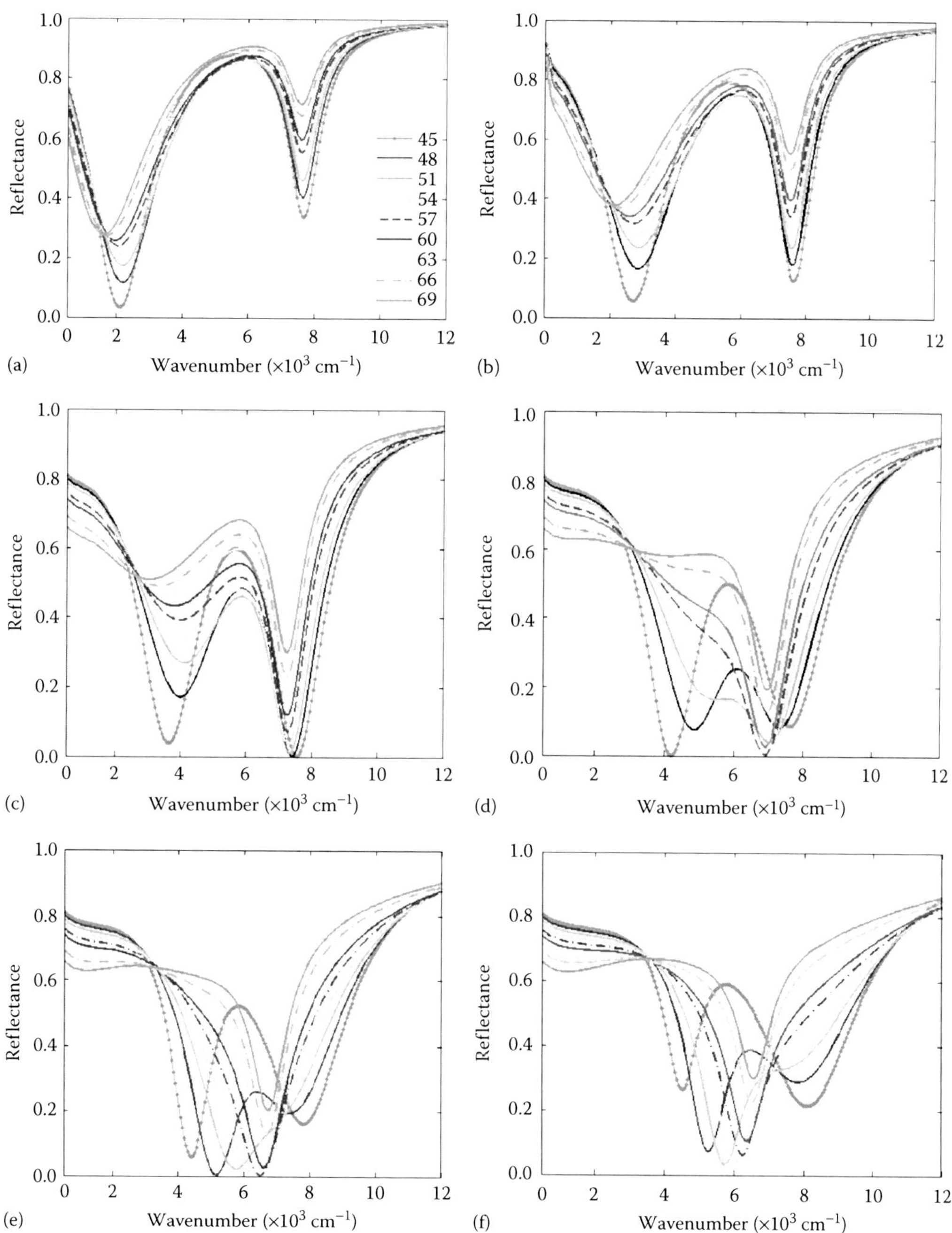

FIGURE 9.44
Reflectance vs. wavenumber curves for different thicknesses of ITO film, with θ varying from 45° to 69° in steps of 3°: (a) 30 nm; (b) 50 nm; (c) 100 nm; (d) 150 nm; (e) 200 nm; (f) 250 nm; (g) 300 nm; (h) 400 nm; (i) 500 nm. Line styles are depicted in the legend of (a).

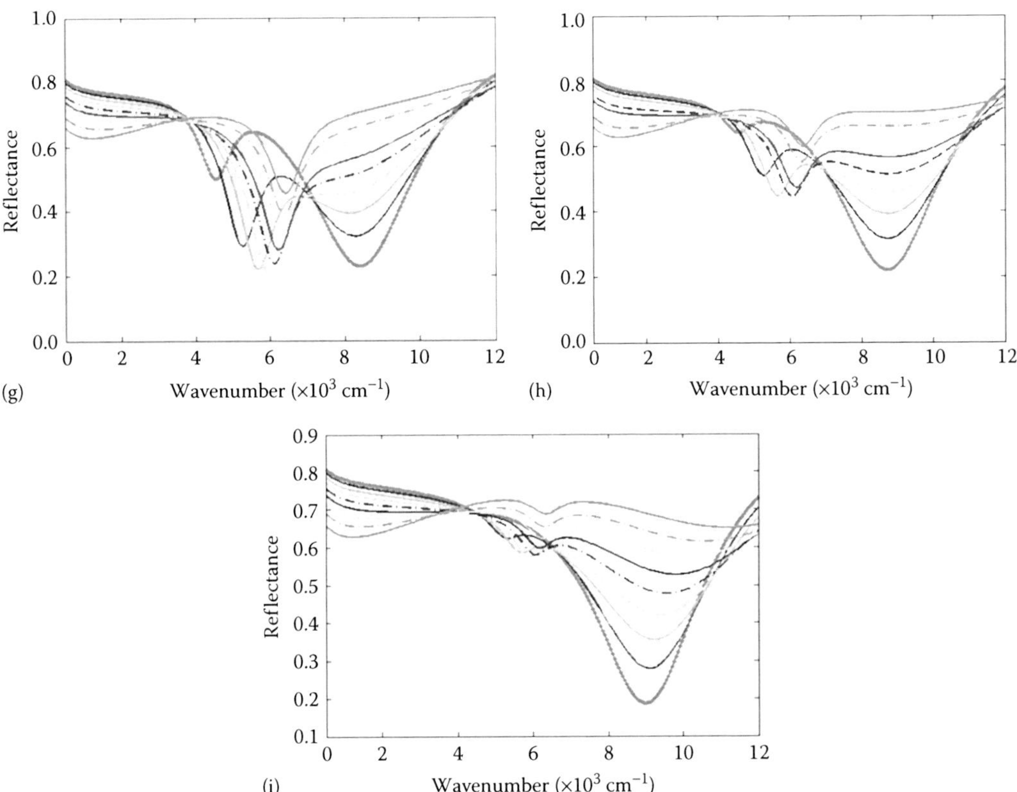

FIGURE 9.44
Continued

of θ, which corresponds to SPP resonance. From the reflectance versus wavenumber curves for different thicknesses, it is shown that SPP coupling is optimum for 200 nm thickness of the ITO film. If the ITO coating is 500 nm thick, it is clearly noticed that only lower angles can support SPP excitation although not much efficient, and for θ above 48°, practically there is no coupling between plasmons and incident radiation.

Figure 9.45 shows the reflectance profile for a particular angle of incidence 45° with different thicknesses of ITO film. Here also, dual resonance dips are noticed with varying reflectance minima values for thickness below 400 nm. For 100–200 nm range, these dual dips attain minimum reflectance, which is quite near to zero reflectance as in Figure 9.46.

9.9.1.2 Doped Zinc Oxide (ZnO:Al and ZnO:Ga) Films

Figure 9.47 depicts the reflectance variation curve with incident angle for three different thicknesses of ITO and ZnO:Al films. From Figure 9.47, it can be concluded that Al-doped ZnO also supports the SPP excitation in the mid-IR wavelength considered here. Figure 9.48 shows the variation of reflectance profile with wavenumber for different thicknesses of ZnO:Al. This type of resonance is sometimes termed as Fano resonance or asymmetric type of dual dips. In our simulation, we have noticed that for $\theta > 51°$, the reflectance profile does not correspond to SPP excitation of interest. This limit of angle is thus lower

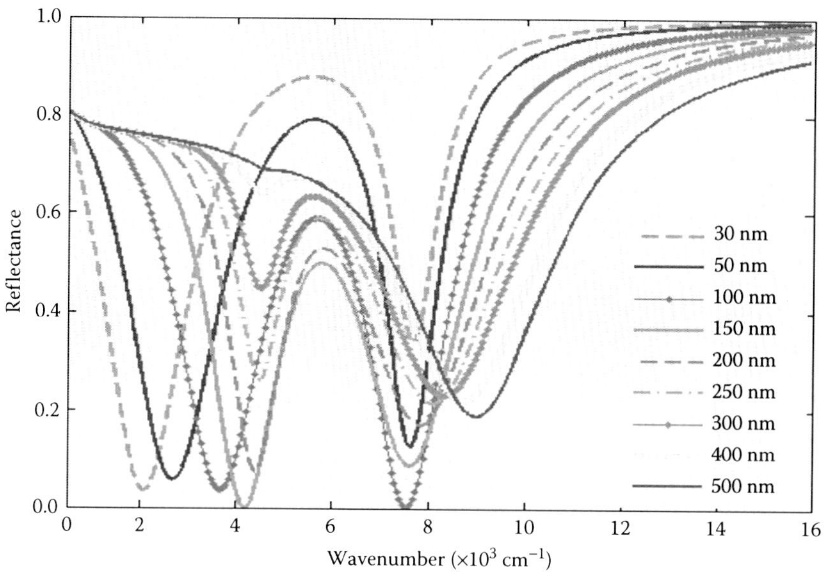

FIGURE 9.45
Variation of reflectance profile with different increasing thicknesses of ITO (from 30 to 500 nm) at 45° incident angle.

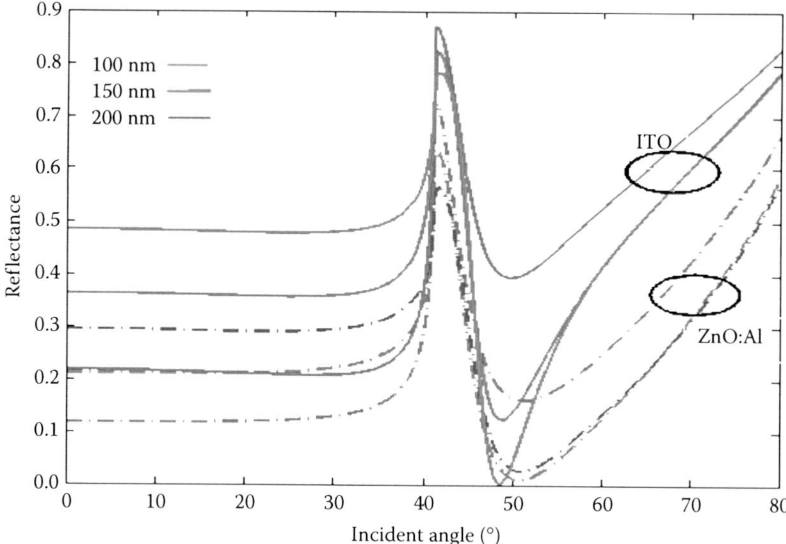

FIGURE 9.46
Reflectance vs. incident angle for ZnO:Al film compared with that for ITO films for three different thicknesses: 100, 150, and 200 nm.

than that for ITO (which is ~70°). For 30 nm ZnO:Al film, dual dips having minima at 954 and 3873 cm^{-1} corresponding to angles 51° and 43°, respectively. The reflectance profile is deeper for lower incident angle for the second dip. Again, for 500 nm thickness, the resonance curve become symmetric type with a single reflecting dip occurring at 4651 cm^{-1} (0.58 eV).

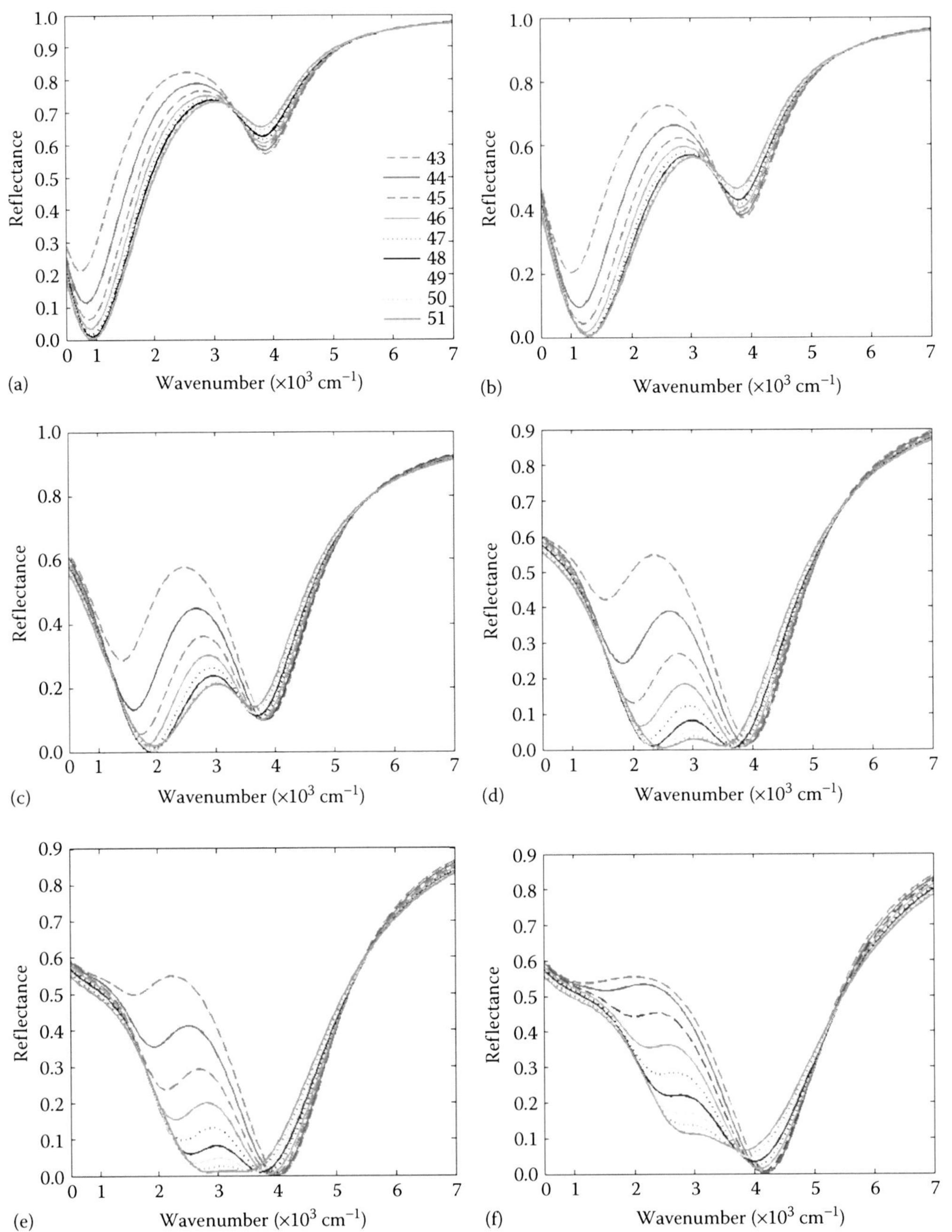

FIGURE 9.47
Reflectance vs. wavenumber curves for different thicknesses of ZnO:Al film, with θ varying from 43° to 69° in steps of 1°: (a) 30 nm; (b) 50 nm; (c) 100 nm; (d) 150 nm; (e) 200 nm; (f) 250 nm; (g) 300 nm; (h) 400 nm; (i) 500 nm.

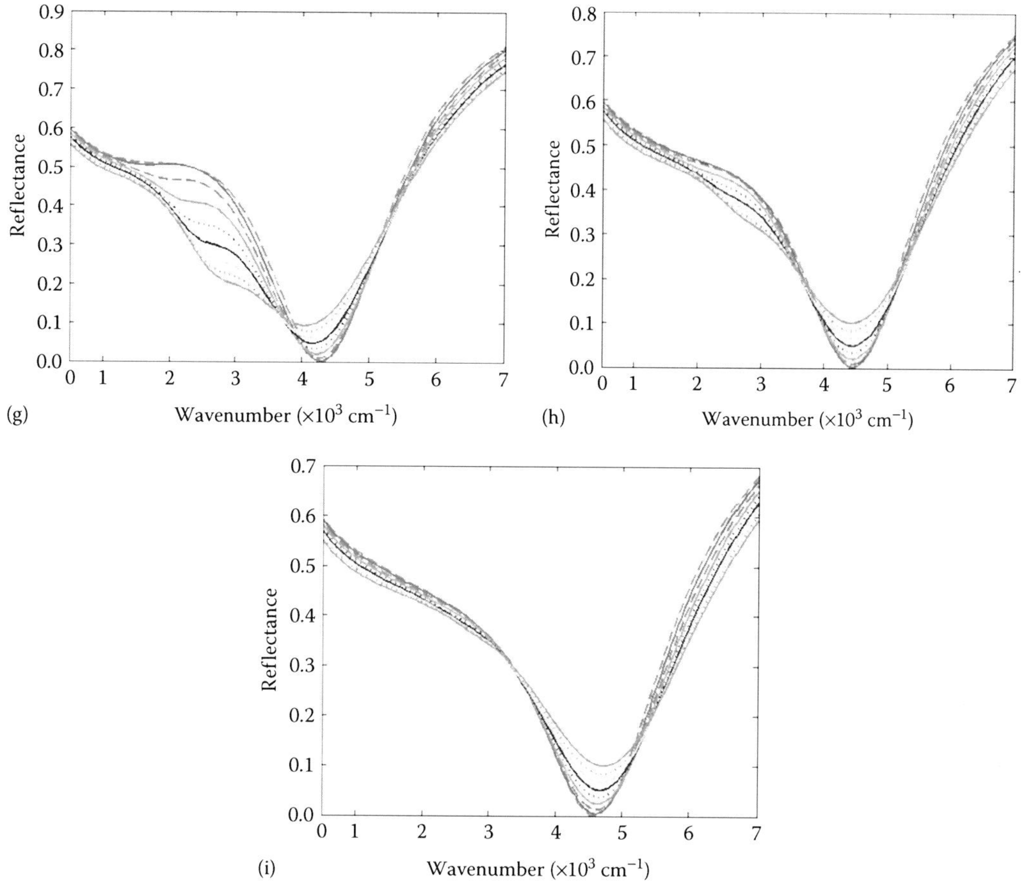

FIGURE 9.47
Continued

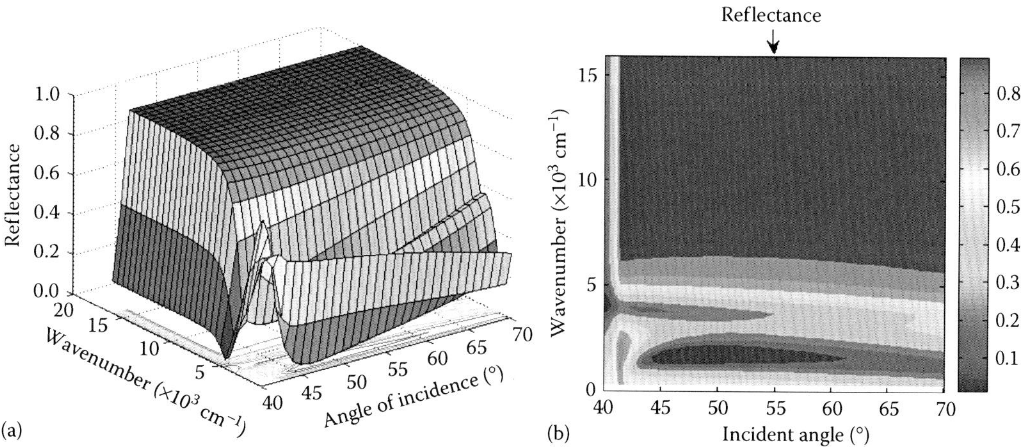

FIGURE 9.48
(See color insert.) (a) 3D SPR spectra and (b) contour plot for ZnO:Al film showing variation with wavenumber and angle of incidence.

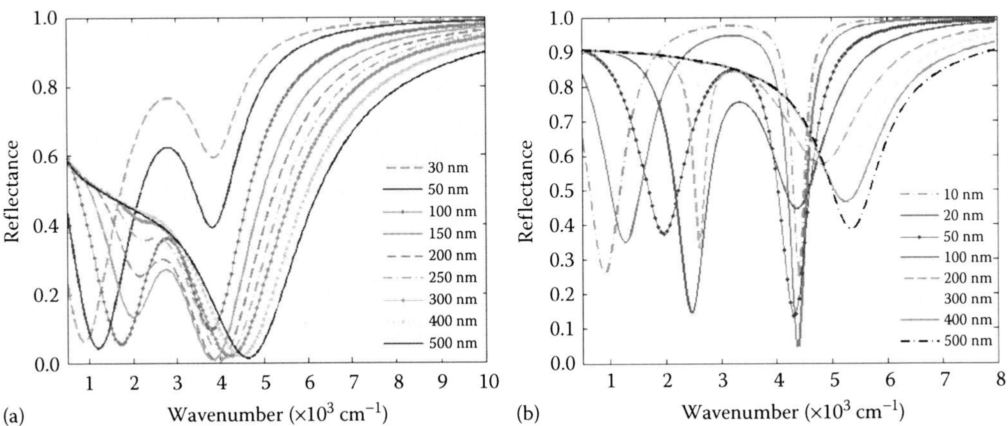

FIGURE 9.49
(See color insert.) Variation of reflectance profile with different increasing thicknesses of (a) ZnO:Al (from 30 to 500 nm) and (b) ZnO:Ga (from 10 to 500 nm) at 45° incident angle.

From Figure 9.47, it can be noticed that the reflectance dip is observed at two values of frequency: the lower wavenumber position, where reflectance drops to 0.23% observed at 1698 cm^{-1} (eV) corresponding to the SP frequency ω_{sp}, and the second dip corresponding to the plasma frequency ω_p at 3820 cm^{-1}(eV), where reflectivity reduced almost 100%. There is an absorption of light resulting in generation of surface wave at the interface.

Figure 9.48 shows the 2D variation of reflectivity with frequency for different thicknesses of both Al- and Ga-doped ZnO when exciting radiation is incident at an angle 45° to the metal oxide layer. This feature follows from the 3D and contour plots shown in Figure 9.49a and b where reflectance is plotted as function of thickness and wavenumber. Compared to ITO, minimum reflectivity position of ZnO:Al is lower or in other words corresponds to higher absorbance of incident light. Above 200 nm thickness, ZnO:Al gives almost 100% absorption, whereas for ITO, it is seen that above 100 nm, almost 20% of incident light gets reflected.

The performance of ZnO:Al for SPR effect based on the variation of thickness of conducting metal oxide (CMO) and wavenumber as 3D plot is demonstrated in Figure 9.50a and b for 45° fixed incident angle. Similar to ITO, for thickness below 100 nm, the reflectivity curves of ZnO:Al and ZnO:Ga show two dips and the minimum reflectivity is not close to zero. But for still thicker layer of both these CMO films, SPR effect can be seen with single resonant dip with a wavenumber shift. This shift is from 3820 cm^{-1} (for 100 nm) to 4934 cm^{-1} (for 500 nm) for ZnO:Al and from 4480 cm^{-1} (second dip of 100 nm) to 5178 cm^{-1} (500 nm) for ZnO:Ga.

When the exciting light incident at an angle 46° on 73 nm-thick ZnO:Al film coated on the BK7 glass substrate, the film becomes transparent and causes SPR effect at 1963 and 3767 cm^{-1} frequencies. This theoretical prediction is supported by the graphical contour plot as shown in Figure 9.50b.

From Figure 9.51a and b, we can see that there may be a chance of dual resonance at $\theta = 43°$. The minimum reflectivity position occurs at frequencies 901 and 2175 cm^{-1} for 50 nm-thick ZnO:Ga film.

3D response of reflectance for 2% ZnO:Ga is shown simultaneously with variation of wavenumber and thickness of the film in Figure 9.52a along with the contour plot in Figure 9.52b. The spectral reflectance curve for 500 nm is symmetric and has single resonant frequency at 2706 cm^{-1}. It can also be observed that above 250 nm value of thickness, the profile becomes symmetric with single dip. But the film remains opaque to some extent

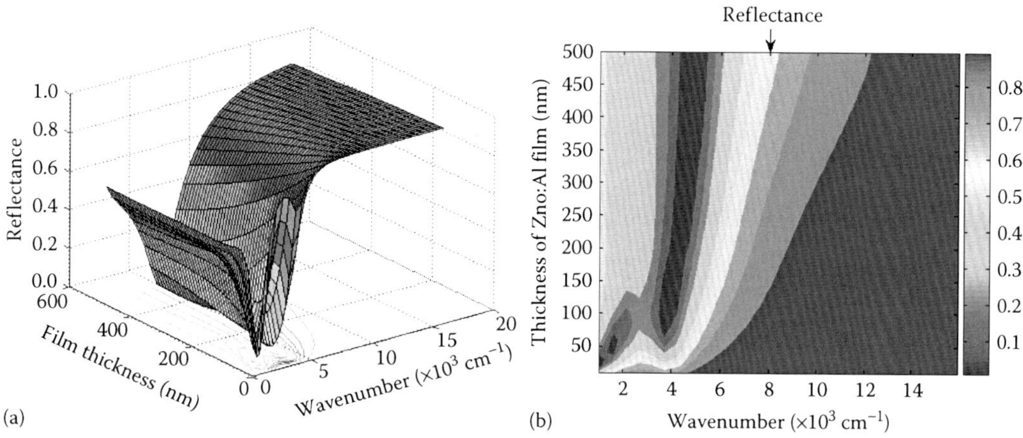

FIGURE 9.50
(See color insert.) (a) 3D SPR spectra and (b) contour plot for ZnO:Al film showing variation with wavenumber and thickness of the film.

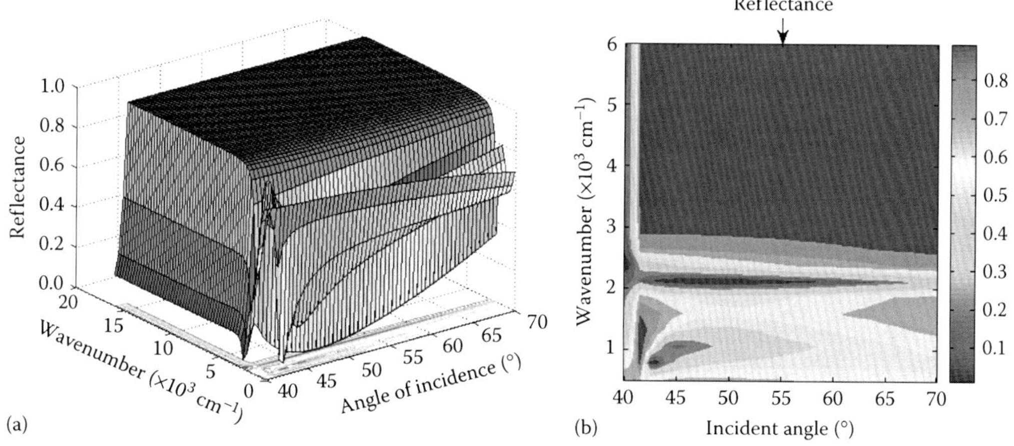

FIGURE 9.51
(See color insert.) (a) 3D SPR spectra and (b) contour plot for ZnO:Ga film showing variation with wavenumber and angle of incidence.

(~15%) to the incident radiation. Ga-doped ZnO, before and after heat treatment, supports SP excitations for thin-film thickness below 200 nm.

Moreover, from Figures 9.49b and 9.51b, the comparison of the films for generating SPWs and performance for resonant condition becomes easy. From these plots, it is seen than Al-doped ZnO gives better sensitivity for the entire thickness range. In the minimum reflectivity position of wavenumber for ZnO:Al, the reflectance minimum is more toward zero than in the other case, although the frequency range of resonance is different for different film materials.

9.9.1.3 Titanium Nitride Film

In this section, the performance of titanium nitride (TiN) under different resonance conditions is discussed. Reflectivity response of TiN with different film thicknesses and angles of incidence is shown in Figure 9.53. The film gives resonance at 35° of the incident angle

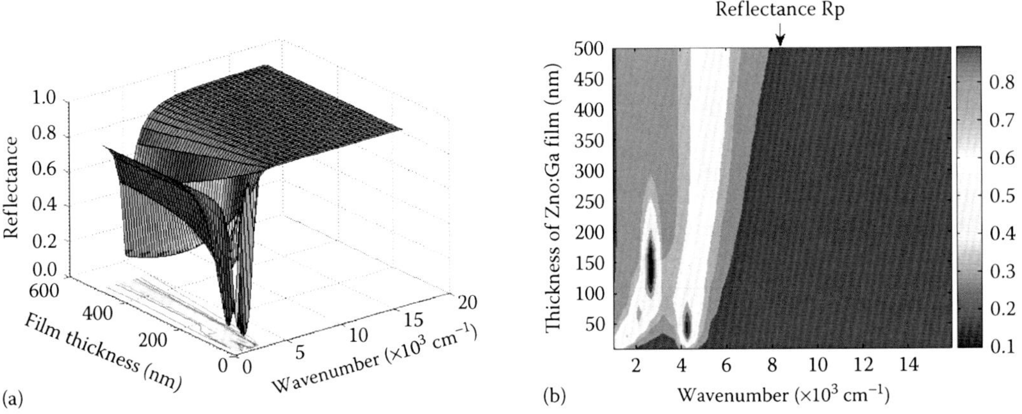

FIGURE 9.52
(a) 3D SPR spectra and (b) contour plot for ZnO:Ga film showing variation with wavenumber and thickness of the film.

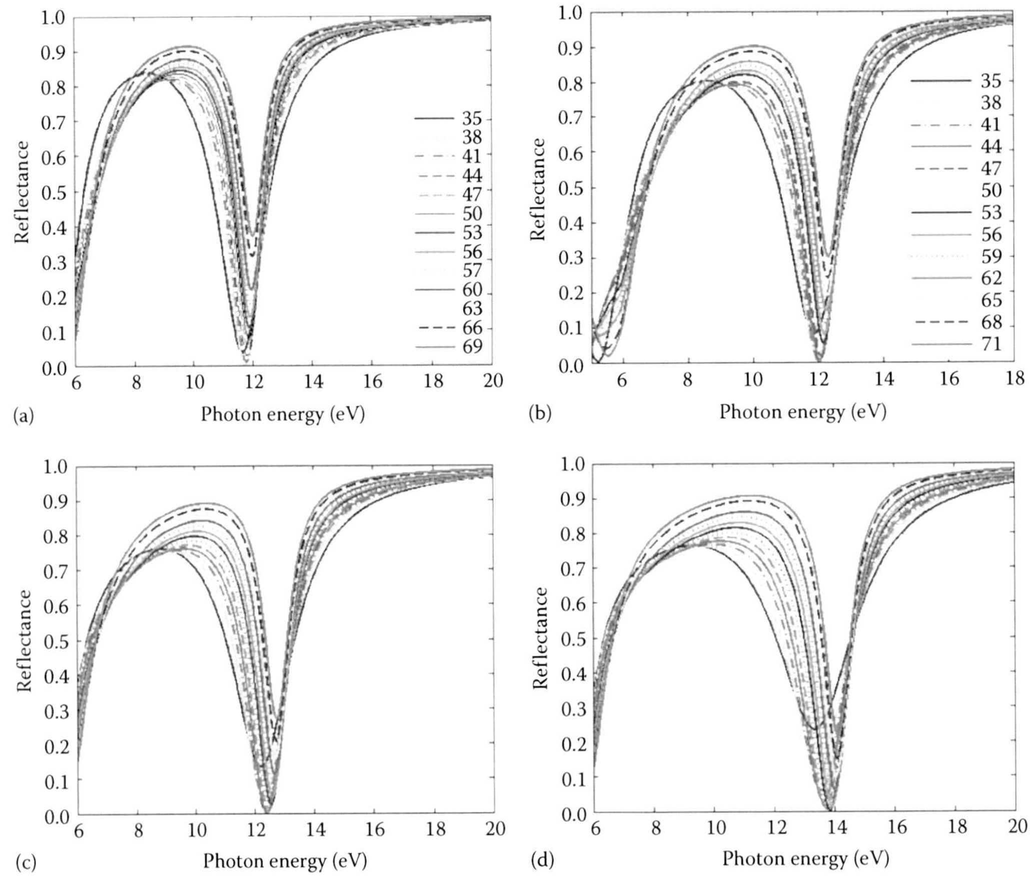

FIGURE 9.53
Reflectance response of TiN with the incident angle varying from 35 to 70° for (a) 25, (b) 35, (c) 55, (d) 70, (e) 90, (f) 100, and (g) 105 nm-thick films.

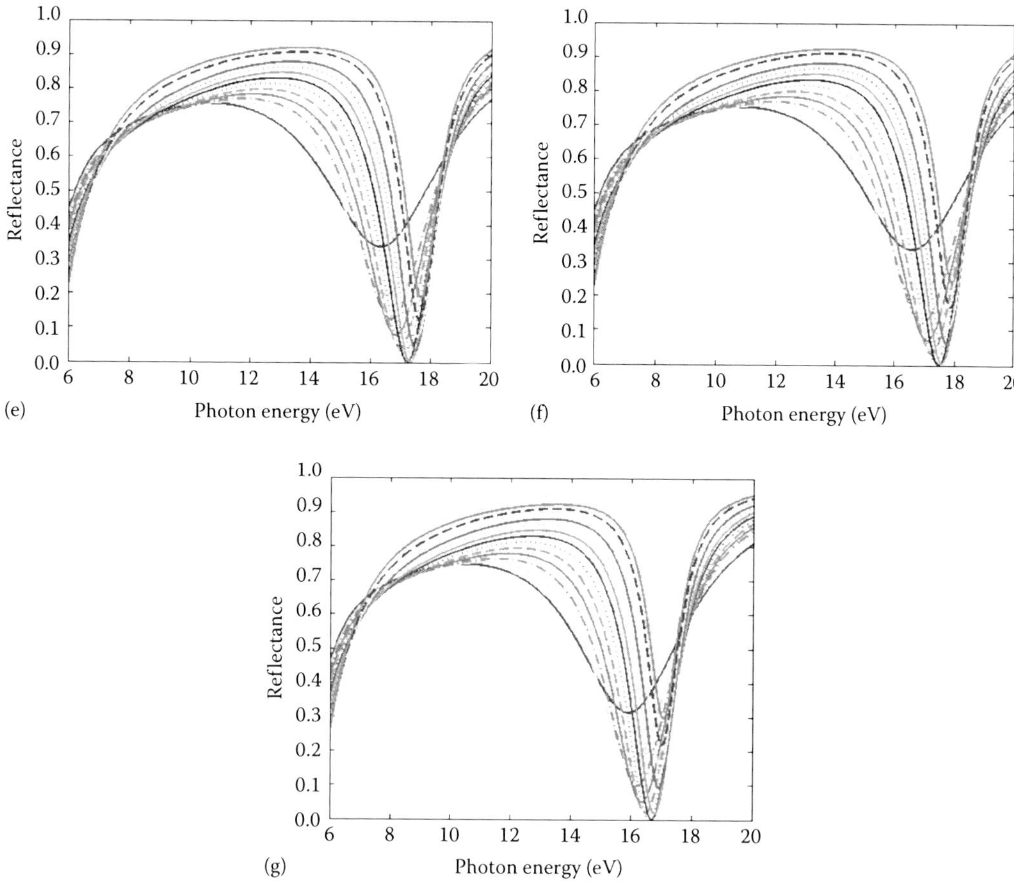

FIGURE 9.53
Continued

and also gives resonance up to 70°. The perfect coupling of incident light with the SPWs occurred at very lower angle than the TCO films. The main advantage of using transition metal nitrides is that it gives response in visible frequency and can be a substitute to metal in this optical frequency for the application of MM and optoelectronic devices. The thickness dependence of the reflectance spectra for TiN is shown in Figure 9.54.

9.10 Conclusions

Earlier we have reported a detailed discussion on computational techniques related to multilayer coupled nanoplasmonic structures [101]. In this chapter, we have discussed about different computational approaches applied to various nanoplasmonic structures for simulation of mainly the resonance response in the spectral domain to be applied to optical biological spectroscopy and imaging.

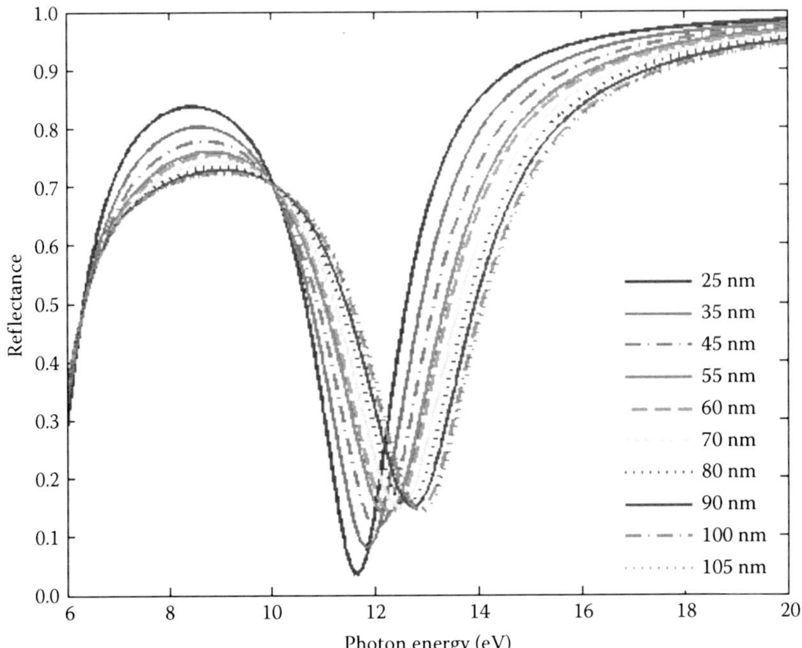

FIGURE 9.54
Reflectance variation for different thicknesses of TiN for 35° incident angle.

Prism material dependency of a bioplasmonic structure operating under angular and wavelength interrogation modes has been theoretically investigated using the admittance loci method for two different prism materials: chalcogenide (2S2G) and silicon. Simulated sensitivity plots recommend the selection of low-index prism materials (in comparison with silicon) such as chalcogenide (2S2G) and suggest to operate the sensor at lower wavelength rather than to operate it at higher wavelength to achieve higher sensitivity. But using low RI prism material reduces the dynamic range of the bioplasmonic structure. Therefore, low-index prism materials are preferable for the applications where higher sensitivity is required and the range of sample refractive indices is narrow and limited, and while designing a bioplasmonic structure, the choice of prism materials should be in accordance with sensitivity, FWHM, and dynamic range requirements.

Modified MSMD-coupled nanoplasmonic structures can be used for biological sensing. Differential phase and reflectance have been investigated for different concentrations of deoxygenated hemoglobin with respect to water and spectroscopic dependence has also been analyzed. Symmetric (LRSPR) and antisymmetric (SRSPR) modes can be utilized for various nanophotonic applications such as all optical signal processing, information processing, optical switching, and second-harmonic generation. Our proposed special layered geometry and tailoring of the thickness of bimetallic block can produce equi- and nonequireflectance dual resonances, which can be further utilized for plasmonic trinary logic. Moreover, coupling of waveguide and plasmonic resonance due to the positional interchange of the metallo-dielectric block provides an improved performance of nanoplasmonic sensor. Also progressive multiple resonance due to the increase of the thickness of waveguide layer in coupled waveguide plasmonic structure can be used to characterize dielectric material in a multilayer nanocomposite.

Phase-dependent resonant behavior of multilayer coupled nanoplasmonic structure can be utilized in SPR-based interferometric imaging.

We have also explored the use of TCO and metal nitrides as plasmon-generating materials, and based on this simulation, we can conclude that although ITO provides better opportunity to be used in many transparent oxide-based device applications, Al-doped ZnO thin films can support SPP resonance in the 2.5 μm range and can be good replacement for ITO [102]. Again, both ZnO:Al and ZnO:Ga support SPR in the frequency range of interest. GZO can also provide crossover wavelength as low as 1.2 μm, but it has higher optical loss compared to AZO. However, it can be a good substitute to ITO for SPR spectroscopy in the near-IR frequency range.

In conclusion, the content of this chapter will be highly useful for the researchers who are interested in taking challenges to explore the vast field of theoretical nanobioplasmonics and associated applications.

Acknowledgments

M. Bera and S. Ghosh acknowledge the Council of Scientific and Industrial Research (CSIR), New Delhi, India, for providing Senior Research Fellowship. K. Brahmachari is grateful to Technical Education Quality Improvement Programme (TEQIP PHASE II), University College of Technology, University of Calcutta, *Kolkata, India*, for awarding Senior Research Assistantship.

References

1. A. Otto, Excitation of nonradiative surface plasma waves in silver by the method of frustrated total reflection, *Zeitschrift fur Physik*, **216**, 398–410, 1968.
2. E. Kretschmann, H. Raether, Radiative decay of non-radiative surface plasmons excited by light, *Zeitschrift fur Naturforschung*, **23A**, 2135–2136, 1968.
3. B. Liedberg, C. Nylander, I. Lunström, Surface plasmon resonance for gas detection and biosensing, *Sensors and Actuators*, **4**, 299–304, 1983.
4. J. Homola, Surface plasmon resonance sensors for detection of chemical and biological species, *Chemical Review*, **108**, 462–493, 2008.
5. J. Homola, I. Koudela, S.S. Yee, Surface plasmon resonance sensors based on diffraction gratings and prism couplers: Sensitivity comparison, *Sensors and Actuators B*, **54**, 16–24, 1999.
6. J. Le Person, F. Colas, C. Compère, M. Lehaitre, M.-L. Anne, C. Boussard-Plèdel, B. Bureau, J.-L. Adam, S. Deputier, M. Guilloux-Viry, Surface plasmon resonance in chalcogenide glass-based optical system, *Sensors and Actuators B*, **130**, 771–776, 2008.
7. E.D. Palik, *Handbook of Optical Constants of Solids*, Academic Press, New York, 1998.
8. K. Brahmachari, M. Ray, Influence of chalcogenide glass on surface plasmon resonance based sensor design in near infrared region using admittance loci method, *Proceedings of the 5th International Conference on Computers and Devices for Communication (CODEC)*, IEEE catalog number: CFP1201I-CDR, December 17–19, 2012, pp. 1–4, Kolkata, India.
9. K. Brahmachari, M. Ray, Modelling of chalcogenide glass based plasmonic structure for chemical sensing using near infrared light, *Optik*, **124**(21), 5170–5176, 2013.

10. K. Brahmachari, M. Ray performance of admittance loci based design of plasmonic sensor at infrared wavelength, *Optical Engineering*, **52**(8), 087112-1-8, 2013.
11. S. Patskovsky, A.V. Kabashin, M. Meunier, J.H.T. Luong, Near-infrared surface plasmon resonance sensing on a silicon platform, *Sensors and Actuators B*, **97**, 409–414, 2004.
12. S. Patskovsky, A.V. Kabashin, M. Meunier, J.H.T. Luong, Properties and sensing characteristics of surface plasmon resonance in infrared light, *Journal of Optical Society of America A*, **20**(8), 1644–1650, 2003.
13. F. Abeles, T. Lopez-rios, Decoupled optical excitation of surface plasmons at the two surfaces of a thin film, *Optics Communications*, **11**(1), 89–92, 1974.
14. D. Sarid, Long-range surface plasma waves on very thin metal films, *Physical Review Letters*, **47**(26), 1927–1930, 1981.
15. J.C. Quail, J.G. Rako, H.J. Simon, Simultaneous excitation of long and short range surface plasmons in an asymmetric structure, *Optics Letters*, **8**(7), 377–379, 1983.
16. K.R. Welford, J.R. Sambles, Coupled surface plasmons in a symmetric system, *Journal of Modern Optics*, **35**(9), 1467–1483, 1988.
17. K. Matsubara, S. Kawata, S. Minami, Multilayer system for a high-precision surface plasmon resonance sensor *Optics Letters*, **15**(1), 75–77, 1990.
18. M. Bera, M. Ray, Precise detection and signature of biological/chemical samples based on surface plasmon resonance, *Journal of Optics*, **38**(4), 232–248, 2009.
19. R. Slavik, J. Homola, Simultaneous excitation of long and short range surface plasmons in an asymmetric structure, *Optics Communications*, **259**, 507–512, 2006.
20. R. Slavik, J. Homola, H. Vaisocherova, Advanced biosensing of short and long range surface plasmons, *Measurement Science & Technology*, **17**, 932–938, 2006.
21. J.T. Hastings, J. Guo, P.D. Keathley, P.B. Kumaresh, Y. Wei, S. Law, L.G. Bachas, Optimal self-referenced sensing using long- and short-range surface plasmons, *Optics Express*, **15**(26), 17661–17672, 2007.
22. J. Guo, D.P. Keathley, J.T. Hastings, Dual mode surface plasmon resonance sensors using angular interrogation, *Optics Letters*, **33**, 512–514, 2008.
23. G. Dyankov, M. Zekriti, M. Bousmina, Plasmon modes management, *Plasmonics*, **6**, 643–650, 2011.
24. M. Bera, M. Ray, Long-range and short-range surface plasmon resonance in coupled plasmonic structure using bimetallic nanofilms, *Proceedings of the 5th International Conference on Computers and Devices for Communication (CODEC)*, IEEE catalog number: CFP1201I-CDR, December 17–19, 2012, pp. 1–4, Kolkata, India.
25. M. Bera, M. Ray, Equi-reflectance dual mode resonance using bimetallic loaded dielectric plasmonic structure, *IEEE Photonic Technology Letters*, **25**(20), 1965–1968, 2013.
26. T. Nagamura, R. Matsumoto, Highly sensitive ultrafast all optical light modulation by complex refractive index changes in guided mode geometry composed of a photoresponsive polymer and a low refractive index polymer, *Applied Physics Letters*, **87**, 041107-1-3, 2005.
27. J.J. Ju, S. Park, M.-S. Kim, J.T. Kim, S.K. Park, M.-H. Lee, Polymer-based long-range surface plasmon polariton waveguides for 10-Gbps optical signal transmission applications, *Journal of Lightwave Technology*, **26**(11), 1510–1518, 2008.
28. N. Mattiucci, G. D'Aguanno, M.J. Bloemer, Second harmonic generation from metallo-dielectric multilayered structures in the plasmonic regime, *Optics Express*, **18**(23), 23698–23710, 2010.
29. H.A. Macleod, *Thin-Film Optical Filters*, 4th ed., CRC Press, Taylor & Francis Group, New York, 2010.
30. C.W. Lin, K.P. Chen, M.C. Su, C.K. Lee, C.C. Yang, Bio-plasmonics: Nano/micro structure of surface plasmon resonance devices for biomedicine, *Optical and Quantum Electronics*, **37**, 1423–1437, 2005.
31. C.W. Lin, K.P. Chen, M.C. Su, T.C. Hsiao, S.S. Lee, S. Lin, X.J. Shi, C.K. Lee, Admittance loci design method for multilayer surface plasmon resonance devices, *Sensors and Actuators B*, **117**, 219–229, 2006.

32. Y.J. Jen, A. Lakhtakia, C.W. Yu, T.Y. Chan, Multilayered structures for p- and s-polarized long-range surface-plasmon-polariton propagation, *Journal of Optical Society of America A*, **26**(12), 2600–2606, 2009.
33. K. Brahmachari, S. Ghosh, M. Ray, Surface plasmon resonance based sensing of different chemical and biological samples using admittance loci method, *Photonic Sensors*, **3**(2), 159–167, 2013.
34. K. Brahmachari, M. Ray, Effect of prism material on design of surface plasmon resonance sensor by admittance loci method, *Frontiers of Optoelectronics*, **6**(2), 185–193, 2013, Erratum to: Effect of prism material on design of surface plasmon resonance sensor by admittance loci method, *Frontiers of Optoelectronics*, **6**(3), 353, 2013.
35. G. Gupta, J. Kondoh, Tuning and sensitivity enhancement of surface plasmon resonance sensor, *Sensors and Actuators B*, **122**, 381–388, 2007.
36. R. Jha, A.K. Sharma, Design of silicon-based plasmonic biosensor chip for human blood group identification, *Sensors and Actuators B*, **145**, 200–204, 2010.
37. A.K. Sharma, R. Jha, H.S. Pattanaik, Design considerations for surface plasmon resonance based detection of human blood group in near infrared, *Journal of Applied Physics*, **107**, 034701-1-7, 2010.
38. K. Brahmachari, M. Ray, Admittance loci based design of a plasmonic structure using Ag-Au bimetallic alloy film, *ISRN Optics*, 7, 2013, Article ID 946832, http://dx.doi.org/10.1155/2013/946832.
39. N.K. Sharma, Performances of different metals in optical fiber-based surface plasmon resonance sensor, *Pramana*, **78**(3), 417–427, 2012.
40. K. Brahmachari, M. Ray, Performance evaluation of admittance loci based design of a ceramic prism based plasmonic structure, *Proceedings of the International Conference on Control, Instrumentation, Energy and Communication (CIEC)*, January 31, 2013–February 2, 2014, pp. 151–155, Kolkata, India.
41. K. Brahmachari, M. Ray, Design of nanocomposite film based plasmonic device for gas sensing, accepted for publication. *Pramana – Journal of Physics*, **83**, 107, 2014.
42. Z. Salamon, H.A. Macleod, G. Tollin, Coupled plasmon waveguide resonator: A new spectroscopic tool for probing proteolipid film structure and properties, *Biophysical Journal* **73**, 2791–2797, 1997.
43. M. Bera, M. Ray, Coupled plasmonic assisted progressive multiple resonance for dielectric material characterization, *Optical Engineering*, **50**(10), 103801-1-8, 2011.
44. F.C. Chein, S.J. Chen, A sensitivity comparison of optical biosensors based on four different surface plasmon resonance mode, *Biosensors and Bioelectronics*, **20**, 633–642, 2004.
45. F.C. Chein, S.J. Chen, Direct determination of the refractive index and thickness of a biolayer based on coupled waveguide-surface plasmon resonance mode, *Optics Letters*, **31**(2), 187–189, 2006.
46. M. Bera, M. Ray, Role of waveguide resonance in coupled plasmonic structures using bimetallic nanofilms, *Optical Engineering*, **51**(10), 103801-1-10, 2012.
47. M. Bera, M. Ray, Parametric analysis of multi-layer metallo-dielectric coupled plasmonic resonant structures using homo and hetero-bimetallic nanofilms, *Optics Communications*, **294**, 384–394, 2013.
48. A. Lahav, M. Auslender, I. Abdulhalim, Sensitivity enhancement of guided-wave surface-plasmon resonance sensors, *Optics Letters*, **33**, 2359–2541, 2008.
49. H. Raether, *Surface Plasmons on Smooth and Rough Surface and on Gratings*, Springer-Verlag, Berlin, Germany, 1988.
50. X. Chen, S. Ma, L. Liu, Z. Liu, Y. He, H. Shi, J. Guo, SPR sensor by method of electro-optic phase modulation and polarization interferometry, *Chinese Optics Letters*, 2012.
51. F. Wooten, *Optical Properties of Solids*, Academic Press, San Diego, CA, 1972.
52. P.B. Johnson, R.W. Christy, Optical constants of the noble metals, *Physical Review B*. **6**, 4370–4379, 1972.
53. J.P. Marton, B.D. Jordan, Optical properties of aggregated metal systems: Interband transitions, *Physical Review B*, **15**, 1719–1727, 1977.

54. E. Feigenbaum, K. Diest, H. Atwater, unity-order index change in transparent conducting oxides at visible frequencies, *Nano Letters*, **10**, 2111–2116, 2010.
55. D.-G. Park, T.-H. Cha, K.-Y. Lim, H.-J. Cho, T.-K. Kim, S.-A. Jang, Y.-S. Suh, V. Misra, I.-S. Yeo, J.-S. Roh, J.W. Park, H.-K. Yoon, Robust ternary metal gate electrodes for dual gate CMOS devices, *Electron Devices Meeting*, 2001, pp. 30.6.1–30.6.4, International Electron Devices Meeting Technical Digest. Washington, DC, December 2–5.
56. B. Rech, H. Wagner, Potential of amorphous silicon for solar cells, *Applied Physics A: Materal Science & Processing*, **69**(2), 155–167, 1999.
57. M. Zeman, R.A.C.M.M. van Swaaij, J.W. Metselaar, R.E.I. Schropp, Optical modeling of *a*-Si:H solar cells with rough interfaces: Effect of back contact and interface roughness, *Journal of Applied Physics*, **88**(11), 6436–6443, 2000.
58. S. Ferlauto, G.M. Ferreira, J.M. Pearce, C.R. Wronski, R.W. Collins, X. Deng, G. Ganguly, Analytical model for the optical functions of amorphous semiconductors from the near-infrared to ultraviolet: Applications in thin film photovoltaics, *Journal of Applied Physics*, **92**(5), 2424–2436, 2002.
59. P. West, S. Ishii, G. Naik, N. Emani, V. Shalaev, A. Boltasseva, Searching for better plasmonic materials, *Laser and Photonics Review*, **4**, 795–808, 2010.
60. D. Bobb, G. Zhu, M. Mayy, A. Gavrilenko, P. Mead, V. Gavrilenko, M. Noginov, Engineering of low-loss metal for nanoplasmonic and metamaterials applications, *Applied Physics Letters*, **95**, 151102, 2009.
61. G. Naik, A. Boltasseva, Ceramic plasmonic components for optical metamaterials, *Quantum Electronics and Laser Science Conference*, 2011, Paper QTuI1, Optical Society of America , Baltimore, MD, May 1–6.
62. G. Zhu, L. Gu, J. Kitur, A. Urbas, J. Vella, M. Noginov, Organic materials with negative and controllable electric permittivity, *Quantum Electronics and Laser Science Conference*, 2011, Paper QThC3, Optical Society of America, Baltimore, MD, May 1–6.
63. S. Ghosh, M. Ray, Investigation of surface plasmon resonance using differential phase jump analysis at metal-dielectric interface, *Journal of Nanoscience Letters*, **4**, 31, 2013.
64. S. Ghosh, M. Ray, Investigation of surface plasmon resonance using cylindrical dielectric-metal-dielectric (C-DMD) plasmonic configuration, *Optik*, **125**, 2642–2646, 2013, http://dx.doi.org/10.1016/j.ijleo.2013.11.033.
65. S. Ghosh, K. Brahmachari, M. Ray, Experimental investigation of surface plasmon resonance using tapered cylindrical light guides with metal-dielectric interface, *Journal of Sensor Technology*, **2**(1), 48–54, 2012, http://dx.doi.org/10.4236/jst.2012.21007.
66. M. Born, E. Wolf, *Principles of Optics*, 7th expanded ed., Cambridge University Press, Cambridge, 1999.
67. F. Abeles, Recherches sur la propagation des ondes electromagnetiques sinusoidales dans les milieux stratifies, *Application aux couches minces*, *Annales de Physique*, **5**, 596–640, 1950.
68. A. Shalabney, I. Abdulhalim, Prism dispersion effects in near guided wave surface plasmon resonance sensors, *Annalen der physik*, 2012.
69. D.B. Hand, The refractivity of protein solutions, *Journal of Biological Chemistry*, 1935, 108:703–707.
70. S.G. Nelson, K.S. Johnston, S.S. Yee, High sensitivity surface plasmon resonance sensor based on phase detection, *Sensors and Actuators B*, **35–36**, 187–191, 1996.
71. A.V. Kabashin, P.I. Nikitin, Surface plasmon resonance interferometer for bio- and chemical-sensors, *Optics Communications*, **150**, 5–8, 1998.
72. P.I. Nikitin, A.A. Beloglazov, E.V. Kochergin, V.M. Valeiko, I.T. Ksenevich, Surface plasmon resonance interferometry for biological and chemical sensing, *Sensors and Actuators B*, **54**, 43–50, 1999.
73. N.A. Grigorenko, P.I. Nikitin, V.A. Kabashin, Phase jumps and interferometric surface plasmon resonance imaging, *Applied Physics Letters*, **75**(25), 3917–3919, 1999.
74. Y. Xinglong, W. Dingxin, Y. Zibo, Simulation and analysis of surface plasmon resonance biosensor based on phase detection. *Sensors and Actuators B*, **91**, 285–290, 2013.

75. G. Nemova, A.V. Kabashin, R. Kashyap, Surface plasmon–polariton Mach-Zehnder refractive index sensor, *Journal of Optical Society of America B*, **25**(10), 1673–1677.
76. H.P. Ho, W. Yuan, C.L. Wong, S.Y. Wu, Y.K. Suen, S.K. Kong, C. Lin, Sensitivity enhancement based on application of multi-pass interferometry in phase-sensitive surface plasmon resonance biosensor, *Optics Communication*, **275**, 491–496, 2007.
77. P.H. Chiang, T.H. Yeh, M.C. Chen, C.J. Wu, Y.S. Su, R. Chang, J.Y. Wu, D. Tsai, U.S. Jen, T.P. Leung, Surface plasmon resonance monitoring of temperature via phase measurement, *Optics Communications*, **241**, 409–418, 2004.
78. Y. Zhang, Study of an absorption-based surface plasmon resonance sensor in detecting the real part of refractive index, *Optical Engineering*, **52**(1), 014405-1-7, 2003.
79. G.A. Notcovich, V. Zhuk, G.S. Lipson, Surface plasmon resonance phase imaging, *Applied Physics Letters*, **76**(13), 1665–1667, 2000.
80. H.K. Chen, C.C. Hsu, C.D. Su, Measurement of wavelength shift by using surface plasmon resonance heterodyne interferometry, *Optics Communications*, **209**, 167–172, 2002.
81. R. Wang, C. Zhang, Y. Yang, S. Zhu, C.X. Yuan, Focused cylindrical vector beam assisted microscopic SPR biosensor, with an ultra wide dynamic range, *Optics Letters*, **37**(11), 2091–2093, 2012.
82. C. Zhang, R. Wang, C. Min, S. Zhu, C.X. Yuan, Experimental approach to the microscopic phase-sensitive surface plasmon resonance biosensor, *Applied Physics Letters*, **102**, 011114-1-5, 2013.
83. H.Y. Huang, P.H. Ho, Y.S. Wu, K.S. Kong, Detecting phase shifts in surface plasmon resonance: A review, *Advances in Optical Technologies*, **2012**, Article ID 471957, 1–12, 2012.
84. M. Bera, M. Ray, Circular phase response based analysis for swapped multilayer metallo-dielectric plasmonic structures, *Plasmonics*, doi:10.1007/s11468-013-9617-8.
85. M. Griot, *The Practical Application of Light*, Vol. X, Barloworld Scientific, Staffondshire, UK, p. 4.8, 2006.
86. J. Homola, On the sensitivity of surface plasmon resonance sensors with spectral interrogation, *Sensors and Actuators B*, **41**, 207–211, 1997.
87. M. Bera, J. Banerjee, M. Ray, Resonance parameters based analysis for metallic thickness optimization of bimetallic plasmonic structure, *Journal of Modern Optics*, doi:10.1080/09500340.2013.878043.
88. O. Zhernovaya, O. Sydoruk, V. Tuchin, A. Douplik, The refractive index of human hemoglobin in the visible range, *Physical Medical Biology*, **56**, 4013–4021, 2011.
89. H. Brewer, S. Franzen, Calculation of the electronic and optical properties of indium tin oxide by density functional theory, *Chemical Physics*, **300**(1–3), 285–293, 2004.
90. S. Franzen, C. Rhodes, M. Cerruti, R.W. Gerber, M. Losego, J.P. Maria, D.E. Aspenes, Plasmonic phenomena in indium tin oxide and ITO-Au hybrid films, *Optics Letters*, **34**(18), 2867–2869, 2009.
91. S. Franzen, Surface plasmon polaritons and screened plasma absorption in indium tin oxide compared to silver and gold, *Journal of Physical Chemistry C*, **112**(15), 6027–6032, 2008.
92. S.H. Brewer, S. Franzen, Optical properties of indium tin oxide and fluorine-doped tin oxide surfaces: Correlation of reflectivity, skin depth, and plasmon frequency with conductivity, *Journal of Alloys & Compounds*, **338**, 73–79, 2002.
93. M.D. Losego, A.Y. Efremenko, C.L. Rhodes, M.G. Cerruti, S. Franzen, J.P. Maria, Conductive oxide thin films: Model systems for understanding and controlling surface plasmon resonance, *Journal of Applied Physics*, **106**(2), 024903–1–024903–8, 2009.
94. C. Rhodes, S. Franzen, J.P. Maria, M. Losego, D.N. Leonard, B. Laughlin, G. Duscher, S. Weibel, Surface plasmon resonance in conducting metal oxides, *Journal of Applied Physics*, **100**, 054905-1–054905-4, 2006.
95. C. Rhodes, M. Cerruti, A. Efremenko, M. Losego, D.E. Aspnes, J.-P. Maria, S. Franzen, Dependence of plasmon polaritons on the thickness of indium tin oxide thin films, *Journal of Applied Physics*, **103**(9), 093108-1–093108-6, 2008.

96. A. Solieman, M.A. Aegerter, Modeling of optical and electrical properties of In_2O_3:Sn coatings made by various techniques, *Thin Solid Films*, **502**(1/2), 205–211, 2006.
97. M. Hiramatsu, K. Imaeda, N. Horio, M. Nawata, Transparent conducting ZnO thin films prepared by XeCl excimer laser ablation, *Journal of Vacum Science Technology A*, **16**(2), 669–673, 1998.
98. Ikehata, T. Itoh, Y. Ozaki, Surface plasmon resonance near-infrared spectroscopy, *Analytical Chemistry*, **76**(21), 6461–6469, 2004.
99. J.J. Ju, S. Park, M.S. Kim, J.T. Kim, S.K. Park, Y.J. Park, M.H. Lee, 40 Gbit/s light signal transmission in long-range surface plasmon waveguides, *Applied Physics Letters*, **91**(17), 171117, 2007.
100. F. Michelotti, L. Dominici, E. Descrovi, N. Danz, F. Menchini, Thickness dependence of surface plasmon polariton dispersion in transparent conducting oxide films at 1.55 μm, *Optics Letters*, **34**(6), 839–841, 2009.
101. M. Ray, M. Bera, Multilayer coupled nanoplasmonic structures and related computational techniques, *Computational Nanophotonics: Modeling and Applications*, S. Musa, Ed. pp. 315–343, CRC Press, London, 2013.
102. S. Rajak and M. Ray, Comparative study of plasmonic resonance in transparent conducting oxides: ITO and AZO, **43**(3), 231–238, 2014. DOI:http://dx.doi.org/10.1007/s12596-014-0215-8.

10

Nanoimaging and Polarimetric Exploratory Data Analysis

Suman Shrestha, Aditi Deshpande, Tannaz Farrahi, Yinan Li, Stefanie Marotta, and George C. Giakos

CONTENTS

10.1 Introduction

Nanoscience is a broad subject that covers a wide range of structures, phenomena, and properties. Although defining nanoscience in a more concise and precise manner is a difficult task, a definition from the National Nanotechnology Initiative is as follows:

> The study of structures, dynamics, and properties of systems in which one or more of the spatial dimensions is nanoscopic (1–100 nm), thus resulting in dynamics and properties that are distinctly different (often in extraordinary and unexpected ways that can be favorably exploited) from both small molecule systems and systems macroscopic in all dimensions. [1]

With the context of this definition, three broad classes of nanoscience systems were considered [1]:

1. Nano building blocks (such as nanotubes, quantum dots, clusters, and nanoparticles)
2. Complex nanostructures and nano-interfaces
3. Dynamics, assembly, and growth of nanostructures

Well-characterized building blocks will be the center of new nanoelectronic or nanomagnetic materials or devices. The best characterized and the most appropriate building blocks are as follows:

- Clusters and molecular nanostructures
- Nanotubes and related systems
- Quantum wells, wires, films, and dots

One of the well-characterized building blocks for nanoscience is shown in Figure 10.1.

Complexity at the nanoscale range plays a vital role in all aspects of nanoscience and nanotechnology. Complexity of the nano-interfaces of the molecules may provide structural stability of the material. A new physics is likely to evolve when materials and devices are dominated by nano-interfaces like the ones that include the electric field at surfaces, luminescence of rare-earth orthosilicates, surface-enhanced Raman scattering, and deformation of nanocrystalline materials. Figure 10.2 shows an atomistic simulation of nanostructure composed of polyhedral oligomeric silsesquioxane cages.

Dynamics, assembly, and growth of nanostructures deal with the time-dependent process required to build nanostructures with processes such as vapor deposition methods. Figure 10.3 shows one particular case of the process required to build Mobil Composition of Matter No. 41 (MCM-41) nanostructure.

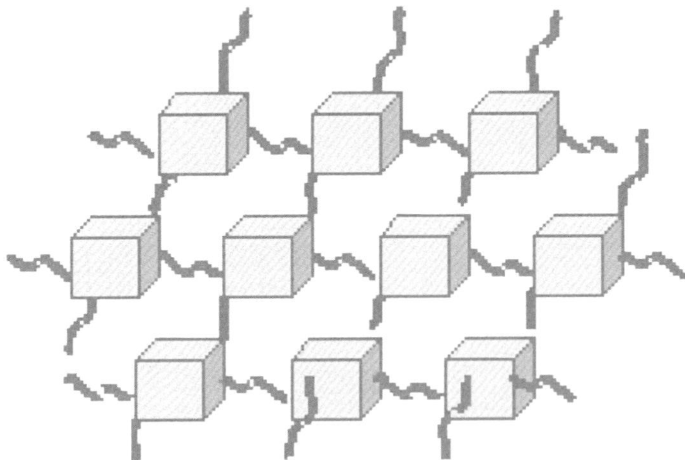

FIGURE 10.1
Schematic of an assembled structure of nanoscopic building blocks tethered by organic *linkers*. The structure is dominated by interfacial connections. (Data from Glotzer, S.C., 2002; McCurdy, C.W. et al., Theory and modeling in nanoscience, Report of Workshop Conducted by the Basic Energy Sciences and Advanced Scientific Computing Advisory Committees to the Office of Science, Lawrence Berkeley National Laboratory, Berkeley, CA, June 28, 2002.)

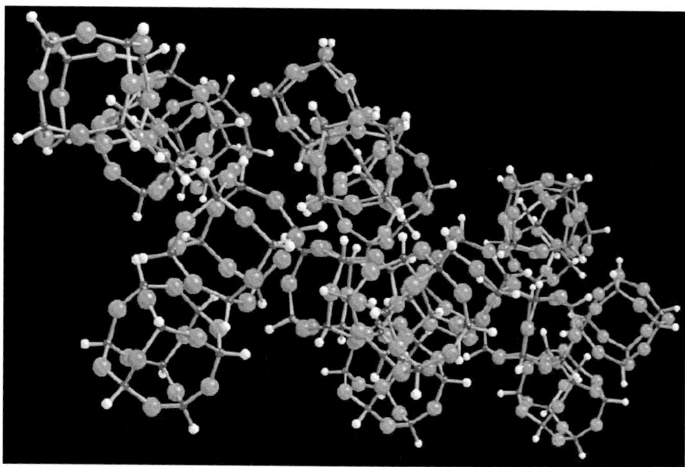

FIGURE 10.2
Atomistic simulation of a nanostructure composed of polyhedral oligomeric silsesquioxane cages. (Data from Kieffer, J., 2002; McCurdy, C.W. et al., Theory and modeling in nanoscience, Report of Workshop Conducted by the Basic Energy Sciences and Advanced Scientific Computing Advisory Committees to the Office of Science, Lawrence Berkeley National Laboratory, Berkeley, CA, June 28, 2002.)

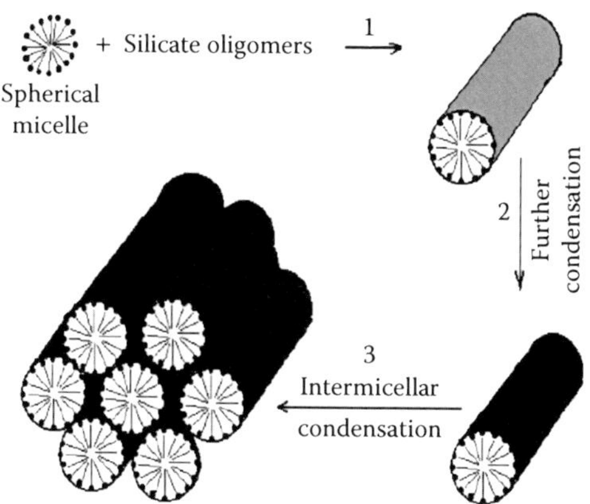

FIGURE 10.3
Templated self-assembly of nanoporous material (in this case, MCM-41). (Data from Zhao, X.S., Synthesis, modification, characterization and application of MCM-41 for VOC Control, PhD Thesis, University of Queensland, Queensland, Australia, 1998; McCurdy, C.W. et al., Theory and modeling in nanoscience, Report of Workshop Conducted by the Basic Energy Sciences and Advanced Scientific Computing Advisory Committees to the Office of Science, Lawrence Berkeley National Laboratory, Berkeley, CA, June 28, 2002.)

10.2 Nanoimaging and Its Techniques

To gain an insight into the complexity of the nanostructures and nanomolecules, imaging is an integral part of biology, allowing researchers to see the detailed orientation of various cellular lives. Particularly, imaging techniques capable of reaching the level of nanoscale spatial resolutions are extremely important to the study of the orientation of nanostructures [3]. Three such techniques that will be discussed are as follows:

1. Light microscopy
2. Electron microscopy
3. Scanning Probe microscopy

10.2.1 Light Microscopy

Light microscopy is the well-used imaging technique in biology; this technique uses visible light to detect nanoparticles and materials. There are a lot of optical factors that determine the power of the microscope used to view nanomaterials. Around 1870, Ernst Abbe formulated his famous sine theory for the resolving power of a light microscope, which demonstrates the importance of the numerical aperture (NA) of the lens used and the wavelength of light. The higher the NA and the shorter the wavelength, the better the resolving power of the microscope. The resolving power of an objective lens is the ability to show two object details separately from each other in the microscope image [4].

By 1900, the theoretical principles of the microscope were well understood and the microscope had become a well-established research tool for professional scientists. Further clever refinements have improved the performance of the instrument and these, together with the development of photographic and now digital imaging techniques, have led us to the modern research microscope used as an essential tool in laboratories all over the world.

There are different techniques used in light microscopy. Some of these techniques will be discussed with examples and figures:

- Transmitted light microscopy
- Bright-field microscopy
- Dark-field microscopy
- Phase contrast
- Polarized light microscopy
- Differential interference microscopy
- Fluorescence microscopy
- Confocal microscopy

When selecting the system for imaging living cells, there are three things that should be considered: detector sensitivity, speed of data acquisition system, and the cell used for imaging. Figure 10.4 shows the comparison of different systems used for imaging with respect to the source used for imaging, detector specifications, and different optical/imaging materials used. The difference between these three systems suggests that a single system may not be feasible for different types of experiments.

Researchers nowadays are more interested in three- and four-dimensional imaging because of the high-resolution images that are produced from these methods. Many researches were performed using different fluorescence imaging systems to detect tumor margins in the cells. Researchers, from the George R. Harrison Spectroscopy Laboratory at the Massachusetts Institute of Technology, designed a portable spectroscopic scanner consisting of a fiber-optic probe for intraoperative margin assessment, based on diffuse reflectance spectroscopy and intrinsic fluorescence spectroscopy [6].

Another technique, which has been applied in margin detection and analysis, relies on the application of terahertz waves. Terahertz waves are capable of discriminating healthy and pathological tissues based on differences in their optical properties [7].

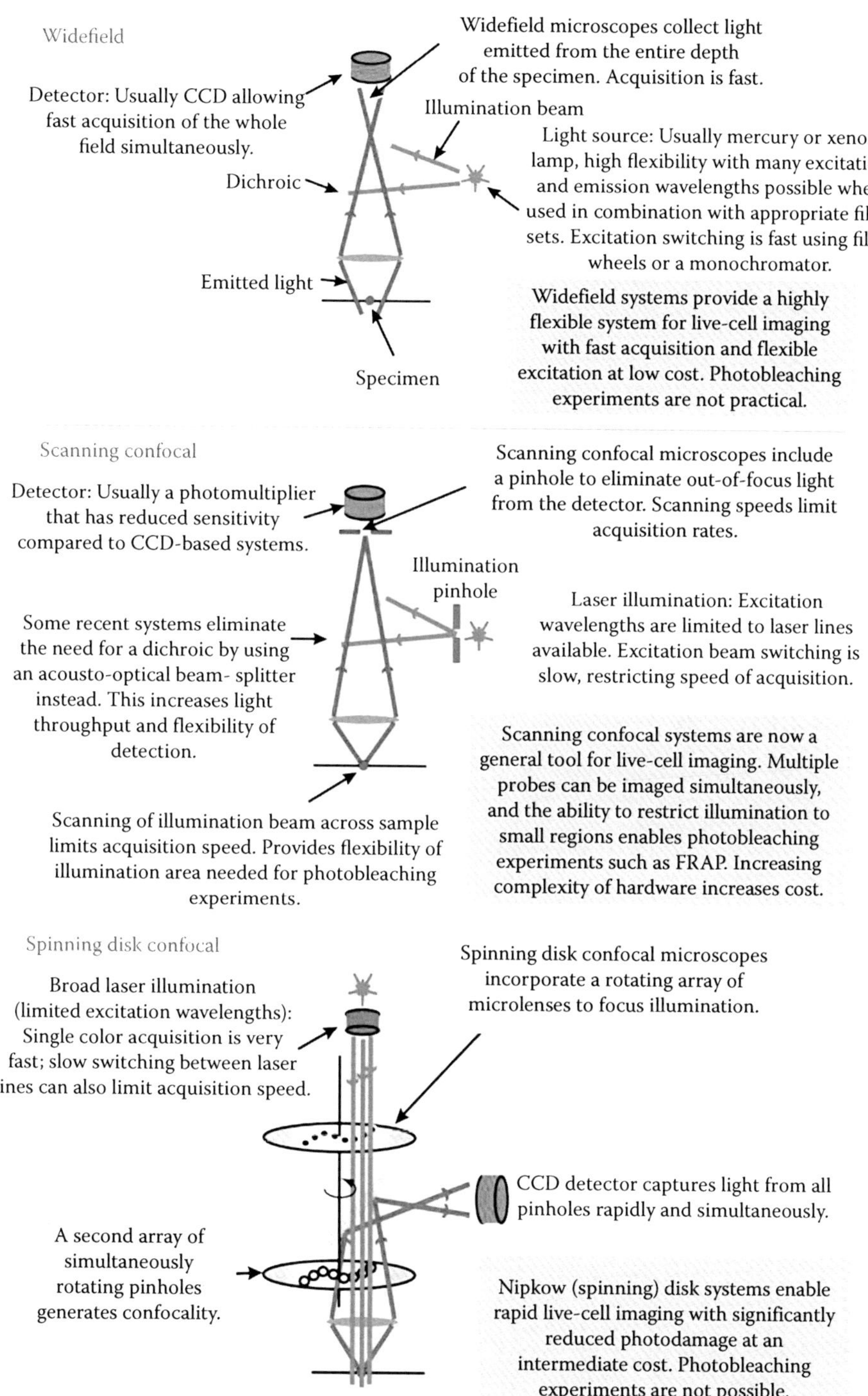

FIGURE 10.4
Comparison of widefield, scanning confocal, and spinning confocal microscopies. CCD, charge-coupled device; FRAP, fluorescence recovery after photobleaching. (Data from Stephens, A., *Sci. Rev.*, 300, 82–86, 2003.)

Recently, photoacoustic imaging has been applied for potential intraoperative tumor margin detection [8]. Fluorescence of ovarian cancer margins was studied using intraoperative tumor-specific fluorescence imaging [9]. In this study, a folate receptor-α-targeted fluorescent agent was used for better staging of cancer and to aid *cytoreductive surgery.*

Giakos et al. (2013) [10] designed an automated digital fluorescence imaging system for detection of tumor margins. The outcome of this study indicated that the fluorescent images demonstrated a clear demarcation of the tumor margins due to the distinctively different fluorescence properties of cancer. The images obtained using the fluorescence microscopy were further improved using a novel image processing system. Figure 10.5 shows the two images of the brain tumor sample obtained at depth of 10 and 30 μm, respectively, [10].

10.2.2 Electron Microscopy

Electron microscopy is a type of microscopic technique that uses a beam of electrons to create an image of the sample. It provides much higher magnifications and a greater resolving power than a light microscopic technique, allowing it to see much smaller objects in finer detail. They are large, expensive pieces of equipment, generally standing alone in a small, specially designed room and requiring trained personnel to operate them [10].

This technique uses electromagnetic/electrostatic lenses to control the path of electrons. Such lens is a solenoid through which current passes and induces an electromagnetic field. The electron beam passes through the center of such lens toward the sample to be imaged. The speed of the electrons determines the wavelength, which in turn determines the resolution of the microscope used. The arrangement of the electron gun, the lenses, the specimen, and the screen is shown in Figure 10.6. Some of the types of electron microscopy techniques commonly used are as follows:

- Transmission electron microscopy (TEM)
- Scanning electron microscopy (SEM)

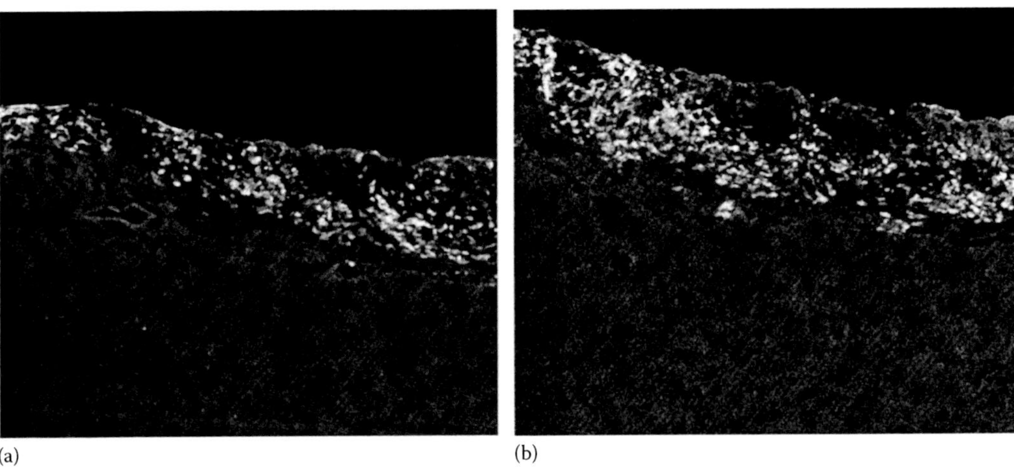

(a) (b)

FIGURE 10.5
Fluorescence images of the brain tumor sample at slice depth 10 (a) and 30 μm (b). (Data from Giakos, G. et al., An automated digital fluorescence imaging system of tumor margins using clustering-based image thresholding, Imaging Systems and Techniques (IST), 2013 IEEE International Conference, Beijing, People's Republic of China, 2013.)

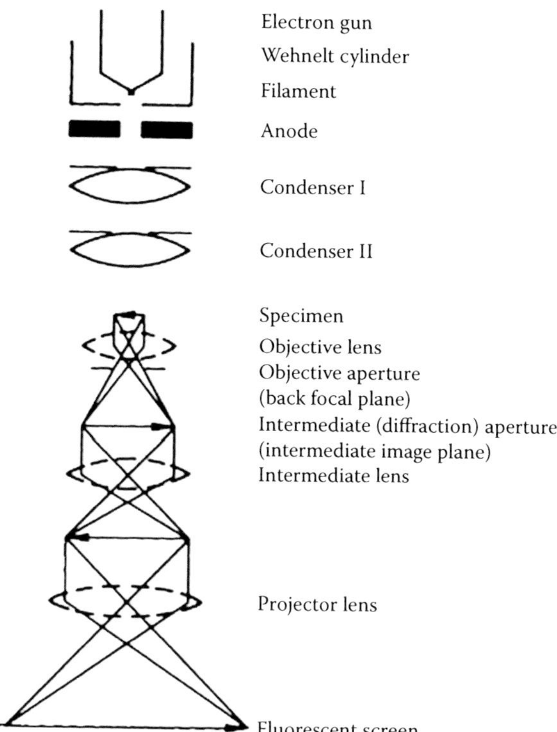

FIGURE 10.6
Basic building block of an electron microscope. (Data from Williams, DB and Carter, CB, *Syst. Mater. Anal.*, 4, Chapter 4, 1978.)

TEM uses a high-voltage beam emitted by the cathode. The beam passes through the sample to be imaged. The information of the specimen is magnified by a number of lenses (depending on the application) and is detected by the screen that is used as an imager. For imaging nanoparticles, the most important application of TEM is atomic resolution real-space imaging of nanoparticles [14]. In TEM, the images of nanoparticles to be imaged are dominated by three types of contrasts:

1. Diffraction/amplitude contrast: It highlights the effects of defects in nanoparticles on the amplitude of the transmitted wave.
2. Phase contrast: It is produced by the phase transfer in the incoming beam when it travels through the material.
3. Atomic number contrast: Different atomic number materials have different powers of scattering.

TEM has also made possible to obtain holograms using electron waves [15–17].

SEM, however, detects the sample and then images it by detecting the secondary electron transmitted through the specimen due to excitation by the primary electron beam. Figure 10.7 shows a schematic diagram of a scanning electron microscope.

All the samples prepared for imaging with SEM must be of an appropriate size to fit into the specimen chamber. The sample is required to be completely dry as the chamber is at high vacuum. Some of the applications of SEM are briefly described in Sections 10.2.2.1 and 10.2.2.2.

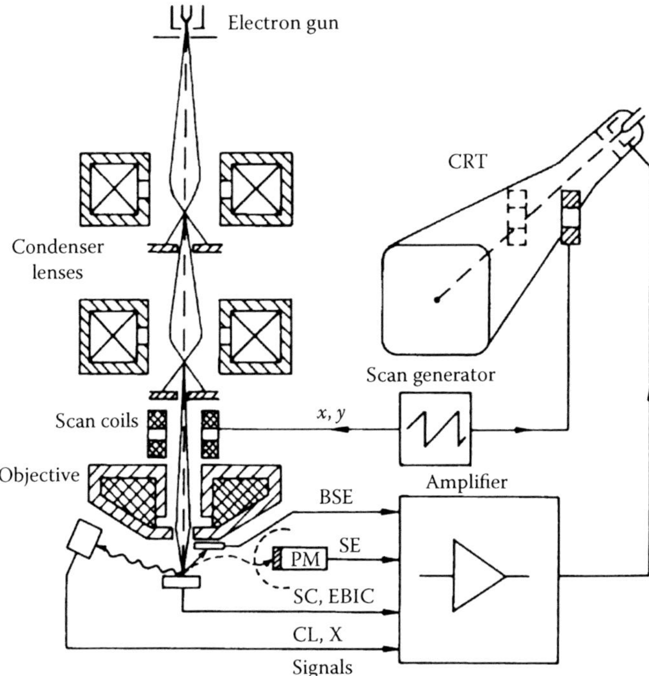

FIGURE 10.7
Scanning electron microscope (SEM). BSE, back-scattered electrons; CL, characteristic X-rays (cathodoluminescence); CRT, cathode ray tube; EBIC, electron beam-induced current; SC, specimen current; SE, secondary electrons; X, X-beam. (Data from Reimer, L. and Kohl, H., *Transmission Electron Microscopy. Physics of Image Formation*, Springer, New York, 2008.)

10.2.2.1 Human Embryo Development Imaging by SEM

Figure 10.8 shows a SEM of a lower limb of a human embryo after 10 weeks' development. The presence of a tactile metatarsal prominence can be seen (with an arrow sign) on the distal tips of the fingers. In addition, one can see how, at this stage of development, the fingers are already independent of each other [18].

10.2.2.2 Microstructure Changes in the Field of Agricultural Products Drying

The scanning electron micrographs on the surface of fresh and dried samples are used to analyze the microstructure changes during drying or dehydration process (Figure 10.9). Fresh apple tissue has a well-organized structure consisting of cells and intercellular spaces; however, the breakdown of cell walls, a decreased intercellular contact, and the collapse of cell structure were found in the dried apple tissues [19,20].

10.2.3 Scanning Probe Microscopy

Scanning probe microscopy is the branch of nanoimaging technique that forms images of surfaces using a physical probe that scans the sample. It is used in a wide variety of disciplines such as fundamental surface science, routine surface roughness analysis, three-dimensional imaging—from atoms of silicon to microsized protrusions on the surface of a living cell. This nanoimaging technique has a vast dynamic range, spanning the realms of optical and electron microscopes. In some cases, it measures the physical properties

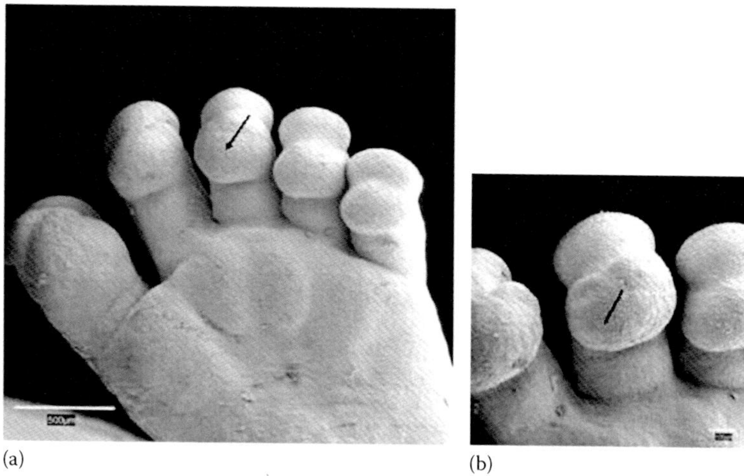

FIGURE 10.8
(a) SEM of a lower limb of a human embryo after 10 weeks' development. (b) Detailed plantar view. (Data from Cortadellas, N. et al., *Handbook of Instrumental Techniques from CCiTUB*. Unitat de Microscòpia Electrònica (Casanova), CCiT-UB, Universitat de Barcelona, Barcelona, Spain.)

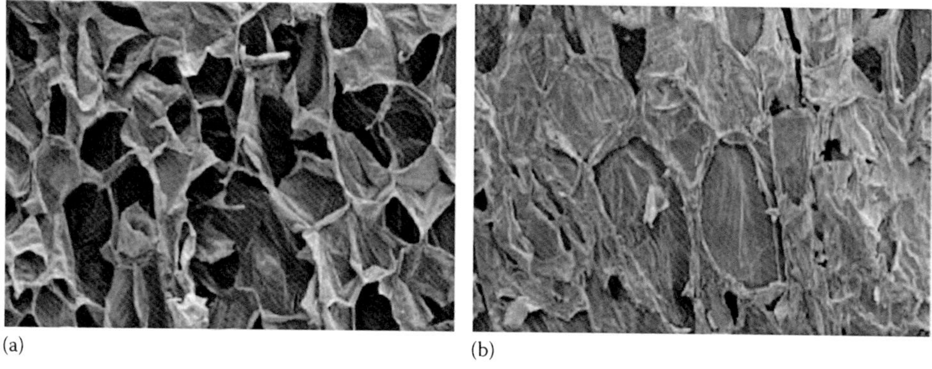

FIGURE 10.9
Scanning electron micrographs of fresh (a) and osmotically treated and freeze-dried apples (b). (Data from Deng, Y. and Zhao, Y.Y., *J. Food Eng.*, 85, 84–93, 2008.)

such as surface conductivity, static charge distribution, localized friction, magnetic fields, and elastic moduli. As a result, the applications of SPMs are very diverse [21]. Figure 10.10 shows the schematics of a general scanning probe microscope.

There are varieties of types of scanning probe microscopes. Some of them are shown in Figure 10.11 [22].

Here, we discuss the operation of two types of scanning probe microscopy: scanning tunnel microscopy (STM) and atomic force microscopy (AFM).

10.2.3.1 Scanning Tunnel Microscopy

This was invented in 1981 by Gerd Binnig and Heinrich Rohrer at IBM Zurich, Rüschlikon, Switzerland, which won them the Nobel Prize in Physics for the invention. This was the first instrument to generate real space images of a surface. Scanning tunnel microscopes (STM) use a sharpened, conducting tip with a bias voltage applied between the tip and

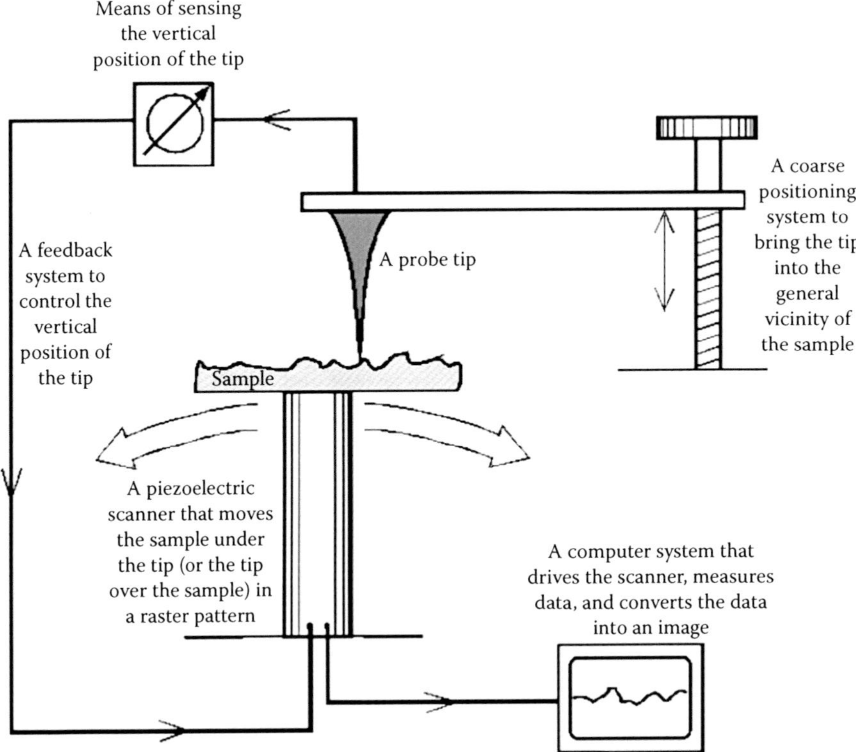

FIGURE 10.10
Schematic of a generalized scanning probe microscope. (Data from Washington State University, *Basic Guide to Scanning Probe Microscopy*. http://public.wsu.edu/~hipps/pdf_files/spmguide.pdf.)

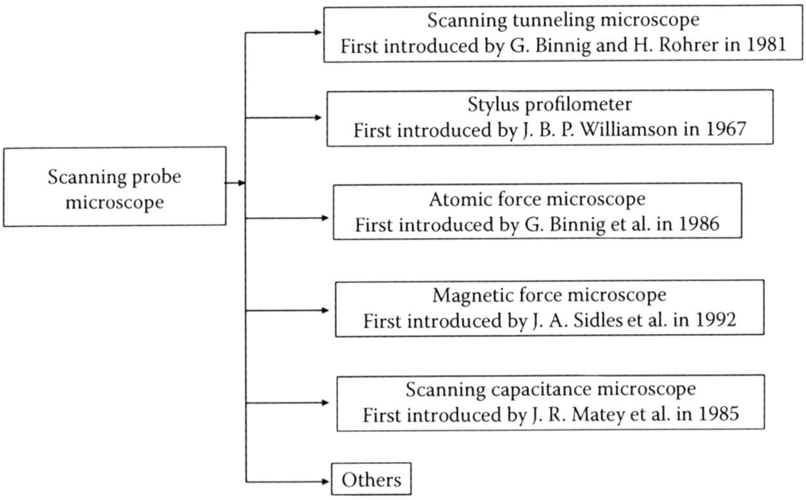

Scanning probe microscope

- Scanning tunneling microscope
  First introduced by G. Binnig and H. Rohrer in 1981
- Stylus profilometer
  First introduced by J. B. P. Williamson in 1967
- Atomic force microscope
  First introduced by G. Binnig et al. in 1986
- Magnetic force microscope
  First introduced by J. A. Sidles et al. in 1992
- Scanning capacitance microscope
  First introduced by J. R. Matey et al. in 1985
- Others

FIGURE 10.11
Varieties of scanning probe microscopes. (Data from Office of Science U.S. Department of Energy, *Basic of Scanning Probe Microscopy*, Pioneering Science and Technology. http://www1.na.infn.it/TIMSI/materialicorsi/iavarone/chapter1.pdf.)

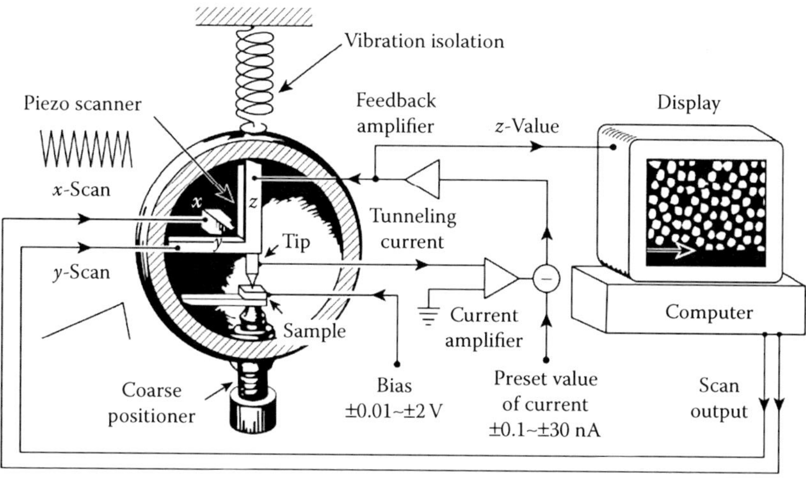

FIGURE 10.12
Basic schematic of scanning tunneling microscope. (Data from Chen, C.J., *Introduction to Scanning Tunneling Microscopy*, Oxford University Press, Columbia University, New York, 2008.)

the sample. For tunneling to take place, both the sample and the tip must be conductors or semiconductors. However, scanning tunnel microscopes cannot image insulating materials. Figure 10.12 shows the basic schematics of scanning tunnel microscope.

The tip of the microscope is virtually grounded. The bias voltage here is the sample voltage. When this voltage is greater than zero, the electrons tunnel from the tip to the sample, and while the voltage is less than zero, the electrons tunnel from the sample to the tip. This current is then converted to a voltage by a current amplifier, which is then compared to the reference voltage. If this value is greater than the reference voltage, the voltage applied to it tends to withdraw the tip from the sample surface, and vice versa. This is then stored and displayed on the computer screen, typically as grayscale image. Figure 10.13 shows the grayscale image and contour plot of Si (111) 7 × 7.

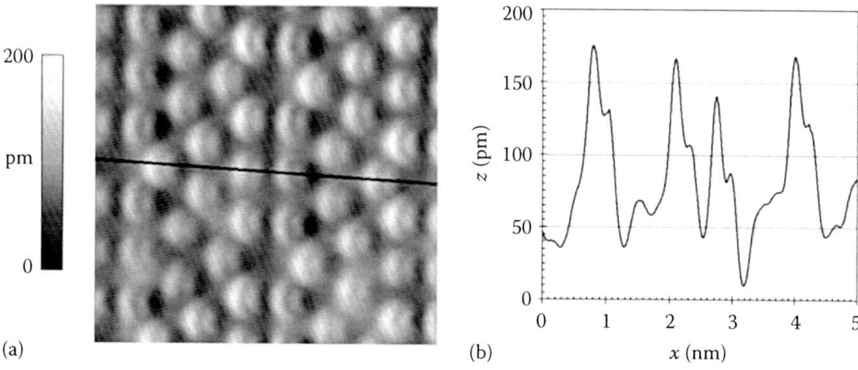

FIGURE 10.13
(a) Grayscale image of Si (111) 7 × 7. The bright spots are the protrusions and the dark are the depressions. The z-values are represented by the scale bar. (b) The contour plot of (a). (Data from Chen, C.J., *Introduction to Scanning Tunneling Microscopy*, Oxford University Press, New York, 2008.)

10.2.3.2 Atomic Force Microscopy

This is a very high-resolution type of microscopic technique that demonstrates the resolution on the order of nanometers. It is a combination of the principles of STM and the stylus profilometer [24]. The experimental setup of the first AFM based on STM sensing is shown in Figure 10.14. The cantilever is sandwiched between the sample and the tip. AFM imaging is performed by sensing the force between a very sharp probe and a sample as shown in Figure 10.15.

In AFM, there are number of imaging modes. The most widely used one is the contact mode that simply records the cantilever deflection when the sample is scanned horizontally. The feedback output is then used to display a true *height image* [25]. Another type of

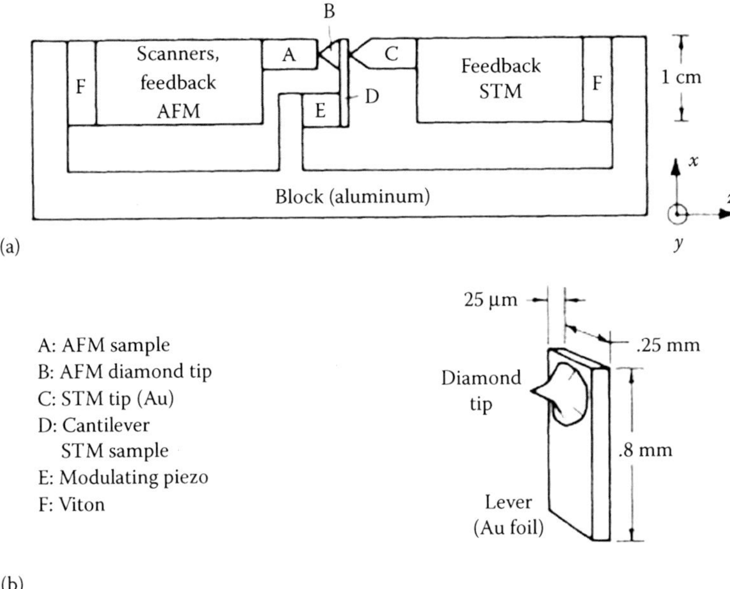

FIGURE 10.14
(a) Experimental setup of AFM; (b) dimensions. (Data from Binnig, G. et al., *Phys. Rev. Lett.*, 56, 1986.)

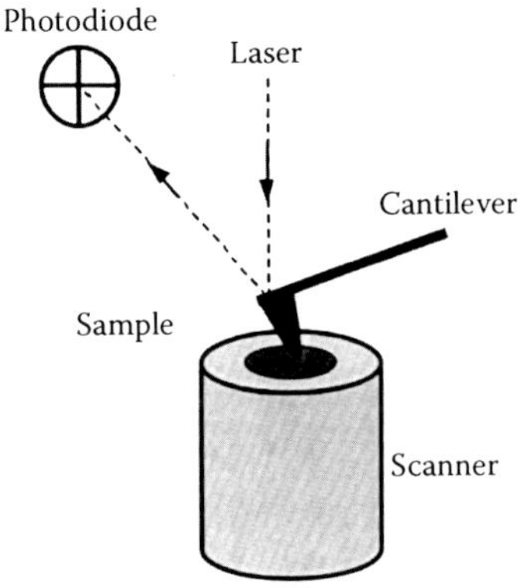

FIGURE 10.15
General principle of AFM. (Data from Dufrene, Y.F., *J. Bacteriol.*, 184, 2002.)

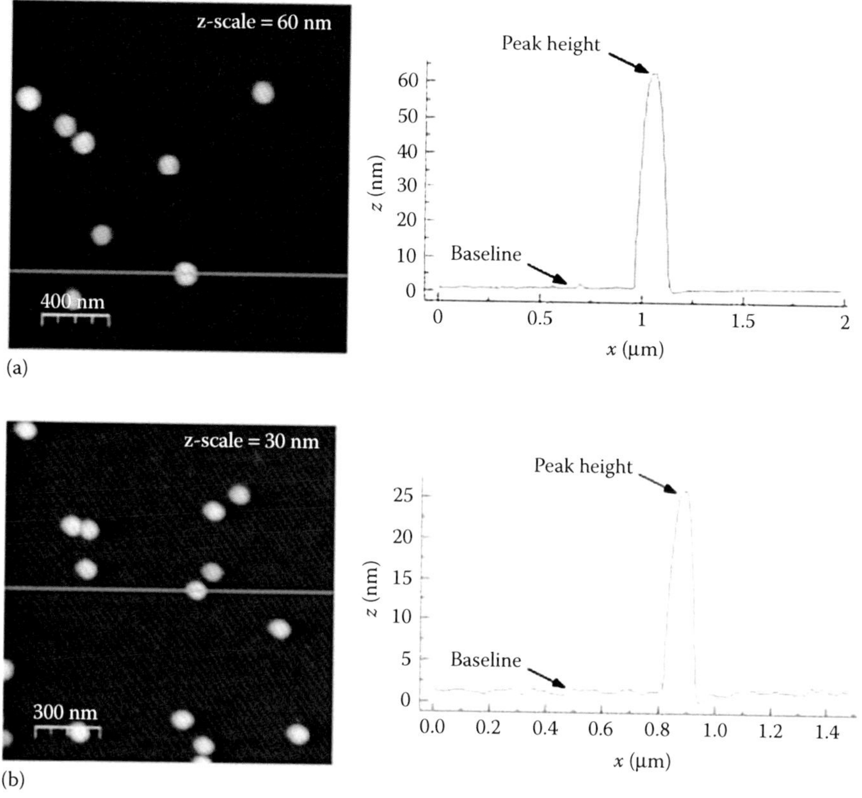

FIGURE 10.16
AFM images and cross sections for 60 (a) and 30 nm (b) nanoparticles (nominal). The difference between the peak height and the average baseline (from both sides of the nanoparticle) is the particle height. (Data from National Institute of Science and Technology, US Department of Commerce, NCI Alliance for Nanotechnology in Cancer, Gaithersburg, MD, 2009.)

imaging mode available in AFM is the tapping mode. AFM can image the surface of a living cell in high resolution in real mode, which is a complementary to electron microscopy. It can be used for the imaging of various kinds of cancer cells and nanoparticles. One such example of AFM imaging of nanoparticle is shown in Figure 10.16 [26].

10.2.3.3 AFM Cantilever

A flexible cantilever exerts less distortion and less damage. It has a spring constant less than 0.1 N/m. Higher the resonance frequency of the cantilever, the better the imaging of the microscopic technique.

$$\text{Resonant frequency} = \frac{1}{2\Pi}\sqrt{\frac{\text{Spring constant}}{\text{Mass}}} \qquad (10.1)$$

Microlithography process is used to make such cantilevers. Silicon (Si) is used for making tips. A think gold layer is deposited on the upper side of the tip for good reflectivity of the beam. AFM is a very versatile technique that has a wide range of resolution (Figure 10.17).

The resolution of AFM is shown in Figure 10.18, which makes use of the resolution of the tip and the sample.

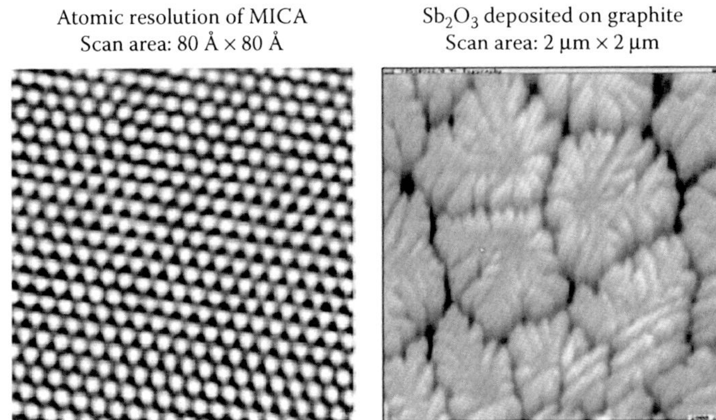

FIGURE 10.17
A wide range of resolution of AFM. (Data from Wang, Z., RHK Technology, http://www1.na.infn.it/TIMSI/materialicorsi/iavarone/chapter1.pdf.)

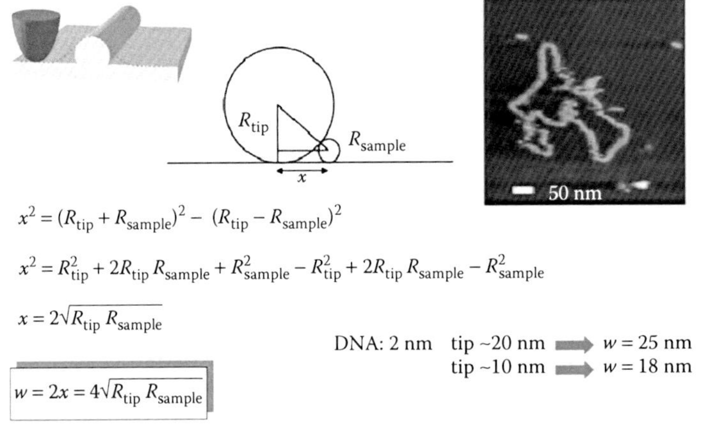

FIGURE 10.18
Resolution of atomic force microscopy. (Data from Office of Science, U.S. Department of Energy, *Basic of Scanning Probe Microscopy*, Pioneering Science and Technology. http://www1.na.infn.it/TIMSI/materialicorsi/iavarone/chapter1.pdf.)

10.3 Nanoscale Biosensors and Their Use in Biomedical Industry

Early detection and sensing techniques can be greatly important for the reduction of cost of diseases. These costs have been estimated to be nearly US$75 billion [27] and US$90 billion [28] for cancer and diabetes, respectively. There have been a large number of such devices but have their own limitations with regard to response time and burden to patients. Thus, there is a great deal of interest in the development of such reliable sensing and detection instruments. A biosensor is commonly defined as an analytical device that uses a biological recognition system to target molecules or macromolecules [29]. Biosensors consist of three components: (1) a detector, (2) a transducer and (3) the output system.

An early example of the use of cells as biosensors occurred in 1977 when Rechnitz et al. [30] coupled living microorganisms (*Streptococcus faecium*) on the surface of an ammonia gas-sensing membrane electrode. Rechnitz's electrode biosensor was capable of detecting the amino acid Arginine. Modern nanoscale biosensors can greatly enhance the detection technique using cheaper, faster, and easy-to-use tools. The class of nanoscale biosensor using the method of biological signaling employs five major mechanisms that are shown in Figure 10.19. Nanoparticles have emerged as widely used materials used for nanoscale biosensing. Quantum dots have mostly been used for *in vitro* experimentation [31,32].

There are a lot of applications of nanoscale biosensors that have been used in health care and treatment of diseases. Some of them are discussed in Sections 10.3.1 and 10.3.2.

10.3.1 Glucose Detection *In Vivo*

This detection technique was originally introduced in the 1980s [33], which has now made possible for patients to self-monitor their glucose level with these instruments. One example of such device is GlucoWatch developed by Cygnus, Inc., Redwood City, California, which utilizes a glucose-containing interstitial fluid that is lured to the skin surface by a small current passing between two electrodes [34]. But the major limitation of these modern devices is the delay between the glucose concentration variations in the interstitial fluid and the corresponding changes in the blood [29]. Thus, there is a need for future glucose sensors to get rid of these limitations.

10.3.2 Bacterial Urinary Tract Infections

Currently, microbial culture techniques are used for the identification of urinary tract pathogens. But this method is a cumbersome one, which has a 2-day lag period between the collection of the sample and the identification of the pathogen. Thus, to reduce this lag period, new techniques are needed for improvement of health care and reduction of cost. Electrochemical DNA biosensors have been used in the literature to detect and identify pathogens [35,36], where the detection is done through the use of an electrochemical transducer.

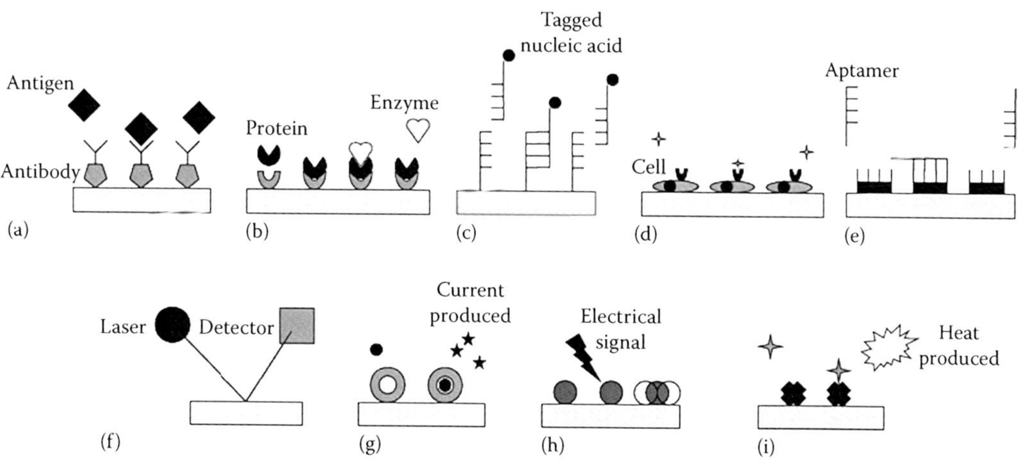

FIGURE 10.19
Biosensing and transduction classes for *in vitro* biosensors. Methods of biosensing: (a) antibody/antigen; (b) enzyme catalyzed; (c) nucleic acid; (d) cell-based; (e) biomimetic. Methods of transduction: (f) optical; (g) electrochemical; (h) mass-sensitive; (i) thermal. (Data from Gonsalves, K.E. et al., *Biomedical Nanostructures*, John Wiley & Sons, Inc., Hoboken, NJ, 2008.)

There are two basic modes to detect DNA with this configuration. The first method requires target immobilization followed by detection with a labeled probe [29,37]. In the second method, the DNA target initially binds to a surface oligonucleotide through hybridization. This is followed by hybridization to a marker probe for signal transduction [29,37].

10.4 Applications of Nanospectroscopy in Cancer Treatment and Imaging of Nanocrystals

Nanospectroscopy has found its applications in the imaging of various cells and tissue surfaces of animals as well as formed a new era of treatment of cancer cells. We discuss some of the applications of nanospectroscopy in this section.

Early cancer detection included works based on interrogation of living epithelial cells based on light scattering spectroscopy, which has been reported in Reference 38. Both single backscattering from the uppermost epithelial cells and multiple scattered light have been obtained, but the index of refraction and the size distribution of the epithelial cells were assessed through single backscattering. Another study on the detection technique was a method introduced by Backman et al. [39] for selective detection of size-dependent scattering characteristics of epithelial cells *in vivo* based on polarized illumination and polarization-sensitive detection of scattered light. It was found that reflectance spectroscopy with polarized light can provide quantitative morphological information, which could be used for noninvasive detection of neoplastic changes. Another study reveals the combination of both wavelength-dependent and light scattering [40]. The result of the study was the difference in scattering attributed to the average dimension of the scatterers, being a few tens of nanometers smaller in the healthy cells compared with the cancerous cells.

A new technique has been developed known as surface-enhanced Raman spectroscopy (SERS), which is due to the near-field enhancement of the electromagnetic field [41]. This application provides a strongly enhanced Raman response from molecular adsorbates on rough metals [42]. However, there is another technique termed as tip-enhanced Raman spectroscopy (TERS), which combines the advantages of SERS with those offered by scanning near-field optical microscopy (SNOM). SNOM is an optical imaging technique that provides access to subwavelength scale spatial resolution. [43,44]. TERS provides ultra-high-sensitivity and nanometer spatial resolution imaging. This has been studied on various materials and molecular systems absorbed on fat and corrugated surfaces [45,46].

The experimental layout of TERS is shown in Figure 10.20. The incident radiation is focused on the tip sample gap and the tip-scattered Raman light is detected. The signal is detected by any photodiode (an avalanche photodiode here). The polarization direction of both incident and scattered light can be controlled. An AFM is used for the experiments. The signal is reduced due to the oscillating tip.

The tip of the probe holds the central function in this technique that provides the enhanced electromagnetic field at the apex. It presents strong plasmon resonance leading to enhanced incident radiation and scattered field at the apex [47]. TERS presents a signal-to-noise ratio of more than 40:1 when probing nearly 100 molecules for 1 s accumulation time [47]. This makes this technique potentially usable in single-molecule sensitivity. In Figure 10.21, it can be seen that TERS measures a time series with 1 s acquisition time for each spectrum. Figure 10.21a shows TERS for submonolayer surface with enhancement

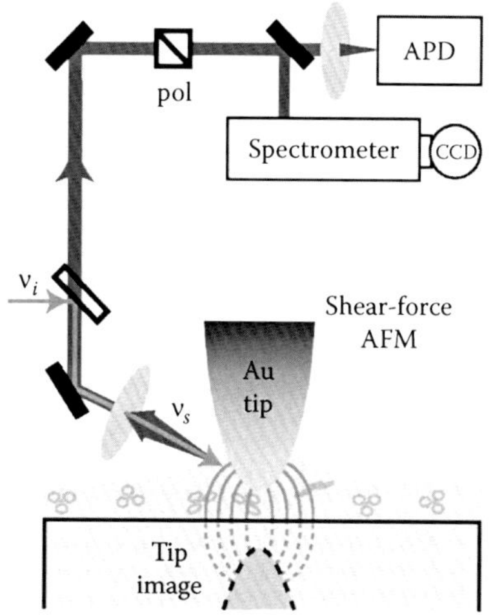

FIGURE 10.20
Experiment arrangement of TERS. CCD, charge-coupled device. APD, avalanchephotodiode; pol, polarizer; v_i, incident radiation; v_s, tip-scattered Raman light. (Data from Neacsu, C.C. et al., *Nanobiotechnology*, 3, 172–196, 2007.)

of 5×10^9. Figure 10.21b in the middle shows comparison of sum spectra of data and tip-enhanced Raman spectra for different enhancements while Figure 10.21c on the right shows time variation of Raman spectroscopy.

On the whole, since TERS offers chemical specificity, nanometer resolution, and single-molecule sensitivity, this technique will certainly come out as an important tool for chemical and structural identification of various nanostructures [47]. With experiments performed under the controlled environmental condition, reproducibility can be obtained in this technique.

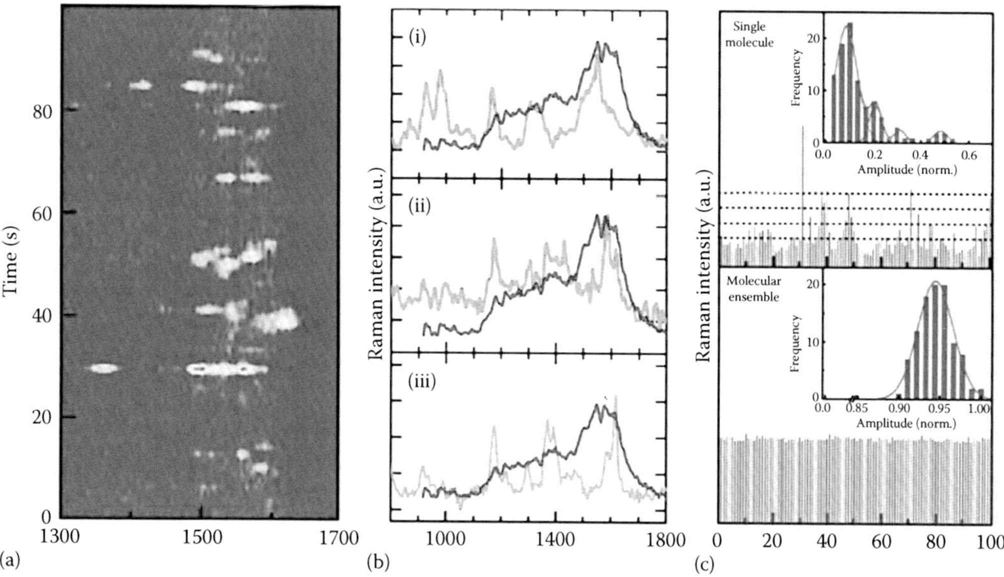

FIGURE 10.21
(a–c) Imaging with TERS. (Data from Neacsu, C.C. et al., *Nanobiotechnology*, 3, 172–196, 2007.)

Nanotechnology-based treatment of cancer provides therapy that could increase efficacy and improve the patient the quality of life [48]. A wide range of gold nanoparticles have been developed for use in imaging and therapy applications. Some of the forms of gold nanoparticle-based therapy applications are discussed here. Silica-based nanoshells have been used *in vitro* as probes for various types of cancers [49–51]. These nanoshells can cause cancer cell deaths by converting light to heat as shown in Figure 10.22.

Like nanoshells, nanorods can be tuned to the near-infrared (NIR) region and are used in cancer therapy applications. Their advantages are small cells, high absorption coefficient, and narrow spectral bandwidths [48]. However, gold nanorods have lower photothermal transduction cross section compared to gold nanoshells [52]. Although both nanorods and nanoshells have a potential for therapy application, these have less than 5% of dose accumulating in the tumors. Thus, to address these issues, smaller particles are applied in therapy applications.

Small NIR-tunable gold nanoparticles are potential to therapy application to address the problems posed by nanoshells and nanorods. Gobin et al. [53] showed that gold–gold sulfide have a larger absorption-to-scattering ratio due to its smaller size. Also hollow gold nanoshell is seen as potential applications in cancer therapy applications. These nanoparticles consist of hollow center with a gold shell. Li et al. showed that these hollow gold nanoparticles could be used to ablate the tumors [54].

Gold nanoparticles are also used as potential probe in treatment in mice (Figure 10.23). The main thing to consider in this treatment is the amount of gold nanoparticle biodistribution in mice, which is influenced by the size of the nanoparticle and its surface characteristic. A study by Perrault et al. showed the size of the nanoparticle and the weights of polyethylene glycol coating that produced increased blood half-life in smaller and larger diameter particles as shown in Figure 10.24.

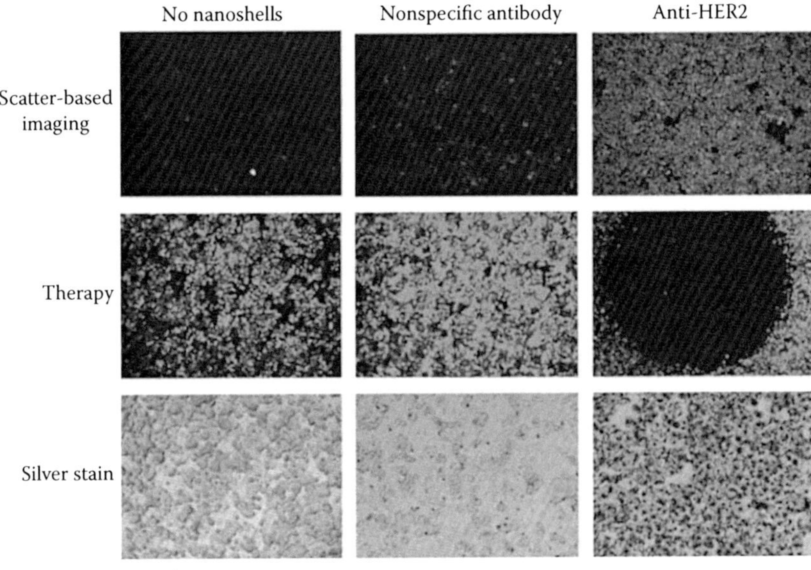

FIGURE 10.22
In vitro imaging of breast cancer cells using gold–silica nanoshells. (Originally from Loo, C. et al., *Nano Lett.*, 5, 709, 2005; Taken from Kennedy, L.C. et al., *Small*, 7, 169–183, 2011.)

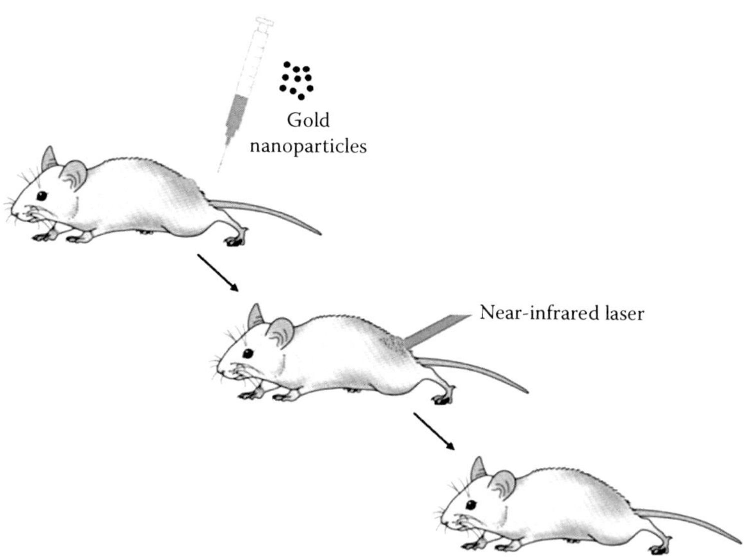

FIGURE 10.23
Gold nanoparticles as potential probe for use in mice. (Data from Kennedy, L.C. et al., *Small*, 7, 169–183, 2011.)

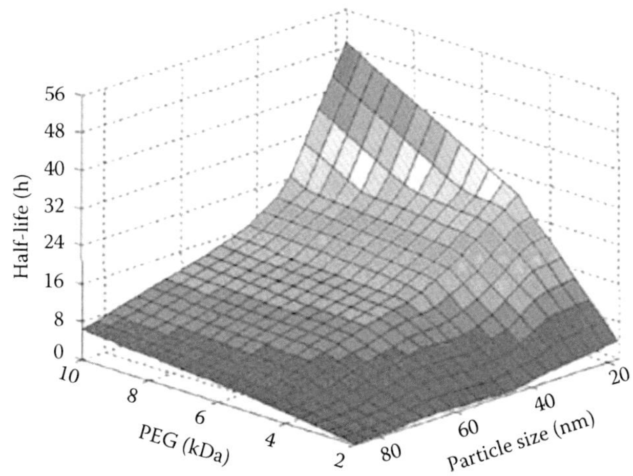

FIGURE 10.24
Effect of nanoparticle size and polyethylene glycol (PEG) weight on half-life in blood. (Originally from Perrault, S.D. et al., *Nano Lett.*, 9, 1909, 2009; Taken from Kennedy, L.C. et al., *Small*, 7, 169–183, 2011.)

For the biodistribution found in mice, there are a variety of ways different groups have presented their results. Some have presented their data as a ratio of gold in the organ to gold in the tissue, whereas others as a percentage of injected dose in the body. Comparing the gold nanoparticle distribution in the liver and spleen, gold nanorods show more percentage of accumulation in the liver than in the spleen, while the nanorods shows more concentration in liver than spleen [48]. Figure 10.25 shows the biodistribution of gold nanoparticles in a mouse model.

Thus, nanoshells, nanorods, and different gold nanoparticles have emerged as potential for use in radiotherapy; however, more research should be focused on improving methods

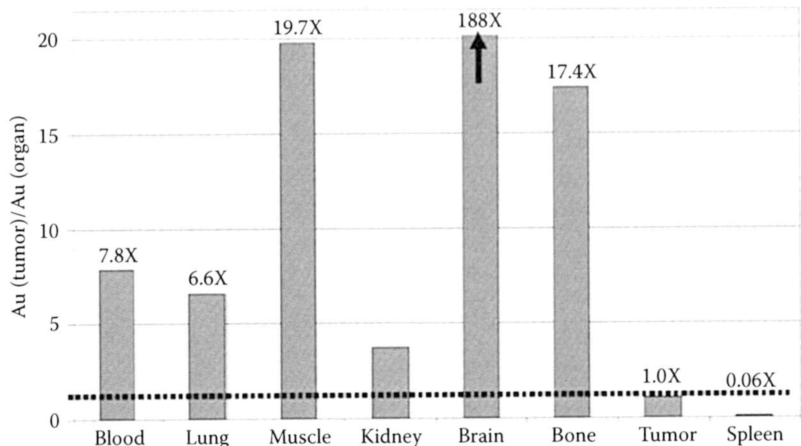

FIGURE 10.25
Biodistribution of gold nanoparticles in a mouse model. (Originally from James, W.D. et al., *J. Radioanal. Nucl. Chem.*, 271, 455, 2007; Taken from Kennedy, L.C. et al., 7, 2, 169–183, 2011.)

for reaching deeper tumors through various radiation modalities, less attenuation in the body parts, and improved delivery methods in order to obtain higher accumulation of particles in the tumor site [48].

10.5 Imaging Mass Spectroscopy

Imaging mass spectroscopy (IMS) is an indispensable technique for use in studying biological research. Mass spectroscopic imaging technique needs a computer-controlled sample stage for the motion in lateral and longitudinal direction of the samples. Also pixel resolution can be maintained by creating a pixel by assigning it to a defined area of a sample. With advancement in instrumentation, this technique is continuing to grow its application range from biology to bacterial thin films, plant leaves and stems, and even fingerprints [55]. Figure 10.26 shows the general workflow of the IMS technique.

The samples to be analyzed can be mounted on a surface. It is then programmed in the samples across different sampling positions. The data is then collected and processed to improve the image quality with combination of other imaging techniques. There are various techniques that could be applied for the visualization of IMS images, which include background correction, normalization, denoising, and contrast enhancement techniques. The summary of these techniques used in the analysis of IMS images is shown in Figure 10.27.

Figure 10.28 shows the proposed workflow of an online public database for published IMS datasets. All the contributors can submit their datasets, which can be used by biologists for their research.

This technique has several underlying technical challenges that must be addressed. For instance, better sample preparation is needed. Poorly prepared samples such as unevenness and thickness may result in deviation of signals. Further improvements can be made in the instrumentation to achieve increased resolution, sensitivity, and spatial resolution. The

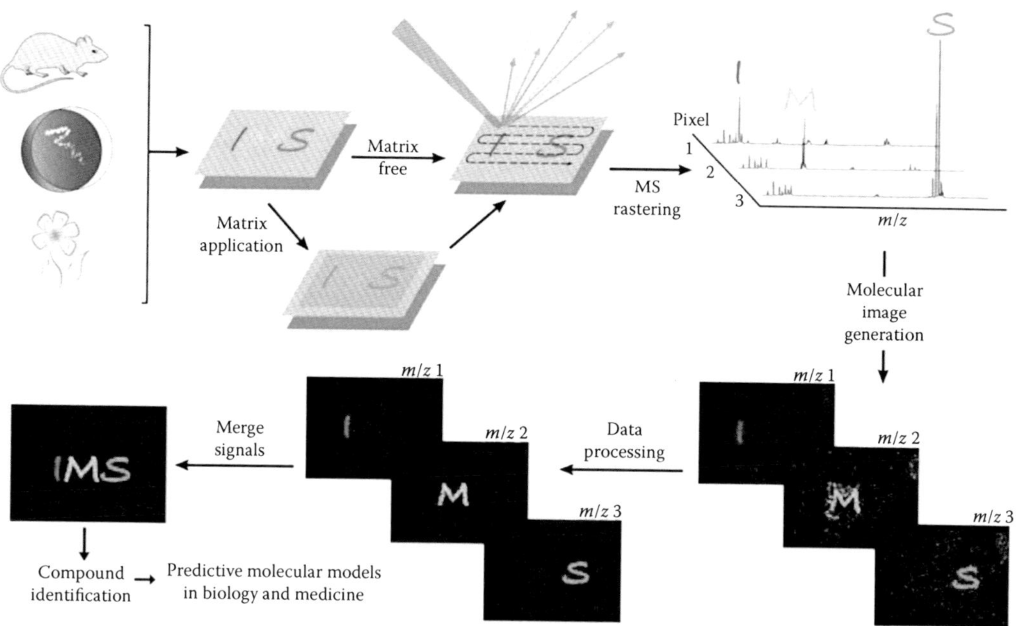

FIGURE 10.26
General workflow of IMS technique. MS, mass spectrometer. (Data from Watrous, J.D. et al., *J. Mass. Spectrom.*, 46, 209–222, 2011.)

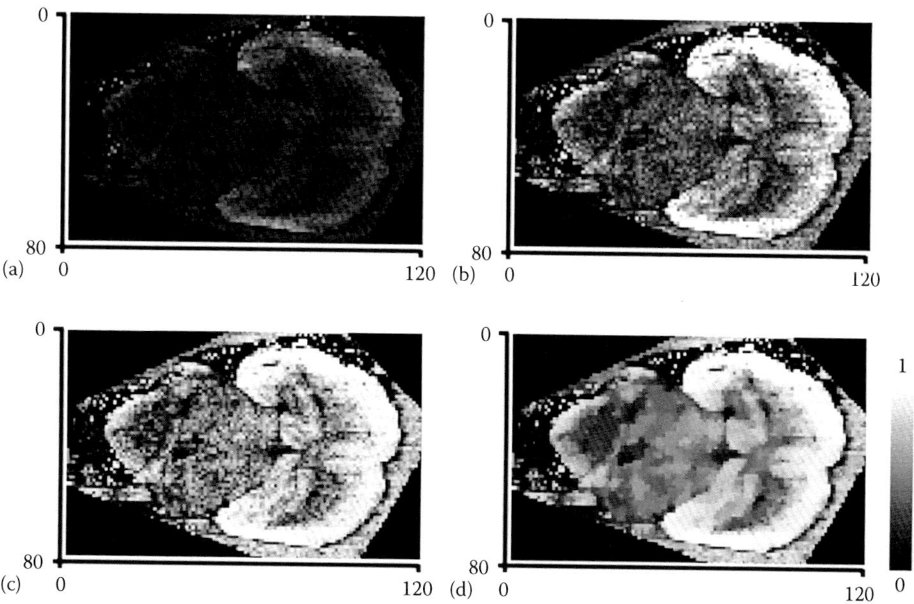

FIGURE 10.27
Techniques used for the processing of IMS images: (a) Original unprocessed image; (b) after suppressing 5% brightest pixels; (c) after histogram equalization; (d) after contrast stretching and denoising. (Data from Watrous, J.D. et al., *J. Mass. Spectrom.*, 46, 209–222, 2011.)

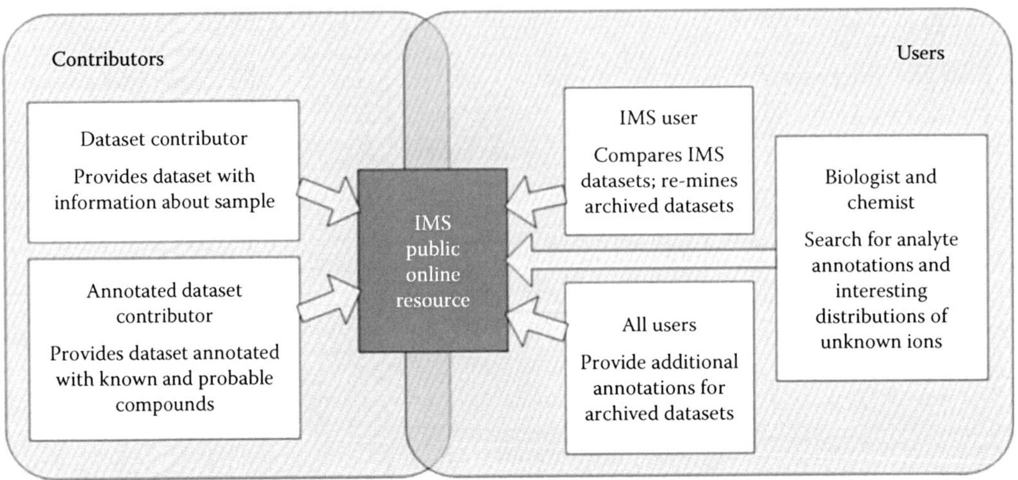

FIGURE 10.28
Proposed workflow for online database for published IMS dataset. (Data from Watrous, J.D. et al., *J. Mass. Spectrom.*, 46, 209–222, 2011.)

spatial resolution of the spectrometer can be achieved by making the pixel sizes smaller. Also improved data mining technique is another method for improvement in the IMS technique. Some of the possible applications of IMS are imaging of live cells, studying of signal interactions and community dynamics, and imaging of tissue biopsies. IMS requires cells to be killed before imaging, but if the source of ionization becomes more sensitive, live-cell imaging will be possible [55]. Thus, IMS has the potential to become a prominent imaging technique in the biological and medical research, provided that further improvements in sample preparation and instrumentation of the spectrometer can be made.

10.6 Polarimetric Exploratory Data Analysis

Recently, a new optical polarimetric metrics has been introduced by Giakos [56], the polarimetric exploratory data analysis (pEDA), which combines polarimetry with histogram data analysis.

The pEDA aims to quantify the signal characteristics of photons transmitted or backscattered through/from optically active media and tissue in terms of enhanced contrast to potentially discriminate optical signatures. Specifically, combining polarimetric analysis with exploratory data analysis offers the opportunity to relate a physical process to categorize signatures by grouping and separating different parts of the histograms, eliminating outliers, and applying thereafter fittings of different statistical curves. Histograms provide a visual discriminator for detecting outliers or unusual behavior from a sample.

Spatial information of the tissue can be obtained from the histograms of backscattered signals. Depending on the physical phenomenology of the sample, Gaussian or other suitable interpolation curves can be used to fit histogram data. A Gaussian distribution is used in the current single-pixel detection experiments to fit the amplitude measurements of the backscattered signals. The Gaussian (normal) probability density function in one dimension is shown in Equation 10.2 [56,57].

$$f(x;\mu,\sigma^2) = \frac{1}{\sqrt{2\pi\sigma^2}} \exp\left(-\frac{(x-\mu)^2}{2\sigma^2}\right) \tag{10.2}$$

where:
- μ is mean
- σ^2 is variance

A form of validation is carried out to determine the goodness of fit between the fitted Gaussian curve and the original data. This is done by creating a probability plot of the original data and the fitted Gaussian curve. Linearity between the two sets of data suggests a good fit. The correlation coefficient between the observed and fitted data is found to further validate the goodness of fit.

The full width at half maximum (FWHM) can be calculated from the Gaussian model used to fit the data. It is used to determine the dynamic range (DR) of the system for a particular sample:

$$\text{FWHM} = 2.354 \times \sigma \tag{10.3}$$

$$\text{DR(dB)} = 20 \times \log\left(\frac{V_{max}}{V_{min}}\right) \tag{10.4}$$

where:
- σ is the standard deviation of the data
- V_{max} and V_{min} are the maximum and minimum voltages, respectively

They correspond to the *x*-values (amplitudes) of the points on the Gaussian curve that are at 50% of the maximum on either side of the peak. The *x*-axis is normalized with respect to the sample mean, μ.

$$V_{max} = +\sigma\sqrt{2\ln 2} + \mu \tag{10.5}$$

$$V_{min} = -\sigma\sqrt{2\ln 2} + \mu \tag{10.6}$$

The *angular dynamic range* of a sample is similar to Equation 10.4 but is based on interrogation of the heterogeneous sample at different aspect angles

$$\text{DR}(\theta) = 20 \times \log\left(\frac{V_{max}}{V_{min}}\right) \tag{10.7}$$

Angular dynamic range can be further broken down into copolarized and cross-polarized angular dynamic ranges defined as follows:

$$\text{DR}(\theta)_{\parallel} = 20 \times \log\left(\frac{V_{max\parallel}}{V_{min\parallel}}\right) \tag{10.8}$$

$$\text{DR}(\theta)_{\perp} = 20 \times \log\left(\frac{V_{max\perp}}{V_{min\perp}}\right) \tag{10.9}$$

In single-pixel detection, a single pixel displays the intensity of a one-dimensional imaging signal. The dynamic range for single-pixel detection expresses the gray-level range associated with this one-dimensional imaging signal. An increase in the backscattered signal will result in an increase in the dynamic range [56].

Dynamic range and angular dynamic range can aid in characterizing each of the 16 elements of the Mueller matrix under different aspect angles. The Mueller matrix M transforms an input Stokes vector, S, into an output Stokes vector, S', according to the following equation:

$$\begin{pmatrix} S_0' \\ S_1' \\ S_2' \\ S_3' \end{pmatrix} = \begin{pmatrix} m_{11} & m_{12} & m_{13} & m_{14} \\ m_{21} & m_{22} & m_{23} & m_{24} \\ m_{31} & m_{32} & m_{33} & m_{34} \\ m_{41} & m_{42} & m_{43} & m_{44} \end{pmatrix} \begin{pmatrix} S_0 \\ S_1 \\ S_2 \\ S_3 \end{pmatrix} \tag{10.10}$$

Rotating the sample in reflection configuration, the rotated Mueller matrix of the sample, M_R, is obtained by [58]

$$M_R(\theta) = \mathbf{R}(\theta) \cdot M \cdot \mathbf{R}(\theta) \tag{10.11}$$

where:

θ is the rotation of the sample

$\mathbf{R}(\theta)$ is the rotational change in the basis matrix for Stokes vectors and Mueller matrices for the optical components of the system

Applying the polar Mueller matrix decomposition [59] allows for the calculation of depolarization, retardance, and diattenuation of the Mueller matrices of the sample. Typically, the Mueller matrix elements are used to provide a polarimetric intensity value that mainly contributes to the image contrast; however, the pEDA would provide both polarimetric contrast and enhanced discrimination potential.

As an example, several lung cancer types were discriminated against each other, using the discrimination potential of the pEDA.

A microphotograph of the normal, lung carcinoma *in situ* (CIS), and stage I nonsmall cell lung carcinoma (NSCLC) tissues is shown in Figure 10.29.

In Figure 10.29a, normal lung tissue is composed of thin-walled alveoli. Alveoli are the lung's air spaces (the large empty areas in the image), which comprise most of the volume of the lung. Each alveolar wall has a simple squamous epithelium (1) lining each exposed surface, with a thin stroma of capillaries (2) and delicate interstitial supporting connective tissue (3) sandwiched in between.

The simple squamous epithelium is a single layer of flattened cells with disc-shaped nuclei, which form a thin boundary that easily allows materials to diffuse through it.

The capillaries running between two adjacent alveoli are gorged with blood and pretty well defined by the erythrocytes in them. Any two neighboring alveoli are separated from each other by a blood vessel, which is therefore exposed to the air on either side, facilitating gas exchange.

Connective tissue functions not only as a mechanical support but also as an avenue for communication and transport. Collagen is the most abundant and pervasive component in connective tissue. Fibroblasts are part of the connective tissue, which secrete collagen and other elements of the extracellular matrix in the lungs.

A large number of free cells in the lumen of the alveoli are macrophages (4). These alveolar macrophages can then crawl over the free surface and scavenge dust particles and

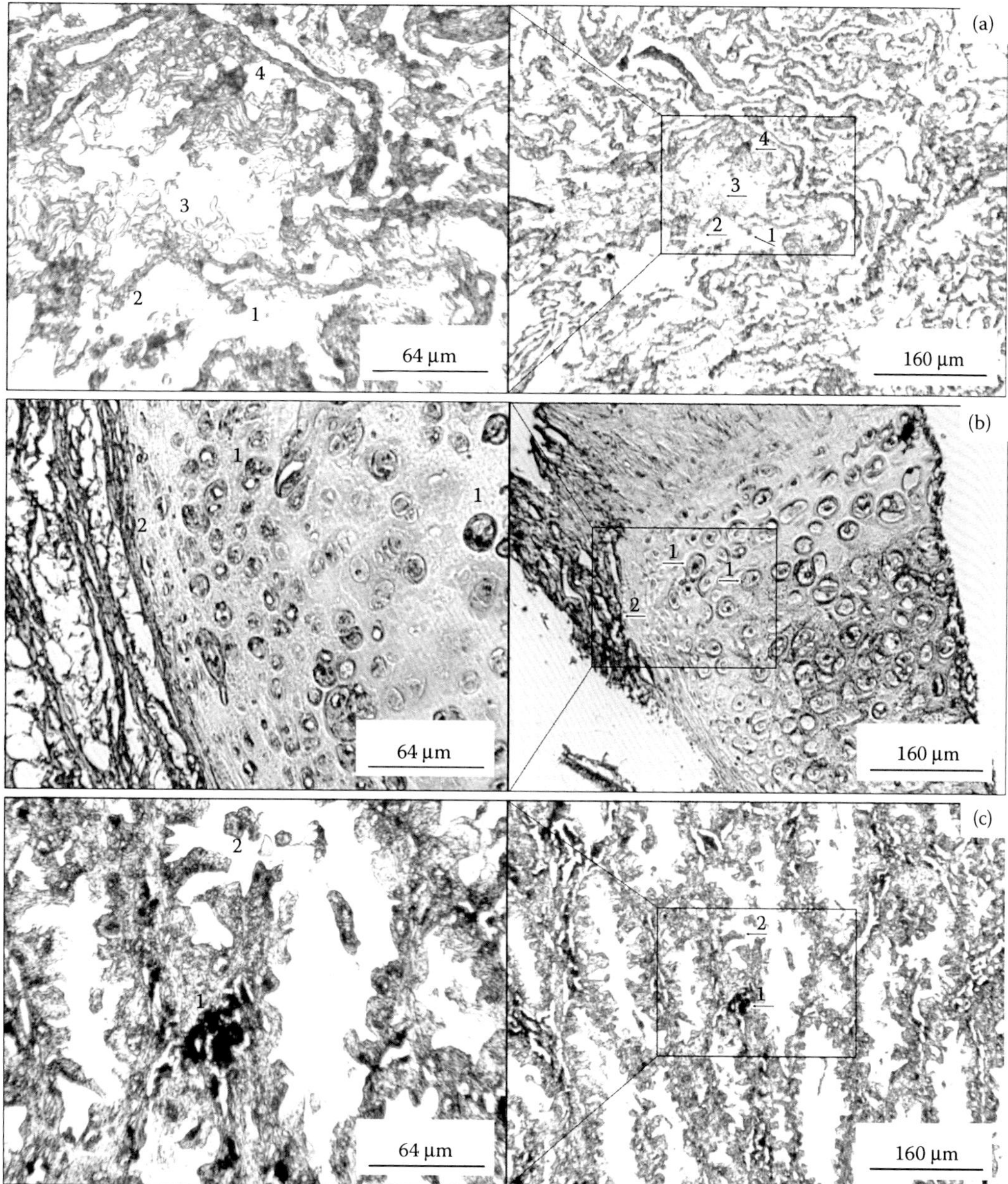

FIGURE 10.29
(a) A histology slide of normal lung tissue with a thickness of 5 μm. (b) A sample of lung CIS with a thickness of 5 μm. (c) A sample consisting of stage I NSCLC adenocarcinoma with a papillary bronchioloalveolar pattern, T1, N0 (without node involvement), G2 (early stage) with a thickness of 5 μm. See text for further details.

bacteria that have been inhaled. Ingested material can accumulate in lysosomal vesicles and become visible as lipofuscin granules.

In Figure 10.29b, the lung CIS tissue was characterized as poorly differentiated non-small cell lung carcinoma, with glandular and squamous features. Respiratory epitheliums are replaced by malignant squamous cells with hyperchromatic and pleomorphic characters (1).

The nuclear atypia and frequent mitoses figures are pretty evident; neoplastic glands in the desmoplastic stroma with cribriform pattern and stromal invasion with severe desmoplasia (2).

In Figure 10.29c, the malignant tumor cells with enlarged and vesicular nuclei and prominent nucleoli (1) are cuboid to columnar in shape, grow along the alveolar walls for support; replace the original pneumocytes; frequently secrete mucin; form a tubular, acinar, or papillary structure; and thicken the alveolar septae (2).

Following the treatment of Chipman [59], histograms were recorded for each sample for all three tests performed and fit with a Gaussian curve. An overlay of histograms obtained from each sample displaying the difference in intensity distributions is shown in Figure 10.30. Table 10.1 reports the average values of all statistical parameters calculated from the histograms.

The dynamic range from histograms of lung tissue samples is plotted in Figure 10.31.

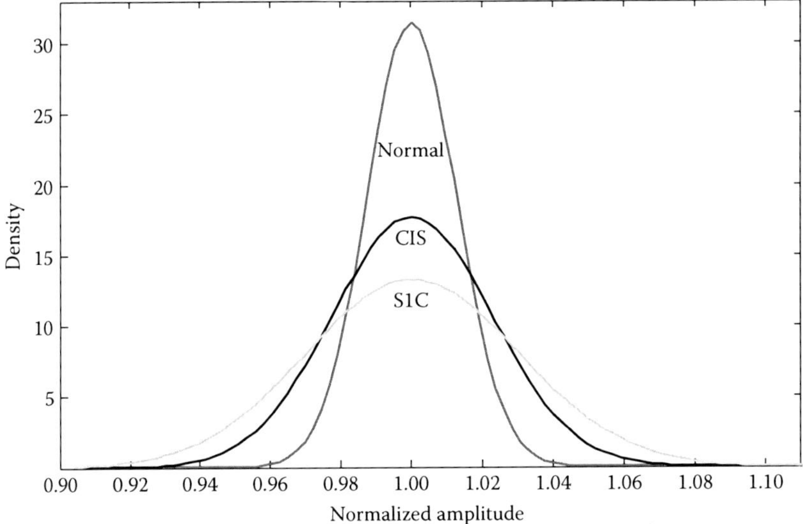

FIGURE 10.30
Gaussian fits to detected amplitudes from normal lung tissue, CIS, and stage I carcinoma (S1C) to display the differences in intensity spreads.

TABLE 10.1
Histogram Statistical Analysis for the Three Lung Samples: Normal, CIS, and Stage I Carcinoma

	Normal Lung Tissue		CIS		Stage I Carcinoma	
	Average	Standard Error of Mean	Average	Standard Error of Mean	Average	Standard Error of Mean
μ (mV)	500.36	2.4700	390.33	1.8244	299.22	5.0230
Standard error of μ	0.2461	0.0041	0.2600	0.0121	0.2816	0.0013
σ (mV)	7.7842	0.1299	8.2231	0.3833	8.9036	0.0441
Standard error of σ	0.1742	0.0029	0.1840	0.0085	0.1993	0.0009
FWHM (mV)	17.5165	0.2535	19.1820	0.9198	20.7055	0.1401
Dynamic range (dB)	0.1521	0.0027	0.1666	0.0080	0.1799	0.0013
Correlation coefficient	0.8563	0.0010	0.8997	0.0132	0.8979	0.0282

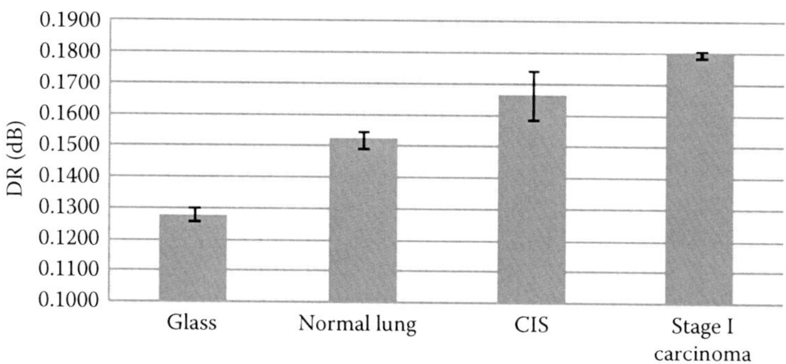

FIGURE 10.31
Dynamic range (DR) in decibels calculated from the histograms of each sample: normal, CIS, stage I carcinoma, and the background (glass). The error bars represent the standard error of the mean.

10.7 Computational Nanophotonics with MATLAB®

In the following example, the Mueller matrix of the object, consisting of nanoparticles or a group of molecules, will be calculated. In this specific example, a Mueller matrix polarimeter is considered; a light source (from a laser) interrogates an object, with a series of generator states g_i and analyzes the light that has interacted with the object with a set of analyzer states a_i recording a set of flux measurements arranged in a vector **P**. As well known from Equation 10.10 and Reference 58, the output Stokes vectors S′ are related to the input S via the Mueller matrix of the sample,

$$\begin{bmatrix} S'_0 \\ S'_1 \\ S'_2 \\ S'_3 \end{bmatrix} = \begin{bmatrix} m_{11} & m_{12} & m_{13} & m_{14} \\ m_{21} & m_{22} & m_{23} & m_{24} \\ m_{31} & m_{32} & m_{33} & m_{34} \\ m_{41} & m_{42} & m_{43} & m_{44} \end{bmatrix} \cdot \begin{bmatrix} S_0 \\ S_1 \\ S_2 \\ S_3 \end{bmatrix} \tag{10.12}$$

A data reduction technique [59] is applied to calculate the Mueller matrix of the object. For this, we must first know the configurations of the polarization state generator and the polarization state analyzer for each of the q intensity measurements. These configurations lead to a $q \times 16$ polarimetric measurement matrix, W. Let G_q denote the Stokes vector for the generator states and A_q denote the Stokes vector of the analyzer states, where q represents the states of the experiment. Also let P_q denote the measured intensity of each of the states.

Typically, the LabVIEW code cycles through all 16 states of the generator–analyzer, which in this case were chosen to be for states $q1$ through $q16$, as shown in Table 10.2 [60].

TABLE 10.2
Polarization States = 16

*q*1	*q*2	*q*3	*q*4	*q*5	*q*6	*q*7	*q*8	*q*9	*q*10	*q*11	*q*12	*q*13	*q*14	*q*15	*q*16
1	2	3	4	5	6	7	8	9	10	11	12	13	14	15	16
HH	HV	HP	HR	VH	VV	VP	VR	PH	PV	PP	PR	RH	RV	RP	RR

Note: H = horizontal, V = vertical, P = +45°, R = right circular.

TABLE 10.3

Mueller Matrix Scattering Elements = 16

m_{11} = HH + HV + VH + VV	m_{22} = HH − HV − VH + VV	m_{13} = PH + PV − MH − MV	m_{14} = RH + RV − LH − LV
m_{21} = HH − HV + VH − VV	m_{22} = HH − HV − VH + VV	m_{23} = PH − PV − MH + MV	m_{24} = RH − RV − LH + LV
m_{31} = HP − HM + VP − VM	m_{32} = HP − HM − VP + VM	m_{33} = PP − PM − MP + MM	m_{34} = RP − RM − LP + LM
m_{41} = HR − HL + VR − VL	m_{42} = HR − HL − VR + VL	m_{43} = PR − PL − MR + ML	m_{44} = RR − RL − LR + LL

Note: H = horizontal, V = vertical, P = +45°, M = −45°, R = right circular, and L = left circular.

The scattering Mueller matrix elements were then estimated by combining the 16 states of the generator–analyzer so that the Mueller matrix of the target was constructed in the form given in Table 10.3.

Examples of representative acquired states and waveforms plotted using MATLAB are shown in Figures 10.32 and 10.33, respectively.

Let M denote the Mueller matrix of the object. Then,

$$P_q = A_q^{\,T}.M.S_q = \begin{bmatrix} A_{q,1} & A_{q,2} & A_{q,3} & A_{q,4} \end{bmatrix} \begin{bmatrix} m_{11} & m_{12} & m_{13} & m_{14} \\ m_{21} & m_{22} & m_{23} & m_{24} \\ m_{31} & m_{32} & m_{33} & m_{34} \\ m_{41} & m_{42} & m_{43} & m_{44} \end{bmatrix} \begin{bmatrix} G_{q,1} \\ G_{q,2} \\ G_{q,3} \\ G_{q,4} \end{bmatrix} \qquad (10.13)$$

$$= \sum_{j=1}^{4} \sum_{k=1}^{4} A_{q,j} m_{j,k} G_{q,k}$$

The analyzer and generator matrix can be ordered as in Equation 10.1 where it can be expressed as a 16 × 1 measurement vector for the qth state [59]:

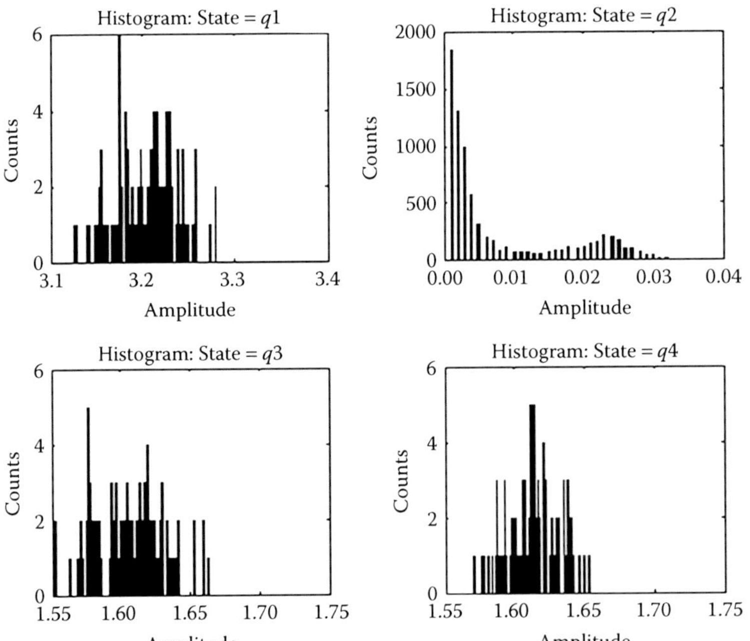

FIGURE 10.32
Example of acquisition of polarization histogram states.

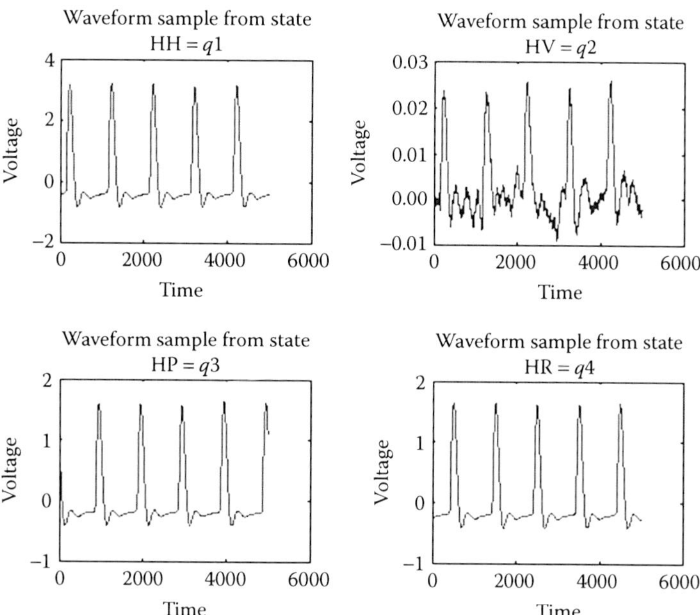

FIGURE 10.33
Example of acquisition of polarization waveform states.

$$W_q = \begin{bmatrix} A_{q,1}G_{q,1} & A_{q,1}G_{q,2} & A_{q,1}G_{q,3} & \ldots & A_{q,4}G_{q,4} \end{bmatrix}^T \tag{10.14}$$

Then the Mueller matrix can be presented as a 16 × 1 vector, thus giving 16 intensities for each state represented by P_q. Thus,

$$P_q = \begin{bmatrix} A_{q,1}G_{q,1} \\ A_{q,1}G_{q,2} \\ A_{q,1}G_{q,3} \\ \cdot \\ \cdot \\ A_{q,4}G_{q,4} \end{bmatrix} \begin{bmatrix} m_{11} \\ m_{12} \\ m_{13} \\ \cdot \\ \cdot \\ m_{44} \end{bmatrix} \tag{10.15}$$

and 16 measurements need to be taken that makes $q = 16$. Thus, W_q turns to be a 16 × 16 matrix, which when multiplied by 16 × 1 Mueller vector gives 16 × 1 intensity vector, each element of a vector representing an intensity for each state. Thus, the Mueller matrix of the object can be determined as

$$M = W^{-1} \bullet P \tag{10.16}$$

If W contains 16 linearly independent rows, all the elements of the Mueller matrix can be determined using Equation 10.6. But if $q > 16$ and M is overdetermined, then the optimal polarimetric data reduction equation uses the pseudoinverse W_p^{-1} of W and is given by Equation 10.7 [59]:

$$M = [W^T W]^{-1} W^T P = W_p^{-1} P \tag{10.17}$$

The advantages of the polarimetric data reduction technique are as follows:

1. This method can be easily understood and implemented.
2. It is independent of the specific type of light source or detector used. If the Stokes vector associated with the generator and analyzer arm are determined, the effect of the nonideal elements are corrected.
3. It can be used for overdetermined measurement states, that is, $q > 16$ using Equation 10.17.

The Mueller matrix of an element consists of various optical properties of an element. In order to know about these properties, the Mueller matrix can be decomposed into three other matrices, which can be used to further characterize the materials. These three matrices are the depolarization, diattenuation, and retardance matrices. This algorithm of decomposing the Mueller matrix into these three matrices is called Lu–Chipman decomposition and is given by Equation 10.8.

$$M = M_{\Delta} \cdot M_{R} \cdot M_{D} \tag{10.18}$$

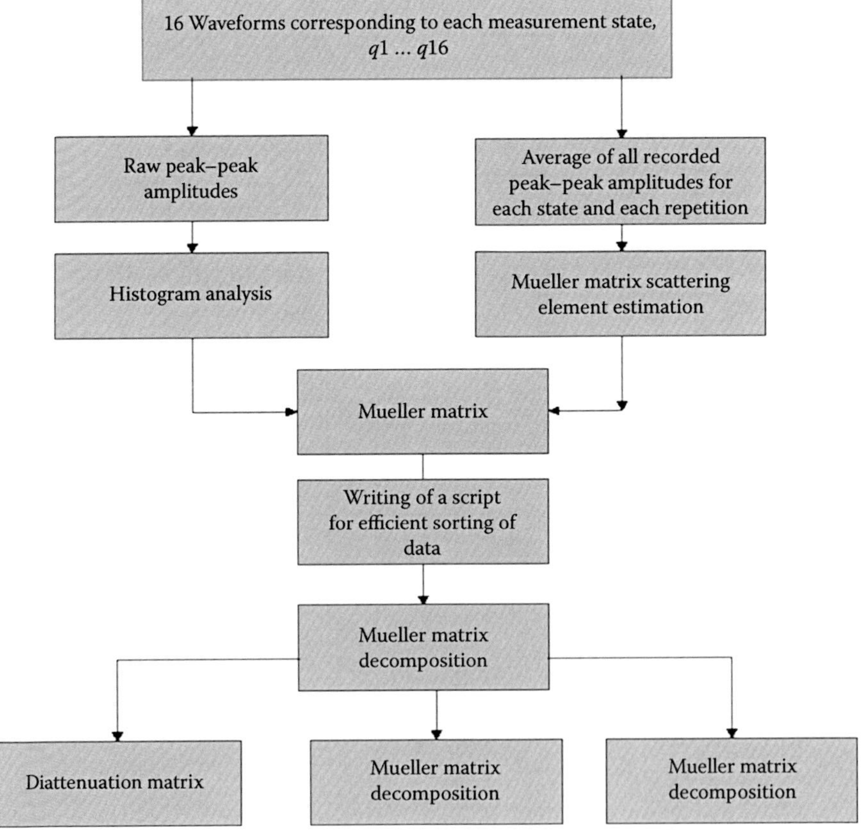

FIGURE 10.34
Block diagram of the applied polarimetric algorithm.

where:

M_Δ is the depolarization matrix
M_R is the retardance matrix
M_D is the diattenuation matrix

For the algorithm of the decomposition of Mueller matrix to the three matrices, refer to Reference 59.

The physical algorithm is shown in Figure 10.34.

10.8 Conclusions

In this chapter, nanoimaging system principles, key designs, and representative techniques addressing applications from the areas relating to bioscience and medical diagnostics have been introduced. Selected nanospectroscopic techniques that are being applied in medical imaging and radiotherapy have been presented. The principles of IMS, one of the newest metabolic imaging techniques, has been introduced and discussed. Finally, optical polarimetric metrics, the pEDA—which combines polarimetry with histogram data analysis aimed at providing enhanced discrimination among cancerous tissues and cells—have been introduced and discussed. A preclinical example of enhanced discrimination among different lung cancer types has been presented, and a computational analysis of the polarimetric principles has been offered and discussed.

References

1. McCurdy CW, Stechel E, Cummings P, Hendrickson B, Keyes D. "Theory and modeling in nanoscience." Report of Workshop Conducted by the Basic Energy Sciences and Advanced Scientific Computing Advisory Committees to the Office of Science, Lawrence Berkeley National Laboratory, Berkeley, CA, June 28, 2002.
2. Zhao XS. "Synthesis, modification, characterization and application of MCM-41 for VOC Control." PhD Thesis, University of Queensland, Australia, 1998.
3. Sousa AA, Kruhlak MJ. "Introduction: Nanoimaging Techniques in Biology: Nanoimaging." *Methods in Molecular Biology*, 950:1–10, 2013. DOI:10.1007/978-1-62703-137-0_1.
4. John Innes Center. "What is light microscopy?" https://www.jic.ac.uk/microscopy/intro_EM.html [Accessed: February 2, 2013].
5. Stephens DJ, Allan VJ. "Light microscopy of live cell imaging." *Science Review*, 300(5616):82–86, 2003.
6. Lue N, Kang JW, Yu C-C, Barman I, Dingari NC, Feld MS, Dasari RR et al. "Portable optical fiber probe-based spectroscopic scanner for rapid cancer diagnosis: A new tool for intraoperative margin assessment." *PLoS One*, 7(1):e30887, 2012.
7. Hassan AM, Hufnagle DC, El-Shenawee M, Pacey GE, "Terahertz imaging for margin assessment of breast cancer tumors." Published in Microwave Symposium Digest (MTT), Montreal, QC, Canada, 2012 IEEE MTT-S International, pp. 1–3, June 17–22, 2012.
8. Xi L, Grobmyer SR, Wu L, Chen R, Zhou G, Gutwein LG, Sun J et al. "Evaluation of breast tumor margins in vivo with intraoperative photoacoustic imaging." *Optics Express*, 20(8):8726, 2012.

9. George Themelis GMD, Crane LMA, Harlaar NJ, Pleijhuis RG, Kelder W, Sarantopoulos A, Jong JS et al. "Intraoperative tumor-specific fluorescence imaging in ovarian cancer by folate receptor-[alpha] targeting: first in-human results." *Nature Medicine*, 10:1315–1319, 2011.
10. Giakos G, Deshpande A, Shrestha S, Quang T. "An automated digital fluorescence imaging system of tumor margins using clustering-based image thresholding." Imaging Systems and Techniques (IST), 2013 IEEE International Conference on, Beijing, People's Republic of China, pp. 116–120, 201. DOI:10.1109/IST.2013.6729674. http://ieeexplore.ieee.org/xpl/login.jsp?tp=&arnumber=6729674&url=http%3A%2F%2Fieeexplore.ieee.org%2Fxpls%2Fabs_all.jsp%3Farnumber%3D6729674.
11. James WD, Hirsch LR, West JL, O'Neal PD, Payne JD. *Journal of Radioanalytical and Nuclear Chemistry*, 271:455, 2007.
12. Williams JC, Patton N. "Transmission Electron Microscopy." In *Systematic Materials Analysis*, Vol. 4, Richardson JH and Peterson RV, Eds., 407–432. Academic Press Inc, New York, 1978.
13. Reimer L, Kohl H. *Transmission Electron Microscopy. Physics of Image Formation*, Springer, Optical Sciences Series, Vol. 36, 5th ed., Helmut, 2008.
14. Wang ZL. "Transmission electron microscopy of shape-controlled nanocrystals and their assemblies." *Journal of Physical Chemistry B*, 2000(104):1153–1175.
15. Tonomura A. *Electron Holography*. Springer-Verlag, New York, 1993.
16. Lichte H. *Advances in Optical and Electron Microscopy*, 12:25, 1991.
17. Tonomura A, Allard LF, Pozzi G, Joy DC, Ono YA, Eds. *Electron Holography*. Elsevier Science, New York, 1995.
18. Cortadellas N, Fernández E, Garcia A. "Biomedical and biological applications of scanning electron microscopy." *Handbook of Instrumental Techniques from CCiTUB*. Unitat de Microscòpia Electrònica (Casanova), CCiT-UB, Universitat de Barcelona. Facultat de Medicina, Barcelona, Spain. http://diposit.ub.edu/dspace/bitstream/2445/31943/1/BT03%20-%20Biomedical%20and%20Biological%20Applications%20of%20SEM%20_ed2.pdf.
19. Deng Y, Zhao YY. "Effects of pulsed-vacuum and ultrasound on the osmodehydration kinetics and microstructure of apples." *Journal of Food Engineering*, 85:84–93, 2008.
20. Deng Y, Zhao YY. "Effect of pulsed vacuum and ultrasound osmopretreatments on glass transition temperature, texture, microstructure and calcium penetration of dried apples." *LWT-Food Science and Technology*, 41:1575–1585, 2008.
21. Washington State University. "Basic guide to scanning probe microscopy." http://public.wsu.edu/~hipps/pdf_files/spmguide.pdf [Accessed: February 17, 2014].
22. Office of Science U.S. Department of Energy. *Basic of Scanning Probe Microscopy*. Pioneering Science and Technology, Chapter 1. http://www1.na.infn.it/TIMSI/materialicorsi/iavarone/chapter1.pdf.
23. Chen CJ. *Introduction to Scanning Tunneling Microscopy*, 2nd Edition. Department of Applied Physics and Applied Mathematics, Columbia University, New York, 2008.
24. Binnig G, Quote CF, Gerber Ch. "Atomic force microscope." *Physical Review Letters*, 56(6):930–934, 1986.
25. Dufrene YF. "Atomic force microscopy, a powerful tool in microbiology." *Journal of Bacteriology*, 184(19):930–934, 2002.
26. National Institute of Science and Technology, US Department of Commerce. "Size measurement of nanoparticles using atomic force microscopy." NCI Alliance for Nanotechnology in Cancer, Gaithersburg, MD, October 2009.
27. American Cancer Society. *Cancer Facts and Figures*. American Cancer Society, Atlanta, GA, 2006.
28. Hogan P, Dall T, Nikolov P. "Economic costs of diabetes in the US in 2002." *Diabetes Care*, 26:917–932, 2003.
29. Gonsalves KE, Laurencin CL, Halberstadt CR, Nair LS. *Biomedical Nanostructures*. John Wiley & Sons, Inc., Hoboken, NJ; New York, 2008.
30. Rechnitz GA, Kobos RK, Riechel SJ, Gebauer CR. "A bio-selective membrane electrode prepared with living bacterial cells." *Analytica Chimica Acta*, 94:357–365, 1977.
31. Alivisatos AP, Gu W, Larabell C. "Quantum dots as cellular probes." *Annual Review of Biomedical Engineering*, 7:55–76, 2005.

32. Mansson A, Sundberg M, Balaz M, Bunk R, Nicholls IA, Omling P, Tågerud S, Montelius L. "In vitro sliding of actin filaments labelled with single quantum dots." *Biochemical and Biophysical Research Communications*, 314:529–534, 2004.
33. Turner AP, Chen B, Piletsky SA. "In vitro diagnostics in diabetes: Meeting the challenge." *Clinical Chemistry*, 45:1596–1601, 1999.
34. Bolinder J, Ungerstedt U, Arner P. "Microdialysis measurement of the absolute glucose concentration in subcutaneous adipose tissue allowing glucose monitoring in diabetic patients." *Diabetologia*, 35:1177–1180, 1992.
35. Wang J. "Electrochemical nucleic acid biosensors." *Analytica Chimica Acta*, 469:63–71, 2002.
36. Drummond TG, Hill MG, Barton JK. "Electrochemical DNA sensors." *Nature Biotechnology*, 21:1192–1199, 2003.
37. Campbell CN, Gal D, Cristler N, Banditrat C, Heller A. "Enzyme-amplified amperometric sandwich test for RNA and DNA." *Analytical Chemistry*, 74:158–162, 2002.
38. Sokolov K, Drezek R, Gossage K, Richards-Kortum R. "Reflectance spectroscopy with polarized light: In-situ sensitive to cellular and nuclear morphology." *Optics Express*, 5(13):302–317, 1999.
39. Backman V, Gurjar R, Badizadegan K, Itzkan I, Dasari RR, Perelman LT, Feld MS. "Polarized light scattering spectroscopy for quantitative measurement of epithelial cellular structures in situ." *IEEE Journal of Selected Topics in Quantum Electronics*, 5(4):1019–1026, 1999.
40. Mourant JR, Hielscher AH, Eick AA, Johnson TM, Freyer, JP. "Evidence of intrinsic differences in the light scattering properties of tumorigenic and nontumorigenic cells." *Cancer Cytopathology*, 84(6):366–374, 1998.
41. Moskovits M. "Surface roughness and the enhanced intensity of Raman scattering by molecules absorbed on metals." *Journal of Chemical Physics*, 69:4159, 1978.
42. Albercht M, Creighton J. "Anomalously intense Raman spectroscopy of pyridine at a silver electrode." *Journal of the American Chemical Society*, 99:5215, 1977.
43. Pohl DW, Denk W, Lanz M. "Optical stethoscopy: Image recording with resolution $\lambda/20$." *Applied Physics Letters*, 44:651–653, 1984.
44. Courjon D. *Near-Field Microscopy and Near-Field Optics*, Imperial College Press, London, 2003.
45. Anderson MS. "Locally enhanced Raman spectroscopy with an atomic force microscope." *Applied Physics Letters*, 76, 3130–3132, 2000.
46. Domke KF, Zhang D, Pettinger B. "Toward Raman fingerprints of single dye molecules at atomically smooth Au(111)." *Journal of the American Chemical Society*, 128(45):14721–14727, 2006.
47. Neacsu CC, Berweger S, Raschke MB. "Tip-enhanced Raman imaging and spectroscopy: Sensitivity, symmetry and selection rules." *Nanobiotechnology*, 3:172–196, 2007. DOI:10.1007/s12030-008-9015-z.
48. Kennedy LC, Bickford LR, Lewinski NA, Coughlin AJ, Hu Y, Day ES, West JL, Drezek RA. "A new era for cancer treatment: Gold-nanoparticle-mediated thermal therapies." *Small*, 7(2):169–183, 2011.
49. Loo C, Hirsch L, Lee MH, Chang E, West J, Halas N, Drezek R. "Gold nanoshell bioconjugates for molecular imaging in living cells." *Optics Letters*, 30:1012, 2005.
50. Loo C, Lin A, Hirsch L, Lee MH, Barton J, Halas N, West J, Drezek R. "Nanoshell-enabled photonics-based imaging and therapy of cancer." *Technology in Cancer Research and Treatment*, 3:33, 2004.
51. Liu SY, Liang ZS, Gao F, Luo SF, Lu GQ. "In vitro photothermal study of gold nanoshells functionalized with small targeting peptides to liver cancer cells." *Journal of Materials Science: Materials in Medicine*, 21:665, 2010.
52. Cole JR, Mirin NA, Knight MW, Goodrich GP, Halas NJ. "Photothermal efficiencies of nanoshells and nanorods for clinical therapeutic applications." *Journal of Physical Chemistry C*, 113:12090–12094, 2009.
53. Gobin AM, Watkins EM, Quevedo E, Colvin VL, West JL. "Near-infrared-resonant gold/gold sulfide nanoparticles as a photothermal cancer therapeutic agent." *Small*, 6:745, 2010.

54. Melancon MP, Lu W, Yang Z, Zhang R, Cheng Z, Elliot AM, Stafford J, Olson T, Zhang JZ, Li C. "In vitro and in vivo targeting of hollow gold nanoshells directed at epidermal growth factor receptor for photothermal ablation therapy." *Molecular Cancer Therapeutics*, 7(6):1730–1739, 2008. doi:10.1158/1535-7163.MCT-08-0016.
55. Watrous JD, Alexandrov T, Dorrestein PC. "The evolving field of imaging mass spectrometry and its impact on future biological research." *Journal of Mass Spectrometry*, 46(2):209–222, 2011.
56. Giakos GC. "Novel biological metamaterials, nanoscale optical devices and polarimetric exploratory data analysis (pEDA)." *International Journal of Signal and Imaging Systems Engineering*, 3(1):3–12, 2010.
57. Marotta S. "Polarimetric exploratory data analysis (pEDA) using dual rotating retarder polarimetry for in vitro detection of early stage lung cancer." Master Thesis, University of Akron, OH, 2011.
58. Goldstein D. *Polarized Light*, 2nd Edition. Marcel Dekker Inc., New York, 2003.
59. Chipman RA. "Polarimetry." *Handbook of Polarimetry*, 2nd Edition, Vol. 2. McGraw-Hill, New York, 1994.
60. Shrestha S. "High resolution polarimetric imaging techniques for space and medical applications." Master Thesis, The University of Akron, OH, May 2013.
61. Wang Z, RHK Technology. http://www1.na.infn.it/TIMSI/materialicorsi/iavarone/chapter1.pdf.
62. Loo C, Lowery A, Halas N, West J, Drezek R. "A new era for cancer treatment: Gold-nanoparticle-mediated thermal therapies." *Nano Letters*, 5:709, 2005.
63. Perrault SD, Walkey C, Jennings T, Fischer HC, Chan WC. *Nano Letters*, 9:1909, 2009.

11

Medical Imaging Instrumentation and Techniques

Viroj Wiwanitkit

CONTENTS

11.1 Introduction

Optics is an important term in science. When one mentions optics, it usually means something related to light and image. Optics also gives the sense of *seeing*. Talking about optics, one can imagine the relationship to magnetoelectricity wave. Indeed, light is a kind of energy and it is related to the vision and further related to optics. It is not possible to talk about optics without referring to light. As already mentioned, vision is directly related to optics.

It can be simply said that light is the basic requirement of any vision. This physical property can be mentioned and considered as a specific property of magnetoelectricity wave. Any object, small or large, has to get light before it can be visualized.

Now, talking about optics is not a new concept. Optics is an actual classical focus in science that is still important. The scientific society turns its current direction to optics property. Light and visualization of objects are also focused. The physical properties of optics light as an *object* become challenges in science. This leads to the consideration of light as wave in the scientific world. In the past, the human ability limited the study of light. Light in the classical science might be visible but not touchable. However, when the development of new theory of electromagnetic wave was complete, the scientific society turned its face to a facet of light as an object. With the use of new scientific tools, light can be visualized and manipulated as objects. This leads to a jump to a new paradigm to deal with optics in the scientific world. Many new things were identified, especially for photon. However, the science continues on its own ways. The scientists continued to seek and found the new things in the field of optics.

Generally, visualization or *ability to see* is directly related to light and this is an actual topic in optics. The science related to optics is called optics science. It mainly deals with light; therefore, it is the *science of light*. Light is the basic requirement for any vision. Without light, visualization cannot be possible.

At present, optics is worldwide studied. Many specific scientific societies are set up corresponding to optics science. The most well-known society is the Optical Society (OSA), which was founded in 1916. According to its declaration in its website (which can be accessible at http://www.osa.org/en-us/about_osa/) [1], the OSA is aiming at "the specific society that will be the leading professional association in optics and photonics, home to accomplished science, engineering, and business leaders from all over the world." The OSA has several members around the world. The OSA continuously sets and launches its publications, meetings, and membership programs. Also, the OSA gives specific knowledge and updated information on the science of light, that is, optics.

Now the scientist successfully reaches a marvelous level, an extremely small scale, at nanolevel. At this level, the nanoscale also covers the scale of wavelength of light. New science focuses and acts on this extremely nanolevel; hence, the optics also becomes a topic that directly deals with nanoscience. The scientists deal with the nanoscale optics and the new approach is under the new science, which is called *nanoscience*. This is a real challenge of optical study. Nanoscience is widely studied and becomes the hot topic. In fact, it nanoscience can be useful in understanding optics. The application of nanoscience for optics can be seen in many aspects. With the use of nanoscience, many new facts of optics can be discovered. Many new things can be detected and become the new knowledge. With the nanoscience application, the manipulation and understanding of the optics become a reality. As already mentioned, nanoscience deals with the study of nanoscale phenomenon that includes optics. The photon theory is the core theory that describes light as a small object. The property of small objects can be assigned to light. Then light will becomes a real object similar to other objects at the nanolevel. However, there are some interesting properties of optics as nanoscale objects to be mentioned.

Basically, when an object decreases in size, new properties of object can be expected. When the size changes, the object poses, plays, and presents new looks. The differences of properties are important, which are widely mentioned in nanoscience. The differences can be seen in several ways covering both new physical and chemical properties. The new properties of nano-objects result in the fact that the nanoscale objects are totally different from their corresponding large-scale companions. The change is according to the change in physical and chemical properties and this leads to many interesting nanoscience-based apparatuses. Being a small thing, the change in surface area leads to many other changes including the presentation of *electricity*. However, the light is a nano-object in sense of photon. Hence, it is no doubt that the electrical property of light or photon can be expected. It is no doubt that the advanced electrical tools and apparatuses can deal with photon. This becomes the new concept and facet in electrical engineering.

In fact, nano-object is not strictly to be a biological or a physical object. Hence, the nanoscience is not a pure but interdisciplinary theme bridging between physical and biological sciences. Nanoscience is an actual hybrid merging both physical and biological sciences. The answer to physical, biological, or hybrid questions can be based on the general concept of nanoscience. At present, nanoscience is worldwide studied and the questions on extremely small objects including light and photon, which are previously difficult to solve, can be and becomes simple thing to answer. Nanoscience can help to answer those difficult questions. Hence, nanoscience acts as a new alternative approach for solving the problem in optics.

In medicine, visualization at both ordinary and extraordinary small-scale things are all within the scope of optical science. This is an important topic to be discussed. Indeed, the talk of the electromagnetic wave including light is not a new thing but very old topic in medicine. The visible light is a basic wave that is required by everyone to fulfill the basic needs for visualization. The visualization has been generally studied and focused in medicine for very long time. There is a well-known medical science that deals with visible light. This medical science is called as *medical optics*. With improvement of medical and scientific knowledge, many new things in medicine have been continuously launched. The focus at present shows paradigm shifts, changing from simple visible light to other nonvisible electromagnetic waves. The use of medical microscope is the best example. It leads to the specific biomedical science, which is called *clinical microscopy*. This specific science has been introduced in the medical society and becomes the basic required knowledge for all physicians and medical personnel. For many years, this science has been included in the biomedical curriculum and becomes a core part in modern medical education. Also, the clinical microscopy is the main medical activity in any medical service unit. Several new microscopes are available in clinics and hospitals. Medical microscope is a tool that is important for *seeing* the small objects. Sometimes, this can be applicable to the nanoscale objects, such as bacteria and virus. To see means to visualize an object. To make the visualized objects a recorded evidence, a new technique, imaging, has to be used. These processes for visualizing objects are simple. Good examples are the processes of getting medical images from electronic electron microscopy and near-field scanning optical microscopy. It is no doubt that novel nanophotonics techniques can be applied in those microscopic techniques in clinical microscopy [2]. The use of spectrum analysis is an important process. Since spectrum is the basic component of light, it is necessary to deal with it if ones deal with optics. In clinical microscopy, dealing with spectrum is also the basic thing. Several new spectrometry tools have been continuously developed. Of those new tools, the good example is the specific microscope spectrometer. The microscope spectrometer is the specific microscope that can determine the ultraviolet (UV)–visible–near-infrared (NIR) spectrum of very small things. This kind of microscope is an actual useful tool that is widely used. Mainly, there are two kinds of microscope spectrometer: the basic microspectrometer, an integrative microscope built for specific microspectrometry work, and the special microspectrometer that is designed as a spectrometer unit, an additional attachment piece to a simple classical open photoport of an optical microscope.

To facilitate the nanoscience work, *in vivo*, *in vitro*, or *in silico* (computational) techniques can be applied [3]. Classically, the two common techniques in medicine are the technique that directly deals with living things (*in vivo*) or the technique that is totally done outside living things in laboratory (*in vitro*). Indeed, these two classical approaches require several things and result in a high cost. The newest concept is the use of computational modeling (*in silico*) approach. This is the basic fundamental concept in computational medicine. The *in silico* approach is the hot issue and a new useful technology. For nanoscience, the two classical approaches, *in vivo* and *in vitro* techniques, can still be used, but they still pose the problems. The basic limitations include high cost and time consuming as already mentioned. Hence, the use of the modern computational technique, *in silico* approach, is widely used in nanomedicine. Focusing on the new *in silico* approach, the scientists can clarify and manipulate many research problems in nanoscience, which are usually dealt with very small objects, including photon, that are hard to study by either simple *in vivo* or *in vitro* techniques.

Computational technology is the widely used concept. It can be used as an *in silico* approach in nanoscience including nanomedicine. The so-called computational nanomedicine is one of the newest biomedical sciences. It is no doubt that the *in silico* approach is not a real life. However, its effectiveness and reliability are confirmed. The newest *in silico* technique can additionally support the classical *in vivo* and *in vitro* techniques. Computational technology can significantly reduce time and cost. As already mentioned, giving salutation to difficult nanomedicine questions can be possible within a short time at low unit cost. Also, since it is a simulating world, there will be no problem for confounding factors or interferences. It can be said that the *in silico* approach is a practical approach that is free from any interference effects, which are common problems in *in vivo* or *in vitro* study. Focusing on the *in silico* approach, *omics* science is a good example that is worldwide used [4]. Several new omics sciences are already in use. The omics is an interesting scientific technique that has just been introduced for a few years with an extremely rapid expansion. Several new omics sciences are available; hence, it is no doubt that these new omics sciences become the rapid progress of science. The present era of omics is already declared [4]. Similar to any field of scientific study, nanomedicine can apply computational technology for usage. With the help of the available omics techniques, many things become simple in medicine. Solutions to many advanced problems in nanomedicine can be simply derived. *In silico* technique can be useful in giving fast answers to the nanoscale problems. Either clarification of the scenario or prediction of imaginary phenomenon in nanomedicine can be done [5,6]. The imaginary approach with *in silico* computational approach in nanomedicine by a specific technique called nanoinformatics is currently in use. A lot of information can be derived from nanoinformatics. This specific subbranch of nanomedicine is an actual modern advent that is the actual hope in medicine. In optics, the medical imaging as well as the optical biomedical spectroscopy has its more advent, with the applied computational approach, at the nanolevel. Novel techniques based on nanotechnology are also introduced for using as the nano-medical imaging and nano-biomedical spectroscopy, which are further discussed in detail in the next headings on nano-medical imaging and nano-biomedical spectroscopy.

11.2 Overview of Optical Biomedical Spectroscopy and Imaging and Its Application in Medicine

Electromagnetic wave can be visible in medicine. One might not be able to imagine for its appearance in medicine. For example, several rays have been mentioned in biomedical science for a long time. Several rays have been applied in medicine for a very long times.

For centuries, the rays have been known and applied for clinical use in biomedicine. The light interaction is an important topic that is widely studied in the medical study. To deal with light interaction, the specific knowledge on spectrometry should be mentioned. Nevertheless, the *spectrum* should be first understood. Spectrometry directly deals with the spectrum; therefore, one has to know the spectrum before dealing with spectrometry. In science, the term spectrum is a condition that presents continuum and shows no boundary. There is also no specific limitation for the spectrum. The basic property of the spectrum includes (1) a set of data (derived from directly measured or not), (2) having specificity, (3) collective as a set, (4) continuum, and (5) really existed. The spectrum is usually mentioned in optics when ones talk about and deals with the light. Indeed, the specific applied science optics directly deals with spectrum. For nanoscience, nano-optics is the specific subbranch that focuses its interests on the behavior of light or optics on the nanometer scale. Applied in medicine, nanophotonics is currently useful and applicable in many ways in biomedicine, including laboratory medicine, ophthalmology, and radiology.

Of several available optical tools, spectrometry is a basic widely used tool. This tool is used for measuring the spectrum. Spectrometry is sometimes called spectrography. Its main function is to measure radiation intensity. The function of wavelength is usually assessed and the final outcome is called spectrum. There are several apparatuses in spectrometry. Spectrometers, spectrophotometers, spectrographs, and spectral analyzers are good examples of those apparatuses, have been used for many years in science. In medicine, these tools have also been used for a long time. Dealing with optics in medicine, spectrometry measurement is the basic requirement. Focusing on spectroscopy, as noted, it is a scientific approach for studying spectrum. Simply, spectrometry is used for getting spectroscopic data that can represent the interaction. Briefly, spectroscopy is a specific assessment of the interaction between the object and the light (or wave). Classically, spectroscopy is a basic knowledge in optics. The old concept is usually on the study of visible light. The spectrum of visible light can be dispersed due to its wavelength with the use of the specific tool, such as a *prism*. However, nowadays, the concept of spectrum is not limited to visible light but extends to other waves as already mentioned. The study extends to nonvisible lights; hence, the scope of spectroscopy is extensively expanded. As already mentioned, the interaction is the main thing to be manipulated in spectrometry. Spectroscopic data are useful in representing the interaction and they are the basic outcome of spectroscopy usage. The result from spectroscopy study is usually written as a simple spectrum. As already mentioned, the characteristics that can be seen in any spectrum are (1) focus on interaction, (2) interrelationship of things and radiations, and (3) assessment using scientific tool. In general, when we talk about radiation, as already mentioned, the ray is within the very small scale, nanoscale. Its main characteristic can be reflected by its specific radiation's property, the wavelength. Generally, the wavelength is usually presented as a nanometer. Hence, it is not surprising that an actual nanoscience study covers optics [7]. The light or photon is a nanoscale thing based on its own characteristic. In addition, general objects can interact with radiation. These objects might be large or small. In case that the object is within the nanoscale, the role of nanoscience cannot be overlooked, as previously mentioned that nano-optics is an important part of nanoscience.

In science, the application of spectrometry can be seen in several ways. It is the basic analytical technique in science. Specifically, several kinds of spectrometry can be applied in several scientific works. The well-known examples include ion-mobility spectrometry [2,7,8], mass spectrometry [9–11], neutron triple-axis spectrometry [12–14], and optical spectrometry [15–17]. Ion-mobility spectrometry [2,7,8] is the specific spectrometry technique that deals with ion motility. It is based on separation and detection of small molecules. Hence, the scope includes nanomolecules, based on the movement of their ions. The use of a buffer gas-phase carrier is generally needed for analytical process in ion-mobility spectrometry. Mass spectrometry [9–11] is another widely used spectrometry method. It is a specific spectrometry technique focusing on mass analysis. Indeed, mass is the basic property and identification of any objects used for is an actual specific property. In addition to mass, several physical properties of substances can be referred to. Mass spectrometry is a specific basic technique that detects mass and its relationship to charge. The charge of the substance is another important specific property to be mentioned. As already mentioned, in nanoscience, when mass becomes very low, the charge will be altered. Hence, the application of mass spectrometry, measuring mass-to-charge ratio-based technique, can be useful in nanoscience. In medicine, mass spectrometry is currently used worldwide in biochemical laboratory. Good examples are the mass spectrometry tools used in pharmacological and toxicological laboratories. Neutron triple-axis spectrometry [12–14] deals with neutrons. Its process starts from activation and the neutron scattering will be further gathered and interpreted in analytical process. The first outcome is inelastic neutral scattering. Then it will be further transferred and manipulated into spectrometry data. A similar spectrometry technique to be mentioned is Rutherford backscattering spectrometry [18–21]. This is another spectrometry technique that measures scattering particles. Rutherford backscattering spectrometry measures backscattering of a high-energy ion beam during analytical process, which is different from neutron triple-axis spectrometry. It can also be useful in several works in science. Finally, optical spectrometry [15–17] is another well-known spectrometry tool. It measures distributing light across optical spectrum. Optical spectrometry covers a wide region ranging from infrared through visible light to ultraviolet. The measurement of this wide spectrum is the basic activity in the analytical process of optical spectrometry.

When spectrometry is applied for microscale work, it will be called microspectrometer. Its characteristics depend on its composition, strength, and configuration. It can be used for determining the spectrum of microscopic samples. The analytical process makes use of transmission, absorbance, reflectance, fluorescence, emission as well as polarization spectrometry. Some specific microspectrometers have increased properties that result in high-resolution digital imaging. The applied computational technology, with additional new software, can increase the ability in determining thin-film thickness and colorimetry. This tool can be applied as a new diagnostic tool in many fields of medicine such as laboratory medicine.

For application of spectroscopy in medicine, the first thing to be mentioned is its application in medical radiology. The radiology is the basic medical science that directly deals with radiation. This science first deals with X-rays, which are also a kind of photons. The photonics study can be applied. In biomedical science, X-ray spectroscopy is a basic tool that can also be applied in radiology [22]. In fact, this technology is accepted as a simple, accurate, and economical analytical method and can be applied for assessing many chemical compositions. First, it was applied in analysis in material sciences, and then it was further applied for usage in medicine. Crystallography study of medical molecule is the best example. It can be applied for studying new drugs, antibodies, biomolecules, hormones, or enzymes. In order to make the reader better understand, some interesting reports on the application of spectroscopy in medical radiology are given as follows:

- Torres-del-Pliego et al. [23] discussed on the use of microradiography, Fourier-transform infrared spectroscopy, or Raman microspectroscopy for measuring the bone quality. They concluded that "finite-element analysis is an image-based method that enables calculation of bone strength" [23].
- Chigerwe et al. [24] reported the use of microscopy, infrared spectroscopy, and X-ray diffractometry in the study of renal calculi.

- Ober et al. [25] reported the development of new protocol for (1)H-magnetic resonance (MR) spectroscopy of the brain at 3 T.
- Lerner et al. [26] reported the use of advanced MR imaging, including (1)H nuclear MR spectroscopy, diffusion-weighted imaging, and MR perfusion imaging for differentiation of neurocysticercosis lesions from metastatic diseases and pyogenic abscesses.
- Bejjani et al. [27] reported the use of *in vivo* single-voxel proton magnetic resonance spectroscopy [(1)H MR spectroscopy] to study pediatric autism spectrum disorder and found elevated glutamatergic compounds in pregenual anterior cingulate.
- Zynda et al. [28] reported on detected disparity between angiographic coronary lesion complexity and lipid core plaques assessed by near-infrared spectroscopy.
- Cobbold et al. [29] reported on assessment of inflammation and fibrosis in nonalcoholic fatty liver disease by MR spectroscopy techniques.
- Robbins et al. [30] reported the use of positron emission tomography (PET), single-photon emission tomography, magnetic resonance imaging (MRI), and MR spectroscopy for evaluating radiation-induced normal tissue injury.

Another important application of spectroscopy in medicine is in opthalmology. The spectrometry can be helpful in assessment of visualization. The good example is the technique that is used in assessing retinal problems. Functional maps based on chromophore spectrum derived from retinal imaging spectroscopy is widely used in medical ophthalmology [11,12]. Recently, the new snapshot spectral camera has been introduced. This is the new opthalmological tool that can be the noninvasive approach for generating retinal vessel oximetry maps [12]. For further referencing, some more interesting reports on the application of spectroscopy in medical opthalmology are given as follows:

- Borchman et al. [31] reported on using heteronuclear single-quantum correlation spectroscopy for confirmation of the presence of squalene in human eyelid lipid.
- Yu-Wai-Man et al. [32] reported on using proton MR spectroscopy to assess extraocular muscle atrophy and central nervous system involvement in chronic progressive external ophthalmoplegia.
- Kaźmierczak et al. [33] reported on the results of atomic absorption spectroscopy and detection for blood levels of zinc (Zn), iron (Fe), copper (Cu), and cadmium (Cd) in patients with optic neuritis.
- Haritoglou et al. reported on the use of atomic force microscopy in the spectrometry mode for evaluating lens stiffness [34].

In addition to optical biomedical spectroscopy, another important topic of medical optics is imaging. As already noted, imaging is a record of things that are visualized by a visualization tool. Image is created by light phenomenon. It generally resembles the picture that is visualized. In medicine, visualization occurs due to the function of receptor (ocular organ) and the brain. The occipital lobe of the cerebral cortex is the main part of the brain that plays a role in receiving visual signal that is transferred via optic nerve from the eye. The brain functions to interpret the signal and translate it into the percept picture. However, picture from visualization is not a record. It cannot be used as referencing to the other. The way to contract a reference is the generation of the image.

In medicine, imaging is very important because it is the media for communication among physicians. To diagnose and follow up the patients, the physician in-charge must use medical images. In medicine, there are several techniques based on several apparatuses for creating medical images or medial photography [35]. Medical imaging is a specific medical technique for generating images. However, in addition to the classical concept, medical imaging currently deals with the manipulation of the medical images as well. Generally, medical imaging deals with the images from the human body. The images can be macro or micro images. Also, the images shift from static to dynamic ones. This allows the observation and follow-up of both anatomy and physiology of the focused part. The data can be useful in pathology work. Annually, there are many medical images produced in medical clinics and hospitals around the world. It can be noted that medical imaging is one of the two important diagnostic tools in medicine (the other one is laboratory investigation). However, compare to laboratory investigation, imaging is considered to less invasive (although some can be invasive such as the imaging technique that requires insertion of catheter into vascular line or injection of contrast substance).

Medical imaging consists of different imaging modalities. There are many available clinical processes for imaging the human body, aiming at diagnostic and treatment purposes. In medicine, the unit that mainly deals with imaging is radiology. However, it is also done in other units such as medicine, surgery, ophthalmology, and obstetrics. There are many techniques in medicine that relate to medical images. Good examples include X-ray radiography, MRI, nuclear medicine, medical ultrasonography or ultrasound, and endoscopy. It can be seen that several disciplines involve in medical imaging. In a medical team, one can imagine the roles of radiologist, obstetricians, medical internists, endoscopist, cardiologist, and so on. In case of paramedical personnel, medical scientists, medical radiologists, and medical technologists should be mentioned. In case of nonmedical personnel, biomedical engineering, optic engineering, and IT personnel must not be forgotten. Therefore, the knowledge of medical imaging becomes an important requirement and has an important role in the improvement of public health around the world. Furthermore, the area of medical imaging is very complex. Collaborative work among different multidisciplinary personnel (medical doctors, medical physicists, biomedical engineers, medical technicians, etc.) as previously mentioned is needed. To improve health-care policy, the concern for the proper use of medical imaging technology is needed. It is also recommended to improve, upgrade, and increase the number of available medical imaging apparatus. As a result, the number of complex medical imaging procedures continuously increases considerably. An effective and good-quality imaging is needed. To reach the up-to-date technology, the integration of advanced computational technology is an important channel. Some interesting reports on the application of computational technology in medical imaging are given as follows:

- Castaneda et al. [36] reported the application of numerical methods to elasticity imaging.
- Urbanik and Chrzan [37] reported the application of computed tomography (CT) examination for forensic medicine.
- Malanchuk et al. [38] reported on the determination of regimes of functional loading in patients with traumatic mandibular fractures after osteosynthesis performance using modern methods of computer modeling.
- Okada et al. [39] reported on the band-limited double-step Fresnel diffraction and its application to computer-generated hologram.

- Chin et al. [40] reported on the ideal starting point and trajectory for C2 pedicle screw placement, a 3D CT analysis using perioperative measurements.
- Wang et al. [41] reported on the application of automatic tube current modulation on image quality and radiation dose at abdominal CT.
- Bai et al. [42] reported on improving the image quality in CT pulmonary angiography with dual-energy subtraction.
- Dai et al. [43] reported on the application value of multislice spiral CT for imaging determination of metastatic lymph nodes of gastric cancer.
- Wang et al. [44] reported on a sparse-projection CT reconstruction method for *in vivo* application of in-line phase-contrast imaging.

11.3 Computational Optical Biomedical Spectroscopy in Medicine

11.3.1 Usefulness of Computational Optical Biomedical Spectroscopy in Medicine

It is no doubt that the computational optical biomedical spectroscopy can be applicable in medicine. There are several advantages of this application. Talking about the specific manipulation technique of optics, a nanoscale electromagnetic wave, several computational techniques can be applied. This is a way to construct bridging between physical and biological sciences. This results in a very useful new complex science. Basically, visualization or ability to see is directly related to visible electromagnetic wave or visible light. Light is an important basic requirement. An object cannot be visible if it does not interrelate with light. Indeed, interrelation has a specific site of interaction and this can be implied as a specific space. The concept of finite-element method can be applied at this point. Here, the space can be gridded when one focuses. Additionally, any interaction has a time period. Hence, in one interrelationship, both space and time have to be dealt with. Theoretically, the time-dependent Maxwell's equations can be used, which are within the scope of finite-difference time–domain (FDTD) method. Space-based central difference estimations and time-based partial derivatives can be matched. With the use of advanced computational technology, the generated finite-difference equations can be solved by the available computational software and hardware.

Focusing on optics, electromagnetic wave must be manipulated into electrical and magnetic issues. All focused electrical components within the studied volume of space will be solved within the instant assigned time. For magnetic field vector components, the same process can be separately done. The simulation of electromagnetic wave interaction with objects including nano-objects, which are the light or optics, can be performed. As an electromagnetic wave, the light can transmit and reflect. Manipulation of information on photonic interaction with objects is an important step. Simply, this can be assessed by nanoscale spectrometry as already mentioned. With the use of nanoscale spectrometry, the light can be interpreted into its corresponding wavelength function. The computational technology can play its role at this stage. In medicine, good examples of computational optical biomedical spectroscopy can be seen in several medical areas, including clinical microscopy, ophthalmology, radiology, oncology, and laboratory medicine, which are further discussed in the next headings on each specific medical field.

11.3.2 A Summarization on Important Reports on Application of Computation Optical Biomedical Spectroscopy in Medicine

As already noted, the application of computational optical biomedical spectroscopy can be seen in several areas in medicine. First, the area of laboratory medicine should be mentioned. The widely used application is clinical microscopy, which is the specific branch of laboratory medicine that deals with optics. In fact, clinical microscopy has been introduced in the medical society for many years and becomes the core facet of the laboratory medicine. Indeed, clinical microscopy directly deals with optics; hence, the application of computational optical biomedical spectroscopy is possible. Several kinds of microscopes (both visible and non—visible wave based types) are now available as important diagnostic tools in clinical microscopy. The application of computational technology to those apparatuses can be seen in medicine.

As already mentioned, the new microscope spectrometer is available and the computational application is possible. In general, microspectrometer is used for measuring the spectrum of microscopic samples. The analysis is directly based on transmission, absorbance, reflectance, fluorescence, polarization, and emission spectrometry. With microspectrometer, high-resolution digital imaging can be derived. The applied computational technology helps better manipulate the microscope data. To help the reader better understand on this application, some interesting reports are given as follows:

- Miernik et al. [45] reported on the computational automated analysis of urinary stone composition using Raman spectroscopy.
- Hayasaka et al. [46] reported on the development of imaging mass spectrometry dataset extractor software, IMS convolution. They concluded that "IMS Convolution with ROIs could automatically extract the meaningful peaks from large-volume IMS datasets for inexperienced users as well as for researchers who have performed the analysis" [46].
- Rich and Wampler [47] reported on a flexible, computer-controlled video microscope capable of quantitative spatial, temporal, and spectral measurements. They concluded that "Flexible use of this system in these various applications is possible because it allows operation with illumination intensities over a dynamic range of 100 000:1" [47].
- Kucera et al. [48] reported on a new computer-controlled double-beam scanning microspectrophotometry for rapid microscopic image reconstructions. They concluded that this technique can be "used for the study of naturally fluorescent intracellular components in living tissue culture" [48].
- Arnold [49] reported on the application of Fourier transformation infrared spectrometry as a method of detection in forensic chemistry and criminal investigation.
- Frank et al. [50] reported on a new white light confocal microscope for spectrally resolved multidimensional imaging.
- Skala et al. [51] reported on a combined hyperspectral and spectral domain optical coherence tomography microscope for noninvasive hemodynamic imaging. They showed an example of using this technique to microscopically study on hemoglobin molecule.

Based on the given examples, it is no doubt that computational technology can be well applied in clinical microscopy. It can be used for manipulation of data, confirmation of microscopic experiment, and improvement of resulted microscopic spectrum.

Focusing on radiology, it is another medical science that deals with optics; hence, the application of computer can be expected. Since X-ray is a kind of electromagnetic wave, posing spectrum, the role of spectroscopy can be seen. Also, the manipulation using the computational technique is possible. The applied computational technology helps better manipulate the investigation data. To help the reader better understand on this application, some interesting reports are given as follows:

- Zimmy et al. [52] reported on the *in vivo* evaluation of brain damage in the course of systemic lupus erythematosus using MR spectroscopy, perfusion-weighted, and diffusion-tensor imaging.
- Runge et al. [53] reported on measuring liver triglyceride content by new noninvasive MR method and proposed this technique as an alternative to histopathology.
- Maldonado et al. [54] reported on using noninvasive characterization of the histopathologic features of pulmonary nodules of the lung adenocarcinoma spectrum using computer-aided nodule assessment and risk yield.
- Jeong et al. [55] reported on computerized analysis of osteoporotic bone patterns using texture parameters characterizing the bone architecture.

11.4 Computational Imaging in Medicine

11.4.1 Usefulness of Computational Imaging in Medicine

Medical imaging becomes an important activity that cannot be lacked in any hospital. There are several kinds of medical imaging. Based on the different spectra of rays, several techniques of medical imaging are introduced. At present, there are many available medical imaging instrumentation and techniques. Good examples are X-ray imaging, utrasonography imaging, and MRI.

11.4.1.1 X-Ray Imaging

X-ray imaging is the most classical technique in medical radiology [56–59]. This practice has been implemented in medicine for many years. The main aim of this technique is the diagnosis. The principle of X-ray imaging is the getting the picture generated from X-ray penetration through body. Different densities of air, solid, and liquid within the body will result in different final appearances in X-ray image. Nowadays, there are many X-ray apparatuses in medicine. The commonly used instruments include medical imaging, projection radiography, and fluoroscopy. The first phase of X-ray imaging is the two-dimensional (2D) technique, and the additional new technique, advanced three-dimensional (3D) technique, is also available at present. For any technique, the basic process starts from X-ray beam generation and then pointing it to the focused site and letting it penetrating through that area.

An image receptor is further required to receive the penetrating beam and converts it into an image. Its main composition includes an image amplifier and a large vacuum tube coated with cesium iodide on one end and covered by mirror or camera on the other end.

Focusing on projectional X-ray tool, it is the classical technique that is widely used in almost all hospitals. Its basic purpose is usually to diagnose bone and lung disorders. Sometimes, the modification of technique using contrast media can be seen. The good example is the use of contrast media projectional X-ray investigation for determining the gastrointestinal problems. For the fluoroscopy, it is a more advanced technique. It is a mobile X-ray. It can be used at the studied site and facilitate the diagnosis. The technique also uses a X-ray beam but at a lower dose rate compared to the projectional X-ray technique. Real-time results can be observed using this technique.

With the advanced computational technology, simple X-rays can be modulated by the computer. The best example is the CT apparatus, which becomes one of the most useful medical imaging instruments in the present day. The CT technique is the continuum of the classical X-ray technique, linear tomography. Conceptually, to get a tomogram, the X-ray tube has to move from one to another point on the patient and the cassette holder, which contains image receptor, simultaneously moves in the counter-direction. To construct the tomogram, the fulcrum pivot point has to be assigned to the studied area. Considering this rotation, the focused area will clearly appear in the final medical image with blurred background. Before the final development of CT apparatus, there are also other techniques of tomography. The details of these techniques are provided in Table 11.1 [60].

11.4.1.2 Ultrasonography Imaging

Ultrasonography imaging is another widely used imaging technique in medicine at present [61–64]. This is a radiation-free technique. The sound wave is the main wave used for construction of image. The main principle is to use high-frequency sound waves (megahertz) to penetrate into the body, which results in reflection. Then detection of reflected sound waves by tissue will be done. The detector collects the reflected waves from various degrees and uses them for further production of medical images. Similar to X-ray imaging, the classical ultrasonography imaging is 2D imaging. However, at present, the new techniques allow 3D and four-dimensional (4D) image construction (Table 11.2).

11.4.1.3 Magnetic Resonance Imaging

MRI is another advanced medical imaging technology [65–67]. Compared to the CT technique, MRI technique is newer. There are many MRI instruments at present. Indeed, there

TABLE 11.1

Tomography Techniques in Medical Imaging

Techniques	Brief Details
Linear tomography	It is a classical original technique and is already out of date.
Polytomography	It is a tomography with a number of complex geometrical movements (hypocycloidic, circular, and elliptical). It is used for studying the inner ear.
Zonography	It is a specific tomography for visualizing the kidney during an intravenous urogram.
Orthopantomography	It is a specific tomography for imaging mandible.
CT	It is a computational application and is widely used in medical practice at present. Compared to projectional X-ray technique, it has a significant greater ionizing radiation.
PET	It is an advanced CT technique that makes use of nuclear medicine application. It uses coincidence detection to image functional processes via determination of positron emission (from short-lived positron emitting isotope).

Source: Pfeiler, M. and Steiner, K., *Morphol Med.*, 3, 117–124, 1983.

TABLE 11.2

Ultrasonography Techniques in Medical Imaging

Techniques	Brief Details
2D sonography	It is a classical technique. Its principles are collection of reflected sound waves from studied tissue and formatting them into the final images.
3D sonography	It allows more dimensional view of the studied tissue due to adjustment of the detection to cover the range from more directions.
4D sonography	It is a continuous 3D sonography that monitors dynamicity of the focused tissue. It is widely used in obstetrics at present.
Elastography	It is a specific application of ultrasonography. It is used for determining the elastic properties of soft tissue.
Doppler ultrasonography	It is a specific application of ultrasonography. It is used for determining the flow such as blood flow.
Echocardiography	It is a specific application of ultrasonography. It is used for studying the heart.

Source: Donald, I., *J. Clin. Ultrasound.*, 4, 323–328, 1976; Gravelle, I.H., *Trans. Med. Soc. Lond.*, 85, 161–166, 1969; Ross, F.G., *Recenti. Prog. Med.*, 62, 262–294, 1977; Berkhout, A.J., *Med. Prog. Technol.*, 11, 197–207, 1986.

are many new applied instruments that are very useful in medicine. Focusing on its principle, MRI does not make use of X-rays. First, the MRI scanner makes use of magnets to polarize and excite the hydrogen nuclei or single proton in water molecules within the human tissue. The MRI machine can emit a pulsatile radio frequency that specifically attacks hydrogen within the studied area of the body. Then the stimulation will result in a detectable signal that is spatially encoded. This signal is used for further construction of image.

Focusing on the signal generation process, when the magnetic impulse hits the photon in the focussed area, the absorption of energy occurs, which can further induce spinning in a different direction and there will be a well-known phenomenon, resonance. The specific frequency of resonance is known as Larmour frequency, which can be calculated based on the particular tissue that is studied and the strength of the magnetic field from the MRI tool. Focusing on the magnetic field, there are three common fields: (1) a very strong or static magnetic, which is used to polarizing the hydrogen nuclei in the tissue; (2) a gradient field (intermediate field); and (3) a weak field, which is used to stimulating the hydrogen nuclei to produce measurable signals. At present, MRI tools are widely used. The main purpose is to examine the brain, the spinal cord, and the internal organs.

11.4.2 A Summary of the Important Reports on Application of Computation Imaging in Medicine

As already noted, the computational technology can be applicable to support and fulfill the medical imaging technology. Nowadays, the computer becomes an important integral part among many medical image instruments. The details will be further discussed. First, for X-ray imaging, it is no doubt that computational technology can be applied to instrumentation. The best example is the construction of the CT scan. The interesting reports on computational technology application in X-ray imaging are given as follows:

- Yamamura et al. [68] reported the use of dynamic CT of locally advanced pancreatic cancer and further mentioned the effect of low tube voltage and a hybrid iterative reconstruction algorithm on image quality.

- Thing et al. [69] reported on patient-specific scatter correction in clinical cone beam CT imaging made possible by the combination of Monte Carlo simulations and a ray tracing algorithm.
- Yu et al. [70] reported on the low tube voltage intermediate tube current liver MDCT and discussed for sinogram-affirmed iterative reconstruction algorithm for detection of hypervascular hepatocellular carcinoma.
- Ay and Zaidi [71] reported on the development and validation of MCNP4C-based Monte Carlo simulator for fan and cone beam X-ray CT. In this work, they used MATLAB® 6.5.1 to create the scanner geometry at different views as MCNP4C's input file.
- Apostolou et al. [72] reported on advanced image fusion algorithms for Gamma Knife treatment planning. They. evaluated these algorithms for MATLAB platform and head images.
- Otton et al. [73] reported on the 4D image processing of myocardial CT perfusion for improved image quality and noise reduction. In this work, they measured the distribution of local Hounsfield values in both time and space with the use of a customized program within the MATLAB software.
- Yu et al. [74] reported on MATLAB-based simulation tools for two-dimensional experiments in X-ray CT using the FORBILD head phantom.

For ultrasound imaging, the computational technology can be applied to the basic ultrasound instrumentation. The computational algorithm and MATLAB can also be applicable . The interesting reports on computational technology application in ultrasonography imaging are given as follows:

- Molinari et al. [75] reported on a new MATLAB toolbox for the simulation and reconstruction of photoacoustic wave fields. They concluded that their tool could make realistic photoacoustic modeling simple and fast.
- Wang et al. [76] reported their success on fusion of color Doppler and MR images of the heart. In this work, MATLAB programming was used for fusion process.
- Jassar et al. [77] reported on quantitative mitral valve modeling using real-time 3D echocardiography. In this work, custom MATLAB algorithms were used to generate fully quantitative 3D mitral valve models of the annular and leaflet point cloud.
- McFarlin et al. [78] reported on ultrasonic attenuation estimation of the pregnant cervix. In this report, they used MATLAB for digital image transforming into radio-frequency pattern.
- Uhercík et al. [79] reported on model fitting using random sample consensus (RANSAC) for surgical tool localization in 3D ultrasound images.

Similarly, the computational technology can be applied to the MRI instrumentation. The use of computational algorithm and MATLAB are also applicable. The interesting reports on computational technology application in MRI imaging are given as follows:

- Tana et al. [80] reported on the new MATLAB toolbox for spectral Granger causality analysis of functional MRI (fMRI) data.
- Tian and Liu [81] reported on depth-compensated diffuse optical tomography enhanced by general linear model analysis and an anatomical atlas of human head.

- Wilson et al. [82] reported on a new platform for multimodal monitoring, data collection, and research in neurocritical care. Their proposed platform is "optimized for real-time analysis of multimodal data using advanced time and frequency domain analyses and is extensible for research development using a combination of C++, MATLAB, and Python languages" [82].
- Clas et al. [83] reported on a semiautomatic algorithm for determining the demyelination load in metachromatic leukodystrophy. In this report, an algorithm called Clusterize was developed and implemented in MATLAB for semiautomatic segmentation.
- Zhang et al. [84] reported on a new toolbox implementing a Bayesian spatial model for brain activation and connectivity.

11.4.3 Important Computational Imaging Tools for Biomedical Work

There are several computational imaging tools for management of biomedical data. Some important computation tools are given as follows:

1. FDTD Solutions Knowledge Base: It is a specific database focusing on the information of the FDTD method that can be well applied for studying optics and imaging.
2. COSBID-M3: It was developed by Wilson et al. [82]. C++, MATLAB, and Python languages were used in development of COSBID-M3. COSBID-M3 is a platform for multimodal monitoring, covering data collection and integration. It can be applied for neuromonitoring in patients with severe brain trauma and stroke [82].
3. Granger multivariate autoregressive connectivity: It is a computational toolbox that was developed by Tana et al. [80]. It is a toolbox for spectral Granger causality analysis of fMRI data [80]. The available features of GMAX include "fMRI data importing/exporting, network nodes definition, time series preprocessing, multivariate autoregressive modeling, spectral Granger causality indexes estimation, statistical significance assessment using surrogate data, network analysis and visualization of connectivity results" [80].
4. BSMac: It is another toolbox using a flexible Bayesian modeling framework [84]. It is used in neurology. BSMac helps perform parameter estimation based on Markov chain Monte Carlo methods and generates plots for activation. It can give interactive 2D maps of voxel and region-level task-related changes in neural activity and animated 3D graphics of the results. The toolbox can be derived from http://www.sph.emory.edu/bios/CBIS/ [84].
5. RANSAC: It is designed for help perform good surgery by fitting the position [79]. It starts with "thresholding and model fitting using random sample consensus for robust localization of the axis," then it will continue "subsequent local optimization refines its position" [79].
6. PANDA: It is a toolbox for analyzing brain diffusion images [85]. Its package includes FMRIB Software Library, Pipeline System for Octave and MATLAB, Diffusion Toolkit and MRIcron [85]. It is available online at http://www.nitrc.org/projects/panda/ [85].
7. iBEAT: It is a toolbox designed for infant brain MRI processing [86]. It helps several analyses including image preprocessing, brain extraction, tissue segmentation,

and brain labeling. It presents as a downloadable Linux-based package available at http://www.nitrc.org/projects/ibeat [86].

8. FocusDET: It is a new toolbox for SPECT Co-registered to MRI analysis [87]. It can help localize the epileptogenic focus in patients with intractable partial epilepsy. filtered backprojection algorithm and an ordered subsets expectation maximization reconstruction method are mainly used in reconstruction process [87].
9. SimTB: It is a simulation toolbox for fMRI data under a model of spatiotemporal separability [88]. It is accessible at http://mialab.mrn.org/software.
10. elastix: It is a toolbox for intensity-based medical image registration [89]. It is a "publicly available computer program for intensity-based medical image registration" [89].
11. PyMVPA: It is a python-based toolbox for multivariate pattern analysis of fMRI data [90].
12. BAX: It is a toolbox for the dynamic analysis of fMRI datasets that are currently widely used in neuroinformatics [91].
13. Template-O-Matic: It is a toolbox for customized pediatric templates [92].
14. ASLtbx: ASLtbx is a toolbox for arterial spin labeling (ASL) data processing [93].
15. LI-tool: It is a toolbox for assessment of lateralization in fMRI data [94].

11.5 Common Applications of Computational Medical Imaging in General Medical Practice

As earlier noted, computational applications can be used in medical imaging work. Some common applications of computational medical imaging in medicine are as follows:

1. Application for clarification: A good example of computational application for clarification is the usage of modeling related to medical imaging. Image processing and image reconstruction are the widely performed activities. Several published reports are already previously discussed in this chapter.
2. Application for prediction: A good example of application of computational technology in this field is the simulation of medical images. Prediction might be done with the help of computational FDTD simulation.

   In sum, it can be seen that the computational method can be applicable for both diagnosis and therapy in medicine.

   a. Diagnosis: The computational domain method in diagnostic medical imaging is the main work at present. The aim is usually to answer the questions on optics as previously discussed. Also, it can be applied to the routine medical imaging work. It can help make and adjust the images. The construction of 3D and 4D medical imaging instruments can be good examples.
   b. Treatment: The use of computational technique in using medical images for treatment can be expected. It is not a direct activity but it is a related topic. A good example is the use of cancer therapy in clinical oncology. Manipulation of medical images and reconstruction can be useful in planning for radiotherapy and surgical management.

To help computational manipulation on medical image from medical imagine instrument, there are several available programs that can be selected. However, MATLAB is a good basic program that is widely used in many publications. The ways how MATLAB can be used are simply discussed as follows:

1. Creating a graphical model: MATLAB can be used for creating a graphical model. For example, this is a case of using MATLAB for creating a model of X-ray coverage at different angles and radii from a focused studied pulmonary vasculature. The writing of MATLAB code in the MABLAB Command window is shown as follows:
```
>>theta = 0:0.3:5*pi;
>>rho = theta/(5*pi);
>>polar(theta, rho);
```
2. Solving the problem of functional equation: Considering the nature of rays that are used by medical imaging instrument, the quantity of exposure is very important in determining the quality of a final medical image. Also, the quality of exposure is also important for specifying the risk of the patients. The dosage is usually varied depending on the distance from the beam or ray focus. Here, there is a case of using MATLAB for solving the function of quantity of exposed radiation in the studied soft tissue during a tomography study. An example of the specific code is shown as follows:
```
>>x = [0: 10: 500];
>>y = [34 22 14 6 2];
>>stairs(x,y);
>>axis([0 500 2 40]);
>>xlabel('Distance (Millimeter)');
>>ylabel('Dosage (MicroRad)');
```

11.5.1 Examples of Medical Researches Based on Computational Medical Imaging Application

The computational application in medical imaging in general medicine is a very interesting topic. There are many new applied techniques that help better create good medical images, which lead to the advent in medical instrumentation technology.

Example 11.1: Data Mining to Find the Features of Medical Imaging Appearance in H1N1 Swine Flu

Basically, the new emerging influenza is the important concern in public health. In the past few years, there are several new emerging influenza infections including H5N1 bird flu, H7N9 bird flu, H6N1 bird flu, and H1N1 swine flu. Of these new cross-species infection, the H1N1 swine seems to be the one that has a worldwide effect [95–98]. This infection results in respiratory tract problems. It shows several respiratory signs and symptoms (as well as other nonrespiratory signs and symptoms [99]. The standard oseltamivir antiviral drug is used in the management of this infection [100]. As a kind of influenza, this infection usually presents with acute high fever accompanied with cough. The more severe infection can result in lower respiratory tract involvement and the outcome is the serious lung infection called pneumonia. It is the need to concern that the rapid progression of pneumonia can be expected and this can lead to the most serious outcome, the acute respiratory distress syndrome and respiratory failure. The affected cases can end up with death. It is the international policy to get the infection control implementation for prevention of the spreading of H1N1 swine flu [101].

A challenge is how to facilitate early diagnosis and treatment of the H1N1 swine flu. The concept of preventive medicine is to provide early diagnosis and prompt treatment. This will result in confining of the infection and successful management of the illness. To diagnose the H1N1 swine flu, the physician needs to first aware of the possibility of infection. The use of investigation to support the first hypothesis that the patient might have the H1N1 swine flu is required. Of several tools, both laboratory testing and X-ray imaging investigation have to be used. Certainly, the interpretation of the X-ray imaging is sometimes a very difficult thing, which needs experience. However, due to the threatening and widening of the infection, there must be a tool to help the general practitioners for interpreting the X-ray imaging in their general practice. The consultation system for the expert might be limited and the computational tool might be a useful tool.

In fact, a heap of information of H1N1 swine flu has influxed into the medical society since its first outbreak. The system to collect the data is needed. Luckily, with the idea of the National Institutes of Health, the database for flu correspondence is developed. Indeed, the most simplified database in health as PubMed (http://www.pubmed.com) can be well used in data mining for the information on flu. The PubMed is the center for collection of medical publications in the area of biomedicine and it has a simple search engine to facilitate the data mining. Here, the author shows an example of data searching using PubMed [102]. The input keyword is "H1N1 swine flu" and the result shows 1762 related publications (see the result in Figure 11.1). Also, within the package of PubMedCentral Image, there is also a specific function to perform specific data mining with a special focus on medical imaging. This is called *PMC image*." For example, the same keyword "H1N1 swine flu" is used and the searching is done; then the computational database tool gives the final outcome of only medical images. There are 102 related medical images that cover the X-ray images and other related medical images (see the result in Figure 11.1).

Using the mentioned data mining tool, the general practitioner can learn the new updated knowledge of the imaging of the H1N1 swine flu and further apply it in clinical practice. Based on the literature search, it can be shown that the medical image of lungs in the H1N1-infected cases can be shown in many forms [103–107]. It is needed to think about the H1N1 problem in any patient presenting with acute febrile respiratory illness and showing abnormal chest X-ray images.

Example 11.2: Finding Insertional Loss of Ultrasonography Energy due to Amniotic Absorption

Basically, ultrasonography is considered a safe medical imaging technology. The ultrasonography instrument is a radiation-free tool. It generates no radiation but sound waves. Hence, it is considered safe and is widely used in obstetrics. In obstetrics, the radiation is a contraindication since the radiation is strongly related to the abnormality of the fetus *in utero* (if fetus exposes to radiation, it is likely to develop mutation or congenital anomaly). In obstetrics, the main medical imaging technique that is worldwide used for investigation is the ultrasonography [108–116]. It is useful in assessing the fetus *in utero*.

To perform the test, the ultrasonography has to be applied to the abdomen of the pregnant subject. The ultrasonography probe will contact with the abdominal wall and the ultrasound wave will transmit through the abdominal wall and enter into the uterus. In the uterus, the important component inside is the fetus and the amniotic fluid. The focused investigation is the fetus *in utero*. However, it is no doubt that the amniotic fluid will interfere the ultrasonography process. The important concern is

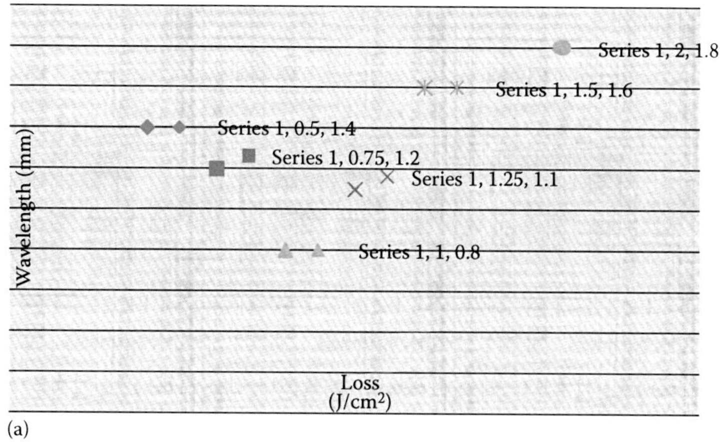

(a)

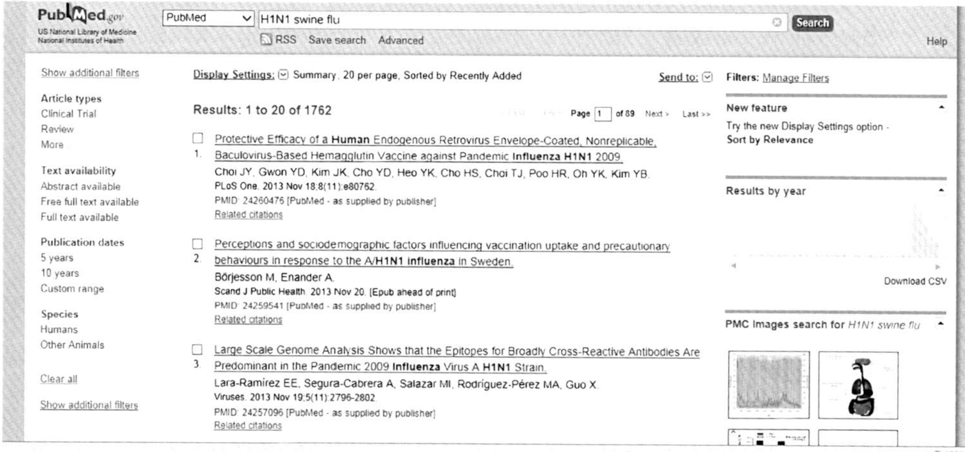

(b)

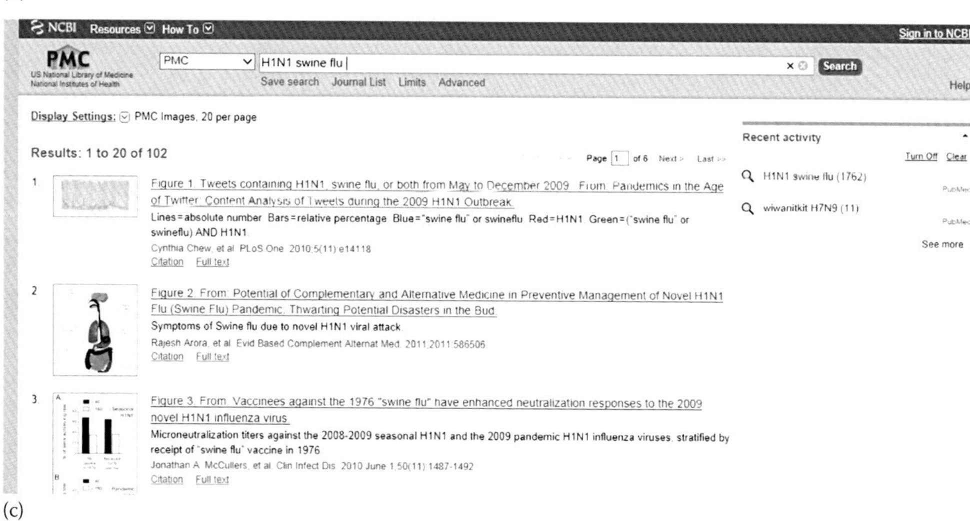

(c)

FIGURE 11.1
Result from data mining (a) using general literature searching (b) and Medical image searching (c) on the PubMED database with the specific input "H1N1 swine flu."

TABLE 11.3

Ultrasonography Wavelength and Insertional Loss

Wavelength (mm)	Loss (J/cm^2)
0.5	1.4
0.75	1.2
1	0.8
1.25	1.1
1.50	1.6
2.0	1.8

on the insertional loss [117,118]. Here, the example of finding the insertional loss of ultrasonography energy due to the amniotic fluid absorption with use of computational technology. The measurement of loss is an important process that can determine the quality of ultrasonography imaging. Basically, loss is usually measured at the focused site to be investigated, the intra-utero area. The loss is generally dependent on the wavelength and frequency (time function). Therefore, the basic algorithm can be presented as follows: concentration = $k \times$ wavelength $\times$ frequency. This is an actual phenomenon dealing with wave and time functions. Hence, the computational FDTD method can be applied. An example of a relationship between change and finalized loss of ultrasonography waves is presented in Table 11.3. In this model, the variation of wavelength is the experimental parameter, the observed loss is the observed parameter, and time is the fixed controlled parameter. It can be seen that the least loss could be observed at the wavelength around 1 mm. This is the wavelength that is widely used for investigation in obstetrics.

Example 11.3: 3D Reconstruction of CT Scan Image of Lung Cancer

Lung cancer is a common malignancy that can be seen in any country around the world [119–123]. This cancer is one of the leading causes of death in the world population. Smoking is mentioned as a risk factor for lung cancer [119–123]. This disease is a deadly disorder and the affected patients usually die within 6 months of the first diagnosis [119–123]. The diagnosis of this cancer is usually difficult. The use of X-ray imaging might help identify the mass within the lung. However, the better investigation is by the use of CT scan instrument. The CT scan can help better find the mass within the lung and can show the interrelationship to other organ [119–123].

In clinical practice, the CT lung is useful in diagnosis and monitoring of lung cancer. Generally, the CT lung is usually performed in any suspected case to give the diagnosis of lung cancer. The result is usually a 2D image. Based on the present advanced computational technology, the reconstruction of the image into a 3D image is possible. Here, the example of such reconstruction is shown in Figure 11.2 using the 3D Anaglyph stereo image reconstruction. It can be seen that the reconstructed image can help the physicians who view the image imagine on the exact volume of the tumor and its relationship to the nearby structure.

11.5.2 Examples of Nanomedicine Researches Based on Computational Nanomedical

11.5.2.1 Imaging Application

As a wave, the wavelength of light is usually present at nanoscale. Hence, any phenomenon related to light and other electromagnetic waves can be explained by nanoscience. The medical imaging can also be explained in nanoscience. It can be said that the medical

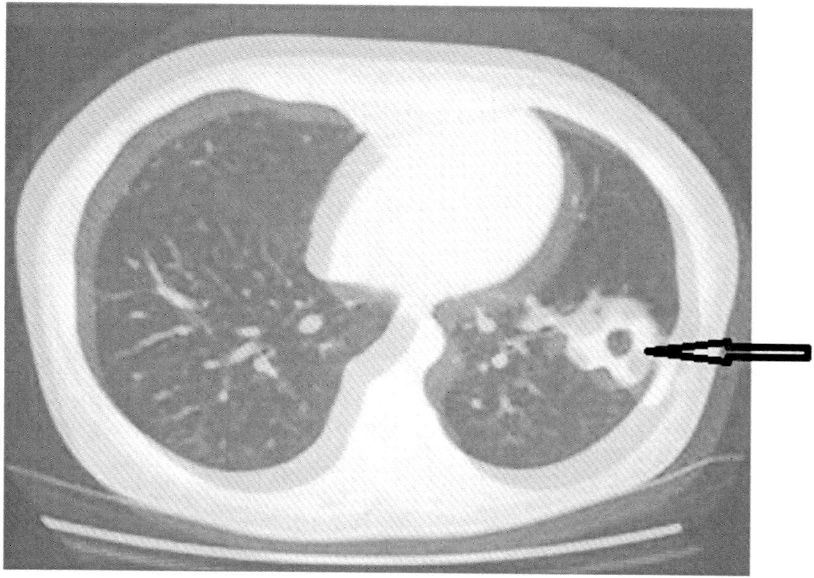

FIGURE 11.2
(See color insert.) Simple reconstruction of 3D image from simple reconstruction from 2D CT lung. The lung cancer can be seen in the figure (black arrow).

imaging at present has its more advent at the nanolevel. Nevertheless, due to extremely small scale, naked eye assessment seems to be not possible to study the medical imaging at the nanoscale. Hence, both *in vivo* and *in vitro* studies are very difficult.

Therefore, in the past, the nanomedical imaging has never been mentioned. However, due to the introduction of the new science, nanoscience, the nanomedical imaging becomes the new topic in medicine. Several tools including the nanophotonics and nanoplasmonics tools are available to study nanomedical imaging. Based on these new tools, the solution of complicated nanomedical imaging work can be probable. To solve the problem, the computational approach plays a very significant role. The *in silico* technique is widely used and the specific informatics technology, nanoinformatics, can be applied for the nanomedical imaging aspect.

Example 11.4: Prediction and Modelling of 3D Structure of Abnormal Coagulation Factor Molecule

Coagulation factor is an important biomolecule seen in the human body. It is a protein that can be synthesized by the liver. This protein is expressed in a tissue-specific manner. The corresponding gene structure and regulatory elements of the coagulation factor have been widely analyzed in detail [124]. There are many studies on the structure and function of the coagulation factor in thrombohemostaseology [124]. Generally, the biological functions of the coagulation factor and its activated form, thrombin, are widely discussed in medicine [124]. Coagulation factor deficiencies are a group of thrombohemostatic disorders that can be seen and can result in both acquired and congenital disorders. Congenital coagulation factor deficiency is a rare bleeding disorder that is genetically considered as an autosomal recessive trait disorder [125]. Some cases are lethal and very serious [125]. Several coagulation factor deficiency disorders can be seen at present.

To analyze the protein structure of the coagulation factor is difficult and needs the complex medical imaging instrumentation. A special microscopy with application of X-ray diffraction technology is required. However, as already noted, this complex tool is limited available. Also, the analysis requires high cost. The application of computational technology can be useful at this point. With the advanced bioinformatics technology, the structural prediction of the abnormal prothrombin can be easily done by the standard structural proteomics technique. Here, an example of using computational bioinformatics evaluation on a coagulation factor disorder is shown.

The example is the case of factor IXHilo disorder. For factor IXHilo disorder, Huang et al. [126] first used a genomic DNA library and an enzymatic DNA amplification technique to discover the factor IX coding sequences and further clarify the basic pathogenesis of this disorder as a substitution mutation. The primary structural disorder of this disorder has a new mutant which is "Glu180 to Arg" [126]. Monroe et al. [127] noted that that the mutation in factor IX Hilo altered the biomolecule resulting in inability to be activated by factor Xla. Further, they suggested that "the mutation resulted in a molecule that interacts with components of the extrinsic pathway to give a prolonged ox brain prothrombin time" [127]. However, no further analysis of the secondary structure of this disorder was successfully done and reported. In this example, which has been previously published in a standard textbook by Wiwanitkit [128], the author generated a predicted medical image of factor IXHilo disorder. Using the NNPREDICT server, the calculation of the secondary structure predictions of factor IX in normal and factor IXHilo disorder was performed (Figure 11.3). According to this study, the secondary structures of factor IX in both normal and factor IXHilo disorder are predicted and presented. From this study, the secondary structure of factor IX is significantly affected by the mutation; a significant defect as an additional helical residual could be observed [128]. This structural aberration might be an important clue for the pathogenesis of this disorder. The predicted image in this study can be a good explanation for mild clinical presentation of factor IXHilo disorder [128].

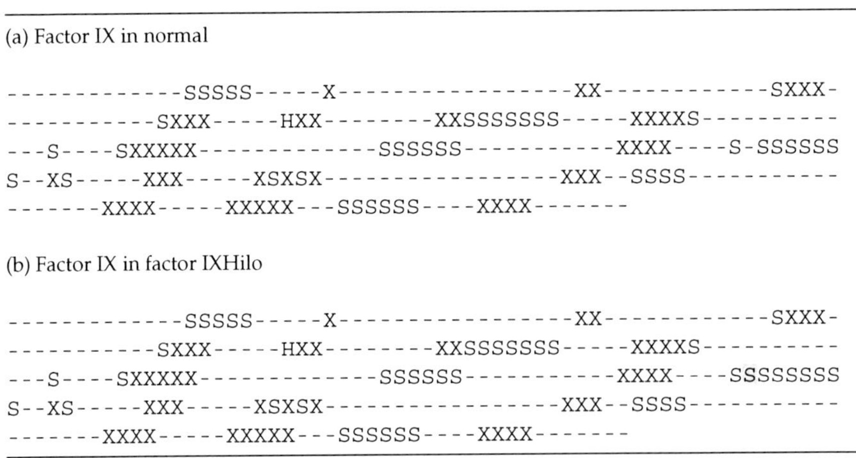

(a) Factor IX in normal

```
-------------SSSSS-----X-----------------XX------------SXXX-
-----------SXXX-----HXX--------XXSSSSSSS-----XXXXS----------
---S----SXXXXX-------------SSSSSS-----------XXXX----S-SSSSSS
S--XS-----XXX-----XSXSX-----------------XXX--SSSS-----------
-------XXXX-----XXXXX---SSSSSS----XXXX-------
```

(b) Factor IX in factor IXHilo

```
-------------SSSSS-----X-----------------XX------------SXXX-
-----------SXXX-----HXX--------XXSSSSSSS-----XXXXS----------
---S----SXXXXX-------------SSSSSS-----------XXXX----SSSSSSSS
S--XS-----XXX-----XSXSX-----------------XXX--SSSS-----------
-------XXXX-----XXXXX---SSSSSS----XXXX-------
```

FIGURE 11.3
Calculated secondary structures of factor IX in normal and factor IXHilo disorder. Additional helix is indicated in highlighting. S indicates helix; X indicates strand; – indicates no prediction. (Adapted from Van Rossum, R., *Brux Med.*, 34, 2301–2319, 1954.)

11.6 Conclusion

Computational technology can be widely applied in several sciences including medicine. Using the computational approach, difficult medical problems can be simple solved. Computational technology application in medicine provides usefulness in this specific filed. Using the computational approach, several new medical apparatuses have been produced and launched for usage in medical practice. Of several computational approaches in medicine, the usage of computational technique to support the work on medical imaging is very interesting. The application of computational technology can serve this purpose. Good examples of roles of computational application include X-ray imaging, utrasonography imaging, MRI, and so on. The medical imaging now turns into a more advanced phase and shifts to the nanolevel. Novel techniques based on nanotechnology are also introduced for using as the nano-medical imaging, which is the new hope in investigative medicine at present.

References

1. The Optical Society. About the Optical Society. Available online at http://www.osa.org/en-us/about_osa/ (Last accessed on October 26, 2014).
2. Colliex C. From electron energy-loss spectroscopy to multi-dimensional and multi-signal electron microscopy. *J Electron Microsc (Tokyo)*. 2011;60(Suppl 1):S161–S171.
3. Gehlenborg N, O'Donoghue SI, Baliga NS, Goesmann A, Hibbs MA, Kitano H, Kohlbacher O et al. Visualization of omics data for systems biology. *Nat Methods*. 2010;7(Suppl 3):S56–S68.
4. Haarala R, Porkka K. The odd omes and omics. *Duodecim*. 2002;118(11):1193–1195.
5. Haddish-Berhane N, Rickus JL, Haghighi K. The role of multiscale computational approaches for rational design of conventional and nanoparticle oral drug delivery systems. *Int J Nanomedicine*. 2007;2(3):315–331.
6. Saliner AG, Poater A, Worth AP. Toward in silico approaches for investigating the activity of nanoparticles in therapeutic development. *IDrugs*. 2008;11(10):728–732.
7. Behari J. Principles of nanoscience: An overview. *Indian J Exp Biol*. 2010;48(10):1008–1019.
8. Meckenstock R. Invited review article: Microwave spectroscopy based on scanning thermal microscopy: Resolution in the nanometer range. *Rev Sci Instrum*. 2008;79(4):041101.
9. Wu J, Gu M. Microfluidic sensing: State of the art fabrication and detection techniques. *J Biomed Opt*. 2011;16(8):080901.
10. Cheung K, Gawad S, Renaud P. Impedance spectroscopy flow cytometry: On-chip label-free cell differentiation. *Cytometry A*. 2005;65(2):124–132.
11. Johnson WR, Wilson DW, Fink W, Humayun M, Bearman G. Snapshot hyperspectral imaging in ophthalmology. *J Biomed Opt*. 2007;12(1):014036.
12. Mordant DJ, Al-Abboud I, Muyo G, Gorman A, Sallam A, Ritchie P, Harvey AR, McNaught AI. Spectral imaging of the retina. *Eye (Lond)*. 2011;25(3):309–320.
13. Duan H, Fernández-Domínguez AI, Bosman M, Maier SA, Yang JK. Nanoplasmonics: Classical down to the nanometer scale. *Nano Lett*. 2012. [Epub ahead of print].
14. Stockman MI. Nanoplasmonics: Past, present, and glimpse into future. *Opt Express*. 2011;19(22):22029–22106.
15. Lin J, Weber N, Wirth A, Chew SH, Escher M, Merkel M, Kling MF, Stockman MI, Krausz F, Kleineberg U. Time of flight-photoemission electron microscope for ultrahigh spatiotemporal probing of nanoplasmonic optical fields. *J Phys Condens Matter*. 2009;21(31):314005.

16. Chen HM, Pang L, Gordon MS, Fainman Y. Real-time template-assisted manipulation of nanoparticles in a multilayer nanofluidic chip. *Small*. 2011;7(19):2750–2757.
17. Sannomiya T, Vörös J. Single plasmonic nanoparticles for biosensing. *Trends Biotechnol.* 2011;29(7):343–351.
18. Fujikawa Y, Sakurai T, Tromp RM. Surface plasmon microscopy using an energy-filtered low energy electron microscope. *Phys Rev Lett.* 2008;100(12):126803.
19. Horiba K, Nakamura Y, Nagamura N, Toyoda S, Kumigashira H, Oshima M, Amemiya K, Senba Y, Ohashi H. Scanning photoelectron microscope for nanoscale three-dimensional spatial-resolved electron spectroscopy for chemical analysis. *Rev Sci Instrum.* 2011;82(11):113701.
20. Tejedor ML, Mizuno H, Tsuyama N, Harada T, Masujima T. In situ molecular analysis of plant tissues by live single cell mass spectrometry. *Anal Chem.* 2012;84(12):5221–5228.
21. Date S, Mizuno H, Tsuyama N, Harada T, Masujima T. Direct drug metabolism monitoring in a live single hepatic cell by video mass spectrometry. *Anal Sci.* 2012;28(3):201–203.
22. Kiser RW, Sullivan RE. Mass spectrometry. *Anal Chem.* 1968;40(5):273R+.
23. Torres-del-Pliego E, Vilaplana L, Güerri-Fernández R, Diez-Pérez A. Measuring bone quality. *Curr Rheumatol Rep.* 2013;15(11):373.
24. Chigerwe M, Shiraki R, Olstad EC, Angelos JA, Ruby AL, Westropp JL. Mineral composition of urinary calculi from potbellied pigs with urolithiasis: 50 cases (1982–2012). *J Am Vet Med Assoc.* 2013;243(3):389–393.
25. Ober CP, Warrington CD, Feeney DA, Jessen CR, Steward S. Optimizing a protocol for (1) H-magnetic resonance spectroscopy of the canine brain at 3T. *Vet Radiol Ultrasound.* 2013;54(2):149–158.
26. Lerner A, Shiroishi MS, Zee CS, Law M, Go JL. Imaging of neurocysticercosis. *Neuroimaging Clin N Am.* 2012;22(4):659–676.
27. Bejjani A, O'Neill J, Kim JA, Frew AJ, Yee VW, Ly R, Kitchen C et al. Elevated glutamatergic compounds in pregenual anterior cingulate in pediatric autism spectrum disorder demonstrated by 1H MRS and 1H MRSI. *PLoS One.* 2012;7(7):e38786.
28. Zynda TK, Thompson CD, Hoang KC, Seto AH, Glovaci D, Wong ND, Patel PM, Kern MJ. Disparity between angiographic coronary lesion complexity and lipid core plaques assessed by near-infrared spectroscopy. *Catheter Cardiovasc Interv.* 2013;81(3):529–537.
29. Cobbold JF, Patel D, Taylor-Robinson SD. Assessment of inflammation and fibrosis in non-alcoholic fatty liver disease by imaging-based techniques. *J Gastroenterol Hepatol.* 2012;27(8):1281–1292.
30. Robbins ME, Brunso-Bechtold JK, Peiffer AM, Tsien CI, Bailey JE, Marks LB. Imaging radiation-induced normal tissue injury. *Radiat Res.* 2012;177(4):449–466.
31. Borchman D, Yappert MC, Milliner SE, Smith RJ, Bhola R. Confirmation of the presence of squalene in human eyelid lipid by heteronuclear single quantum correlation spectroscopy. *Lipids.* 2013;48(12):1269–1277.
32. Yu-Wai-Man C, Smith FE, Firbank MJ, Guthrie G, Guthrie S, Gorman GS, Taylor RW et al. Extraocular muscle atrophy and central nervous system involvement in chronic progressive external ophthalmoplegia. *PLoS One.* 2013;8(9):e75048.
33. Kaźmierczak K, Malukiewicz G, Lesiewska-Junk H, Laudencka A, Szady-Grad M, Klawe J, Nowicki K. Blood plasma levels of microelements in patients with history of optic neuritis. *Curr Eye Res.* 2014;39(1):93–98.
34. Haritoglou C, Mauell S, Schumann RG, Henrich PB, Wolf A, Kernt M, Benoit M. Increase in lens capsule stiffness caused by vital dyes. *J Cataract Refract Surg.* 2013;39(11):1749–1752.
35. Hansell P. Medical photography: A review. *Lancet.* 1946;2(6418):296–299.
36. Castaneda B, Ormachea J, Rodríguez P, Parker KJ. Application of numerical methods to elasticity imaging. *Mol Cell Biomech.* 2013;10(1):43–65.
37. Urbanik A, Chrzan R. Application of computed tomography (CT) examination for forensic medicine. *Przegl Lek.* 2013;70(5):229–242.

38. Malanchuk VO, Kopchak AV, Kryshchuk MH. Determination of regimes of functional loading in the patients with traumatic mandibular fractures after osteosynthesis performance using modern methods of computer modeling. *Klin Khir*. 2013;3:53–58.
39. Okada N, Shimobaba T, Ichihashi Y, Oi R, Yamamoto K, Oikawa M, Kakue T, Masuda N, Ito T. Band-limited double-step Fresnel diffraction and its application to computer-generated holograms. *Opt Express*. 2013;21(7):9192–9197.
40. Chin KR, Mills MV, Seale J, Cumming V. Ideal starting point and trajectory for C2 pedicle screw placement: A 3D computed tomography analysis using perioperative measurements. *Spine J*. 2014;14(4):615–618. doi:10.1016/j.spinee.2013.06.077.
41. Wang Q, Zhao X, Song J, Guo N, Zhu Y, Liu J, Qi W et al. The application of automatic tube current modulation (ATCM) on image quality and radiation dose at abdominal computed tomography (CT): A phantom study. *J Xray Sci Technol*. 2013;21(4):453–464.
42. Bai A, Sun Y, Qi L, Yang Y, Hua Y. Improving the image quality in computed tomographic pulmonary angiography with dual-energy subtraction: A new application of spectral computed tomography. *J Comput Assist Tomogr*. 2013;37(5):718–724.
43. Dai CL, Yang ZG, Xue LP, Li YM. Application value of multi-slice spiral computed tomography for imaging determination of metastatic lymph nodes of gastric cancer. *World J Gastroenterol*. 2013;19(34):5732–5737.
44. Wang L, Li X, Wu M, Zhang L, Luo S. A sparse-projection computed tomography reconstruction method for in vivo application of in-line phase-contrast imaging. *Biomed Eng Online*. 201330;12:75.
45. Miernik A, Eilers Y, Bolwien C, Lambrecht A, Hauschke D, Rebentisch G, Lossin PS et al. Automated analysis of urinary stone composition using Raman spectroscopy: Pilot study for the development of a compact portable system for immediate postoperative ex vivo application. *J Urol*. 2013;190(5):1895–1900.
46. Hayasaka T, Goto-Inoue N, Ushijima M, Yao I, Yuba-Kubo A, Wakui M, Kajihara S, Matsuura M, Setou M. Development of imaging mass spectrometry (IMS) dataset extractor software, IMS convolution. *Anal Bioanal Chem*. 2011;401(1):183–193.
47. Rich ES Jr, Wampler JE. A flexible, computer-controlled video microscope capable of quantitative spatial, temporal, and spectral measurements. *Clin Chem*. 1981;27(9):1558–1568.
48. Kucera P, de Ribaupierre Y, de Ribaupierre F. Computer-controlled double-beam scanning microspectrophotometry for rapid microscopic image reconstructions. *J Microsc*. 1979;116(2):173–184.
49. Arnold W. Fourier transformation infrared spectrometry—A new (old) method of detection in forensic chemistry and criminal investigation. *Beitr Gerichtl Med*. 1989;47:123–147.
50. Frank JH, Elder AD, Swartling J, Venkitaraman AR, Jeyasekharan AD, Kaminski CF. A white light confocal microscope for spectrally resolved multidimensional imaging. *J Microsc*. 2007;227(Pt 3):203–215.
51. Skala MC, Fontanella A, Hendargo H, Dewhirst MW, Izatt JA. Combined hyperspectral and spectral domain optical coherence tomography microscope for noninvasive hemodynamic imaging. *Opt Lett*. 2009;34(3):289–291.
52. Zimny A, Szmyrka-Kaczmarek M, Szewczyk P, Bladowska J, Pokryszko-Dragan A, Gruszka E, Wiland P, Sasiadek M. In vivo evaluation of brain damage in the course of systemic lupus erythematosus using magnetic resonance spectroscopy, perfusion-weighted and diffusion-tensor imaging. *Lupus*. 2014;23(1):10–19.
53. Runge JH, Bakker PJ, Gaemers IC, Verheij J, Hakvoort TB, Ottenhoff R, Nederveen AJ, Stoker J. Measuring liver triglyceride content in mice: Non-invasive magnetic resonance methods as an alternative to histopathology. *MAGMA*. 2014;27(4):317–327.
54. Maldonado F, Boland JM, Raghunath S, Aubry MC, Bartholmai BJ, Deandrade M, Hartman TE et al. Noninvasive characterization of the histopathologic features of pulmonary nodules of the lung adenocarcinoma spectrum using computer-aided nodule assessment and risk yield (CANARY)—A pilot study. *J Thorac Oncol*. 2013;8(4):452–460.

55. Jeong H, Kim J, Ishida T, Akiyama M, Kim Y. Computerised analysis of osteoporotic bone patterns using texture parameters characterising bone architecture. *Br J Radiol*. 2013;86(1021): 20101115.
56. Cornwell WS. Radiography and photography in problems of identification: A review. *Med Radiogr Photogr*. 1956;32(1):1.
57. Staff C. Automatic x-ray processing machines: A review of present-day equipment. *Radiography*. 1958;24(283):161–167.
58. Svoboda M. Technical aspects and new trends in roentgenologic diagnosis: Critical review. *Cas Lek Cesk*. 1957;96(30):49–58.
59. Zappoli F, Lavaroni A, Leonardi M. Computed tomography. *Eur Neurol*. 1989;29(Suppl 2):30–32.
60. Pfeiler M, Steiner K. Computed tomography—Basic principles and outlook for related procedures. *Morphol Med*. 1983;3(3):117–124.
61. Donald I. The ultrasonic boom. *J Clin Ultrasound*. 1976;4(5):323–328.
62. Gravelle IH. Ultrasonography and thermography in medical diagnosis. *Trans Med Soc Lond*. 1969;85:161–166.
63. Ross FG. Use of ultrasonics in medical diagnosis. *Recenti Prog Med*. 1977;62(3):262–294.
64. Berkhout AJ. Ultrasonic medical imaging, current techniques and future developments. *Med Prog Technol*. 1986;11(4):197–207.
65. Budinger TF, Lauterbur PC. Nuclear magnetic resonance technology for medical studies. *Science*. 1984;226(4672):288–298.
66. Margulis AR, Hricak H, Crooks L. Medical applications of nuclear magnetic resonance imaging. *Q Rev Biophys*. 1987;19(3/4):221–237.
67. Council on Scientific Affairs. Fundamentals of magnetic resonance imaging. Council on Scientific Affairs. *JAMA*. 1987;258(23):3417–3423.
68. Yamamura S, Oda S, Utsunomiya D, Funama Y, Imuta M, Namimoto T, Hirai T, Chikamoto A, Baba H, Yamashita Y. Dynamic computed tomography of locally advanced pancreatic cancer: Effect of low tube voltage and a hybrid iterative reconstruction algorithm on image quality. *J Comput Assist Tomogr*. 2013;37(5):790–796.
69. Thing RS, Bernchou U, Mainegra-Hing E, Brink C. Patient-specific scatter correction in clinical cone beam computed tomography imaging made possible by the combination of Monte Carlo simulations and a ray tracing algorithm. *Acta Oncol*. 2013;52(7):1477–1483.
70. Yu MH, Lee JM, Yoon JH, Baek JH, Han JK, Choi BI, Flohr TG. Low tube voltage intermediate tube current liver MDCT: Sinogram-affirmed iterative reconstruction algorithm for detection of hypervascular hepatocellular carcinoma. *AJR Am J Roentgenol*. 2013;201(1):23–32.
71. Ay MR, Zaidi H. Development and validation of MCNP4C-based Monte Carlo simulator for fan- and cone-beam x-ray CT. *Phys Med Biol*. 2005;50(20):4863–4885.
72. Apostolou N, Papazoglou T, Koutsouris D. Advanced image fusion algorithms for Gamma Knife treatment planning. Evaluation and proposal for clinical use. *Technol Health Care*. 2006;14(3):143–156.
73. Otton JM, Kühl JT, Kofoed KF, McCrohon J, Feneley M, Sammel N, Yu CY, Chiribiri A, Nagel E. Four-dimensional image processing of myocardial CT perfusion for improved image quality and noise reduction. *J Cardiovasc Comput Tomogr*. 2013;7(2):110–116.
74. Yu Z, Noo F, Dennerlein F, Wunderlich A, Lauritsch G, Hornegger J. Simulation tools for two-dimensional experiments in x-ray computed tomography using the FORBILD head phantom. *Phys Med Biol*. 2012;57(13):N237–N252.
75. Molinari F, Meiburger KM, Zeng G, Acharya UR, Liboni W, Nicolaides A, Suri JS. Carotid artery recognition system: A comparison of three automated paradigms for ultrasound images. *Med Phys*. 2012;39(1):378–391.
76. Wang C, Chen M, Zhao JM, Liu Y. Fusion of color Doppler and magnetic resonance images of the heart. *J Digit Imaging*. 2011;24(6):1024–1030.
77. Jassar AS, Brinster CJ, Vergnat M, Robb JD, Eperjesi TJ, Pouch AM, Cheung AT et al. Quantitative mitral valve modeling using real-time three-dimensional echocardiography: Technique and repeatability. *Ann Thorac Surg*. 2011;91(1):165–171.

78. McFarlin BL, Bigelow TA, Laybed Y, O'Brien WD, Oelze ML, Abramowicz JS. Ultrasonic attenuation estimation of the pregnant cervix: A preliminary report. *Ultrasound Obstet Gynecol.* 2010;36(2):218–225.
79. Uhercík M, Kybic J, Liebgott H, Cachard C. Model fitting using RANSAC for surgical tool localization in 3-D ultrasound images. *IEEE Trans Biomed Eng.* 2010;57(8):1907–1916.
80. Tana MG, Sclocco R, Bianchi AM. GMAC: A Matlab toolbox for spectral Granger causality analysis of fMRI data. *Comput Biol Med.* 2012;42(10):943–956.
81. Tian F, Liu H. Depth-compensated diffuse optical tomography enhanced by general linear model analysis and an anatomical atlas of human head. *Neuroimage.* 2014;14(4):615–618. doi:10.1016/j.spinee.2013.06.077.
82. Wilson JA, Shutter LA, Hartings JA. COSBID-M3: A platform for multimodal monitoring, data collection, and research in neurocritical care. *Acta Neurochir Suppl.* 2013;115:67–74.
83. Clas P, Groeschel S, Wilke M. A semi-automatic algorithm for determining the demyelination load in metachromatic leukodystrophy. *Acad Radiol.* 2012;19(1):26–34.
84. Zhang L, Agravat S, Derado G, Chen S, McIntosh BJ, Bowman FD. BSMac: A MATLAB toolbox implementing a Bayesian spatial model for brain activation and connectivity. *J Neurosci Methods.* 2012;204(1):133–143.
85. Cui Z, Zhong S, Xu P, He Y, Gong G. PANDA: A pipeline toolbox for analyzing brain diffusion images. *Front Hum Neurosci.* 2013;7:42.
86. Dai Y, Shi F, Wang L, Wu G, Shen D. iBEAT: A toolbox for infant brain magnetic resonance image processing. *Neuroinformatics.* 2013;11(2):211–225.
87. Martí Fuster B, Esteban O, Planes X, Aguiar P, Crespo C, Falcon C, Wollny G et al. FocusDET, a new toolbox for SISCOM analysis. Evaluation of the registration accuracy using Monte Carlo simulation. *Neuroinformatics.* 2013;11(1):77–89.
88. Erhardt EB, Allen EA, Wei Y, Eichele T, Calhoun VD. SimTB, a simulation toolbox for fMRI data under a model of spatiotemporal separability. *Neuroimage.* 2012;59(4):4160–4167.
89. Klein S, Staring M, Murphy K, Viergever MA, Pluim JP. elastix: A toolbox for intensity-based medical image registration. *IEEE Trans Med Imaging.* 2010;29(1):196–205.
90. Hanke M, Halchenko YO, Sederberg PB, Hanson SJ, Haxby JV, Pollmann S. PyMVPA: A python toolbox for multivariate pattern analysis of fMRI data. *Neuroinformatics.* 2009;7(1):37–53.
91. Bagarinao E, Matsuo K, Nakai T, Tanaka Y. BAX: A toolbox for the dynamic analysis of functional MRI datasets. *Neuroinformatics.* 2008;6(2):109–115.
92. Wilke M, Holland SK, Altaye M, Gaser C. Template-O-Matic: A toolbox for creating customized pediatric templates. *Neuroimage.* 2008;41(3):903–913.
93. Wang Z, Aguirre GK, Rao H, Wang J, Fernández-Seara MA, Childress AR, Detre JA. Empirical optimization of ASL data analysis using an ASL data processing toolbox: ASLtbx. *Magn Reson Imaging.* 2008;26(2):261–269.
94. Wilke M, Lidzba K. LI-tool: A new toolbox to assess lateralization in functional MR-data. *J Neurosci Methods.* 2007;163(1):128–136.
95. Dhama K, Verma AK, Rajagunalan S, Deb R, Karthik K, Kapoor S, Mahima M, Tiwari R, Panwar PK, Chakraborty S. Swine flu is back again: A review. *Pak J Biol Sci.* 2012;15(21):1001–1009.
96. Yang S, Zhu WF, Shu YL. An overview on swine influenza viruses. *Bing Du Xue Bao.* 2013;29(3):330–336.
97. Chauhan N, Narang J, Pundir S, Singh S, Pundir CS. Laboratory diagnosis of swine flu: A review. *Artif Cells Nanomed Biotechnol.* 2013;41(3):189–195.
98. Krueger WS, Gray GC. Swine influenza virus infections in man. *Curr Top Microbiol Immunol.* 2013;370:201–225.
99. Wiwanitkit V. Non respiratory manifestations of swine flu. *Clin Ter.* 2009;160(6):499–501.
100. Wiwanitkit V. Antiviral drug treatment for emerging swine flu. *Clin Ter.* 2009;160(3):243–245.
101. Iskander J, Strikas RA, Gensheimer KF, Cox NJ, Redd SC. Pandemic influenza planning, United States, 1978–2008. *Emerg Infect Dis.* 2013;19(6):879–885.
102. Heine MH, Tague JM. An investigation of the optimization of search logic for the MEDLINE database. *J Am Soc Inf Sci.* 1991;42(4):267–278.

103. Wiwanitkit V. Chest imaging in H1N1 influenza. *Acta Radiol.* 2011;52(9):969.
104. Wiwanitkit V. Re: Novel influenza A (H1N1) virus infection in children: Chest radiographic and CT evaluation. *Korean J Radiol.* 2011;12(2):266.
105. Lee JE, Choe KW, Lee SW. Clinical and radiological characteristics of 2009 H1N1 influenza associated pneumonia in young male adults. *Yonsei Med J.* 2013;54(4):927–934.
106. Shim SS, Kim Y, Ryu YJ. Novel influenza A (H1N1) infection: Chest CT findings from 21 cases in Seoul, Korea. *Clin Radiol.* 2011;66(2):118–124.
107. Guo WL, Wang J, Zhou M, Sheng M, Eltahir YM, Wei J, Ding YF, Zhang XL. Chest imaging findings in children with influenza A H1N1. *Saudi Med J.* 2011;32(1):50–54.
108. Dewhurst CJ. Advances in obstetrics and cynaecology. *Practitioner.* 1969;203(216):428–437.
109. Thompson HE. The clinical use of pulsed echo ultrasound in obstetrics and gynecology. *Obstet Gynecol Surv.* 1968;23(10):903–932.
110. Khentov RA. The use of ultrasonic diagnosis in obstetrics and gynecology (review of the literature). *Vopr Okhr Materin Det.* 1967;12(7):70–73.
111. Khadzhiev A. Characteristics of ultrasound used in obstetrics and gynecology. *Akush Ginekol (Sofiia).* 1981;20(6):475–481.
112. Neilson JP, Hood VD. Ultrasound in obstetrics and gynaecology. Recent developments. *Br Med Bull.* 1980;36(3):249–255.
113. Kratochwil A. Ultrasonic diagnosis in obstetrics and gynecology. *Med Prog Technol.* 1980;7(4):157–167.
114. Ylöstalo P, Jouppila P, Kirkinen P. The use of ultrasound in obstetrics. *Ann Clin Res.* 1979;11(5):222–232.
115. Azimi F, Bryan PJ, Marangola JP. Ultrasonography in obstetrics and gynecology: Historical notes, basic principles, safety considerations, and clinical applications. *CRC Crit Rev Clin Radiol Nucl Med.* 1976;8(2):153–253.
116. Gottesfeld KR. Ultrasound in obstetrics and gynecology. *Semin Roentgenol.* 1975;10(4):305–313.
117. Duck FA. Nonlinear acoustics in diagnostic ultrasound. *Ultrasound Med Biol.* 2002;28(1):1–18.
118. Starritt HC, Duck FA, Humphrey VF. Forces acting in the direction of propagation in pulsed ultrasound fields. *Phys Med Biol.* 1991;36(11):1465–1474.
119. Mason GA. Cancer of the lung: Review of 1,000 cases. *Lancet.* 1949;2(6579):587–591.
120. Hastings DR. Primary neoplasms of the lung; review of Minneapolis Public Health Center cases. *Minn Med.* 1953;36(6):594–597.
121. Rigdon RH. Cancer of the lung before 1900: A histologic review. *Tex Rep Biol Med.* 1955;13(4):993–1009.
122. Ochsner A, Ray CJ, Acree PW. Cancer of the lung: A review of experiences with 1,457 cases of bronchogenic carcinoma. *Am Rev Tuberc.* 1954;70(5):763–783.
123. Van Rossum R. Primary cancer of the lung: Review. *Brux Med.* 1954;34(47):2301–2319.
124. Sun WY, Degen SJ. Gene targeting in hemostasis. Prothrombin. *Front Biosci.* 2001;6:D222–D238.
125. Strijks E, Poort SR, Renier WO, Gabreels FJ, Bertina RM. Hereditary prothrombin deficiency presenting as intracranial haematoma in infancy. *Neuropediatrics* 1999;30:320–324.
126. Huang MN, Kasper CK, Roberts HR, Stafford DW, High KA. Molecular defect in factor IXHilo, a hemophilia Bm variant: Arg→Gln at the carboxyterminal cleavage site of the activation peptide. *Blood.* 1989;73:718–721.
127. Monroe DM, McCord DM, Huang MN, High KA, Lundblad RL, Kasper CK, Roberts HR. Functional consequences of an arginine180 to glutamine mutation in factor IX Hilo. *Blood.* 1989;73:1540–1544.
128. Wiwanitkit V. *New Developments in Thrombohemostatic Diseases.* New York: Nova, 2007.

Appendix A: Material and Physical Constants

A.1 Common Material Constants

TABLE A.1

Approximate Conductivity at 20°C

Material	Conductivity (S/m)
Conductors	
Silver	6.3×10^7
Copper (standard annealed)	5.8×10^7
Gold	4.5×10^7
Aluminum	3.5×10^7
Tungsten	1.8×10^7
Zinc	1.7×10^7
Brass	1.1×10^7
Iron (pure)	10^7
Lead	5×10^7
Mercury	10^6
Carbon	3×10^7
Water (sea)	4.8
Semiconductors	
Germanium (pure)	2.2
Silicon (pure)	4.4×10^{-4}
Insulators	
Water (distilled)	10^{-4}
Earth (dry)	10^{-5}
Bakelite	10^{-10}
Paper	10^{-11}
Glass	10^{-12}
Porcelain	10^{-12}
Mica	10^{-15}
Paraffin	10^{-15}
Rubber (hard)	10^{-15}
Quartz (fused)	10^{-17}
Wax	10^{-17}

TABLE A.2

Approximate Dielectric Constant and Dielectric Strength

Material	Dielectric Constant (or Relative Permittivity) (Dimensionless)	Strength, E (V/m)
Barium titanate	1200	7.5×10^6
Water (sea)	80	–
Water (distilled)	8.1	–
Nylon	8	–
Paper	7	12×10^6
Glass	5–10	35×10^6
Mica	6	70×10^6
Porcelain	6	–
Bakelite	5	20×10^6
Quartz (fused)	5	30×10^6
Rubber (hard)	3.1	25×10^6
Wood	2.5–8.0	–
Polystyrene	2.55	–
Polypropylene	2.25	–
Paraffin	2.2	30×10^6
Petroleum oil	2.1	12×10^6
Air (1 atm)	1	3×10^6

TABLE A.3

Relative Permeability

Material	Relative Permeability, μ_r
Diamagnetic	
Bismuth	0.999833
Mercury	0.999968
Silver	0.9999736
Lead	0.9999831
Copper	0.9999906
Water	0.9999912
Hydrogen (STP)	~1.0
Paramagnetic	
Oxygen (STP)	0.999998
Air	1.00000037
Aluminum	1.000021
Tungsten	1.00008
Platinum	1.0003
Manganese	1.001
Ferromagnetic	
Cobalt	250
Nickel	600
Soft iron	5000
Silicon iron	7000

STP, Standard Temperature and Pressure.

TABLE A.4

Approximate Conductivity for Biological Tissue

Material	Conductivity (S/m)	Frequency
Blood	0.7	0 (DC)
Bone	0.01	0 (DC)
Brain	0.1	10^2–10^6 Hz
Breast fat	0.2–1	0.4–5 GHz
Breast tumor	0.7–3	0.4–5 GHz
Fat	0.1–0.3	0.4–5 GHz
	0.03	10^2–10^6 Hz
Muscle	0.4	10^2–10^6 Hz
Skin	0.001	1 kHz
	0.1	1 MHz

TABLE A.5

Approximate Dielectric Constant for Biological Tissue

Material	Dielectric Constant (Relative Permittivity)	Frequency
Blood	10^5	1 kHz
Bone	3,000–10,000	0 (DC)
Brain	10^7	100 Hz
	10^3	1 MHz
Breast fat	5–50	0.4–5 GHz
Breast tumor	47–67	0.4–5 GHz
Fat	5	0.4–5 GHz
	10^6	100 Hz
	10	1 MHz
Muscle	10^6	1 kHz
	10^3	1 MHz
Skin	10^6	1 kHz
	10^3	1 MHz

A.2 Physical Constants

Quantity	Best Experimental Value	Approximate Value for Problem Work
Avogadro's number (per kg mol)	6.0228×10^{26}	6×10^{26}
Boltzmann's constant (J/k)	1.38047×10^{-23}	1.38×10^{-23}
Electron charge (C)	-1.6022×10^{-19}	-1.6×10^{-19}
Electron mass (kg)	9.1066×10^{-31}	9.1×10^{-31}
Permittivity of free space (F/m)	8.854×10^{-12}	$10^{-9}/36\pi$

(Continued)

Quantity	Best Experimental Value	Approximate Value for Problem Work
Permeability of free space (H/m)	$4\pi \times 10^{-7}$	12.6×10^{-7}
Intrinsic impedance of free space (O)	376.6	120π
Speed of light in free space or vacuum (m/s)	2.9979×10^{8}	3×10^{8}
Proton mass (kg)	1.67248×10^{-27}	1.67×10^{-27}
Neutron mass (kg)	1.6749×10^{-27}	1.67×10^{-27}
Planck's constant (J s)	6.6261×10^{-34}	6.62×10^{-34}
Acceleration due to gravity (m/s^2)	9.8066	9.8
Universal constant of gravitation ($m^2/kg\ s^2$)	6.658×10^{-11}	6.66×10^{-11}
Electron volt (J)	1.6030×10^{-19}	1.6×10^{-19}
Gas constant (J/mol K)	8.3145	8.3

Appendix B: Photon Equations Index of Refraction, Electromagnetic Spectrum, and Wavelength of Commercially Available Lasers

B.1 Photon Energy, Frequency, and Wavelength

Photon energy (J)	Planck's constant × frequency
Photon energy (eV)	Planck's constant × frequency/Electron charge
Photon energy (cm^{-1})	Frequency/Speed of light in vacuum
Photon frequency (Hz)	1(cycle)/Period (s)
Photon wavelength (μm)	Speed of light in free space/Frequency

B.2 Index of Refraction for Common Substances

Substance	Index of Refraction
Air	1.00
Diamond	2.24
Ethyl alcohol	1.36
Fluorite	1.43
Fused quartz	1.46
Crown glass	1.52
Flint glass	1.66
Glycerin	1.47
Ice	1.31
Polystyrene	1.49
Rock salt	1.54
Water	1.33

B.3 Electromagnetic Spectrum

See Figure B.1.

B.4 Wavelengths of Commercially Available Lasers

See Figure B.2 and Table B.1.

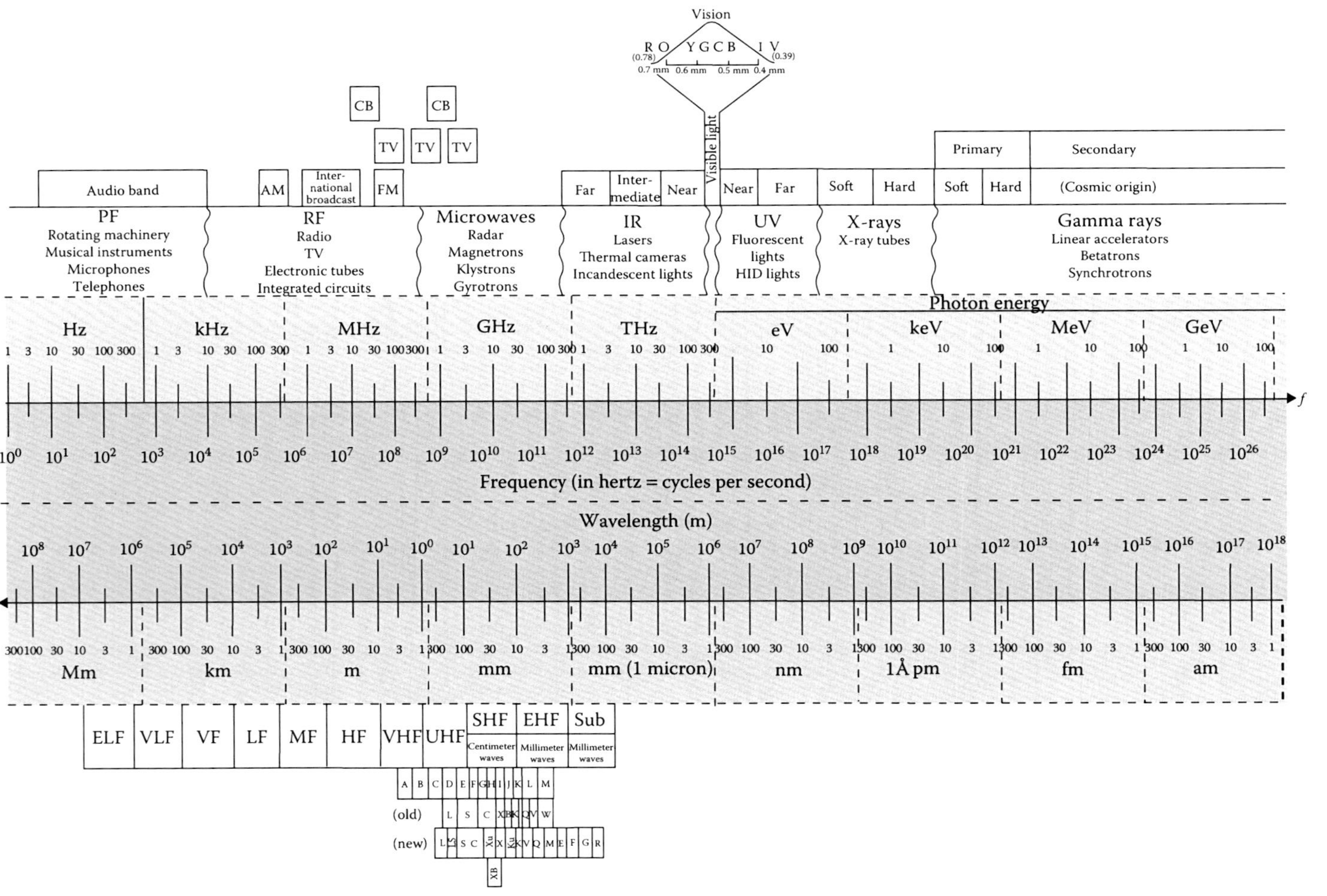

FIGURE B.1
Simplified chart of the electromagnetic spectrum. (Data from Whitaker, J.C., *The Electronics Handbook*, CRC Press, Boca Raton, FL, 1996.)

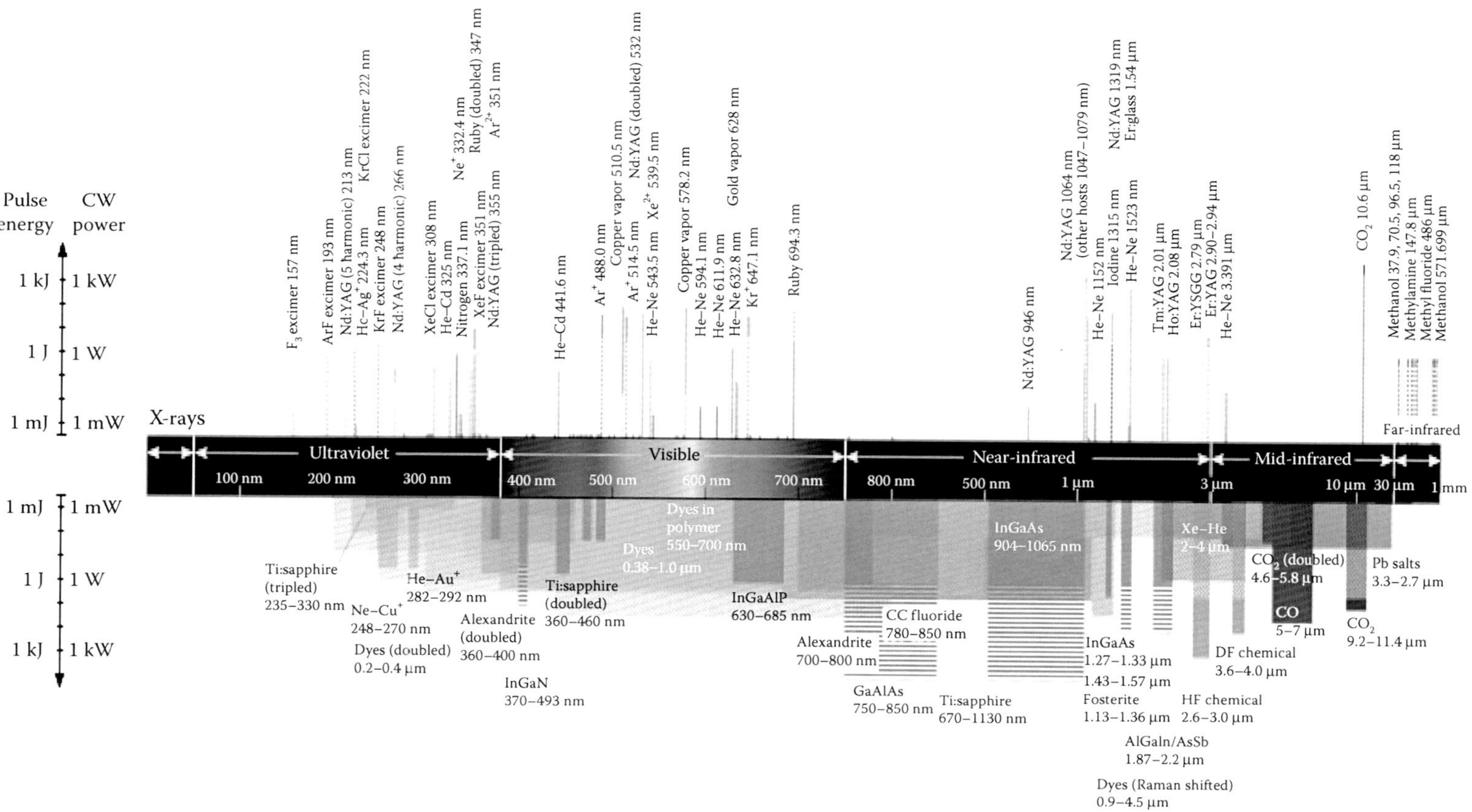

FIGURE B.2
(See color insert.) Wavelengths of commercially available lasers. (Data from Weber, M.J., *Handbook of Laser Wavelengths*, CRC Press, Boca Raton, FL, 1999.)

TABLE B.1
Approximate Common Optical Wavelength Ranges of Light

Color	Wavelength
Ultraviolet region	10–380 nm
Visible region	380–750 nm
Violet	380–450 nm
Blue	450–495 nm
Green	495–570 nm
Yellow	570–590 nm
Orange	590–620 nm
Red	620–750 nm
Infrared	750 nm–1 mm

References

1. Weber, M.J., *Handbook of Laser Wavelengths*, CRC Press, Boca Raton, FL, 1999.
2. Whitaker, J.C., *The Electronics Handbook*, CRC Press, Boca Raton, FL, 1996.

Appendix C: Symbols and Formulas

C.1 Greek Alphabet

Uppercase	Lowercase	Name
Α	α	Alpha
Β	β	Beta
Γ	γ	Gamma
Δ	δ	Delta
Ε	ε	Epsilon
Ζ	ζ	Zeta
Η	η	Eta
Θ	θ, ϑ	Theta
Ι	ι	Iota
Κ	κ	Kappa
Λ	λ	Lambda
Μ	μ	Mu
Ν	ν	Nu
Ξ	ξ	Xi
Ο	ο	Omicron
Π	π	Pi
Ρ	ρ	Rho
Σ	σ	Sigma
Τ	τ	Tau
Υ	υ	Upsilon
Φ	φ, ϕ	Phi
Χ	χ	Chi
Ψ	ψ	Psi
Ω	ω	Omega

C.2 International System of Units Prefixes

Power	Prefix	Symbol	Power	Prefix	Symbol
10^{-35}	Stringo	–	10^{0}	–	–
10^{-24}	Yocto	y	10^{1}	Deca	da
10^{-21}	Zepto	z	10^{2}	Hecto	h
10^{-18}	Atto	A	10^{3}	Kilo	k
10^{-15}	Femto	f	10^{6}	Mega	M
10^{-12}	Pico	p	10^{9}	Giga	G

(Continued)

Power	Prefix	Symbol	Power	Prefix	Symbol
10^{-9}	Nano	n	10^{12}	Tera	T
10^{-6}	Micro	μ	10^{15}	Peta	P
10^{-3}	Milli	m	10^{18}	Exa	E
10^{-2}	Centi	c	10^{21}	Zetta	Z
10^{-1}	Deci	d	10^{24}	Yotta	Y

C.3 Trigonometric Identities

$$\cot\theta = \frac{1}{\tan\theta}, \quad \sec\theta = \frac{1}{\cos\theta}, \quad \mathrm{cosec}\,\theta = \frac{1}{\sin\theta}$$

$$\tan\theta = \frac{\sin\theta}{\cos\theta}, \quad \cot\theta = \frac{\cos\theta}{\sin\theta}$$

$$\sin^2\theta + \cos^2\theta = 1, \quad \tan^2\theta + 1 = \sec^2\theta, \quad \cot^2\theta + 1 = \csc^2\theta$$

$$\sin(-\theta) = -\sin\theta, = \cos(-\theta) = \cos\theta, \tan(-\theta) = -\tan\theta$$

$$\csc(-\theta) = -\csc\theta, \quad \sec(-\theta) = \sec\theta, \quad \cot(-\theta) = -\cot\theta$$

$$\cos(\theta_1 \pm \theta_2) = \cos\theta_1\cos\theta_2 \pm \sin\theta_1\sin\theta_2$$

$$\sin(\theta_1 \pm \theta_2) = \sin\theta_1\cos\theta_2 \pm \cos\theta_1\sin\theta_2$$

$$\tan(\theta_1 \pm \theta_2) = \frac{\tan\theta_1 \pm \tan\theta_2}{1 \mp \tan\theta_1 \pm \tan\theta_2}$$

$$\cos\theta_1\cos\theta_2 = \frac{1}{2}\left[\cos(\theta_1 + \theta_2) + \cos(\theta_1 - \theta_2)\right]$$

$$\sin\theta_1\sin\theta_2 = \frac{1}{2}\left[\sin(\theta_1 + \theta_2) + \sin(\theta_1 - \theta_2)\right]$$

$$\sin\theta_1\cos\theta_2 = \frac{1}{2}\left[\sin(\theta_1 + \theta_2) + \sin(\theta_1 - \theta_2)\right]$$

$$\cos\theta_1\sin\theta_2 = \frac{1}{2}\left[\sin(\theta_1 + \theta_2) + \sin(\theta_1 - \theta_2)\right]$$

$$\sin\theta_1 + \sin\theta_2 = 2\sin\left(\frac{\theta_1 + \theta_2}{2}\right)\cos\left(\frac{\theta_1 - \theta_2}{2}\right)$$

$$\sin\theta_1 - \sin\theta_2 = 2\cos\left(\frac{\theta_1 + \theta_2}{2}\right)\sin\left(\frac{\theta_1 - \theta_2}{2}\right)$$

$$\cos\theta_1 + \cos\theta_2 = 2\cos\left(\frac{\theta_1 + \theta_2}{2}\right)\cos\left(\frac{\theta_1 - \theta_2}{2}\right)$$

$$\cos\theta_1 - \cos\theta_2 = 2\sin\left(\frac{\theta_1+\theta_2}{2}\right)\sin\left(\frac{\theta_1-\theta_2}{2}\right)$$

$$a\sin\theta - b\cos\theta = \sqrt{a^2+b^2}\cos(\theta+\phi), \quad \text{where } \phi = \tan^{-1}\left(\frac{b}{a}\right)$$

$$a\sin\theta - b\cos\theta = \sqrt{a^2+b^2}\sin(\theta+\phi), \quad \text{where } \phi = \tan^{-1}\left(\frac{b}{a}\right)$$

$$\cos(90^\circ-\theta) = \sin\theta, \quad \sin(90^\circ-\theta) = \cos\theta, \quad \tan(90^\circ-\theta) = \cot\theta$$

$$\cos(90^\circ-\theta) = \tan\theta, \quad \sec(90^\circ-\theta) = \operatorname{cosec}\theta, \quad \operatorname{cosec}(90^\circ-\theta) = \sec\theta$$

$$\cos(\theta\pm 90^\circ) = \mp\sin\theta, \ \sec(\theta\pm 90^\circ) = \pm\sin\theta, \ \tan(\theta\pm 90^\circ) = -\cot\theta$$

$$\cos(\theta\pm 180^\circ) = -\cos\theta, \quad \sin(\theta\pm 180^\circ) = -\sin\theta, \quad \tan(\theta\pm 180^\circ) = \tan\theta$$

$$\cos 2\theta = \cos^2\theta - \sin^2\theta, \quad \cos 2\theta = 1-2\sin^2\theta, \quad \cos 2\theta = 2\cos^2\theta - 1$$

$$\sin 2\theta = 2\sin\theta\cos\theta, \quad \tan 2\theta = \frac{2\tan\theta}{1-\tan^2\theta}$$

$$\cos 3\theta = 4\cos^3\theta - 3\sin\theta$$

$$\sin 3\theta = 3\sin\theta - 4\sin^3\theta$$

$$\sin\frac{\theta}{2} = \pm\sqrt{\frac{1-\cos\theta}{2}}, \quad \cos\frac{\theta}{2}\pm\sqrt{\frac{1+\cos\theta}{2}},$$

$$\sin\theta = \frac{e^{j\theta}-e^{j\theta}}{2j}, \quad \cos\theta = \frac{e^{j\theta}+e^{j\theta}}{2}\left(j=\sqrt{-1}\right), \quad \tan\theta = \frac{e^{j\theta}+e^{-j\theta}}{j\left(e^{j\theta}+e^{-j\theta}\right)}$$

$$e^{\pm j\theta} = \cos\theta \pm j\sin\theta \quad \text{(Euler's identity)}$$

$$1\,\text{rad} = 57.296^\circ$$

$$\pi = 3.1416$$

C.4 Hyperbolic Functions

$$\cosh x = \frac{e^x+e^{-x}}{2}, \quad \sinh x = \frac{e^x-e^{-x}}{2}, \quad \tanh x = \frac{\sinh x}{\cosh x}$$

$$\cosh x = \frac{1}{\tanh x}, \quad \operatorname{sech} x = \frac{1}{\cosh x}, \quad \operatorname{cosech} x = \frac{1}{\sinh x}$$

$$\sin jx = j\sinh x, \quad \cos jx = \cosh x$$

$$\sinh jx = j\sin x, \quad \cosh jx = \cos x$$

$$\sin(x \pm jy) = \sin x \cosh y \pm j\cos x \sinh y$$

$$\cos(x \pm jy) = \cos x \cosh y \pm j\sin x \sinh y$$

$$\sinh(x \pm y) = \sinh x \cosh y \pm \cosh x \sinh y$$

$$\cosh(x \pm y) = \cos x \cosh y \pm j\sin x \sinh y$$

$$\sinh(x \pm jy) = \sinh x \cos y \pm j\cosh x \sin y$$

$$\cosh(x \pm jy) = \cosh x \cos y \pm j\sinh x \sin y$$

$$\tanh(x \pm jy) = \frac{\sinh 2x}{\cosh 2x + \cos 2y} \pm j\frac{\sin 2y}{\cosh 2x + \cos 2y}$$

$$\cosh^2 x - \sinh^2 x = 1$$

$$\operatorname{sech}^2 x + \tanh^2 x = 1$$

C.5 Complex Variables

A complex number can be written as

$$z = x = jy = r\angle\theta = re^{j\theta} = r(\cos\theta + j\sin\theta),$$

where:

$x = \text{Re } z = r\cos\theta$

$y = \text{Im } z = r\sin\theta$

$$r = |z| = \sqrt{x^2 + y^2}, \quad \theta = \tan^{-1}\left(\frac{y}{x}\right)$$

$$j = \sqrt{-1}, \quad \frac{1}{j} = -j, \quad j^2 = -1$$

The complex conjugate of $z = z^* = x - jy = r\angle{-\theta} = re^{-j\theta} = r(\cos\theta - j\sin\theta)$

$$(e^{j\theta})^n = e^{jn\theta} = \cos n\theta + j\sin n\theta \quad \text{(deMoivre's theorem).}$$

If $z_1 = x_1 + jy_1$ and $z_2 = x_2 + jy_2$, then only if $x_1 = x_2$ and $y_1 = y_2$,

$$z_1 \pm z_2 = (x_1 + x_2) \pm j(y_1 + y_2)$$

$$z_1 z_2 = (x_1 x_2 - y_1 y_2) + j(x_1 y_2 + x_2 y_1) = r_1 r_2 e^{j(\theta_1 + \theta_2)} = r_1 r_2 \angle \theta_1 + \theta_2$$

$$\frac{z_1}{z_2} = \frac{(x_1 + jy_1)}{(x_2 + jy_2)} \times \frac{(x_2 + jy_2)}{(x_2 + jy_2)} = \frac{x_1 x_2 + y_1 y_2}{x_2^2 + y_2^2} + j\frac{x_2 y_1 - x_1 y_2}{x_2^2 + y_2^2} = \frac{r_1}{r_2} e^{j(\theta_1 + \theta_2)} = \frac{r_1}{r_2} \angle \theta_1 - \theta_2$$

$$\ln(re^{j\theta}) = \ln r + \ln e^{j\theta} = \ln r + j\theta + j2m\pi \quad (m = \text{integer})$$

$$\sqrt{z} = \sqrt{x + jy} = \sqrt{r}\left(e^{j\theta/2}\right) = \sqrt{r} \angle \theta/2$$

$$z^n = (x + jy)^n = r^n e^{jn\theta} = r^n \angle n\theta \quad (n = \text{integer})$$

$$z^{1/n} = (x + jy)^{1/n} = r^{1/n} e^{jn\theta} = r^{1/n} \angle \frac{\theta}{2} + \frac{2\pi m}{n} \quad (m = 0, 1, 2, \ldots, n-1)$$

C.6 Table of Derivatives

$y=$	$\frac{dy}{dx} =$
C (constant)	0
cx^n (n any constant)	cnx^{n-1}
e^{ax}	ae^{ax}
a^x ($a > 0$)	$a^x \ln a$
In x ($x > 0$)	$\frac{1}{x}$
$\frac{c}{x^a}$	$\frac{-ca}{x^{a+1}}$
$\log_a x$	$\frac{\log_a e}{x}$
$\sin ax$	$a \cos ax$
$\cos ax$	$-a \sin ax$
$\tan ax$	$-a\sec^2 ax = \frac{a}{\cos^2 ax}$
$\cot ax$	$-a\,\text{cosec}^2\, ax = \frac{-a}{\sin^2 ax}$
$\sec ax$	$\frac{a \sin ax}{\cos^2 ax}$
$\text{cosec}\, ax$	$\frac{-a \cos ax}{\sin^2 ax}$
$\arcsin ax = \sin^{-1} ax$	$\frac{a}{\sqrt{1 - a^2x^2}}$
$\arccos ax = \cos^{-1} ax$	$\frac{-a}{\sqrt{1 - a^2x^2}}$
$\arctan ax = \tan^{-1} ax$	$\frac{a}{1 + a^2x^2}$
$\text{arccot}\, ax = \cot^{-1} ax$	$\frac{-a}{1 + a^2x^2}$

(Continued)

$y=$	$\frac{dy}{dx}=$
$\sinh ax$	$a\cosh ax$
$\cosh ax$	$a\sinh ax$
$\tanh ax$	$\frac{a}{\cosh^2 ax}$
$\sinh^{-1} ax$	$\frac{a}{\sqrt{1-a^2x^2}}$
$\cosh^{-1} ax$	$\frac{a}{\sqrt{a^2x^2-1}}$
$\tanh^{-1} ax$	$\frac{a}{1-a^2x^2}$
$u(x)+\upsilon(x)$	$\frac{du}{dx}+\frac{d\upsilon}{dx}$
$u(x)\,\upsilon(x)$	$u\frac{d\upsilon}{dx}+\upsilon\frac{du}{dx}$
$\frac{u(x)}{\upsilon(x)}$	$\frac{1}{\upsilon^2}\left(\upsilon\frac{du}{dx}-u\frac{d\upsilon}{dx}\right)$
$\frac{1}{\upsilon(x)}$	$\frac{-1}{\upsilon^2}\frac{d\upsilon}{dx}$
$y(\upsilon(x))$	$\frac{dy}{d\upsilon}\frac{d\upsilon}{dx}$
$y(\upsilon(u(x)))$	$\frac{dy}{d\upsilon}\frac{d\upsilon}{du}\frac{du}{dx}$

C.7 Table of Integrals

$$\int a\,dx = ax + c \quad (c \text{ is an arbitrary constant})$$

$$\int x\,dy = xy - \int y\,dx$$

$$\int x^n dx = \frac{x^{n+1}}{n+1} + c, \quad (n \neq -1)$$

$$\int \frac{1}{x} dx = \ln|x| + c$$

$$\int e^{ax} dx = \frac{e^{ax}}{a} + c$$

$$\int a^x dx = \frac{a^x}{\ln a} + c \quad \text{for} \quad (a > 0)$$

$$\int \ln x\, dx = x\ln x - x + c \quad \text{for} \quad (x > 0)$$

$$\int \sin ax \, \mathrm{d}x = \frac{-\cos ax}{a} + c$$

$$\int \cos ax \, \mathrm{d}x = \frac{\sin ax}{a} + c$$

$$\int \tan ax \, \mathrm{d}x = \frac{-\ln|\cos ax|}{a} + c$$

$$\int \cot ax \, \mathrm{d}x = \frac{-\ln|\cos ax|}{a} + c$$

$$\int \sec ax \, \mathrm{d}x = \frac{-\ln\left(\dfrac{1-\sin ax}{1+\sin ax}\right)}{2a} + c$$

$$\int \operatorname{cosec} ax \, \mathrm{d}x = \frac{-\ln(1-\cos ax / 1+\cos ax)}{2a} + c$$

$$\int \frac{1}{x^2+a^2} \mathrm{d}x = \frac{\tan^{-1}(x/a)}{a} + c$$

$$\int \frac{1}{x^2-a^2} \mathrm{d}x = \frac{\ln(x-a/x+a)}{2a} + c \text{ or } \frac{\tanh^{-1}(x/a)}{a} + c$$

$$\int \frac{1}{a^2-x^2} \mathrm{d}x = \frac{\ln(x+a/x-a)}{2a} + c$$

$$\int \frac{1}{\sqrt{a^2-x^2}} \mathrm{d}x = \sin^{-1}(x/a) + c$$

$$\int \frac{1}{\sqrt{a^2-x^2}} \mathrm{d}x = \frac{\sinh^{-1}(x/a)}{a} + c \text{ or } \ln\left(x+\sqrt{x^2+a^2}\right) + c$$

$$\int \frac{1}{\sqrt{x^2-a^2}} \mathrm{d}x = \ln\left(x+\sqrt{x^2+a^2}\right) + c$$

$$\int \frac{1}{x\sqrt{x^2-a^2}} \mathrm{d}x = \frac{\sec^{-1}(x/a)}{a} + c$$

$$\int x\mathrm{e}^{ax} \mathrm{d}x = \frac{(ax-1)\mathrm{e}^{ax}}{a^2} + c$$

$$\int x\cos ax \, \mathrm{d}x = \frac{\cos ax + ax\sin ax}{a^2} + c$$

$$\int x\sin ax \, \mathrm{d}x = \frac{\sin ax + ax\cos ax}{a^2} + c$$

$$\int x\ln x \, \mathrm{d}x = \frac{x^2}{2}\ln x - \frac{x^2}{4} + c$$

$$\int x\mathrm{e}^{ax} \, \mathrm{d}x = \frac{\mathrm{e}^{ax}(ax-1)}{a^2} + c$$

$$\int e^{ax}\cos bx\,dx = \frac{e^{ax}(a\cos bx + b\sin bx)}{a^2+b^2} + c$$

$$\int e^{ax}\sin bx\,dx = \frac{e^{ax}(-b\cos bx + a\sin bx)}{a^2+b^2} + c$$

$$\int \sin^2 x\,dx = \frac{x}{2} - \frac{\sin 2x}{4} + c$$

$$\int \cos^2 x\,|dx = \frac{x}{2} - \frac{\sin 2x}{4} + c$$

$$\int \tan^2 x\,dx = \tan x - x + c$$

$$\int \cot^2 x\,dx = \cot x - x + c$$

$$\int \sec^2 x\,dx = \tan x + c$$

$$\int \operatorname{cosec}^2 x\,dx = -\cot x + c$$

$$\int \sec x\tan x\,dx = -\sec x + c$$

$$\int \operatorname{cosec} x\cot x\,dx = -\operatorname{cosec} x + c$$

C.8 Table of Probability Distributions

1. Discrete Distribution	Probability $P(X = x)$	Expectation (Mean) μ	Variance σ^2
Binomial $B(n,p)$	$\binom{n}{r}p^r(1-p)^{n-r} = \frac{n!p^rq^{n-1}}{r!(n-r)!'}np$ r = 0, 1,…, n	np	$np(1-p)$
Geometric $G(p)$	$(1-p)^{r-1}p$	$\frac{1}{p}$	$\frac{1-p}{p^2}$
Poisson $p(\lambda)$	$\frac{\lambda^n e^{-\lambda}}{n!}$	λ	λ
Pascal (negative binomial) $NB(r,p)$	$\begin{pmatrix} x & -1 \\ r & -1 \end{pmatrix}p^r(1-p)^{x-r},$ $x = r, r+1, \ldots$	$\frac{r}{p}$	$\frac{r(1-p)}{p^2}$
Hypergeometric $H(N,n,p)$	$\frac{\binom{Np}{r}\binom{N-Np}{n-r}}{\binom{N}{n}}$	np	$np(1-p)\frac{N-n}{N-1}$

(Continued)

2. Discrete Distribution	Density $f(x)$	Expectation (Mean) μ	Variance σ2
Exponential $E(\lambda)$	$\begin{cases} \lambda e^{-\lambda x}, & x \geq 0 \\ 0, & x < 0 \end{cases}$	$\frac{1}{\lambda}$	$\frac{1}{\lambda^2}$
Uniform $U(a,b)$	$\begin{cases} \frac{1}{b-a}, & a < x < b \\ 0, & \text{elsewhere} \end{cases}$	$\frac{a+b}{2}$	$\frac{(b-a)^2}{12}$
Standardized normal $N(0,1)$	$\varphi(x) = \frac{e^{-x^2/2}}{\sqrt{2\pi}}$	0	1
General normal	$\frac{1}{\sigma}\varphi\left(\frac{x-\mu}{\sigma}\right)$	μ	σ^2
Gamma $\Gamma(n,\lambda)$	$\frac{\lambda^n}{\Gamma(n)} x^{n-1} e^{-\lambda x}$	$\frac{n}{\lambda}$	$\frac{n}{\lambda^2}$
Beta $\beta(p,q)$	$a_{p,q} x^{p-1}(1-x)^{q-1}, 0 \leq x \geq 1$ $a_{p,q} = \frac{\Gamma(p+q)}{\Gamma(p)\Gamma(q)}, \ p > 0, \ q > 0$	$\frac{p}{p+q}$	$\frac{pq}{(p+q)^2(p+q+1)}$
Weibull $W(\lambda,\beta)$	$\lambda^\beta \beta x^{\beta-1} e^{-(\lambda x)\beta}, \ x \geq 0$ $F(x) = 1 - e^{-(\lambda x)\beta}$	$\frac{1}{\lambda}\Gamma\left(1+\frac{1}{\beta}\right)$	$\frac{1}{\lambda^2}(A-B)$ $A = \Gamma^2\left(1+\frac{2}{\beta}\right)$ $B = \Gamma^2\left(1+\frac{1}{\beta}\right)$
Rayleigh $R(\sigma)$	$\frac{x}{\sigma^2} e^{-x^2/2\sigma^2}, \ x \geq 0$	$\sigma\sqrt{\frac{\pi}{2}}$	$2\sigma^2\left(1-\frac{\pi}{4}\right)$

C.9 Summations (Series)

1. Finite element of terms

$$\sum_{n=0}^{N} a^n = \frac{1-a^{N+1}}{1-a}; \ \sum_{n=0}^{N} na^n = a\left[\frac{1-(N+1)a^N + Na^{N+1}}{(1-a)^2}\right]$$

$$\sum_{n=0}^{N} n = \frac{N(N+1)}{2}; \ \sum_{n=0}^{N} n^2 = \frac{N(N+1)(2N+1)}{6}$$

$$\sum_{n=0}^{N} n(n+1) = \frac{N(N+1)(N+2)}{3};$$

$$(a+b)^N = \sum_{n=0}^{N} NC_n a^{N-n} b^n, \text{ where } NC_n = NC_{N-n} = \frac{NP_n}{n!} = \frac{N!}{(N-n)!n!}$$

2. Infinite element of terms

$$\sum_{n=0}^{\infty} x^n = \frac{1}{1-x}, (|x|<1); \quad \sum_{n=0}^{\infty} nx^n = \frac{1}{(1-x)^2}, \ (|x|<1)$$

$$\sum_{n=0}^{\infty} n^k x^n = \lim_{a\to 0}(-1)^k \frac{\partial^k}{\partial a^k}\left(\frac{x}{x-e^{-a}}\right), \ (|x|<1); \quad \sum_{n=0}^{\infty} \frac{(-1)^n}{2n+1} = 1-\frac{1}{3}+\frac{1}{5}-\frac{1}{7}+\cdots = \frac{1}{4}\pi$$

$$\sum_{n=0}^{\infty} \frac{1}{n^2} = 1+\frac{1}{2^2}+\frac{1}{3^2}+\frac{1}{4^2}+\cdots = \frac{1}{6}\pi^2$$

$$e^x = \sum_{n=0}^{\infty} \frac{x^n}{n!} = 1+\frac{1}{1!}x+\frac{1}{2!}x^2+\frac{1}{3!}x^3+\cdots$$

$$a^x = \sum_{n=0}^{\infty} \frac{(\ln a)^n x^n}{n!} = 1+\frac{(\ln a)x}{1!}+\frac{(\ln a)^2 x^2}{2!}+\frac{(\ln a)^3 x^3}{3!}+\cdots$$

$$\ln(1\pm x) = \sum_{n=0}^{\infty} \frac{(\pm 1)^n x^x}{n} = \pm x - \frac{x^2}{2} \pm \frac{x^3}{3} - \cdots, \ (|x|<1)$$

$$\sin x = \sum_{n=0}^{\infty} \frac{(-1)^n x^{2n+1}}{(2n+1)!} = x-\frac{x^3}{3!}+\frac{x^5}{5!}-\frac{x^7}{7!}+\cdots$$

$$\cos x = \sum_{n=0}^{\infty} \frac{(-1)^n x^{2n}}{(2n)!} = 1-\frac{x^2}{2!}+\frac{x^4}{4!}-\frac{x^6}{6!}+\cdots$$

$$\tan x = x+\frac{x^3}{3}+\frac{2x^5}{15}+\cdots, \ (|x|<1)$$

$$\tan^{-1} x = \sum_{n=0}^{\infty} \frac{(-1)^n x^{2n+1}}{(2n+1)} = x-\frac{x^2}{3}+\frac{x^5}{5}-\frac{x^7}{7}+\cdots \ (|x|<1)$$

C.10 Logarithmic Identities

$$\log_e a = \ln a \text{ (natural logarithm)}$$

$$\log_{10} a = \log a \text{ (common logarithm)}$$

$$\log ab = \log a + \log b$$

$$\log \frac{a}{b} = \log a - \log b$$

$$\log a^n = n \log a$$

C.11 Exponential Identities

$$e^x = 1+x+\frac{x^2}{2!}+\frac{x^3}{3!}+\frac{x^4}{4!}+\cdots, \text{ where } e \simeq 2.7182$$

$$e^x e^y = e^{x+y}$$

$$(e^x)^n = e^{nx}$$

$$\ln e^x = x$$

C.12 Approximations for Small Quantities

$$\text{If } |a| << 1, \text{then}$$

$$\ln(1+a) \simeq a$$

$$e^a \simeq 1+a$$

$$\sin a \simeq a$$

$$\cos a \simeq 1$$

$$\tan a \simeq a$$

$$(1 \pm a)^n \simeq 1 \pm na$$

C.13 Matrix Notation and Operations

1. Matrices. A *matrix* is a rectangular array of elements arranged in rows and columns. The array is commonly enclosed in brackets. Let a matrix A (expressed in boldface as **A** or in bracket as $[A]$) have m rows and n columns; then the matrix can be expressed as follows:

$$A = [A] = \begin{bmatrix} a_{11} & a_{12} & . & . & . & a_{1j} & . & . & . & a_{1n} \\ a_{21} & a_{22} & . & . & . & a_{2j} & . & . & . & a_{2n} \\ . & . & . & . & . & . & . & . & . & . \\ . & . & . & . & . & . & . & . & . & . \\ . & . & . & . & . & . & . & . & . & . \\ a_{i1} & a_{i2} & . & . & . & a_{ij} & . & . & . & a_{in} \\ . & . & . & . & . & . & . & . & . & . \\ . & . & . & . & . & . & . & . & . & . \\ . & . & . & . & . & . & . & . & . & . \\ a_{m1} & a_{m2} & . & . & . & a_{mj} & . & . & . & a_{mm} \end{bmatrix}$$

where the element a_{ij} has two subscripts: the first refers to the row position of the element in the array and the second refers to the column position. A matrix with m rows and n columns, $[A]$, is defined as a matrix of order or size $m \times n$ (m by n) or an $m \times n$ matrix. A vector is a matrix that consists of only one row or one column.

Location of an element in a matrix:

$$\text{Let } A = \begin{bmatrix} a_{11} & a_{12} & a_{13} & a_{14} \\ a_{21} & a_{22} & a_{23} & a_{24} \\ a_{31} & a_{32} & a_{33} & a_{34} \\ a_{41} & a_{42} & a_{43} & a_{44} \end{bmatrix} \text{ be the matrix with size } 4 \times 4$$

where:

a_{11} is the element a at row 1 and column 1
a_{12} is the element a at row 1 and column 2
a_{32} is the element a at row 3 and column 2

2. Special common types of matrices
   a. If $m \pm n$, then the matrix $[A]$ is called a *rectangular matrix.*
   b. If $m = n$, then the matrix $[A]$ is called a *square matrix of order n.*
   c. If $m = 1$ *and* $n > 1$, then the matrix $[A]$ is called *row matrix or row vector.*
   d. If $m > 1$ *and* $n = 1$, then the matrix $[A]$ is called *column matrix or column vector.*
   e. If $m = 1$ *and* $n = 1$, then the matrix $[A]$ is called *a scalar.*
   f. A *real matrix* is a matrix whose elements are all real.
   g. A *complex matrix* is a matrix whose elements may be complex.
   h. A *null matrix* is a matrix whose elements are all zero.
   i. An *identity* (or *unit*) *matrix*, $[I]$ *or* **I**, *is a square matrix* whose elements are equal to zero except those located on its main diagonal elements, which are unity (or one). The *main diagonal* elements have equal, row and column subscripts. The main diagonal runs from the upper-left corner to the lower-right corner. If the elements of an identity matrix are denoted as e_{ij}, then

$$e_{ij} = \begin{cases} 1 & i = j \\ 0 & i \neq j \end{cases}$$

   j. A *diagonal matrix* is a square matrix that has zero elements everywhere except on its main diagonal. That is, for a diagonal matrix, $a_{ij} = 0$ when $i \pm j$ and not all a_{ii} are zero.
   k. A *symmetric matrix* is a square matrix whose elements satisfy the condition $a_{ij} = -a_{ji}$ for $i \neq j$, and $a_{ii} = 0$.
   l. An *antisymmetric* (or *skew symmetric*) *matrix* is a square matrix whose elements satisfy the condition $a_{ij} = -a_{ji}$ for $i \neq j$, and $a_{ii} = 0$.
   m. A *triangular matrix* is a square matrix whose all elements on one side of the diagonal are zero. There are two types of triangular matrices: first, an upper triangular **U** whose elements below the diagonal are zero, and second, a lower triangular **L** whose elements above the diagonal are all zero.
   n. A *partitioned* (or *block*) *matrix* is a matrix that is divided by horizontal and vertical lines into smaller matrices called submatrices or blocks.

3. Matrix operations

a. Transpose of a matrix. The *transpose* of a matrix $\mathbf{A} = [a_{ij}]$ is denoted as $\mathbf{A}^T = [a_{ji}]$ and is obtained by interchanging the rows and columns in matrix **A**. Thus, if a matrix **A** is of order $m \times n$, then $\mathbf{A}^T$ will be of order $n \times$ m.

b. Addition and subtraction. Addition and subtraction can only be performed for matrices of the same size. The addition is accomplished by adding the corresponding elements of each matrix. For addition, $\mathbf{C} = \mathbf{A} + \mathbf{B}$ implies that $c_{ij} = a_{ij} + b_{ij}$.

Now, the subtraction is accomplished by subtracting the corresponding elements of each matrix. For subtraction, $\mathbf{C} = \mathbf{A} - \mathbf{B}$ implies that $c_{ij} = a_{ij} - b_{ij}$, where c_{ij}, a_{ij}, and b_{ij} are typical elements of the **C**, **A**, and **B** matrices, respectively.

If matrices **A** and **B** are both of the same size $m \times n$, the resulting matrix **C** is also of size $m \times n$.

Matrix addition and subtraction are associative:

$$\mathbf{A} + \mathbf{B} + \mathbf{C} = (\mathbf{A} + \mathbf{B}) + \mathbf{C} = \mathbf{A} + (\mathbf{B} + \mathbf{C})$$

$$\mathbf{A} + \mathbf{B} - \mathbf{C} = (\mathbf{A} + \mathbf{B}) - \mathbf{C} = \mathbf{A} + (\mathbf{B} - \mathbf{C})$$

Matrix addition and subtraction are commutative:

$$\mathbf{A} + \mathbf{B} = \mathbf{B} + \mathbf{A}$$

$$\mathbf{A} - \mathbf{B} = -\mathbf{B} + \mathbf{A}$$

c. Multiplication by scalar. A matrix is multiplied by a scalar by multiplying each element of the matrix by the scalar. The multiplication of a matrix **A** by a scalar c is defined as

$$cA = [ca_{ij}]$$

The scalar multiplication is commutative.

d. Matrix multiplication. The product of two matrices is $\mathbf{C} = \mathbf{AB}$ if and only if the number of columns in A is equal to the number of rows in **B**. The product of matrix **A** of size $m \times n$ and matrix **B** of size $n \times r$ results in matrix **C** of size $m \times r$. Then, $c_{ij} = \sum_{k=1}^{n} a_{ik} b_{kj}$.

That is, the (ij)th component of matrix **C** is obtained by taking the dot product:

$$c_{ij} = (i\text{th row of } \mathbf{A}) \cdot (j\text{th column of } \mathbf{B})$$

Matrix multiplication is associative:

$$\mathbf{ABC} = (\mathbf{AB})\mathbf{C} = \mathbf{A}(\mathbf{BC})$$

Matrix multiplication is distributive:

$$\mathbf{A}(\mathbf{B} + \mathbf{C}) = \mathbf{AB} + \mathbf{AC}$$

Matrix multiplication is not commutative:

$$\mathbf{AB} \neq \mathbf{BA}$$

e. Transpose of matrix multiplication. The transpose of matrix multiplication is usually denoted $(\mathbf{AB})^T$ and is defined as

$$(\mathbf{AB})^T = \mathbf{B}^T \mathbf{A}^T$$

f. Inverse of square matrix. The inverse of a matrix **A** is denoted by $\mathbf{A}^{-1}$. The inverse matrix satisfies

$$\mathbf{AA}^{-1} = \mathbf{A}^{-1}\mathbf{A} = \mathbf{I}$$

A matrix that possesses an inverse is called a *nonsingular matrix* (or *invertible matrix*). A matrix without an inverse is called a *singular matrix.*

g. Differentiation of a matrix. The differentiation of a matrix is differentiation of every element of the matrix separately. To emphasize, if the elements of the matrix **A** are a function of t, then

$$\frac{d\mathbf{A}}{dt} = \left[\frac{da_{ij}}{dt}\right]$$

h. Integration of a matrix. The integration of a matrix is integration of every element of the matrix separately. To emphasize, if the elements of the matrix **A** are a function of t, then

$$\int \mathbf{A}dt = \left[\int a_{ij}dt\right]$$

i. Equality of matrices. Two matrices are equal if they have the same size and their corresponding elements are equal.

4. Determinant of a matrix. The determinant of a square matrix **A** is a scalar number denoted by $|\mathbf{A}|$ or det **A**.

The value of a second-order determinant is calculated from

$$\det\begin{bmatrix} a_{11} & a_{12} \\ a_{21} & a_{22} \end{bmatrix} = \begin{vmatrix} a_{11} & a_{12} \\ a_{21} & a_{22} \end{vmatrix} = a_{11}a_{22} - a_{12}a_{21}$$

By using the sign rule of each term, the determinant is determined by the first row in the following diagram:

$$\begin{vmatrix} + & - & + \\ - & + & - \\ + & - & + \end{vmatrix}$$

The value of a third-order determinant is calculated as

$$\det\begin{bmatrix} a_{11} & a_{12} & a_{13} \\ a_{21} & a_{22} & a_{23} \\ a_{31} & a_{32} & a_{33} \end{bmatrix} = \begin{bmatrix} a_{11} & a_{12} & a_{13} \\ a_{21} & a_{22} & a_{23} \\ a_{31} & a_{32} & a_{33} \end{bmatrix} =$$

$$a_{11}\begin{vmatrix} a_{22} & a_{23} \\ a_{32} & a_{33} \end{vmatrix} - a_{12}\begin{vmatrix} a_{21} & a_{23} \\ a_{31} & a_{33} \end{vmatrix} + a_{13}\begin{vmatrix} a_{21} & a_{22} \\ a_{31} & a_{32} \end{vmatrix}$$

C.14 Vectors

1. Vector derivatives
   a. Cartesian coordinates

Coordinates	(x, y, z)
Vector	$A = A_x a_x + A_y a_y + A_z a_z$
Gradient	$\nabla A = \frac{\partial A}{\partial x} a_x + \frac{\partial A}{\partial y} a_y + \frac{\partial A}{\partial z} a_z$
Divergence	$\nabla \cdot A = \frac{\partial A_x}{\partial x} + \frac{\partial A_y}{\partial y} + \frac{\partial A_z}{\partial z}$
Curl	$\nabla \times A = \begin{vmatrix} a_x & a_y & a_z \\ \frac{\partial}{\partial x} & \frac{\partial}{\partial y} & \frac{\partial}{\partial z} \\ A_x & A_y & A_z \end{vmatrix}$
	$= \left(\frac{\partial A_z}{\partial y} - \frac{\partial A_y}{\partial z}\right) a_x + \left(\frac{\partial A_x}{\partial z} - \frac{\partial A_z}{\partial x}\right) a_y + \left(\frac{\partial A_y}{\partial x} - \frac{\partial A_x}{\partial y}\right) a_z$
Laplacian	$\nabla^2 A = \frac{\partial^2 A}{\partial x^2} + \frac{\partial^2 A}{\partial y^2} + \frac{\partial^2 A}{\partial z^2}$

   b. Cylindrical coordinates

Coordinates	(ρ, ϕ, z)
Vector	$A = A_\rho a_\rho + A_\phi a_\phi + A_z a_z$
Gradient	$\nabla A = \frac{\partial A}{\partial \rho} a_\rho + \frac{1}{\rho}\frac{\partial A}{\partial \phi} a_\phi + \frac{\partial A}{\partial z} a_z$
Divergence	$\nabla \cdot A = \frac{1}{\rho}\frac{\partial}{\partial \rho}(\rho A_\rho) + \frac{\partial A_\phi}{\partial \phi} + \frac{\partial A_z}{\partial z}$

(Continued)

Coordinates	(ρ, ϕ, z)
Curl	$\nabla \times A = \begin{vmatrix} a_\rho & \rho a_\phi & a_z \\ \frac{\partial}{\partial \rho} & \frac{\partial}{\partial \phi} & \frac{\partial}{\partial z} \\ A_\rho & \rho A_\phi & A_z \end{vmatrix}$
	$= \left(\frac{1}{\rho}\frac{\partial A_z}{\partial \phi} - \frac{\partial A_\phi}{\partial z}\right) a_\rho + \left(\frac{\partial A_\rho}{\partial z} - \frac{\partial A_z}{\partial \rho}\right) a_\phi + \frac{1}{\rho}\left[\frac{\partial}{\partial x}(\rho A_\phi) - \frac{\partial A_\rho}{\partial \rho}\right] a_z$
Laplacian	$\nabla^2 A = \frac{1}{\rho}\frac{\partial}{\partial \rho}\left(\rho \frac{\partial A}{\partial \rho}\right) + \frac{1}{\rho^2}\frac{\partial^2 A}{\partial \phi^2} + \frac{\partial^2 A}{\partial z^2}$

c. Spherical coordinates

Coordinates	(r, θ, ϕ)
Vector	$A = A_r a_r + A_\theta a_\theta + A_\phi a_\phi$
Gradient	$\nabla A = \frac{\partial A}{\partial r} a_r + \frac{1}{r}\frac{\partial A}{\partial \theta} a_\theta + \frac{1}{r\sin\theta}\frac{\partial A}{\partial \phi} a_\phi$
Divergence	$\nabla \cdot A = \frac{1}{r^2}\frac{\partial}{\partial r}(r^2 A_r) + \frac{1}{r\sin\theta}\frac{\partial}{\partial \theta}(A_\theta \sin\theta) + \frac{1}{r\sin\theta}\frac{\partial A_\phi}{\partial \phi}$
Curl	$\nabla \times A = \frac{1}{r^2 \sin\theta}\begin{vmatrix} a_r & r a_\theta & (r\sin\theta) a_\phi \\ \frac{\partial}{\partial r} & \frac{\partial}{\partial \theta} & \frac{\partial}{\partial \phi} \\ A_r & r A_\theta & (r\sin\theta) A_\phi \end{vmatrix}$
	$= \frac{1}{r\sin\theta}\left[\frac{\partial}{\partial \theta}(A_\phi \sin\theta) - \frac{\partial A_{\theta\phi}}{\partial \phi}\right] a_r + \frac{1}{r}\left[\frac{1}{\sin\theta}\frac{\partial A_r}{\partial \phi} - \frac{\partial}{\partial r}(r A_\phi)\right] a_\theta$
	$+ \frac{1}{r}\left[\frac{\partial}{\partial r}(r A_\theta) - \frac{\partial A_r}{\partial \theta}\right] a_\phi$
Laplacian	$\nabla^2 A = \frac{1}{r^2}\frac{\partial}{\partial r}\left(r^2 \frac{\partial A}{\partial r}\right) + \frac{1}{r^2 \sin\theta}\frac{\partial}{\partial \theta}\left(\sin\theta \frac{\partial A}{\partial \theta}\right) + \frac{1}{r^2 \sin\theta}\frac{\partial^2 A}{\partial \phi^2}$

2. Vector identities

a. Triple products

$$\mathbf{A}(\mathbf{B} \times \mathbf{C}) = \mathbf{B}(\mathbf{C} \times \mathbf{A}) = \mathbf{C}(\mathbf{A} \times \mathbf{B})$$

$$\mathbf{A}(\mathbf{B} \times \mathbf{C}) = \mathbf{B}(\mathbf{A} \cdot \mathbf{C}) - \mathbf{C}(\mathbf{A} \cdot \mathbf{B})$$

b. Product rules

$$\nabla(fg) = f(\nabla g) + g(\nabla f)$$

$$\nabla(\mathbf{A} \cdot \mathbf{B}) = \mathbf{A} \times (\nabla \times \mathbf{B}) + \mathbf{B} \times (\nabla \times \mathbf{A}) + (\mathbf{A} \cdot \nabla)\mathbf{B} + (\mathbf{B} \times \nabla)\mathbf{A}$$

$$\nabla \cdot (f\mathrm{A}) = f(\nabla \cdot \mathrm{A}) + \mathrm{A} \cdot (\nabla f)$$

$$\nabla(\mathbf{A} \times \mathbf{B}) = \mathbf{B} \cdot (\nabla \times \mathrm{A}) - \mathrm{A} \cdot (\nabla \times \mathbf{B})$$

$$\nabla\times(f\mathbf{A}) = f(\nabla\times \mathrm{A}) - \mathrm{A}\times(\nabla f) = \nabla\times(f\mathrm{A}) = f(\nabla\times \mathrm{A}) + (\nabla f)\times \mathrm{A}$$

$$\nabla\times(\mathbf{A}\times\mathbf{B}) = (\mathbf{B}\cdot\nabla)\mathrm{A} - (\mathrm{A}\cdot\nabla)\mathbf{B} + \mathrm{A}(\nabla\cdot\mathbf{B}) - (\nabla\cdot \mathrm{A})$$

c. Second derivatives

$$\nabla\cdot(\nabla\times\mathbf{A}) = 0$$

$$\nabla\times(\nabla f) = 0$$

$$\nabla\cdot(\nabla f) = \nabla^2 f$$

$$\nabla\times(\nabla\times A) = \nabla(\nabla\cdot\mathbf{A}) - \nabla^2\mathbf{A}$$

d. Addition, division, and power rules

$$\nabla(f+g) = \nabla f + \nabla g$$

$$\nabla\cdot(\mathbf{A}+\mathbf{B}) = \nabla\cdot\mathbf{A} + \nabla\cdot\mathbf{B}$$

$$\nabla\times(\mathbf{A}\times\mathbf{B}) = \nabla\times\mathbf{A} + \nabla\times\mathbf{B}$$

$$\nabla\left(\frac{f}{g}\right) = \frac{g(\nabla f) - f(\nabla g)}{g^2}$$

$$\nabla f^n = nf^{n-1}\nabla f(n = \text{integer})$$

3. Fundamental theorems

a. Gradient theorem

$$\int_a^b (\nabla f)\mathrm{d}l = f(b) - f(a)$$

b. Divergence theorem

$$\int_{\text{volume}} (\nabla\cdot\mathbf{A})\mathrm{d}v = \oint_{\text{surface}} \mathbf{A}\cdot\mathrm{d}s$$

c. Curl (Stokes) theorem

$$\int_{\text{surface}} (\nabla\times\mathbf{A})\mathrm{d}s = \oint_{\text{line}} \mathbf{A}\cdot\mathrm{d}l$$

d. $\oint_{\text{line}} f\mathrm{d}l = -\oint_{\text{surface}} \nabla f\times\mathrm{d}s$

e. $\oint_{\text{surface}} f\cdot\mathrm{d}ls = -\oint_{\text{volume}} \nabla f\cdot\mathrm{d}v$

f. $\oint_{\text{surface}} \mathbf{A}\cdot\mathrm{d}s = -\int_{\text{volume}} \nabla\cdot\mathbf{A}\mathrm{d}v$

Index

Note: Locators followed by "*f*" and "*t*" denote figures and tables in the text

D

E

F

N

Q

R